Timing Channels in Cryptography

Chester Rebeiro • Debdeep Mukhopadhyay
Sarani Bhattacharya

Timing Channels in Cryptography

A Micro-Architectural Perspective

Springer

Chester Rebeiro
Columbia University
New York
New York
USA

Debdeep Mukhopadhyay
Indian Institute of Technology
Kharagpur
West Bengal
India

Sarani Bhattacharya
Department of Computer Science and Engin
IIT Kharagpur
Kharagpur
India

ISBN 978-3-319-12369-1 ISBN 978-3-319-12370-7 (eBook)
DOI 10.1007/978-3-319-12370-7

Library of Congress Control Number: 2014954350

Springer Cham Heidelberg New York Dordrecht London

Printed on acid-free paper

Springer is part of Springer Science+Business Media (www.springer.com)

Foreword

With the publication of the Data Encryption Standard DES in 1977 and the invention of public key cryptography in the 1980s, cryptography has moved into civil applications. This was pushed by the electronic revolution, which needs cryptography to create an electronic equivalent of the real world into the digital world. Cryptography is used to provide confidentiality, digital signatures, anonymity, payments, electronic transactions, elections, and many more.

Cryptography and cryptographic protocols by themselves do not provide security: they need a digital platform that executes the algorithms and protocols. In the early days, these platforms were computers and servers sitting in well protected computer rooms and offices. The main concern when implementing the algorithms and protocols was efficiency, as cryptographic algorithms are typically very computationally demanding. The attacker model assumed that the computer platforms themselves were well protected and that only the result of the encryption, the ciphertext, has to travel over insecure communication channels such as cables or satellite communication or tapes that are transported between bank offices and the bank headquarters.

However, as the electronic revolution expanded, electronics became cheaper, more widespread, and integrated into day to day applications. Examples are pay TV systems, banking cards, identity cards, or access control systems. This omnipresence of electronic gadgets changes the attacker model! Now, the implementation becomes the weak link in the chain: the attacker will attack *while* the electronic device is performing cryptographic operations using the secret key operating on possible sensitive data. During calculations, the devices leak information. These are the so-called side channel attacks. The attacker will observe the device while it performs its calculations. From timing or power consumption variations, or from electromagnetic radiations, the attacker is able to guess which operations and which sensitive data is being handled.

The contributions of this book are situated in this context. One highly important source of side channel information leakage is the timing variations of calculations. Cryptographic algorithms and protocols are written in high level languages and compiled onto sophisticated digital platforms. Both the software stack and the hardware platform include a large variety of optimizations to improve performance. Memory accesses and branches are expensive in execution time, therefore processors include caches and branch predictors. Unfortunately, these optimizations result in timing variations when executing a program. Caches are considered one of the most important sources of side channel information leakage. This book therefore gives a comprehensive overview and an in depth analysis of many flavors of attacks using caches and branch predictors. It ends with a set of countermeasures to address these timing attacks at different abstraction layers.

This book is a fine read for anyone interested in the many possibilities of cache attacks. A software engineer should not write any line of a cryptographic routine without knowing how powerful timing attacks are. This book gives a clear insight on how these attacks work and how they can be mitigated. Enjoy reading!

Leuven, Belgium
December 2014

Ingrid Verbauwhede

Preface

Cryptography plays a vital role in securing e-business and e-commerce transactions. The algorithms used have been rigorously analyzed and tested for their ability to conceal information. However, these algorithms are needed to be realized in systems that are used in a variety of applications. As the saying says, *there's many a slip 'twixt the cup and the lip*, these implementations may leak sensitive information via unintended timing channels. Over the past 20 years, several attacks have been developed that use these *side-channels* to reveal secrets from ciphers. These attacks, known as timing attacks, are powerful enough to break a mathematically robust cipher in few minutes on standard computing platforms.

Starting from its inception to the present day, timing attacks and the underneath statistics have evolved. Enhancements made to computer architecture over the years have also influenced timing channels. To develop a system secure against these threatening channels, one needs to be abreast of the interplay between cryptographic algorithms and computer architecture. Here is where this book steps in. It brings on a single platform aspects of computer architecture and cryptography that are essential to understand *how* timing attacks work and *why* they work. It describes the attacks and analyzes the relationship between the cipher implementation, system micro-architecture, and the attack threat. Various timing channels arising due to cache memories and branch prediction units are presented. The book would help engineers and researchers understand timing attacks and thereafter develop platforms that can tolerate these attacks.

The following topics are covered.

- **Modern Cryptography.** Attackers make use of the cipher's structure to extract the secret key from execution time. The attacks are made even more powerful when ideas from classical cryptanalysis are used with the timing information. The book therefore begins with a review of a variety of modern ciphers ranging from symmetric algorithms like the Advanced Encryption Standard (AES) to asymmetric algorithms like the Rivest-Shamir-Adleman (RSA), and some popular classical cryptanalysis techniques.

- **Superscalar Processor Architectures.** To address how timing channels arise in a system, the book provides a background of the internal micro-architecture of processors. An introduction to modern superscalar architectures has been provided, with an emphasis on the components in them that affect the execution time of ciphers.
- **Time-Driven Cache Attacks on Block Ciphers.** Memory load instructions that cause a cache miss takes considerably longer than those that result in a cache hit. Attackers utilize this difference to build a variety of attacks on block ciphers. A detailed discussion on how these attacks are implemented is presented, with emphasis on various accurate time measurement techniques and analysis methods to observe the cache hit miss phenomenon.
- **Advanced Time-Driven Cache Attacks on Block Ciphers.** Combination of classical cryptanalysis with side-channel attacks paves way to powerful attack methodologies. A combination of classical differential cryptanalysis with cache timing attacks has been discussed in the book to describe this threat.
- **Formal Analysis of Time-Driven Cache Attacks.** Developing suitable metrics for comparison of secured implementations and using them to guide the design flow is an important aspect in security engineering. In this pursuit, the book provides formal modeling of time-driven cache attacks on block ciphers and suggests metrics for evaluation and comparison of such implementations. A discourse on how micro-architectural features, such as pipelining and out-of-order execution, affect cache attacks is presented. These guidelines form a framework for developing ideal implementations of the block ciphers with respect to time-driven cache attacks.
- **Profiled Time-Driven Cache Attacks on Block Ciphers.** There are various types of cache attacks, of which profiled cache attacks are arguably the most powerful. The book provides a detailed description of such profiled attacks, which rely on building timing profiles in a learning phase before mounting the attack. The book also provides insights into developing metrics for estimating the information leakage and relates the leakage to micro-architectural features in modern computers, such as hardware prefetchers.
- **Access-Driven Cache Attacks on Block Ciphers.** Cache attacks come in different flavors, one of them, called access attacks, relies on a spy program running along with the encryption program. The spy uses timing measurements to determine memory access patterns made by the encryption. By monopolizing the OS scheduler, the spy program can accurately determine every memory access made, thereby leading to powerful attacks. The book provides a comprehensive overview on the topic to understand the theory and develop the procedure of such access attacks.
- **Branch Prediction Attacks.** Branch prediction units are artifacts in computer architectures to reduce performance penalties due to branch instructions. They predict whether a branch instruction will be taken or not taken, thereby reducing processor stalls. Athough this can boost performance of an application significantly, it can also lead to timing channels that are catastrophic. Attackers have used these channels to break mathematically strong asymmetric key ciphers such

as the RSA. Several of the attack strategies developed are discussed in detail in the book.

- **Countermeasures**. To circumvent timing attacks, a variety of countermeasures have been proposed at various levels in the system: from the application level and architectural level to the Operating System. The book tries to provide an overview on these countermeasures, how they are applied, their effectiveness, and the consequent overheads involved.

IIT Kharagpur, India
November 2014

Chester Rebeiro
Debdeep Mukhopadhyay
Sarani Bhattacharya

Acknowledgements

This preface would be incomplete without a mention of the people who made this effort possible. The authors would like to express their sincere gratitude to the Computer Science and Engineering Department of IIT Kharagpur, which provides an ideal environment for research on information security. The authors would like to thank all their colleagues, students, collaborators, and funding agencies without whose support this work would not be possible. Special thanks to Abhijit Das, Partha Pratim Chakrabarty, Rajat Subhra Chakraborty, Dipanwita Roy Chowdhury, Phuong Ha Nguyen, and Indranil Sengupta from IIT Kharagpur, and Junko Takahashi (NTT Labs), Toshinori Fukunaga (NTT Labs), and Sobha PM (CDAC Bangalore) for their help and support during the project.

Chester Rebeiro would like to thank his family members especially his parents, wife Sharon, and son Tristan for their love and prayers.

Debdeep Mukhopadhyay would like to thank in particular all his family members and friends. He would like to make a special mention of his parents, wife, and brother for their bonding and wishes. He would like to dedicate this book to his daughter Debanti, who has indeed made his life more interesting.

Sarani Bhattacharya would like to thank her parents and friends for their continuous support and encouragement. She would also like to specially thank her mentor Prof. Debdeep Mukhopadhyay for his invaluable guidance and Dr. Chester Rebeiro for his advice and suggestions.

Acknowledgement

Contents

List of Abbreviations

AES	Advanced Encryption Standard
BEEA	Binary Extend Euclidean Algorithm
BPU	Branch Prediction Unit
BTB	Branch Target Buffer
CFS	Completely Fair Scheduler
CCA	Chosen Ciphertext Attack
CPA	Chosen Plaintext Attack
CPU	Central Processing Unit
CMOS	Complementary Metal Oxide Semiconductor
CRT	Chinese Remainder Theorem
DOM	Difference of Means
DRAM	Dynamic Random Access Memory
GCD	Greatest Common Divisor
GF	Galois Field
GFN	Generalized Feistel Networks
IPC	Inter Process Communication
ISA	Instruction Set Architecture
L1	Level 1
L2	Level 2
L3	Level 3
MM	Montgomery Multiplication
NIST	National Institute Standards and Technology
NSS	Network Security Services
NTT	Nippon Telegraph and Telephone Corporation
PC	Program Counter
RAM	Random Access Memory
RSA	Rivest Shamir Adleman
SCA	Side Channel Attacks
SMT	Symmetrical Multithreading
SNR	Signal to Noise Ratio
SPN	Substitution Permutation Network
SRAM	Static Random Access Memory
TSC	Time Stamp Counter
VTSC	Virtual Time Stamp Counter

Chapter 1
An Introduction to Timing Attacks

With the ever-increasing proliferation of e-business practices, great volumes of business transactions and data transmission are routinely carried out in devices ranging in scale from personal smartcards to business servers. Cryptographic algorithms are employed in order to secure these transactions and to ensure that the communicated data is only accessible by the intended parties. Exclusive access to the data (referred to as *plaintext*) is achieved by encrypting it with a *key* that is known only to the authorized parties.

Since Claude Shannon's benchmark paper on the communication theory of secrecy systems in [1], significant scientific advancements have been made in the development of cryptographic algorithms. The algorithms in use today are supported by extensive mathematical evaluations, which provide guarantees for secrecy of the plaintext and key. The guarantees for instance, ensure that an attacker would find it extremely difficult to get any information about the secret key from the ciphertext even if the plaintext and cryptographic algorithm are known. To take an example, the best known cryptanalytic attack against the popular Advanced Encryption Standard (AES) cipher would require around 2^{126} operations of the cipher before any information about the key is obtained. This would take several centuries to run even with the best computing resources.

To circumvent the strict security guarantees that cryptographic algorithms comply with, attackers target the physical implementation of the algorithms rather than the algorithms itself. Irrespective of the device that the algorithm is implemented in, its execution would leave a trace in the system and the surrounding medium. The trace may reveal information about the internal state of the cipher that classical cryptanalytic attackers do not have. The mathematical proofs that guarantee security of the cipher fail when such additional information is known. This results in attacks on the cipher that are considerably more practical. For instance, when the cryptographic algorithm executes on a processor, it leaves a trace in the power consumed by the device. This trace, which essentially is a modulation of the dynamic power consumption, contains information about the data being processed and has been used by attackers to retrieve the secret key of ciphers in a few hours. Such channels that leak information about the executing cipher are called *side channels*. The class of attacks that use these side channels to obtain secret information are called

C. Rebeiro et al., *Timing Channels in Cryptography*,
DOI 10.1007/978-3-319-12370-7_1

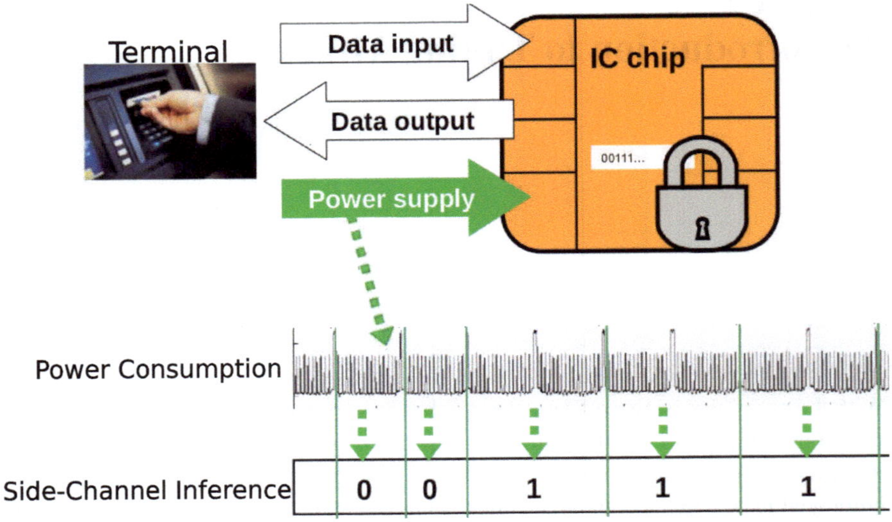

Fig. 1.1 The *power consumption* trace of a smart card reader may reveal the secret. The glitch in the trace can identify a 1 in the secret

side-channel attacks [2]. Figure 1.1 shows how the power consumed by a smart card reader is affected by the secret key stored in the smart card. An attacker monitoring this side can infer the smart card's secret from the glitches in the power trace.

Besides the power consumption, several other side channels have been discovered over the years. For instance, secret keys were recovered from electromagnetic radiation that emanates from a device when executing a cipher [3]. More recently, high-pitched acoustics from the vibration of electronic components have been used to recover cryptographic keys [4]. Side channels based on time have also been used to build attacks against ciphers. Information in timing channels is conveyed by the variation in the execution time of a piece of code [6]. As an example, consider the function Divide in Listing 1.1. The function generally returns the quotient when a is divided by b. However, if b happens to be 0, then the function returns an error value.

Listing 1.1 Function divide returns the quotient of a/b

```
unsigned int Divide(unsigned int a, unsigned int b) {
  if (b== 0)
    return ERROR;
  else
    return a/b;
}
```

Since no division is performed when $b = 0$, the function finishes a bit earlier compared to when $b \neq 0$. An observer who can only monitor the execution time of Divide, would therefore be able to identify invocations when $b = 0$. Further information about b may be obtained from the execution time if the hardware used for division depends on the divisor length. A division where the divisor is large would complete earlier than a division with a small divisor.

Variation in the execution time of Divide is due to the effect of the input b on the conditional statements and underlying hardware. Conditional statements alter the execution path of the function. This may affect execution time if, for instance, one path is substantially more longer than another. Hardware components in the system could also affect execution time. For example, data present in the cache memory would take considerably less time to load compared to when the data is accessed from the RAM. Other components in the system such as branch prediction units, memory interface units, and ALU blocks may also affect execution time of the program. Microarchitectural features in the processor such as symmetrical multithreading, prefetching, parallelization, pipelining, input–output ports, also influence execution time, albeit in a subtle manner.

To a neophyte it may seem trivial to overcome side-channel attacks by adding noise, randomizing, or fuzzing the side channels of the system. While these additions may successfully stymie certain side-channel attacks, there are still considerable number of attacks that are capable of dealing with such naïve countermeasures. In many of these cases, surprisingly, little extra effort is required to overcome the countermeasures. This book takes a look at how various features in the architecture and program structure are exploited to build time-based side-channel attacks against cryptographic ciphers. It determines how the program design and the choice of architectural components in the system can affect the success of the attacks. In this chapter we provide a brief introduction to side-channel attacks, before introducing timing side channels and the various categories of timing attacks.

1.1 Side-Channel Attacks

The idea of using side channels to steal information has been used for over a century. In World War I, the telephones used in battle fields had just one wire and used the earth to carry the return current. Spies would insert rods in the ground and connect them to amplifiers in order to pick up enemy conversation. During World War II, Bell Labs were the first to discover that electromagnetic emissions from devices could be used for spying. They were able to reveal 75 % of the plaintext that was sent in a secured fashion from a distance of over 80 ft. During the 1950s, the Americans used radiations from encoding machines to spy on encrypted Russian message transmission. By building an appropriate device, it was possible to rebuild messages without decryption. These attacks were studied by American scientists under the code name Tempest, in order to identify shields for equipment. It was initially thought that side-channel attacks would require sophisticated tools that were available only to governments. In 1985, Win van Eck published the first unclassified report which showed how low cost equipment could be used to eavesdrop on messages from a distance of a few hundred meters using the emanations from cathode ray tube monitors. The equipment required an antenna tuned to receive radio transmissions from the monitor, and could efficiently reconstruct the images that were displayed on the monitor.

More recent studies show how emissions from cables of LCD monitors, wireless keyboards, and LED indicators can be picked up and decoded from several feet away. In the mid-1990s two seminal papers by Paul Kocher showed how execution time and power consumption could be used to easily retrieve secret keys from naïve implementations of cryptographic ciphers. Subsequently, substantial research activities have been directed to understanding side-channel attacks and implementing defenses. The challenge of side-channel attacks comes from the fact that it violates classical notions of cryptography. Side channels expose the fragility of a mathematically sound cipher, making it unsuitable for security applications. Hence a deeper, thorough, and theoretical study of the topic is required. In this section, we provide an overview of the side-channel attacks on cryptographic ciphers.

1.1.1 Side-Channel Attack Requirements

There are three essential requirements in order that a side-channel attack be successfully carried out. These are a perturbation, manifestation, and an observation. We discuss each of these requirements below.

1. **Perturbation :** A perturbation occurs when the secret that the attacker wants to reveal alters the behavior of the system or its state. For instance, in Listing 1.1, the divisor *b* alters the program execution. Depending on the value of *b*, the result of conditional statements are altered, thus affecting the instructions that are executed. The behavior of the hardware divider may vary depending on the divisor *b*. Additionally, hardware registers that hold intermediate results in the divider may be perturbed during the computations. These perturbations in the behavior or system state contain information about the secret divisor *b*.
2. **Manifestation :** Most often the perturbations in the system, brought about by the execution, cannot be directly accessible by an attacker. In many cases however, they are manifested through side channels, which are generally more accessible. For instance, changes in register values are not always visible to a user executing an application; however, the changes are manifested in the power consumption and electromagnetic radiation of the device. The instructions executed by the program are manifested in the execution time as well as the radiation and power consumption. These manifestations provide information about the program's internals.
3. **Observation :** An attacker would have to observe the side channels in order to obtain the required information about the secret. For example, the power consumption channel would require the attacker to obtain a physical device and monitor its dynamic power consumption through an oscilloscope. For radiation measurements, the attacker would need a receiver and be close enough to pick up emissions from the device. Time-based side channels would require monitoring

of execution time. The clock used to measure time would need to be highly precise in order to distinguish between microarchitectural events in the execution. In certain cases the attacker could create an environment which forces the program or system to behave in a specified way.

1.1.2 The Attacker's Success

Even if the three requirements for a side-channel attack are fulfilled, the attacker may not always be successful in discovering the secret. The amount of success the attacker achieves would depend on additional factors. First, the success would depend on how effectively perturbations convey the secret. For instance, in the function Divide in Listing 1.1, the perturbations due to the conditional statement result in only one value of the divisor to be detected (i.e. when $b = 0$). An attacker would be able to identify cases only when the divisor (b) is zero. All other divisors cannot be distinguished. On the other hand, perturbations in the execution time due to a nonconstant time hardware divider would be able to convey more information about the divisor.

The manifestation of the perturbations through side channels also affects the success of the attack. Perturbations which are well manifested would be more easy to detect. Considering the Divide function again, the presence or absence of the divide operation due to $b = 0$ or $b \neq 0$ has a higher impact on the execution time of the function compared to the subtle variations in time inflicted by the divisor on the divide hardware. As a result, it is easier for an attacker to detect a division compared to detecting the differential behavior of the divider.

A third factor affecting the attack's success is the noise in side channel. Most side-channels are extremely noisy. Further, the amount of noise and their characteristics vary between systems, processors, technologies used, and even depends on the time when measurements are made. An attacker could eliminate a significant amount of noise by signal processing, filtering, and applying other heuristics. Noise, which are uniformly distributed can be eliminated by averaging, provided the attacker is capable of making sufficient number of measurements.

1.1.3 Side-Channel Attack Suppression

Side-channel attacks can be prevented by suppressing or preventing at least one of the afore mentioned requirements. While it is difficult to completely eliminate the perturbations in the system caused when a program executes, the information they convey can be hidden—for instance by masking the secret. Systems could be designed so that manifestations of the perturbations can be eliminated. Take for example, Listing 1.1. The Divide function can be rewritten so that a division is always carried out irrespective of the value of b, for instance, by performing a dummy division when $b = 0$. Further, the division hardware could be redesigned to ensure that the time taken is independent of the value of the divisor b, thereby preventing any timing channels that occur in the divide operation. Registers in the system can be manufactured in such a way that the power consumed does not depend on the data it stores. As a final resort, the systems could be shielded to ensure that side-channel observations are not made. Electromagnetic shields for instance could be used to prevent radiation-based side-channel attacks. Precision clocks in the system could be disabled to ensure high resolution timers are not available to an attacker, thereby mitigating timing side channels.

Such suppression techniques do not completely protect and in many cases come with large overheads. Consider for instance, huge noise generation circuits added to a mobile phone in order to hide side-channel leakage. This would not only make the phone more bulky but also increase its cost. Worse still would be a large electromagnetic shield around the phone to prevent leakage through radiation. Building efficient and low overhead countermeasures for side-channel attacks is still an open research problem.

1.2 Timing Attacks

Time has always been known to contain information but it was not until 1977 that Schaefer, Gold, Linde, and Scheid published a research paper showing how time could be used to carry information [6]. They showed how a timing channel could be established between two programs in a system to communicate covertly. Information can be transferred from one program to another by the amount of time a sender holds the CPU. For instance, holding the CPU for 10, 20, or 30 μs, can be used to represent a 0, 1, or 2 respectively. Over the years several covert timing channels have been discovered. The rate at which a program performs paging, schedules disks, utilizes cache memory, and scheduled to execute in a CPU have been used to transfer information between two collaborating programs.

In 1996, Kocher published the first timing attack against a cryptographic cipher. He used the variance in the execution time of a cipher to predict secret information. This attack kindled interest in the cryptography community and spurred research in the area, resulting in several papers published on timing attacks and defenses on a variety of ciphers. Timing attacks use the fact that the time required to perform an operation varies depending on the cipher's inputs. The inputs, comprising of a plaintext and key, undergo a series of mathematical operations before the encrypted message is obtained. The mathematical operations performed during encryption affects results of conditional statements, branch destinations, the number and type of instructions executed, instruction operands, and the memory locations accessed by the program. These are the perturbations in the system and could potentially contain information about the secret plaintext and key.

There are two ways by which a timing attacker could monitor these perturbations. The first is by timing executions of the cipher. This works since many of the perturbations in the system made during the execution of the cipher directly affects the execution time. For instance, the pattern with which cache hits and misses occur during execution of the cipher would affect its execution time. Alternatively, measuring the perturbations made by a cipher can be done by another program (called the *spy*).

Algorithm 1.1: Modular Exponentiation

```
   Input: y, x, n
   Output: s = y^x mod n
1  begin
2  |  s ← 1
3  |  i ← w − 1
4  |  while i >= 0 do
5  |  |  s ← s^2 mod n
6  |  |  if x[i] = 1 then
7  |  |  |  s ← s * y mod n
8  |  |  end
9  |  |  i ← i − 1
10 |  end
11 |  return s
12 end
```

The spy would determine perturbations by monitoring the usage of various hardware components in the system during or soon after the spy executes. This indirect monitoring is generally done with timing channels. For instance, the spy could identify the memory accesses made by the cipher, by using timing to determine how cache memories are utilized.

Just like in any other side-channel attack, a timing attack has two phases: an online phase, followed by an offline phase. The *online phase* monitors the time required for certain operations to be carried out, while the *offline phase* predicts the secret key from the timing information collected. We illustrate how the two phases work by taking a look at Kocher's timing attack.

1.2.1 Kocher's Timing Attack

Paul Kocher's attack was on a naïve implementation of a modular exponentiation algorithm. The algorithm takes as input three numbers y, x, and n, and determines the remainder when y^x is divided by n. This is represented as $y^x \ mod \ n$. The algorithm (depicted in Algorithm 1.1) is an integral part of ciphers such as RSA (Rivest Shamir and Adleman) and DH (Diffie Hellman) based ciphers. Typically, n is public and y can be found by an eavesdropper, while the exponent x is secret. Each iteration of the loop considers a bit of x (denoted $x[i]$ in the algorithm) starting from its most significant bit. Assuming x to be of length w bits, there are w iterations of the loop. The conditional statement in line 6 results in a multiplication being performed only if $x[i] = 1$. Thus, a value of $x[i] = 1$ results in a modular squaring followed by a modular multiplication, while a value of $x[i] = 0$ only has a modular squaring.

The attacker knows the value of n and wants to get x. We assume the attacker is powerful enough to be able to invoke the program (or function) which implements

Algorithm 1.1 and can choose or monitor different values of y. She can also replace x with her own exponent, which we denote x^*.

The attacker triggers the execution of the function with secret x and monitors its execution time. The execution time for the algorithm varies depending on the length of x (*i.e.,* w) and its value. This can be denoted as

$$T = e + \sum_{i=0}^{w-1} t_i \,, \tag{1.1}$$

where t_i is the time required for the i-th iteration of the loop (i.e., corresponding to the bit $x[i]$) and e includes the measurement error, loop overhead, and all other sources of inaccuracies. For a fixed x, time distributions can be obtained by observing the execution time corresponding to different values of y. The variance of the distribution is given by

$$Var(T) = Var(e) + w \cdot Var(t). \tag{1.2}$$

The mathematical formalism for the encryption time and its variance is used to recover the secret x. The attack is iterative in the sense that the attacker discovers x starting from its most significant bit. She discovers bit $x[b]$ (where $0 \le b < w$) only after recovering all bits from $x[w-1]$ to $x[b+1]$.

The attacker makes a guess of $x^*[b]$ (either 0 or 1), chooses $x^* = (x[w-1] \parallel x[w-2] \parallel x[w-3] \parallel \cdots \parallel x[b+1] \parallel x^*[b])$, triggers modular exponentiation, and obtains the execution time (T^*). The difference between the time for the two executions is

$$\begin{aligned} T - T^* &= [e + \sum_{i=0}^{w-1} t_i] - [t_b^* + \sum_{i=b+1}^{w-1} t_i] \\ &= [e + \sum_{i=0}^{b-1} t_i] + t_b - t_b^* \end{aligned} \tag{1.3}$$

If the guess is correct ($x[b] = x^*[b]$), then $t_b = t_b^*$ and the difference between the time for the two executions is $e + \sum_{i=0}^{b-1} t_i$. Further, on a correct guess, the difference in the variance obtained after measuring the execution time of several exponentiation is given by

$$Var[T - T^*] = Var(e) + b \cdot Var(t). \tag{1.4}$$

If the guess is wrong

$$Var[T - T^*] = Var(e) + (b+2) \cdot Var(t). \tag{1.5}$$

In other words, a correct guess causes a reduction in variance, while a wrong guess causes the variance to increase. This distinguisher can be used to validate the guess for $x[b]$.

The accuracy with which a bit in x is determined depends on how a multiplication operation can impact the execution time of a single iteration of the loop. Modern microprocessors have hardware multipliers capable of multiplying two inputs of 32 or 64 bit in a single clock cycle. If the length of s and y are small, a single microprocessor instruction, typically taking less than a nano second would be required to perform the multiplication. In such a case, the impact of the multiplication in the execution time of an iteration is negligible. In spite of this, the multiplication could be detected with reasonable accuracy, provided sufficient amount of time measurements are collected and averaged. As the length of s and y increase beyond the range of the available hardware multiplier, multiple clock cycles would be required to perform the multiplication. Each clock cycle would typically produce a partial product. In this case, the impact of the multiplier in the execution time would increase and the multiplication will be identified with a higher accuracy.

Subsequent to Kocher's work, several timing attacks have been proposed. The attacks target a variety of ciphers, utilize different techniques, and make different assumptions about the attacker. The next part of this section attempts to classify the different timing attacks on ciphers.

1.2.2 Taxonomy of Timing Attacks

Classification Based on the Type of Cipher: Ciphers are broadly classified as symmetric-key and asymmetric-key ciphers. The attack technique varies in each case. Timing attacks on asymmetric-key ciphers are iterative. One bit of the key is recovered at a time just as in Kocher's timing attack on the modular exponentiation algorithm. Timing attacks on symmetric key ciphers target implementations that use look-up tables. Leakage from these ciphers is due to perturbations in the cache memory present in the system.

Classification Based on the Perturbed Component: This classification is based on the component in the system that is perturbed by the cipher's execution. Kocher's timing attack for instance falls into a category based on the execution path, since different execution paths are taken based on the secret. The requirement for these attacks is that the conditional branch depends on the value of the cipher's secret. Further, the paths should have different execution time requirements. These attacks are generally relevant to public-key ciphers.

A closely related category of timing attacks target perturbations in the processor's branch predictor. These predictors are present in most modern processors in order to reduce the performance overheads when a branch occurs. These units automatically predict the branches during the execution of the program and fetch instructions from the new location. Branch predictors can substantially influence execution time and have been exploited by attackers to determine the secret from public key ciphers.

Perturbations made to cache memories have been used by attackers to discover secret keys from ciphers, especially block ciphers. Attacks that use cache memories require that the cipher accesses memory at locations which depend on the secret key.

Timing attacks based on cache memories use the fact that a cache miss takes considerably longer time compared to a cache hit. This is manifested in the execution time.

Classification Based on the Attacker's Capabilities: Attackers capabilities may differ. The most powerful attacker is one who can choose the plaintext to be encrypted, trigger encryptions, accurately monitor the side channel after every instruction executed. On the other side of the spectrum, the weakest attacker is one who can only monitor execution time remotely, for instance over a network. The capabilities of the attacker would therefore dictate the attack's success.

1.3 Organization

A primary goal of this book is to introduce and analyze timing attacks on block ciphers due to information leaking from cache memories in the system. The book discusses how microarchitectural features in the cache memory such as automatic hardware prefetching, nonblocking, parallel, and pipelined servicing of cache misses can influence leakage. Mathematical analysis is used to quantify this leakage. Another goal of the book is to introduce and analyze timing attacks on asymmetric key ciphers due to information leakage in the processor's branch prediction units. The instruction flow of the asymmetric ciphers can be revealed using the information leakage from the branch predictors. In this book, using timing as side channel, we discuss how the microarchitectural features can be monitored efficiently, and due to their deterministic nature they eventually result in leaking the secret instruction flow. The organization is as follows.

In Chapter 2, a brief review of modern cryptography is given. Topics addressed include an overview of encryption algorithms, their implementation, and classical cryptanalysis. Covert timing channels used for stealthy communication is introduced, and a brief introduction to formal analysis of side-channel attacks is presented.

The aim of Chapter 3 is to provide an overview of superscalar processor architecture, modern cache memories, and branch predictors used in modern computer systems. Subtle changes in these components can significantly affect information leakage from timing channels.

Chapter 4 discusses how time measurements can be accurately made on systems. Accurate time measurements are critical for the success of most timing attacks. The chapter then discusses how cache hits and misses can be distinguished from the execution time. These hits and misses are used to build a time-driven attack on a block cipher.

The information obtained from the attack presented in Chapter 4 is restricted by the size of the cache line. Chapter 5 shows how this restriction can be overcome by advanced attack strategies. The chapter shows how block cipher algorithms and their differential properties can be exploited to build powerful time-driven cache attacks.

Chapter 6 develops a framework to analyze information leakage in time-driven cache attacks. The framework is capable of analyzing modern cache memories that are equipped with microarchitectural acceleration features such as nonblocking, parallel, pipelined, and out-of-order servicing of cache misses. The framework is used to evaluate the leakage of popular block implementations and also identify the ideal way a block cipher should be implemented.

Chapter 7 introduces profiled time-driven cache attacks. These attacks require a powerful adversary who can characterize the system behavior when the cipher is executed. The adversary's success would differ depending on the way the block cipher is implemented. Further, automatic hardware prefetching can abet the attack.

Chapter 8 provides an illustrative description of access-driven attacks on block ciphers. Just as in the attacks described in Chap. 4, these attacks track cache trace patterns made by an executing block cipher. The tracing is done by a malicious process which schedules itself in such a manner, so as to observe the cache traces of individual accesses made by the cipher. The chapter explains various strategies used by the malicious process in order to make fine-grained timing measurements to observe several microarchitectural events.

Chapter 9 starts with an introduction to the algorithms used in the implementation of the public key cipher RSA. The branch predictors play a major role in predicting results of branch conditions in the cipher. In this chapter we provide several implementations where the leakage in the conditional instructions present in RSA are exploited using time as a side channel.

In Chap. 10, we highlight some of the important countermeasures that have been proposed for time-driven attacks. These countermeasures differ in how they eliminate leakage, the location in the system that they are applied, and the performance overhead.

Reference

1. Shannon CE (1949) Communication theory of secrecy systems. Bell Syst Tech J 28, 656–715
2. Kocher PC, Jaffe J, Jun B (1999) Differential power analysis, CRYPTO'99 Proceedings of the 19th Annual International Cryptology Conference on Advances in Cryptology, pp 388–397. Springer-Verlag London
3. Agrawal D, Archambeault B, Rao JR, Rohatgi P (2003) The EM Side-Channel(s), Cryptographic Hardware and Embedded Systems – CHES 2002, pp 29–45, Springer, Berlin
4. Genkin D, Shamir A, Tromer E (2013) RSA Key Extraction via Low-Bandwidth Acoustic Cryptanalysis, Cryptology ePring Archive, Report 2013/857
5. Schaefer M, Gold B, Linde R, Scheid J (n.d.) Program Confinement in KVM/370, ACM'77 Proceedings of the 1977 annual conference, pp 404–410, ACM New York, NY, USA
6. Kocher P (1996) Timing Attacks on Implementations of Diffie-Hellman, RSA, DSS, and other Systems. CRYPTO'96 Proceedings of the 16th Annual International Cryptology Conference on Advances in Cryptology, pp 104–113, Springer-Verlag, London, UK

Chapter 2
Modern Cryptography

The art of cryptographic algorithms is an ever evolving field. Initiating from prehistoric times, the main objective of cryptographic algorithms have been to protect and allow usage of information in a legal manner. Encryption is a process of converting a plaintext message into a piece of random-looking text, often called ciphertext. The ciphertext, ideally, should contain or transfer no information to the curious third party, often referred to as the adversary. In order to allow the intended receiver to obtain back the message from the ciphertext, every encryption algorithm is reversible. Thus, the inverse operation of obtaining back the plaintext from the ciphertext, is called decryption. According to the wisdom in cryptography, the algorithms for encryption and decryption are always published and known to even the adversary. Then what does the security rely on? The mappings (from plaintext to ciphertext and vice versa) depends on an information called the key, which is hidden from the attacker. While the goal of cryptography is to design and construct such ciphering algorithms, the objective of cryptanalysis is to develop techniques to obtain the key more efficiently than making a random guess on the key. There are different classes of cryptographic algorithms, depending on their objectives. While secrecy is the obvious requirement, there are other important goals too; for example, algorithms to guarantee the integrity of information, or methods to ensure that one cannot deny a commitment to a transaction (often called the property of nonrepudiation). Then there are algorithms which ensures that authenticity is maintained in a communication, and legal parties can trust with whom they are communicating. In this book we mainly target cryptographic algorithms, with respect to secrecy of data. But often these constructions can be used for achieving the other requirements, namely authentication and nonrepudiation. Hash functions, which are used for integrity checks are not in the scope of this book.

2.1 Types of Encryption Algorithms

For ciphers, there is an encryption key (K_a) and a decryption key (K_b), which are equal for certain classes of algorithms (called symmetric-key ciphers) and different for the other class (called asymmetric-key ciphers). The scenario of a cryptographic

C. Rebeiro et al., *Timing Channels in Cryptography*,
DOI 10.1007/978-3-319-12370-7_2

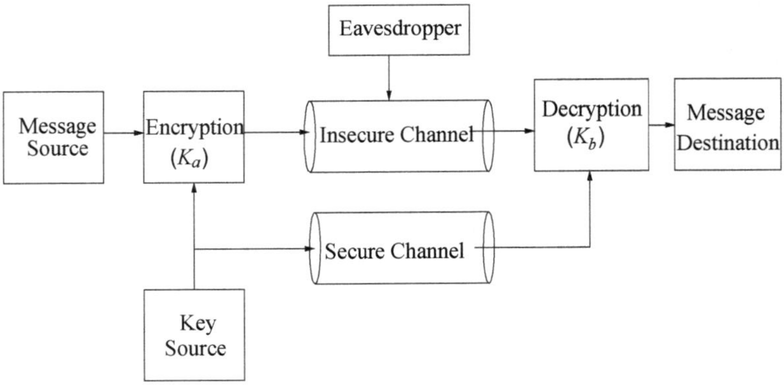

Fig. 2.1 Secret key cryptosystem model

communication is illustrated in Fig. 2.1. The encryptor uses a key K_a and the decryptor a key K_b, where depending on the equality of K_a and K_b there are two important classes of cryptographic algorithms. More precisely, the two classes of ciphers are:

- Private-key (or symmetric) ciphers: These ciphers have the same key shared between the sender and the receiver. Thus, referring to Fig. 2.1 $K_a = K_b$.
- Public-key (or asymmetric) ciphers: In these ciphers we have $K_a \neq K_b$. The encryption key and the decryption keys are different.

As in symmetric-key or private-key algorithms both the encryptor and decryptor use the same key, it must somehow be securely exchanged before secret key communication can begin. The key exchange is a major bottleneck and for n-parties in a network, the number of key exchanges required can grow quite fast (${}^nC_2$ ways). However, these algorithms are often fast and are used for bulk data encryption. Two important subclasses of symmetric-key algorithms are block ciphers and stream ciphers. Block ciphers, as the name suggest operates on fixed blocks or chunks of data, while stream ciphers operate on bits or few bits of data (thus the encryption takes place like a stream!). The Advanced Encryption Standard (AES) is a very popular block cipher, while Trivium is a popular stream cipher.

Public-key algorithms, on the other hand, provide a nice solution to the key-exchange problem. In such algorithms, as we discussed the encryption and decryption keys are different. The algorithms have a key pair, consisting of (i) Public key, which can be freely distributed and is used to encrypt messages. In Fig. 2.1, this is denoted by the key K_a, and (ii) Private key, which must be kept secret and is used to decrypt messages. The decryption key is denoted by K_b in the Fig. 2.1.

In the public key or asymmetric ciphers, the two parties—often called Alice and Bob—are communicating with each other and have their own key pair. They distribute their public keys freely. Mallory (or the adversary) has the knowledge of not only the encryption function, the decryption function, and the ciphertext, but also has the capability to encrypt the messages using Bob's public key. However, she is unaware of the secret decryption key, which is the private key of the algorithm. The security of these classes of algorithms rely on the assumption that it is computationally hard or complex to obtain the private key from the public information. Doing so would imply that the adversary solves a mathematical problem which is widely believed to be difficult. It may be noted that we do not have any proofs for their hardness; however, we are unaware of any efficient techniques to solve them. The elegance of constructing these ciphers lies in the fact that the public keys and private keys still have to be related in the sense, that they perform the invertible operations to obtain the message back. This is achieved through a class of magical functions, which are called *one-way* functions. These functions are easy to compute in one direction, while computing the inverse from the output is believed to be a difficult problem. RSA is a famous public-key algorithm for this class of ciphers.

Example 2.1 This cipher is called the famous RSA algorithm (Rivest Shamir Adleman). Let $n = pq$, where p and q are properly chosen and large prime numbers. Here the proper choice of p and q are to ensure that factorization of n is mathematically complex. The plaintexts and ciphertexts are $P = C = Z_n$, the keys are $K_a = \{n, a\}$ and $K_b = \{b, p, q\}$, such that $ab \equiv 1 \bmod \phi(n)$. The encryption and decryption functions are defined as, $\forall x \in P$, $e_{K_a}(x) = y = x^a \bmod n$ and $d_{K_b}(y) = y^b \bmod n$.

Both symmetric-key ciphers and asymmetric-key ciphers are widely studied. We provide a quick overview on some facts, which we use in the book.

2.2 Block Ciphers: An Important Family of Symmetric-Key Ciphers

The book develops a general theory for time-driven cache attacks on block ciphers. The theory can be applied to any cipher with an iterative structure that is implemented with look-up tables. To test the theory we selected few ciphers, like AES, CAMELLIA, and CLEFIA. AES was chosen because it is the world wide standard. It has a substitution permutation network (SPN) structure and the implementation generally used has 5 large tables of 1024 bytes. The first four tables are invoked 36 times per 128-bit encryption, while the last table is invoked 16 times. Many of the latest CPUs support dedicated instructions for AES [1], on which cache attacks fail. Besides SPN, most cipher designs follow the Feistel structure (Fig. 2.2). The figure shows the rth round of a Feistel block cipher. The block is divided into two parts, L_r (Left) and R_r (Right), which are recursively computed as $L_r = R_{r-1} \oplus \mathbf{F}(L_{r-1}, k_r)$, and $R_r = L_{r-1}$. The transformation $\mathbf{F}$ is composed of several nonlinear transformations or S-Boxes, and combine the round key k_r with a portion of the block. The reversibility of the round does not depend on whether the nonlinear layer $\mathbf{F}$ is invertible or not. The rounds are repeated for certain number of iterations to ensure sufficient security margin against known attacks.

We chose CAMELLIA as a representative of these ciphers. The third cipher we chose is CLEFIA, which has a generalized Feistel structure [2] and a round function which is conceptually similar to that of AES. Further, unlike the AES implementation considered, CLEFIA and CAMELLIA implementations used small tables of 256 bytes. This section provides a brief description of each cipher algorithm.

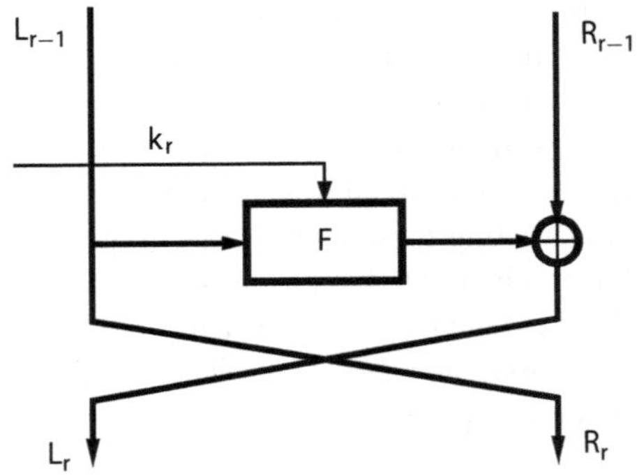

Fig. 2.2 Feistel structure of a block cipher

2.2.1 AES

In 2001, the National Institute of Standards and Technology (NIST) recommended the use of Rijndael as the AES [3]. AES is a symmetric-key block cipher and can use key sizes of either 128, 192, or 256 bits to encrypt and decrypt 128-bit blocks. We summarize the AES-128 standard, which uses a key size of 128 bits. The input to AES-128 is arranged in a 4×4 matrix of bytes called *state*. The state undergoes a series of transformations in ten rounds during the encryption process.

Algorithm 2.1 presents the AES-128 algorithm. The first operation on the input is the AddRoundKeys, which serves to provide the initial randomness by mixing the input key. The state is then subjected to nine rounds to further increase the diffusion and confusion in the cipher [4]. Each round comprises four operations on the state: SubBytes, ShiftRows, MixColumns, and AddRoundKeys. The state is then subjected to a final round, which has all operations except the MixColumns operation.

The four AES operations are defined as follows:

Algorithm 2.1: AES-128

Input: 128 bit plaintext input block $\mathbf{x}$ and 128 bit round keys $\mathbf{k}^{(0)}, \mathbf{k}^{(2)}, \cdots, \mathbf{k}^{(10)}$ generated from a secret key using the AES-128 key generation algorithm
Output: 128 bit ciphertext

begin
  $r \leftarrow 1$
  $\mathbf{s}^{(0)} \leftarrow \mathsf{AddRoundKeys}(\mathbf{x}, \mathbf{k}^{(0)})$
  while $r <= 9$ **do**
    $\mathbf{s}^{(r)} \leftarrow \mathsf{SubBytes}(\mathbf{s}^{(r-1)})$
    $\mathbf{s}^{(r)} \leftarrow \mathsf{ShiftRows}(\mathbf{s}^{(r)})$
    $\mathbf{s}^{(r)} \leftarrow \mathsf{MixColumns}(\mathbf{s}^{(r)})$
    $\mathbf{s}^{(r)} \leftarrow \mathsf{AddRoundKeys}(\mathbf{s}^{(r)}, \mathbf{k}^{(r)})$
  end
  $\mathbf{s}^{(10)} \leftarrow \mathsf{SubBytes}(\mathbf{s}^{(9)})$
  $\mathbf{s}^{(10)} \leftarrow \mathsf{ShiftRows}(\mathbf{s}^{(10)})$
  $\mathbf{y} \leftarrow \mathsf{AddRoundKeys}(\mathbf{s}^{(10)}, \mathbf{k}^{(10)})$
  return y
end

- AddRoundKeys : Each element in the state is subjected to a bitwise ex-or with a 128-bit round key. The round key is generated from the secret key by a key expansion algorithm as in Algorithm 2.2 [4].
- SubBytes : Each element in the state is replaced by an affine transformation of its inverse in the field $GF(2^8)$. For a byte s_i in the state, this operation is denoted by $S(s_i)$.
- ShiftRows : Provides a cyclic shift of the ith row in the state by i bytes toward the left (where $0 \leq i \leq 3$). That is, each byte in the ith row is cyclically shifted to the left by i bytes.
- MixColumns : Provides a column-wise linear transformation of the state matrix. Each column of the state matrix is considered as a polynomial of degree 3 with coefficients in $GF(2^8)$ and multiplied by the polynomial $\{03\}\alpha^3 + \{01\}\alpha^2 + \{01\}\alpha + \{02\} \bmod (\alpha^4 + 1)$. The combination of ShiftRows and MixColumns provide the necessary diffusion for the cipher.

Key Expansion Algorithm [3] takes the secret key as input and generates round keys for 11 AddRoundKeys operations performed in AES. Key expansion as in Algorithm 2.2 uses two operations ROTWORD and SUBWORD which performs cyclic shift and substituition of four bytes (B_0, B_1, B_2, B_3) as

$$ROTWORD(B_0, B_1, B_2, B_3) = (B_1, B_2, B_3, B_0) \tag{2.1}$$

and

$$SUBWORD(B_0, B_1, B_2, B_3) = (SubBytes(B_0), SubBytes(B_1), SubBytes(B_2), SubBytes(B_3)) \tag{2.2}$$

Algorithm 2.2: Key Expansion of AES-128

Input: 128 bit secret key
Output: 11 round keys each of 4 words as $w[0], \cdots, w[43]$

```
begin
    RCon[1] ← 01000000
    RCon[2] ← 02000000
    RCon[3] ← 04000000
    RCon[4] ← 08000000
    RCon[5] ← 10000000
    RCon[6] ← 20000000
    RCon[7] ← 40000000
    RCon[8] ← 80000000
    RCon[9] ← 1B000000
    RCon[10] ← 36000000
    for i <= 0 to 3 do
        w[i] ← (k[4i], k[4i+1], k[4i+2], k[4i+3])
    end
    for i <= 4 to 43 do
        temp ← w[i−1]
        if i ≡ 0( mod 4) then
            temp = SUBWORD(ROTWORD(temp)) ⊕ RCon[i/4]
        end
        w[i] ← w[i−4] ⊕ temp
    end
    return (w[0], ··· , w[43])
end
```

In addition to this, AES involves a round constant term that is defined as $RCon[1], \cdots, RCon[10]$ as constants in hexadecimal before the key expansion Algorithm 2.2.

Starting from the 4×4 byte state, Fig. 2.3 shows the transformation it undergoes in a round (for $1 \leq r \leq 9$).

2.2.1.1 Software Implementations of AES

Of all operations, the SubBytes is the most difficult to implement. On 8-bit micro-controllers, a 256-byte look-up table is ideal to perform this operation. The table provides the necessary flexibility in terms of content, small footprint, and speed. For 32-bit platforms, more efficient implementations can be built using larger tables. We give a brief description of this method, which is known as T-table implementations. T-table implementations were first proposed in [5] and has been adopted by several crypto-libraries such as OpenSSL[1].

Consider four look-up tables defined as follows:

$$\mathcal{T}_0[z] = \begin{bmatrix} 02 \bullet S(z) \\ S(z) \\ S(z) \\ 03 \bullet S(z) \end{bmatrix}; \mathcal{T}_1[z] = \begin{bmatrix} 03 \bullet S(z) \\ 02 \bullet S(z) \\ S(z) \\ S(z) \end{bmatrix}; \mathcal{T}_2[z] = \begin{bmatrix} S(z) \\ 03 \bullet S(z) \\ 02 \bullet S(z) \\ S(z) \end{bmatrix};$$

$$\mathcal{T}_3[z] = \begin{bmatrix} S(z) \\ S(z) \\ 03 \bullet S(z) \\ 02 \bullet S(z) \end{bmatrix} \tag{2.3}$$

Each table is of 1024 bytes mapping a byte z of the state to a 32-bit value. Using these tables, the first nine AES rounds can be expressed as follows

$$\begin{aligned} \mathbf{s}^{(r+1)} =& \mathcal{T}_0[s_0^{(r)}] \oplus \mathcal{T}_1[s_5^{(r)}] \oplus \mathcal{T}_2[s_{10}^{(r)}] \oplus \mathcal{T}_3[s_{15}^{(r)}] \oplus [k_0^{(r)}\ k_1^{(r)}\ k_2^{(r)}\ k_3^{(r)}]^T \ \| \\ & \mathcal{T}_0[s_4^{(r)}] \oplus \mathcal{T}_1[s_9^{(r)}] \oplus \mathcal{T}_2[s_{14}^{(r)}] \oplus \mathcal{T}_3[s_3^{(r)}] \oplus [k_4^{(r)}\ k_5^{(r)}\ k_6^{(r)}\ k_7^{(r)}]^T \ \| \\ & \mathcal{T}_0[s_8^{(r)}] \oplus \mathcal{T}_1[s_{13}^{(r)}] \oplus \mathcal{T}_2[s_2^{(r)}] \oplus \mathcal{T}_3[s_7^{(r)}] \oplus [k_8^{(r)}\ k_9^{(r)}\ k_{10}^{(r)}\ k_{11}^{(r)}]^T \ \| \\ & \mathcal{T}_0[s_{12}^{(r)}] \oplus \mathcal{T}_1[s_1^{(r)}] \oplus \mathcal{T}_2[s_6^{(r)}] \oplus \mathcal{T}_3[s_{11}^{(r)}] \oplus [k_{12}^{(r)}\ k_{13}^{(r)}\ k_{14}^{(r)}\ k_{15}^{(r)}]^T \end{aligned} \tag{2.4}$$

A byte of the state in the current round ($\mathbf{s}^{(r)}$) is denoted $s_i^{(r)}$ and the next round state is denoted $\mathbf{s}^{(r+1)}$, where $0 \leq r \leq 9$ and $0 \leq i \leq 15$. The final round cannot use these tables due to the absence of the MixColumns operation.

[1] http://www.openssl.org.

Fig. 2.3 Transformations of the state in a round of AES ($1 \le r \le 9$)

s_0	s_4	s_8	s_{12}
s_1	s_5	s_9	s_{13}
s_2	s_6	s_{10}	s_{14}
s_3	s_7	s_{11}	s_{15}

s_i for $0 \le i \le 15$ is the elements of the state

↓ SubByte

$S(s_0)$	$S(s_4)$	$S(s_8)$	$S(s_{12})$
$S(s_1)$	$S(s_5)$	$S(s_9)$	$S(s_{13})$
$S(s_2)$	$S(s_6)$	$S(s_{10})$	$S(s_{14})$
$S(s_3)$	$S(s_7)$	$S(s_{11})$	$S(s_{15})$

↓ ShiftRows

$S(s_0)$	$S(s_4)$	$S(s_8)$	$S(s_{12})$
$S(s_5)$	$S(s_9)$	$S(s_{13})$	$S(s_1)$
$S(s_{10})$	$S(s_{14})$	$S(s_2)$	$S(s_6)$
$S(s_{15})$	$S(s_3)$	$S(s_7)$	$S(s_{11})$

↓ MixColumns

$2s'_0 \oplus 3s'_5 \oplus s'_{10} \oplus s'_{15}$	$2s'_4 \oplus 3s'_9 \oplus s'_{14} \oplus s'_3$	$2s'_8 \oplus 3s'_{13} \oplus s'_2 \oplus s'_7$	$2s'_{12} \oplus 3s'_1 \oplus s'_6 \oplus s'_{11}$
$s'_0 \oplus 2s'_5 \oplus 3s'_{10} \oplus s'_{15}$	$s'_4 \oplus 2s'_9 \oplus 3s'_{14} \oplus s'_3$	$s'_8 \oplus 2s'_{13} \oplus 3s'_2 \oplus s'_7$	$s'_{12} \oplus 2s'_1 \oplus 3s'_6 \oplus s'_{11}$
$s'_0 \oplus s'_5 \oplus 2s'_{10} \oplus 3s'_{15}$	$s'_4 \oplus s'_9 \oplus 2s'_{14} \oplus 3s'_3$	$s'_8 \oplus s'_{13} \oplus 2s'_2 \oplus 3s'_7$	$s'_{12} \oplus s'_1 \oplus 2s'_6 \oplus 3s'_{11}$
$3s'_0 \oplus s'_5 \oplus s'_{10} \oplus 2s'_{15}$	$3s'_4 \oplus s'_9 \oplus s'_{14} \oplus 2s'_3$	$3s'_8 \oplus s'_{13} \oplus s'_2 \oplus 2s'_7$	$3s'_{12} \oplus s'_1 \oplus s'_6 \oplus 2s'_{11}$

where $s'_i = S(s_i)$ for $0 \le i \le 15$

↓ AddRoundKey

$2s'_0 \oplus 3s'_5 \oplus s'_{10} \oplus s'_{15} \oplus k_0$	$2s'_4 \oplus 3s'_9 \oplus s'_{14} \oplus s'_3 \oplus k_4$	$2s'_8 \oplus 3s'_{13} \oplus s'_2 \oplus s'_7 \oplus k_8$	$2s'_{12} \oplus 3s'_1 \oplus s'_6 \oplus s'_{11} \oplus k_{12}$
$s'_0 \oplus 2s'_5 \oplus 3s'_{10} \oplus s'_{15} \oplus k_1$	$s'_4 \oplus 2s'_9 \oplus 3s'_{14} \oplus s'_3 \oplus k_5$	$s'_8 \oplus 2s'_{13} \oplus 3s'_2 \oplus s'_7 \oplus k_9$	$s'_{12} \oplus 2s'_1 \oplus 3s'_6 \oplus s'_{11} \oplus k_{13}$
$s'_0 \oplus s'_5 \oplus 2s'_{10} \oplus 3s'_{15} \oplus k_2$	$s'_4 \oplus s'_9 \oplus 2s'_{14} \oplus 3s'_3 \oplus k_6$	$s'_8 \oplus s'_{13} \oplus 2s'_2 \oplus 3s'_7 \oplus k_{10}$	$s'_{12} \oplus s'_1 \oplus 2s'_6 \oplus 3s'_{11} \oplus k_{14}$
$3s'_0 \oplus s'_5 \oplus s'_{10} \oplus 2s'_{15} \oplus k_3$	$3s'_4 \oplus s'_9 \oplus s'_{14} \oplus 2s'_3 \oplus k_7$	$3s'_8 \oplus s'_{13} \oplus s'_2 \oplus 2s'_7 \oplus k_{11}$	$3s'_{12} \oplus s'_1 \oplus s'_6 \oplus 2s'_{11} \oplus k_{15}$

where k_i for $0 \le i \le 15$ is the round key

2.2.2 CLEFIA

CLEFIA is a 128-bit block cipher developed by Sony [6] and is currently incorporated in ISO/IEC 29192-2 as a light-weight block cipher standard. The specification [6, 7] defines three key lengths of 128, 192, and 256 bits. This book considers 128-bit keys though the results are valid for the other key sizes also. The structure of CLEFIA is shown in Fig. 2.4. The input has 16 bytes, P_0 to P_{15}, grouped into four byte words. There are 18 rounds, and in each round, the first and third words are fed into nonlinear functions $F0$ and $F1$ respectively. The output of $F0$ and $F1$ are ex-ored with the second and fourth words. Additionally, the second and fourth words are also whitened at the beginning and end of the encryption. The F functions take four input bytes and four round keys. The nonlinearity in the F functions are due to two 256 element s-boxes $S0$ and $S1$. Matrices $M0$ and $M1$ diffuse the outputs of the s-boxes. They are defined as follows:

$$M0 = \begin{pmatrix} 1 & 2 & 4 & 6 \\ 2 & 1 & 6 & 4 \\ 4 & 6 & 1 & 2 \\ 6 & 4 & 2 & 1 \end{pmatrix} \qquad M1 = \begin{pmatrix} 1 & 8 & 2 & A \\ 8 & 1 & A & 2 \\ 2 & A & 1 & 8 \\ A & 2 & 8 & 1 \end{pmatrix} \tag{2.5}$$

The design of the s-boxes $S0$ and $S1$ differs. $S0$ is composed of four s-boxes $SS0$, $SS1$, $SS2$, and $SS3$; each of 16 bytes. The output of $S0$ is given by :

$$\begin{aligned} \beta_l &= SS2[SS0[\alpha_l] \oplus 2 \cdot SS1[\alpha_h]] \\ \beta_h &= SS3[SS1[\alpha_h] \oplus 2 \cdot SS0[\alpha_l]] \end{aligned} \tag{2.6}$$

where $\beta = (\beta_h|\beta_l)$, $\alpha = (\alpha_h|\alpha_l)$, and $\beta = S0[\alpha]$. The output of $S1$ for the input byte α is given by $g((f(\alpha))^{-1})$, where g and f are affine transforms and the inverse is found in the field $GF(2^8)$.

The CLEFIA encryption algorithm has four whitening keys WK_0, WK_1, WK_2, and WK_3; and 36 round keys $RK_0, \ldots, RK_{35}$. Key expansion is a two-step process. First, a 128-bit intermediate key L is generated from the secret key K using a GFN function [7]. From this, the round keys and whitening keys are generated as shown below:

Step 1: $WK_0|WK_1|WK_2|WK_3 \leftarrow K$

Step 2: For $i \leftarrow 0$ to 8

$T \leftarrow L \oplus (CON_{24+4i}|CON_{24+4i+1}|CON_{24+4i+2}|CON_{24+4i+3})$

$L \leftarrow \Sigma(L)$

if i is odd: $T \leftarrow T \oplus K$

$RK4i|RK4i+1|RK4i+2|RK4i+3 \leftarrow T$

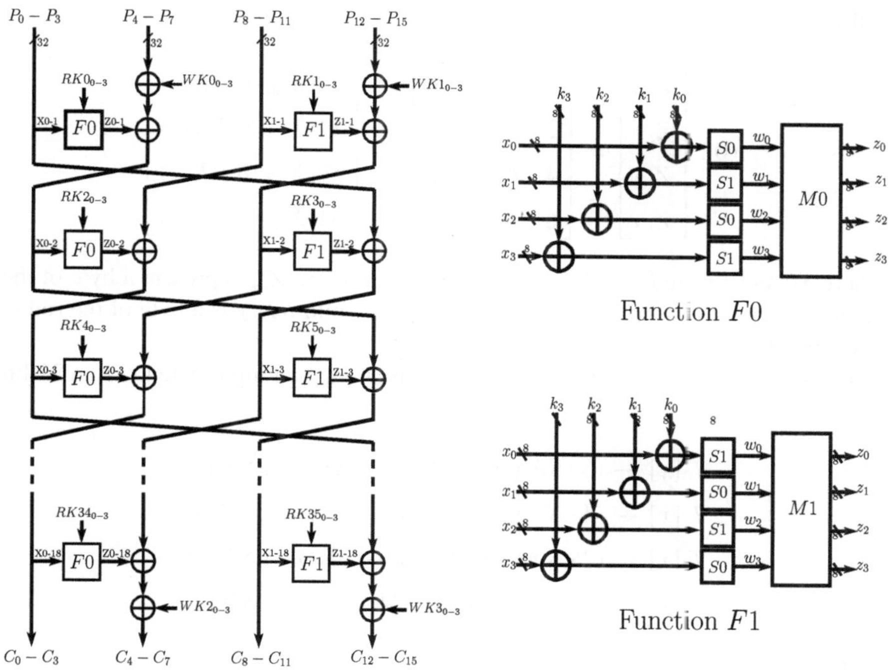

Fig. 2.4 CLEFIA block diagram

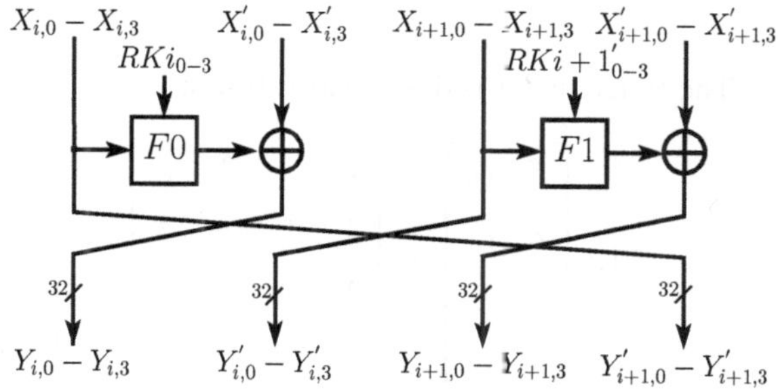

Fig. 2.5 A round of CLEFIA

The function Σ, known as the *double swap* function, rearranges the bits of L.

$$\Sigma(L) \leftarrow L_{(7\cdots 63)}|L_{(121\cdots 127)}|L_{(0\cdots 6)}|L_{(64\cdots 120)} \tag{2.7}$$

In the later part, thirty-six 32-bit constant values CON_i $(24 \le i < 60)$ are used.

Just like AES, CLEFIA can be implemented with T-tables. The T-table implementation for CLEFIA is in the Sect. 2.2.2.1.

2.2.2.1 *T*-Table Implementation of CLEFIA

A round of CLEFIA has 128-bit input and produces 128-bit output. Each input and output is grouped as four words as shown in Fig. 2.5. The function $F0$ is defined as follows,

$$F0: \begin{bmatrix} Z_{i,0} \\ Z_{i,1} \\ Z_{i,2} \\ Z_{i,3} \end{bmatrix} = \begin{bmatrix} 1 & 2 & 4 & 6 \\ 2 & 1 & 6 & 4 \\ 4 & 6 & 1 & 2 \\ 6 & 4 & 2 & 1 \end{bmatrix} \times \begin{bmatrix} S0(X_{i,0} \oplus RK_{i,0}) \\ S1(X_{i,1} \oplus RK_{i,1}) \\ S0(X_{i,2} \oplus RK_{i,2}) \\ S1(X_{i,3} \oplus RK_{i,3}) \end{bmatrix}$$

where $S0$ and $S1$ are CLEFIA's 8×8 s-boxes. $X_{i,k}$ and $Z_{i,k}$ represent a byte of the input and output word respectively $(0 \leq k \leq 3)$, while $RK_{i,k}$ is a byte of the round key. The matrix multiplication is in a finite field.

Consider four look-up tables as follows which map an input byte (x) to a 32-bit word.

$$T_0[x] \leftarrow (6 \cdot S0(x)|4 \cdot S0(x)|2 \cdot S0(x)|1 \cdot S0(x))$$
$$T_1[x] \leftarrow (4 \cdot S1(x)|6 \cdot S1(x)|1 \cdot S1(x)|2 \cdot S1(x))$$
$$T_2[x] \leftarrow (2 \cdot S0(x)|1 \cdot S0(x)|6 \cdot S0(x)|4 \cdot S0(x))$$
$$T_3[x] \leftarrow (1 \cdot S1(x)|2 \cdot S1(x)|4 \cdot S1(x)|6 \cdot S1(x)).$$

Then,

$$(Z_{i,3}|Z_{i,2}|Z_{i,1}|Z_{i,0}) = T_0[X_{i,0} \oplus RK_{i,0}] \oplus T_1[X_{i,1} \oplus RK_{i,1}] \oplus T_2[X_{i,2} \oplus RK_{i,2}] \oplus T_3[X_{i,3} \oplus RK_{i,3}]$$

The function $F1$ is defined as follows:

$$F1: \begin{bmatrix} Z_{i+1,0} \\ Z_{i+1,1} \\ Z_{i+1,2} \\ Z_{i+1,3} \end{bmatrix} = \begin{bmatrix} 1 & 8 & 2 & A \\ 8 & 1 & A & 2 \\ 2 & A & 1 & 8 \\ A & 2 & 8 & 1 \end{bmatrix} \times \begin{bmatrix} S1(X_{i+1,0} \oplus RK_{i+1,0}) \\ S0(X_{i+1,1} \oplus RK_{i+1,1}) \\ S1(X_{i+1,2} \oplus RK_{i+1,2}) \\ S0(X_{i+1,3} \oplus RK_{i+1,3}) \end{bmatrix}$$

The function F1 can be implemented using four tables as shown below. Each table maps the 8-bit input x to a 32-bit output.

$$T_4[x] \leftarrow (A \cdot S1(x)|2 \cdot S1(x)|8 \cdot S1(x)|1 \cdot S1(x))$$
$$T_5[x] \leftarrow (2 \cdot S0(x)|A \cdot S0(x)|1 \cdot S0(x)|8 \cdot S0(x))$$
$$T_6[x] \leftarrow (8 \cdot S1(x)|1 \cdot S1(x)|A \cdot S1(x)|2 \cdot S1(x))$$
$$T_7[x] \leftarrow (1 \cdot S0(x)|8 \cdot S0(x)|2 \cdot S0(x)|A \cdot S0(x))$$

Then,

$$\begin{aligned}(Z_{i+1,3}|Z_{i+1,2}|Z_{i+1,1}|Z_{i+1,0}) =& T_4[X_{i+1,0} \oplus RK_{i+1,0}]\\ &\oplus T_5[X_{i+1,1} \oplus RK_{i+1,1}]\\ &\oplus T_6[X_{i+1,2} \oplus RK_{i+1,2}]\\ &\oplus T_7[X_{i+1,3} \oplus RK_{i+1,3}].\end{aligned}$$

The output of the round is

$$\begin{aligned}(Y_{i,0}\cdots Y_{i,3}) &\leftarrow (Z_{i,0}\cdots Z_{i,3}) \oplus (X'_{i,0}\cdots X'_{i,3})\\ (Y'_{i,0}\cdots Y'_{i,3}) &\leftarrow (X_{i+1,0}\cdots X_{i+1,3})\\ (Y_{i+1,0}\cdots Y_{i+1,3}) &\leftarrow (Z_{i+1,0}\cdots Z_{i+1,3}) \oplus (X'_{i+1,0}\cdots X'_{i+1,3})\\ (Y'_{i+1,0}\cdots Y'_{i+1,3}) &\leftarrow (X_{i,0}\cdots X_{i,3}).\end{aligned}$$

In a similar manner, all rounds of CLEFIA can be implemented. Thus reducing the implementation to a series of memory accesses intertwined with ex-ors.

2.2.3 *CAMELLIA*

CAMELLIA is the 128-bit block cipher that was jointly developed by Mitsubishi and NTT in 2000 [8]. Since the cipher has been made available under a royalty-free license, it has been certified for use by the European Union and Japan. It has also become part of the OpenSSL project, and incorporated in Mozilla's Network Security Services (NSS modules). Support for CAMELLIA has been added to several security libraries as well as Mozilla's popular web browser, Firefox.

The 128-bit block cipher CAMELLIA has a Feistel structure as shown in Fig. 2.6. The 16 bytes plaintext input is grouped in two words of 8 bytes each: $\mathbf{x} = (x_1\|x_2\|\cdots\|x_8)$ and $\mathbf{y} = (y_1\|y_2\|\cdots\|y_8)$. There are 18 rounds in all, broken up into groups of 6 each. After the 6th and the 12th rounds, there are two FL/FL^{-1} function layers. In each round, there is an F function, which is a combination of key addition, substitution (S), and permutation (P). The substitution is done by using four s-boxes, whereas, the P function is implemented by using a diffusion matrix. Figure 2.7 shows the permutation operation and the diffusion matrix. The diffusion matrix also has an inverse depicted in the figure.

Each round has an addition of a round key. The ith round uses the round key $\mathbf{k}^{(i)}$. Each of these round keys are of 64 bits. Additionally, whitening keys $kw1$ and $kw2$ are applied at the start of encryption, while $kw3$ and $kw4$ are applied at the end of encryption.

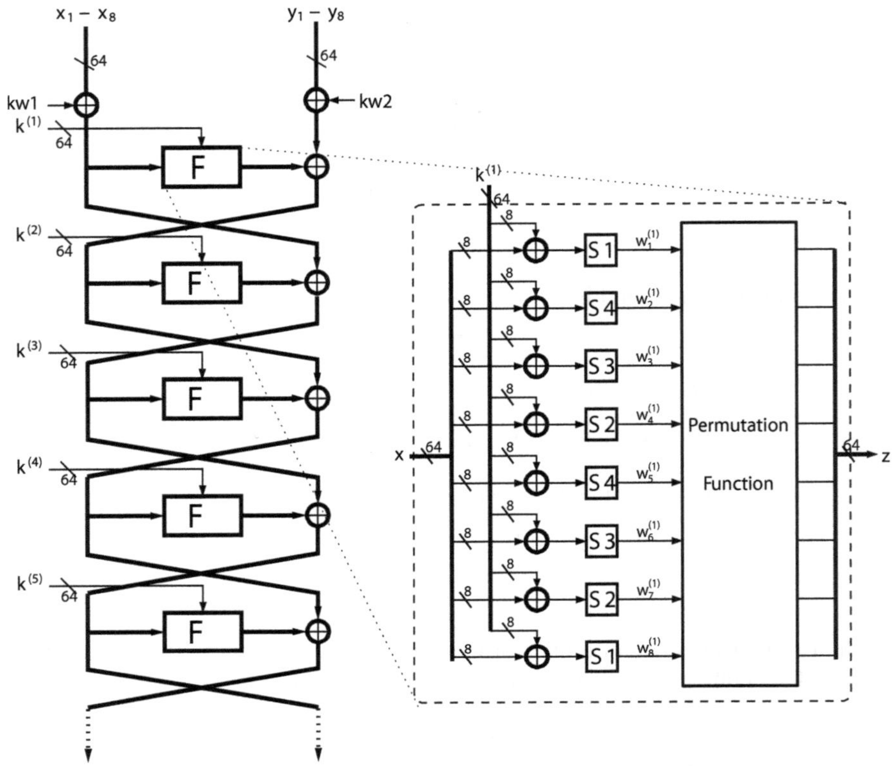

Fig. 2.6 Structure of CAMELLIA

Fig. 2.7 CAMELLIA's diffusion layer and its inverse

$$\begin{pmatrix} z_8 \\ z_7 \\ \cdot \\ \cdot \\ z_1 \end{pmatrix} \leftarrow P \bullet \begin{pmatrix} w_8 \\ w_7 \\ \cdot \\ \cdot \\ w_1 \end{pmatrix}$$

(a) Permutation Function

$$\begin{pmatrix} 0&1&1&1&1&0&0&1 \\ 1&0&1&1&1&1&0&0 \\ 1&1&0&1&0&1&1&0 \\ 1&1&1&0&0&0&1&1 \\ 0&1&1&1&1&1&1&0 \\ 1&0&1&1&0&1&1&1 \\ 1&1&0&1&1&0&1&1 \\ 1&1&1&0&1&1&0&1 \end{pmatrix}$$

(b) Diffusion Matrix P

$$\begin{pmatrix} 1&1&1&0&1&0&0&1 \\ 0&1&1&1&1&1&0&0 \\ 1&0&1&1&0&1&1&0 \\ 1&1&0&1&0&0&1&1 \\ 0&1&1&1&0&1&1&1 \\ 1&0&1&1&1&0&1&1 \\ 1&1&0&1&1&1&0&1 \\ 1&1&1&0&1&1&1&0 \end{pmatrix}$$

(c) Inverse Diffusion Matrix P^{-1}

2.3 Classical Cryptanalysis

A cryptanalytic attack is a procedure through which an attacker gains information about the secret decryption key. Attacks are classified according to the level of a priori knowledge available to the attacker.

A *ciphertext-only attack* is an attack where the cryptanalyst has access to ciphertexts generated using a given key but has no access to the corresponding plaintexts or the key. A *known-plaintext attack* is an attack where the cryptanalyst has access to both ciphertexts and the corresponding plaintexts but not the key.

A **chosen-plaintext attack** (CPA) is an attack where the cryptanalyst can choose plaintexts to be encrypted and has access to the resulting ciphertexts, again their purpose being to determine the key.

A **chosen-ciphertext attack** (CCA) is an attack in which the cryptanalyst can choose ciphertexts, apart from the challenge ciphertext and can obtain the corresponding plaintext. The attacker has access to the decryption device.

In case of CPA and CCA, adversaries can make a bounded number of queries to its encryption or decryption device. The encryption device is often called an oracle; meaning it is like a black-box without details like in an algorithm of how an input is transformed or used to obtain the output. Although this may seem a bit hypothetical, but there are enough real life instances where such encryption and decryption oracles can be obtained.

In the next section, we discuss one form of classical cryptanalysis of block ciphers, namely differential attacks, which is used in developing some of the cache attacks described in the book.

2.3.1 Classical Cryptanalysis of Block Ciphers

Block ciphers have been subjected to several forms of classical cryptanalysis; namely linear cryptanalysis, differential cryptanalysis, impossible differential attacks, related key attacks, boomerang attacks, square attacks are some popular cryptanalysis techniques. In this book, we study how microarchitectural features like cache memories and the accompanying hardware increases leakage when an encryption program runs on this platform. These attacks, often called as side-channel attacks (SCA) exploit additional leakage through timing information, and combine with classical methods, like differential attacks to provide efficient attack techniques. In this section, we provide a quick overview on the ideas of differential cryptanalysis, which will be useful for other attacks. Interested readers may refer [9] for ideas on other forms of cryptanalysis, which may (or may not) lead to more efficient cache attacks.

2.3.2 The Idea of Differential in Block Ciphers

Differential cryptanalysis is a chosen plain text attack and the attack is based on the information of the ex-or of two inputs, and the ex-or of corresponding outputs. Consider, a cipher expressed as $c = m \oplus k$, where m, k, and c are the plaintext, key, and the ciphertext blocks respectively, each of size b-bits. We know this is a secured cipher if the key is randomly and uniquely chosen for every encryption. The attacker has no information about the plaintext from the ciphertext. However, if the key is used twice for two encryptions there is a leakage of information. It is trivial to note, $c_0 \oplus c_1 = m_0 \oplus m_1$, thus showing that the *differential* (or, the difference which is computed useing ex-ors between two blocks) does not depend on the key, and thus the key (although unknown) does not hide the plaintext from the attacker! The differential is often denoted by $\Delta(c) = c_0 \oplus c_1$. Thus, in brevity we can state

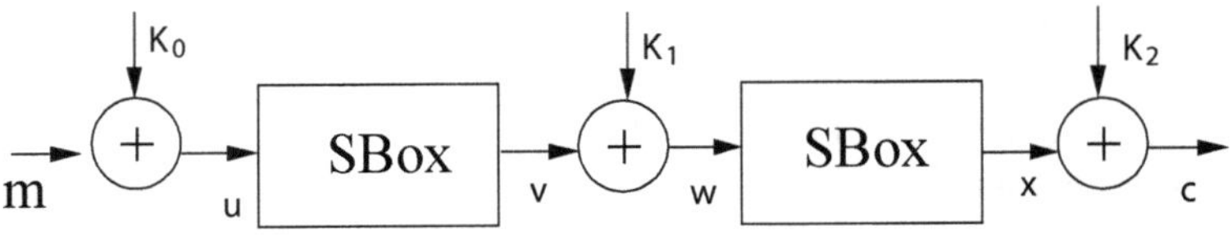

Fig. 2.8 Illustration of a differential attack on a toy cipher

Table 2.1 The S-Box description

x	0	1	2	3	4	5	6	7	8	9	A	B	C	D	E	F
$s[x]$	C	5	6	B	9	0	A	D	3	E	F	8	4	7	1	2

$\Delta(c) = \Delta(m)$ for the above equality. It may be noted that if the key mixing was done by some other reversible process (like modular integer addition), the differential would be accordingly modified (by using modular integer subtraction). Let us see how we can use this fact to cryptanalyze block ciphers with S-Boxes.

In the Fig. 2.8, a two round encryption is depicted where the plaintext m is transformed by mixing (ex-oring) with the keys, k_0, k_1, and k_2 along with substitution which occurs due to the nonlinear S-Boxes. Let us denote the S-Box mapping by S, and thus an input x is transformed to an output which is $S[x]$. The mapping we choose for illustration is that of the block cipher PRESENT[10]. For convenience we present the mapping in Table 2.1.

From the notion of differentials we know that ex-oring with the key has no effect on the differential. However, the S-Box being a nonlinear layer prevents passing of the ex-or differential. Thus, the attacker can guess k_2 and obtain the value of x, for two different encryptions. We denote this by saying that c_0 gives x_0, while c_1 gives x_1. One can perform the inverse S-Boxes (which have to be invertible!), and obtain w_0 and w_1. However, the uncertainty of k_1 prevents us from computing the values of v_0 and v_1. However, we know $\Delta(w) = w_0 \oplus w_1 = v_0 \oplus v_1 = \Delta(v)$. Likewise, if the attacker chooses $\Delta(m) = m_0 \oplus m_1$, she can compute $\Delta(u)$, however the S-Box prevents from determining $\Delta(v)$.

So, differential cryptanalysis performs a differential analysis of the S-Box. Consider, the Table 2.2, where the input differential $i \oplus j$, for two inputs i and j, is *chosen* to be F. The output differential, $S[i] \oplus S[j]$ is computed, and its frequency is observed. It may be seen that some values do not occur in the output differential. Good design practice in the PRESENT S-Box ensures that the probability distribution is uniform, meaning the differentials, namely E, 4, 1, and F which occurs in the output occurs with the same probability, i.e., $\frac{1}{4}$. The attacker can thus choose the values of m_0 and m_1, such that $\Delta(m) = F$, and thus $\Delta(u) = F$. Due to the differential property of the S-Box we know that in 4 out of 16 cases, the output differentials can be E, 4, 1, and F. Thus, $\Delta(w)$ is also any one of the above choices. The attacker also guesses k_2, and obtains the value of $\Delta(w)$ by inverting the S-Box on the values x_0 and x_1, obtained by guessing the key k_3. It is clear that those guesses which provide $\Delta(w)$ any other value, but E, 4, 1, F, can be eliminated as wrong keys. However, the differential property of the S-Box cannot distinguish the keys which lead to $\Delta(w)$

Table 2.2 Differential analysis of the S-Box

i	j	$S[i]$	$S[j]$	$S[i] \oplus S[j]$
0	F	C	2	E
1	E	5	1	4
2	D	6	7	1
3	C	B	4	F
4	B	9	8	1
5	A	0	F	F
6	9	A	E	4
7	8	D	3	E
8	7	3	D	E
9	6	E	A	4
A	5	F	0	F
B	4	8	9	1
C	3	4	B	F
D	2	7	6	1
E	1	1	5	4
F	0	2	C	E

equal to the above four values. This approximately eliminates 3/4 of the keys! This simple example shows the power of differential cryptanalysis, and can be used to distinguish the wrong keys from the correct ones.

In real life the block ciphers are cascades of substitution and permutations, and thus often referred to as SPN (Substitution Permutation Network) ciphers. However, the above discussion of differential cryptanalysis can be conceptually applied even for such a construction. In such a case, for a four-round cipher as shown in Fig. 2.9, an input–output differential is found for three rounds of the cipher which occurs with a very high probability. It is often defined as a differential trail, $\Delta_i \rightarrow \Delta_o$, where Δ_i and Δ_o are the input differential and output differential after the third round. The output differential also should ensure it is of low weight, meaning that it disturbs or affects less number of S-Boxes of the final round. The attacker can then guess a part of the last round key, corresponding to the disturbed S-Boxes, and then decrypt a portion of the ciphertext to check the differential at the output of the third round. It is expected that for the correct key the differential should be equal to Δ_o, with a very high probability. It may be noted that the efficiency of such an attack compared to a brute force attack is because it is a divide-and-conquer strategy, and thus can be successful by guessing portions of the key.

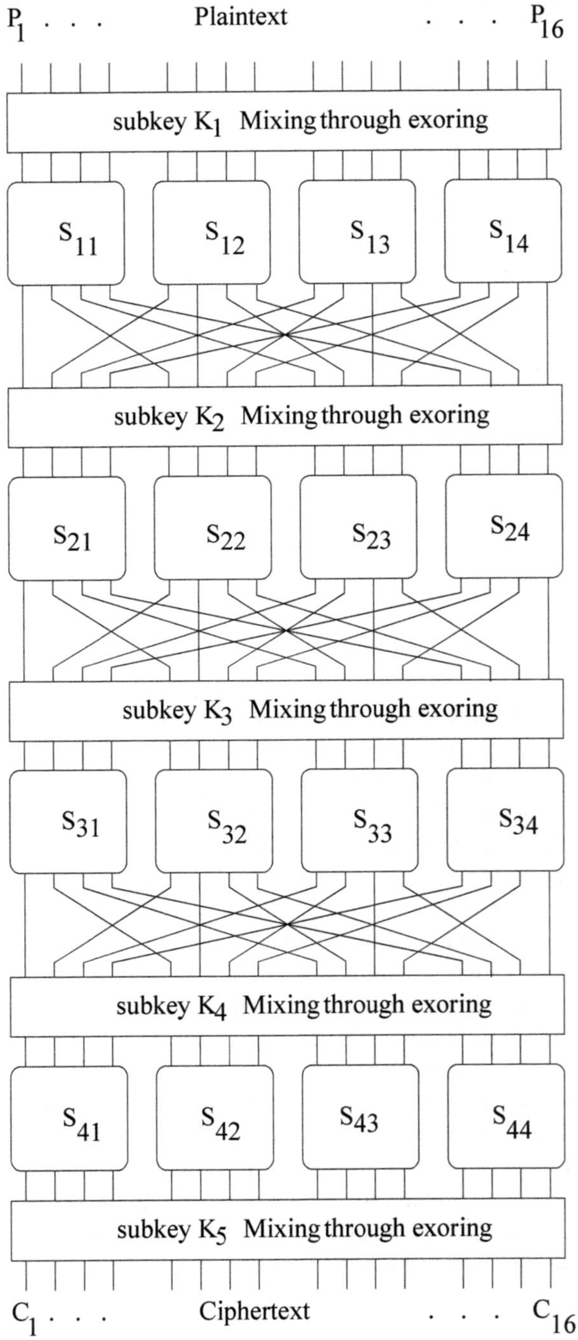

Fig. 2.9 The SPN block cipher

2.4 Asymmetric-Key Ciphers

The book also deals with attacks on asymmetric-key ciphers also known as public-key ciphers. These algorithms are often computation intensive and operate on large finite fields. For efficient design several field operation algorithms have been developed for supporting multiplication, inverse, exponentiation etc., and have been implemented in software libraries. The next section provides a quick overview on some of the commonly known techniques which are employed to realize the ciphers, and are hence targeted for demonstrating timing attacks. We focus on RSA, as it is still one of the most popular and utilized public-key algorithms. Further more, many of the attack techniques that we discuss in the book can be extended to other well known public-key algorithms, like elliptic curve cryptosystems.

2.5 RSA: An Asymmetric-Key Algorithm

RSA works by considering two keys: a *public key* is known to every one whereas a *private key* is *secret*. Encryption of a message is performed using the public key, but decryption requires the knowledge of the private key. All the operations are done mod n, where n is the product of two large distinct prime numbers p and q. The values of p and q are, however, private and hence not disclosed to all. The encryption key, which is public is a value b, where $1 \leq b \leq \phi(n)$ and $\phi(n)$ is the Euler's totient function. Very simply put, Euler's totient function or phi function, $\phi(n)$, is an arithmetic function that counts the positive integers less than or equal to n that are relatively prime to n. The decryption key is a private value a, which is selected such that $ab \equiv 1 \, mod \, \phi(n)$. The owner of the private key (p, q, a) publishes the value (b, n), which is the public key.

The encryptor chooses a message x, where $x \in Z_n$. It may be mentioned that $Z_n = \{0, 1, \ldots, n-1\}$. The encryption process is computing the cipher as $y \equiv x^b \, mod \, n$ using the public key b. Since the decryptor knows the value of a, which is the private key, he computes the value of x from y by computing $y^a \equiv (x^b)^a \, mod \, n \equiv x \, mod \, n$.

Algorithm 2.3: Square and Multiply algorithm for Exponentiation

Input: Ciphertext: C, Modulus: N and Private key: $k = (k_{m-1}, k_{m-2} \cdots k_0)$, where $k_{m-1} = 1$
Output: Plaintext: $M \equiv C^k \bmod N$

```
R = C
for i = m-2 to 0 do
    R = R^2 mod N
    if k_i = 1 then
        R = R × C mod N
    end
end
return R
```

The security of *RSA* is based on the assumption that decryption can be performed only by the knowledge of the private key b. However, to obtain the private information from the public value a requires one to compute the modular inverse of a modulo $\phi(n)$. It is believed that to obtain $\phi(n)$ from n requires the knowledge of the prime factors of n, namely p and q. The security of RSA is thus based on the hardness assumption of factorization of large n. Thus, the underlying operation to perform RSA encryption and decryption is modular exponentiation, i.e., $y \equiv x^b \, mod \, n$, where b is typically 1024 bits or 2048 bits large. This is achieved by the popular square and multiply algorithm.

2.5.1 Square and Multiply Algorithm to Perform Exponentiation

In this section, we present the square and multiply algorithm to perform modular exponentiation. We present the decryption algorithm, which uses the decryption key, denoted as $k = a$. We focus on the decryption algorithm as that is a natural target for an attacker, as the secret key is involved in the computation. The key is an m-bit key, denoted as $k = (k_{m-1}, k_{m-2} \cdots k_0)$, where $k_{m-1} = 1$.

This algorithm, as we later elaborate in the book, is naturally vulnerable against SCA. The reason being that the operations, namely squaring and multiplication, have different fingerprints on the side-channel leakage. For example, with respect to a timing attack, both requires different number of clock cycles to execute. We can observe that if a key bit is one, then a multiplication is performed, else not. The attacker hence tries to exploit this conditional property on the key bits to devise an attack. In the literature, there are several ways of performing a square and multiply algorithm to achieve exponentiation. One of the most popular techniques is called Montgomery Ladder, which is explained in the Algorithm 2.4.

It may be observed that in the Montgomery's Ladder, irrespective of the key bit, a multiplication and squaring is always performed. Thus, the design is naturally resistant against some SCA, which are possible over the naïve square and multiply algorithm.

Algorithm 2.4: Montgomery's Ladder for Exponentiation

Input: Ciphertext: C, Modulus: N and Private key: $k = (k_{m-1}, k_{m-2} \cdots k_0)$, where $k_{m-1} = 1$
Output: Plaintext: $M \equiv C^k \text{mod } N$

$P_1 = C, P_2 = C^2 \text{ mod } N$
for $i = m-2$ **to** 0 **do**
  if $k_i = 1$ **then**
    $P_1 = P_1 \times P_2 \text{ mod } N, P_2 = P_2^2 \text{ mod } N$
  end
  else
    $P_2 = P_1 \times P_2 \text{ mod } N, P_1 = P_1^2 \text{ mod } N$
  end
end
return P_1

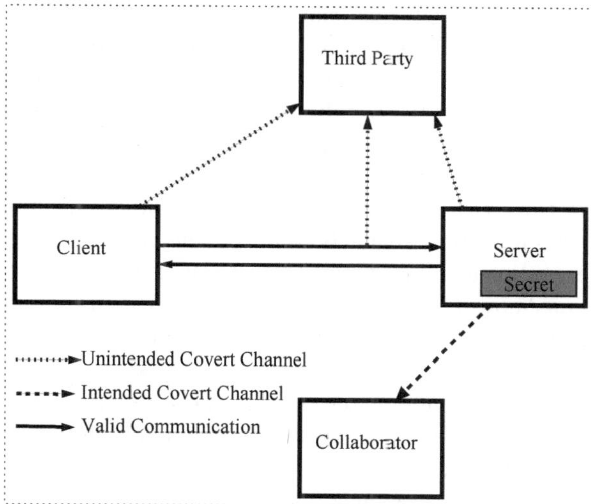

Fig. 2.10 Lampson's confinement problem

2.6 Confinement Problem and Covert Channels

Now that we have reviewed the essential components in cryptography, we move to side channels and how they covertly leak information about an executing application.

Consider a client using the services of a server. The client provides an input which is operated on by the server using some stored secret. In an untrusted environment, the client should ensure that the server does not communicate its input to a collaborator (the attacker), while the secret stored in the server should not be revealed to the client. Additionally no third-party should gain any information about the transactions between the client and server. This is the *confinement problem* as defined by Lampson in 1973 [10] and pictorially represented in Fig. 2.10.

Generally, the system can ensure that the server does not communicate with the collaborator by disabling writing to files, shared memories, and other inter-process communication (IPC) protocols. The server's secret can be protected by implementing memory protection using schemes such as paging and access control. In spite of these protection schemes, there still exists indirect paths by which data can be communicated. These indirect paths are not meant for communication, hence known as *covert channels*. They are of interest in the domain of computer security because they allow programs to bypass security policies of the system. They are all the more relevant in the cloud computing environments where such untrusting parties sharing resources are common. Covert channels can either be *intended* (where a nexus exists between the server and the collaborator) or *unintended* (where a third party obtains information about the computations). Covert channels are generally noisy, but information theory can be used to devise an encoding, which will allow data to get through reliably no matter how small the signal is, provided it is not zero [10].

An example of a covert channel was illustrated by Schaefer et al. in [11]. Consider the server and collaborator sharing a CPU. The client can signal information to the collaborator by the amount of time it holds the CPU. For instance, holding the CPU for 10, 20, or 30 μs can be used to represent a 0, 1, or 2 respectively [11]. This forms an *intentional* covert channel between the two cooperating processes (the client and the collaborator). Over the years several covert channels have been discovered such as the rate at which a program performs paging [10, 12], CPU scheduling [13], disk scheduling [13, 14], and cache memory usage [15, 16].

Covert channels do not always have to be intentional. There can also be unintentional covert channels, which results in transfer of information about a program execution to a third party. The program is unaware of these transfers. Such unintentional covert channels occur due to physical attributes of a device such as the power consumption and electromagnetic radiation, as well as execution time. These unintentional covert channels are commonly known as *side channels*. In 1996, Kocher showed that a timing side channel can be used to reveal the secret key of a cipher to an adversary [17]. Later, power consumption of the device was used for the same purpose [18]. These works that came to be known as SCA (side-channel attacks), kindled interest in the academic community and led to a new domain of cryptographic research; targeting implementations of ciphers rather than just the algorithm. The next section surveys SCA against ciphers that use unintentional covert timing channels. This class of attacks is called *timing attacks*.

2.7 Formal Analysis of Side-Channel Attacks

Since Shannon's benchmark paper in [19], there has been significant progress in the mathematical treatment given to cryptography. Ciphers, such as the three discussed in the previous section, underwent rigorous analysis with strong attack models such as adversaries who know the cipher algorithm, the input, as well as the output. The only secret being the cipher's key. All ciphers standardized and in use today have bounded security against these attack models.

The advent of SCA in [17] introduced a stronger (yet practical) attack model. Here, the adversary has access to side-channel information leakage of the encrypting device in addition to the input, output, and cipher algorithm. The mathematical tools developed so far failed to model these attack classes. It was not until 2004 that new models were developed by Micali and Reyzin to analyze cryptography in the presence of side-channel leakage [20]. This they called *physical observable cryptography*. The new theory was enhanced by the works of Standaert [21–24] and by Backes and Kopf [25, 26, 27] and used to provide a fair evaluation and comparison of ciphers in the presence of side-channel leakage. This has led to the development of a new class of cryptographic algorithms and countermeasures with provable security in the presence of leakages (for example [28, 29, 30]). The algorithms currently available are inefficient to implement. The hope is that the tools and models will lead to practically realizable ciphers with provable resistance against SCA. In this section we present briefly the formal notions in side-channel analysis.

Let $E_{\mathbf{k}}$ be a cipher having a secret key $\mathbf{k}$ chosen uniformly from the set $\mathscr{K}^m$ for some positive integer m. Let $\widetilde{E}_{\mathbf{k}}$ be an implementation of $E_{\mathbf{k}}$ on a device. A *q-limited side-channel key recovery adversary* is a statistical program, which can make at-most q queries to $\widetilde{E}_{\mathbf{k}}$ and monitor the leakage through channels such as power-consumption, timing, or electromagnetic radiation of the device. To quantify leakage, a *leakage function* $L(\cdot)$ is defined that mathematically abstracts the characteristics of the side channel and the measurement setup. For example, it is well-known that power consumption can be modeled in terms of Hamming weight or Hamming distances for CMOS devices.

The q-limited side-channel key recovery adversary follows a divide-and-conquer strategy by splitting $\mathbf{k}$ into smaller parts for example $\mathbf{k} = (k_1 \| k_2 \| k_3 \| \cdots \| k_m)$, where $k_1, k_2, \cdots, k_m \in \mathscr{K}$, and each key part is targeted independently. Further, the leakage partitions the key space $\mathscr{K}$ into equivalence classes such that the SCA cannot distinguish between two keys in the same class. The goal of the adversary is to guess a key class with nonnegligible probability. To do so, typically the attack has two phases. An online phase in which the q side-channel leakages are collected and an offline phase in which the leakages are analyzed using statistical distinguishers in order to obtain a ranking of the keys in $\mathscr{K}$ in an order of their likelihood.

There are two metrics by which the success of the attack can be measured. The first defines the oth order *success rate* as the probability that the correct key is ranked among the top o keys. For example, if $\mathscr{G} = (g_1, g_2, \cdots g_o, g_{o+1}, \cdots, g_{|\mathscr{K}|})$ is the ordered sequence of keys ranked from most likely to least likely, then the probability that the correct value of the key is in $g_1, g_2 \cdots, g_o$ is the oth order success rate of the attack. Alternatively, *guessing entropy* [31] can be used as a metric to evaluate the success of an attack. This measures the average number of guesses that are required before obtaining the correct key. For example, if the correct key is present in the jth location in the ranking $\mathscr{G}$ then the guessing entropy is j. The second metric uses information-theoretic metrics to determine for example $\mathbb{H}[K|L]$, i.e., the entropy of the key given the leakage. This should be much less than $\mathbb{H}[K]$ for a strong attack.

This book shows how timing attacks can be modeled by analyzing the cipher's memory access patterns. The mathematical framework thus developed is used to evaluate ciphers for their resistance against timing attacks and choose implementation strategies for the ciphers.

2.8 Conclusion

This chapter provided an overview of various symmetric and asymmetric encryption schemes. Classical cryptanalytic techniques were dwelt upon, and side channels and their formal analysis were introduced. To a side-channel attacker, it is not just the enciphering algorithm and their implementations that are important, but the system and CPU architecture is also crucial. The next chapter provides an overview of modern CPUs and the few components in them that have been used to develop SCA.

References

1. Gueron S (2010) Intel Advanced Encryption Standard (AES) instructions set (Rev : 3.0)
2. Zheng Y, Matsumoto T, Imai H (1989) On the construction of block ciphers provably secure and not relying on any unproved hypotheses. In: Brassard G (ed) CRYPTO. Lecture notes in computer science, vol 435. Springer, Berlin, pp 461–480
3. Federal Information Processing Standards Publication 197 (2001) Announcing the Advanced Encryption Standard (AES)
4. Stinson D (2002) Cryptography: theory and practice, 2nd edn. Chapman and Hall, London, pp 117–154
5. Daemen J, Rijmen V (2002) The design of Rijndael: AES—the Advanced Encryption Standard. Springer, Berlin
6. Shirai T, Shibutani K, Akishita T, Moriai S, Iwata T (2007) The 128-bit blockcipher CLEFIA (extended abstract). In: Biryukov A (ed) FSE Lecture notes in computer science, vol 4593. Springer, Berlin, pp 181–195
7. Sony Corporation (2007) The 128-bit blockcipher CLEFIA : algorithm specification
8. Aoki K, Ichikawa T, Kanda M, Matsui M, Moriai S, Nakajima J, Tokita T (2000) Camellia: A 128-bit block cipher suitable for multiple platforms—design and analysis. In: Stinson DR, Tavares SE (eds) Selected areas in cryptography. Lecture notes in computer science, vol 2012. Springer,Berlin, pp 39–56
9. Knudsen LR, Robshaw MJB (2011) The block cipher companion. Springer, Berlin
10. Lampson BW (1973) A note on the confinement problem. Commun ACM 16(10):613–615. http://doi.acm.org/10.1145/362375.362389
11. Schaefer M, Gold B, Linde R, Scheid J(1977) Program confinement in KVM/370. In: Proceedings of the 1977 annual conference, ser. ACM '77. ACM: New York, pp. 404–410. http://doi.acm.org/10.1145/800179.1124633. Accessed Dec 2013
12. Van Vleck T (1990) Timing channels. Multics, technical report, 1990
13. Kemmerer RA (1983) Shared resource matrix methodology: an approach to identifying storage and timing channels. ACM Trans Comput Syst 1(3):256–277. http://doi.acm.org/10.1145/357369.357374
14. Karger PA, Wray JC (1991) Storage channels in disk arm optimization. IEEE symposium on security and privacy. IEEE, Oakland, pp 52–63
15. Wray JC (1991) An analysis of covert timing channels. In: Research in security and privacy, 1991. Proceedings, 1991 IEEE computer society symposium. May 1991, pp 2–7
16. Hu W-M (1992) Lattice scheduling and covert channels. In: Proceedings of the 1992 IEEE symposium on security and privacy. SP '92. IEEE Computer Society, Washington, DC, pp 52–61
17. Kocher PC (1996) Timing attacks on implementations of Diffie-Hellman, RSA, DSS, and other systems. In: Koblitz N (ed) CRYPTO '96: proceedings of the 16th annual international cryptology conference on advances in cryptology. Lecture notes in computer science, vol 1109. Springer-Verlag, London, pp 104–113
18. Kocher PC, Jaffe J, Jun B (1999) Differential power analysis. In: Wiener MJ (ed) CRYPTO. Lecture notes in computer science, vol 1666. Springer, Berlin, pp 388–397
19. Shannon CE (1949) Communication theory of secrecy systems. Bell Syst Tech J 28:656–715
20. Micali S, Reyzin L (2004) Physically observable cryptography (extended abstract). In: Naor M (ed) TCC. Lecture notes in computer science, vol 2951. Springer, Berlin, pp 278–296
21. Standaert F-X, Peeters E, Archambeau C, Quisquater J-J (2006) Towards security limits in side-channel attacks. In: Goubin L, Matsui M (eds) CHES. Lecture notes in computer science, vol 4249. Springer, Berlin, pp 30–45
22. Standaert F-X, Pereira O, Yu Y, Quisquater J-J, Yung M, Oswald E (2009c) Leakage resilient cryptography in practice. Cryptology ePrint archive, Report 2009/341. http://eprint.iacr.org/. Accessed June 2010

23. Standaert F-X, Malkin T, Yung M (2009b) A unified framework for the analysis of side-channel key recovery attacks. In: Joux A (ed) EUROCRYPT. Lecture notes in computer science, vol 5479. Springer, Berlin, pp 443–461
24. Standaert F-X, Koeune F, Schindler W (2009a) How to compare profiled side-channel attacks? In: Abdalla M, Pointcheval D, Fouque P-A, Vergnaud D (eds) ACNS. Lecture notes in computer science, vol 5536, pp 485–498
25. Köpf B, Basin DA (2007) An information-theoretic model for adaptive side-channel attacks. In: Ning P, di Vimercati SDC, Syverson PF (eds) ACM conference on computer and communications security. ACM, Alexandria, pp 286–296
26. Backes M, Köpf B (2008) Formally bounding the side-channel leakage in unknown-message attacks. In: Jajodia S, 6pez JL (eds) ESORICS. Lecture notes in computer science, vol 5283. Springer, Berlin, pp 517–532
27. Backes M, Köpf B, Rybalchenko A (2009) Automatic discovery and quantification of information leaks. In: IEEE symposium on security and privacy. IEEE Computer Society, 2009, pp 141–153
28. Dziembowski S, Pietrzak K (2008) Leakage-resilient cryptography. In: FOCS. IEEE Computer Society, 2008, pp 293–302
29. Naor M, Segev G (2009) Public-key cryptosystems resilient to key leakage. In: Halevi S (ed) CRYPTO. Lecture notes in computer science, vol 5677. Springer, Berlin, pp 18–35
30. Pietrzak K (2009) A leakage-resilient mode of operation. In: Joux A (ed) EUROCRYPT. Lecture notes in computer science, vol 5479. Springer, Berlin, pp 462–482
31. Massey J (1994) Guessing and entropy. In: Information theory, 1994. Proceedings, 1994 IEEE international symposium, 1994, p 204

Chapter 3
Superscalar Processors, Cache Memories, and Branch Predictors

Memory accesses and branches are arguably the biggest performance bottlenecks in a program. To mitigate their effect on the performance, modern superscalar microprocessors incorporate cache memories and branch predictors in their architecture. While cache memories bridge the performance gap between the processor and the main memory, branch predictors make predictions about branch destinations to reduce the overhead of branches in the program. The pitfall of these components is that they result in information leakage through side channels, which have been used to break crypto-systems. This book provides an analysis of the attacks that occur due to these components. This chapter provides a brief overview of superscalar processors, its memory organization, cache memories, and branch predictors [1–3].

3.1 Superscalar Processors

In this section, superscalar processors and features such as speculation and out-of-order execution are discussed. Figure 3.1 shows a block diagram of a superscalar processor. A superscalar processor can be thought of as a pipeline with several stages; each stage doing a specific job. In every clock cycle, multiple instructions are fetched from the instruction cache into an instruction/decode buffer. The dispatch unit then scrutinizes the instructions in the buffer and attempts to issue them to an appropriate functional unit (such as integer arithmetic logic units (ALU), floating point units (FPU), branch unit, etc.). The issue is done provided the functional unit is available and the operands used by the instruction are up-to-date. It may be noted that the instructions can be issued to the functional units out-of-order (i.e., in an order that does not strictly follow the program flow). Consequently, the result from the functional unit is obtained out-of-order. This is stored in a reorder buffer before it is retired (i.e., the processor registers updated). An instruction is retired if and only if all instructions preceding it have retired. Instructions can even also be executed speculatively, that is, before a branch can be taken.

Figure 3.2 shows an example of how instructions generally flow in a processor having a five stage pipeline. In first stage (IF), instructions are fetched from the

C. Rebeiro et al., *Timing Channels in Cryptography*,
DOI 10.1007/978-3-319-12370-7_3

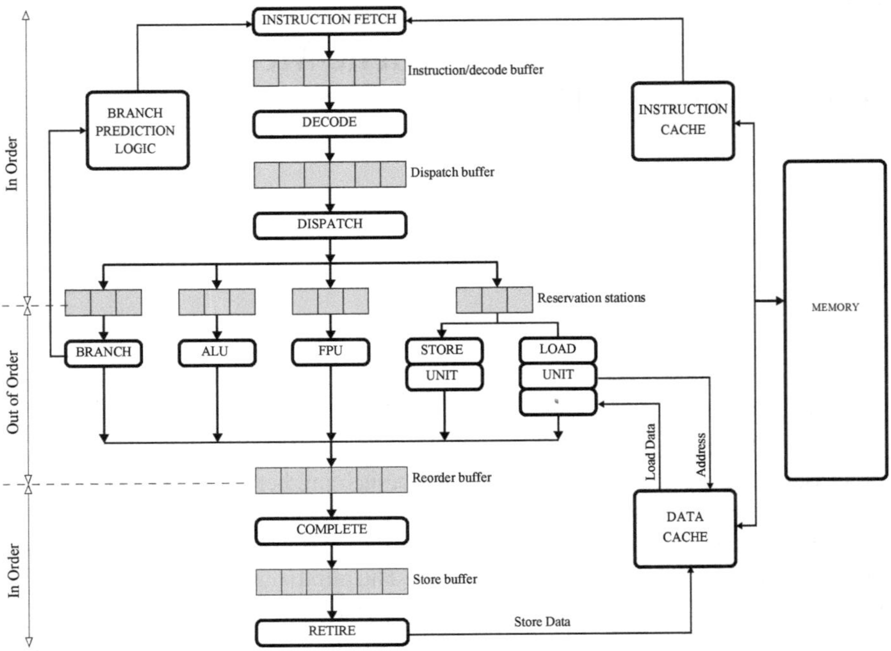

Fig. 3.1 Superscalar processor architecture

instruction cache. The second stage (ID) decodes instructions. The third stage (OF), fetches operands required, while the fourth stage (IE) executes the instructions. The result of the instructions are written back in final stage (WB). Several instructions are simultaneously operated upon in the processor. While one is in the IE stage, others are in the IF, ID, OF, and WB stages. Each stage is small and completes its task fast. This allows the processor to operate at high clock frequencies.

Modern processors can have over 30 pipeline stages. This deeply pipelined execution is possible because instructions are generally executed in a predictable sequence. The processor can, therefore, fetch and decode instructions much ahead of time. This nice flow in the instruction stream is broken by branches in the program. Branch instructions break the sequential flow requiring instructions to be executed from some arbitrary branch destination. A branch instruction requires the entire pipeline to be flushed. All instructions in various stages of execution would need to be aborted and new instructions from the branch destination fetched. This has significant overheads because as many as 100 instructions could concurrently be present in different execution stages of the pipeline. Branch predictors are used to reduce this overhead by predicting branch target destinations and avoid flushing the pipeline. Section 3.3 has more details.

A functional unit of interest in timing attacks is the load-store unit (especially the load part), which handles all memory accesses made by the program. Each load or store has three operations. The first is an address generation (which generates the

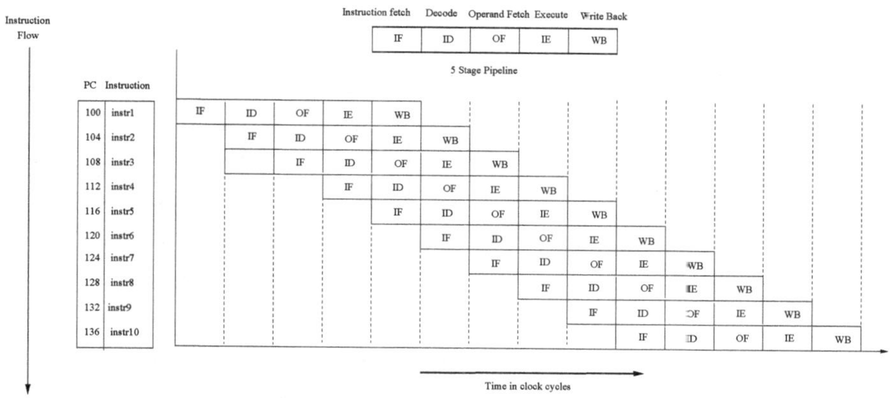

Fig. 3.2 Instruction flow in a five-stage pipeline. The program counter (*PC*) points the instruction being executed

effective address from where data are to be loaded or stored). Next, the address is translated from a virtual address to a physical one using a translation look aside-buffer (TLB). The final operation either loads or stores data to or from the memory. For a load instruction, the loaded data are updated into a register only after the ordering has completed. Load instructions can be pipelined, executed speculatively, and out-of-order. Additionally, multiple load instructions can be issued in parallel if the processor supports multiple load-store units.

In order to extract the best performance out of a superscalar processor, the memory subsystem must match the performance of the processor. The next section briefly describes memory subsystems and branch prediction logic in superscalar processors.

3.2 Memory Hierarchy and Cache Memory

Modern superscalar CPU architectures are plagued with the *von Neumann bottleneck* due to the high speed with which CPUs process instructions and the comparative low speed of memory access. The possibility of using high-speed memory technologies that match CPU processing speeds is restricted by cost factors. To mitigate this speed difference, superscalar CPU architectures are built with memory as a hierarchy of levels, with small and fast memories close to the processor and large, slower, less expensive memories below it. A memory stores a subset of the data that is present in the memory below it, while the lowest memory level (generally main memory that uses low-cost capacitive DRAMs) contains all the data. The aim is to provide the user with large cheap memories, while providing access at processor speeds.

Memories close to the processor are called *cache memories* and are built with SRAM technology, which offers high access speeds, but are large and expensive. Depending on the data it contains, cache memories are categorized as either *data*, *instruction*, or *unified* (that store both data and instructions). Cache memories are

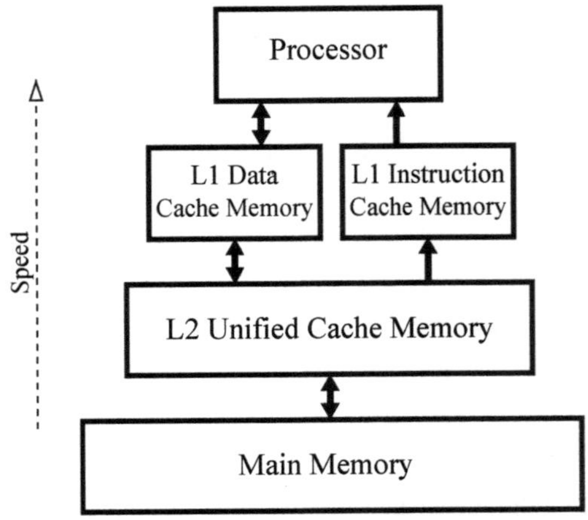

Fig. 3.3 Memory hierarchy structure

also categorized based on their closeness to the processor as L1, L2, and L3, with L1 being the closest to the processor and L3 the farthest. Most modern desktop and server processors have dedicated L1 data and L1 instruction caches, and a larger unified L2 cache memory. Some servers also contain an L3 cache before the main-memory (Fig. 3.3). The L1 caches data present in L2, while L2 caches data present in L3 or the main memory.

Cache memories work by exploiting the *principle of locality*, which states that programs access a small portion of their address space in any given time instant. The entire memory space of the processor is divided into *blocks*, typically of 64 or 128 contiguous bytes. A memory access results in a *cache hit* if the data are available in the cache memory. For example, an *L1-data cache hit* is obtained if the data accessed are available in the L1-data cache memory. In this case, the data are available to the processor quickly and there are no delays. A *cache miss* occurs if the data accessed are not available in the cache memory. In this case the block containing the data is loaded into the cache from a memory lower in the hierarchy. This is a slow process during which the processor may stall. There are three types of cache misses: compulsory misses, capacity misses, and conflict misses. *Compulsory misses* are cache misses caused by the first access to a block that has never been used in the cache. *Capacity misses* occur when blocks are evicted and then later reloaded into the cache. This occurs when the cache cannot contain all the blocks needed during execution of a program. *Conflict misses* occur when one block replaces another in the cache.

3.2.1 Organization of Cache Memory

The cache memory is organized into lines with each line capable of holding a block of memory. Suppose each block has a capacity to store 2^δ words, and there are 2^b lines in the cache, then at most $2^{b+\delta}$ words can be stored in the cache. An address

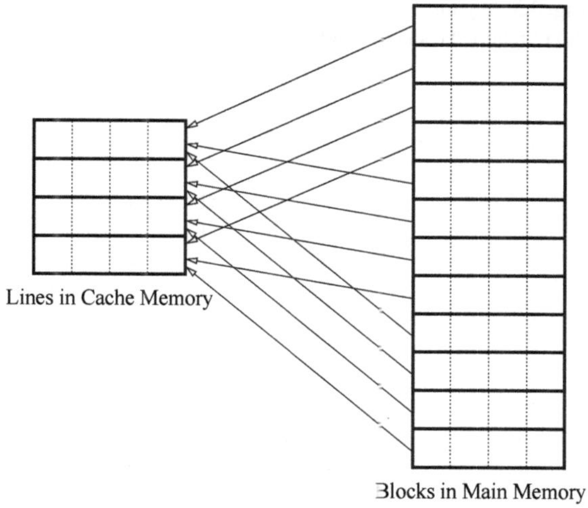

Fig. 3.4 Direct mapped cache memory with $2^b = 4$

translation mechanism is required to determine which cache line a block should get stored into. In the most simple approach, every 2^b-th memory block gets mapped into the same cache line. This mapping is called *direct-mapping* and is depicted in Fig. 3.4.

The address translation mechanism computes two components: the word address A_{word} and the line address A_{line}. If A is the address of the data that is to be accessed then

$$A_{word} = A \bmod 2^{\delta}$$

$$A_{line} = \lfloor A/2^{\delta} \rfloor \bmod 2^b .$$

Note that the size of the address A_{word} is δ bits while the size of the line address is b bits. Problems arise due to the many-to-one mapping from blocks to lines (as seen in Fig. 3.4). The cache controller needs to know which block is present in a cache line before it can decide if a hit or a miss has occurred. This is done by using an identifier called the *tag*. The identifier denoted A_{tag} for the address A is given by

$$A_{tag} = \lfloor \lfloor A/2^{\delta} \rfloor / 2^b \rfloor .$$

Thus, every line in the cache has an associated tag as shown in Fig. 3.5. For every memory access, the tag for the address (A_{tag}) is compared with the tag stored in the cache. A match results in a cache hit, otherwise a cache miss occurs.

Time-driven cache attacks on block ciphers monitor these hits and misses that occur due to look-up tables used in the implementation. A look-up table is defined as a global array in a program. For example, the C construct for a look-up table is as follows.

```
const unsigned char T0[256]={0x63, 0x7C, 0x77, ... };
```

Fig. 3.5 Organization of direct mapped cache memory

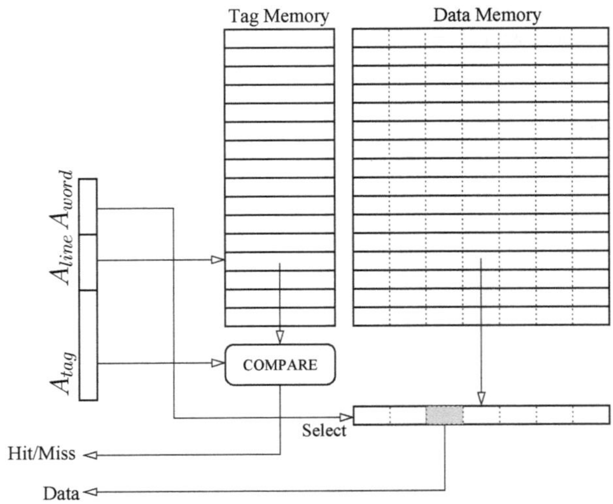

Assuming that the base address for `T0` is `0x804af40`, Table 3.1 shows how the table gets mapped for a $4KB$ direct-mapped cache with a cache line size of 64 bytes. Every block in the table maps to a distinct cache line provided that the table has lesser blocks than the lines in the cache.

Table 3.1 Mapping of Table `T0` to a direct-mapped cache of size $4KB$ ($2^{\delta} = 64$ and $2^b = 64$)

Elements	Address	line	Tag
`T0[0]` to `T0[63]`	`0x804af40` to `0x804af7f`	61	`0x804a`
`T0[64]` to `T0[127]`	`0x804af80` to `0x804afbf`	62	`0x804a`
`T0[128]` to `T0[191]`	`0x804afc0` to `0x804b0ff`	63	`0x804a`
`T0[192]` to `T0[255]`	`0x804b000` to `0x804b03f`	0	`0x804b`

The drawback of the direct mapped scheme is poor cache utilization and *cache thrashing*. Consider a program which continuously reads data from addresses A and then A'. The addresses are such that they map into the same line in the cache (i.e., $A_{line} = A'_{line}$). Thus every memory access would result in a cache miss, considerably slowing down the program even though the other lines in the cache are unused. This is called cache thrashing.

An improved address translation scheme divides the cache into 2^s sets. Each set groups $w = 2^b/2^s$ cache lines. A block now maps to a set instead of a line and can be present in any of the w cache lines in the set. A cache that uses such an address translation scheme is called a *w−way set associative cache*. The address translation for such a scheme is defined as follows,

$$A_{word} = A \bmod 2^{\delta}$$
$$A_{set} = \lfloor A/2^{\delta} \rfloor \bmod 2^{s}$$
$$A_{tag} = \lfloor \lfloor A/2^{\delta} \rfloor / 2^{s} \rfloor$$

If $w = 2$, the thrashing in the previous example is eliminated as the data corresponding to the addresses A and A' can simultaneously reside in the cache. However, the problem arises again if a program is executed that makes continuous accesses to three or more data from different addresses that share the same cache set.

3.2.2 *Improving Cache Performance for Superscalar Processors*

As seen in the previous section, a poorly designed cache can slow down a program instead of accelerating it. Therefore, it is important to understand the effect of the cache memory in a program's execution. The average memory access time is often used as a metric for the purpose. For a system with a single level of cache memory, the metric is defined as follows,

$$\text{Average memory access time} = \text{Hit time} + \text{Miss Rate} \times \text{Miss Penalty},$$

where, Hit time is the time to hit in the cache, while Miss Penalty is the time required to replace the block from memory (i.e., the miss time) Miss Rate is the fraction of the memory accesses that miss the cache.

Improving cache performance can be done by reducing hit time, reducing or hiding miss penalty, and/or reducing miss rate. To certain extent basic parameters in the cache design can be tuned to improve cache performance. The cache size ($2^{b+\delta}$), for example, should be large enough to provide temporal locality for the program, yet be small enough to prevent extensive degradation of hit time. Similarly, other cache parameters such as the associativity (w), block size (2^{b}), and number of levels can be tuned.

Besides tuning the basic parameters in the cache, other micro-architectural techniques are available to improve performance of cache memories. These are generally incorporated in modern superscalar microprocessors. The remaining of the section summarizes some of the important techniques.

- *Wide cache interfaces.* The data bus interface in these caches is capable of transferring multiple contiguous words at a time. For example, two words to an entire line can be transferred simultaneously. This helps in reducing multicycle accesses, thus reducing on average the hit time. Wide cache interfaces are especially suited for instruction caches and L2 and L3 level cache memories, where this feature allows the entire cache line to be transferred efficiently.
- *Multiaccess cache.* Single access cache memories can handle one read or one write per clock cycle. This becomes a bottleneck in superscalar processors, which are capable of multiple loads or stores in a clock cycle. In such cases multiaccess cache

memories are an advantage as they allow multiple accesses in a clock cycle, thus preventing stalls in the processor. There are various ways in which multi-access cache memories can be realized. We enumerate some of the methods.

  - *True Multiporting.* Uses dual-ported memory to store the tags and data. However the size of the cache increases significantly and the hit time is also affected.
  - *Multiple Cache Copies.* Mirrors cache memories so that each mirror holds an identical copy of the data and can handle a single access. Thus n mirrors can support n accesses simultaneously. In addition to the large size, a significant drawback is that stores require all mirrors to be updated in order to maintain coherency.
  - *Overclocking.* The cache is clocked x times faster than the processor. Therefore, a single cycle of the processor can service x accesses to the cache.
  - *Multibanked caches.* The address space is partitioned into multiple banks. Bank 1 caches addresses from partition 1, Bank 2 caches addresses from partition 2, and so on. Each bank can service one memory access per clock cycle. Thus, if there are n partitions, n memory accesses can be serviced together provided they go to different banks. Two simultaneous accesses to the same bank causes a *conflict* and results in a stall. Additionally a crossbar switch is required to channel requests to the appropriate bank and another crossbar switch is required to channel the result to the appropriate port.

- *Sub-blocks.* The cache line is broken into several sub-blocks. Each sub-block has a separate valid bit. On a cache miss, only the required sub-block needs to be fetched. Similarly, for stores, only the valid sub-blocks need to be stored. The others can be ignored thereby reducing the miss penalty.
- *Critical word first.* During a cache miss, the required word in the block is fetched first, thus allowing the processor to restart early. This reduces the miss penalty.
- *Write Buffers.* A write buffer is used to hold data before it is written back into the cache. This reduces the latency for the write and also allows the cache to service other read requests while the store is taking place.
- *Non-blocking or lockup free caches.* In conventional caches while a miss is being serviced, a second access to the cache is stalled. One remedy is to buffer future misses, while allowing the cache to continue servicing hits. A better alternative is to allow multiple misses to be overlapped. This is a nonblocking cache that helps increasing the cache bandwidth.
- *Pipelined caches.* This allows a new request to be serviced before the previous is completed. Thus increasing the cache bandwidth.
- *Victim Caches.* These are small fully associative caches that store blocks recently evicted from the L1 cache. They are accessed on a miss in parallel with the lower level cache. This avoids lower level cache access and reduces the number of cache misses as it is fully associative.
- *Prefetching Cache.* If programs follow strictly the principle of locality, then cache misses can be considerably reduced. However, most programs occasionally access data that are not spatially local. This will create additional cache misses. One

remedy is to load data into the cache before it is actually required, thereby reducing the miss rate. This is called prefetching and can be done either in software or automatically by the hardware. *Software prefetching* requires that prefetch instructions be inserted into the program that would initiate a load of a block into the cache. However, unlike normal loads, the instruction does not stall.
Hardware prefetching uses a prefetcher that monitors the memory access patterns of the executing program and predicts the future data the program will access. This datum is loaded into the cache before the program accesses the data, thus reducing the miss latency. Hardware prefetchers are categorized as *sequential prefetchers* and *stride prefetchers* depending on the sequence of memory accesses. A sequential prefetcher tracks consecutive blocks accessed by the program, while a stride prefetcher looks for accesses with regular strides that are not necessarily consecutive.
Implementing a hardware prefetcher requires a memory reference prediction table, comprising of three fields. The first is the instruction address of the load, which is used as a tag field for selecting an entry of the table. The second field contains the previous instruction address of the load, while the third has the stride value. In order to prefetch, the previous address is added to the stride value in order to obtain a predicted address. If the data corresponding to the predicted address is not in the cache, it is loaded from the memory lower in the hierarchy.

3.3 Branch Prediction Unit

A branch in a program causes its control flow to deviate from a sequential execution. Branch instructions can either be unconditional or conditional. In an unconditional branch instruction, the control flow of the program changes from its sequential flow unconditionally. The next instruction executed is not the next in the instruction sequence. In the case of a conditional branch instruction, the control flow changes only if the condition specified in the instruction is satisfied. The conditions are evaluated only when the branch instruction reaches the execute stage of the processor. By then, several more instructions have entered the processor pipeline and are in various stages of execution. If the branch is not taken, execution flows normally and no overheads are incurred. However, if the branch is taken, all instructions in the pipeline would need to be flushed and new instructions from the branch destination be fetched. This leads to significant overheads.

The Branch Prediction Unit (BPU) plays an important role in reducing this overhead. Much before the branch condition is evaluated, the BPU guesses the probable execution path of the program and causes the processor to fetch instructions from the branch target. This avoids the pipeline flush, saving critical clock cycles. The prediction is made based on historic information about the branches in the program. Figure 3.6 shows the effect of the branch predictor in saving clock cycles.

Like all predictors, the BPU's prediction is not always correct. It could result in a misprediction, where the target address is not the same as the predicted target.

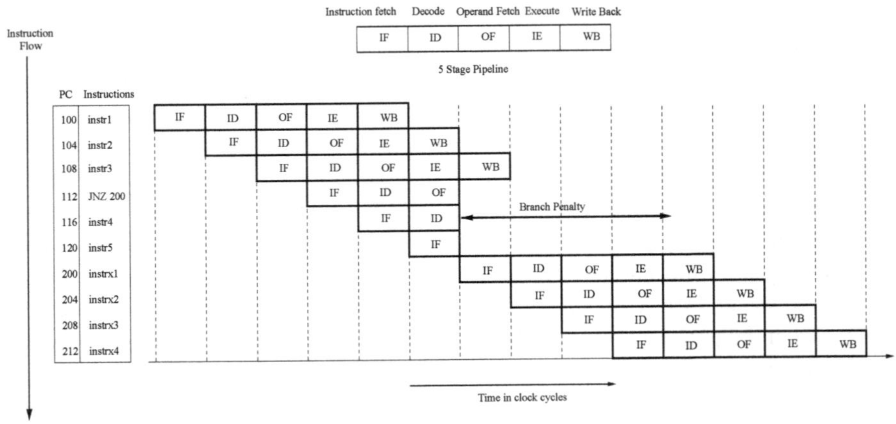

a Without a Branch Predictor

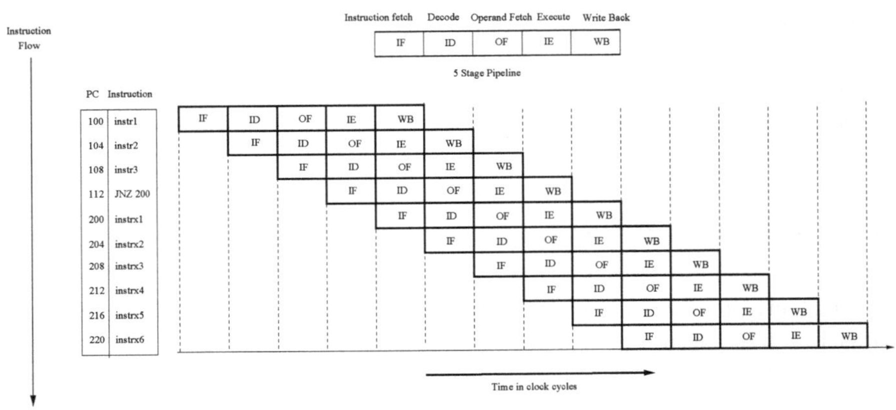

b With a Branch Predictor

Fig. 3.6 Branch predictors can reduce clock cycles by predicting the branch target. The figures show an example of the working of a branch predictor in a processor with a five-stage pipeline. The first column in the figures (100,104,...) are the memory addresses, while the second are the instructions (instr1, instr2, ... instrx1, instrx2,....). Without the predictor, the *JNZ* (jump if not zero) instruction causes the pipeline to be flushed and four clock cycles wasted. On the other hand, a processor with a branch predictor that predicts the branch target (PC = 200) correctly would save these clock cycles

This would cause overheads as the pipeline needs to be flushed. An efficient branch prediction algorithm is one where such mispredictions are minimized.

There has been considerable research in developing the "ideal" branch prediction algorithm; one which has no mispredictions. The algorithms developed are based on either static or dynamic schemes. The next section discusses these schemes.

3.3.1 *Static Branch Prediction*

Some of the static branch prediction schemes are listed below:

- Whenever a branch instruction is present, predict that the branch is always taken. Alternatively, the prediction could always be not taken.
- Predict depending on the opcode. For instance, opcode *X* would always result in a prediction that the branch is taken, while opcode *Y* would always result in a prediction that the branch is not taken.
- Branches that jump forward in the program are always predicted not taken, while branches that jump back are always taken.

The drawback of these approaches is that they do not perform well when the executing program has hierarchical loop statements or when there is an irregularity in the branch behavior. In such situations, the dynamic branch prediction strategies perform better.

3.3.2 *Dynamic Branch Prediction Schemes*

Dynamic prediction schemes store significant amount of information on the run-time instruction execution and use this information to predict the outcome of branch instructions. This section describes some of the most common dynamic predictors.

3.3.2.1 1-bit Branch predictor

This predictor is also known as *last-time* predictor and is arguably the simplest form of dynamic branch prediction. Every time a branch instruction is executed, the predictor saves the result—either taken or not taken. This stored result is used to make a prediction the next time the branch instruction is fetched.

The predictor hardware requires a table containing the address of the branch instruction and a bit, which signifies branch taken or not taken. This bit is updated every time the branch instruction is executed. The automaton for the one bit predictor is shown in Fig. 3.7. The automaton has two states—one predicts branches to be taken and the other predicts branches to be not taken. The transitions are marked by the output of the last branch statement.

3.3.2.2 Bimodal Predictor

In the 1-bit branch predictor discussed above, every misprediction would change the result of the prediction. In a bimodal predictor on the other hand, two consecutive mispredictions are required to change the result of the prediction. Figure 3.8 shows

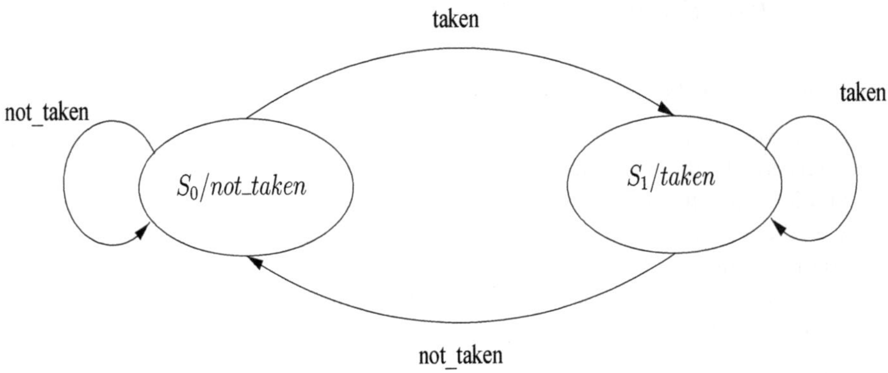

Fig. 3.7 Dynamic 1-bit predictor state machine

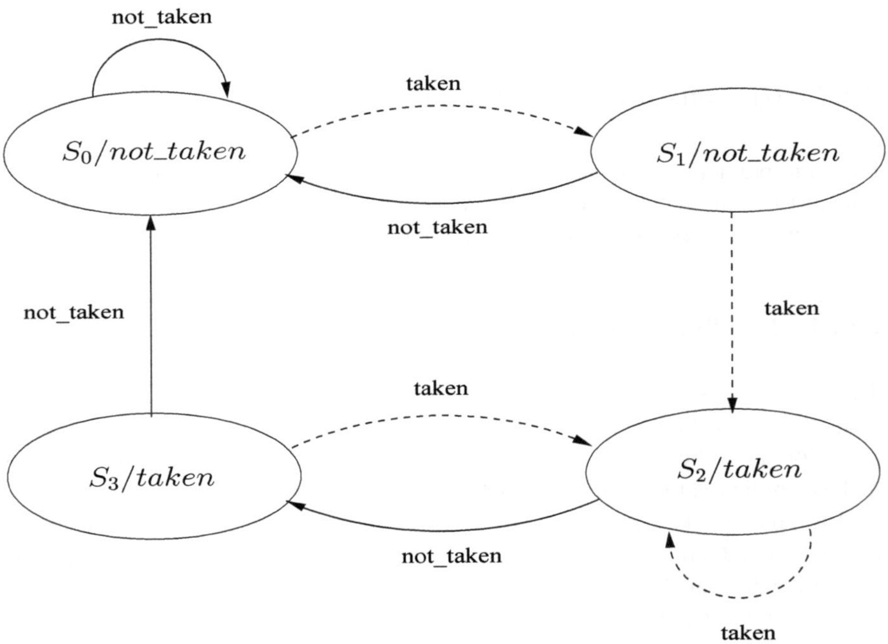

Fig. 3.8 Dynamic 2-bit predictor state machine

the state machine for a bimodal predictor (also called a 2-bit predictor), which predicts branches as either taken or not taken.

The state machine has four states S_0, S_1, S_2, and S_3, each having a predicted output. The arrows represent whether a branch is taken or not, while the label inside the state denotes the prediction (which is the output of the predictor). The original Intel Pentium used this form of prediction.

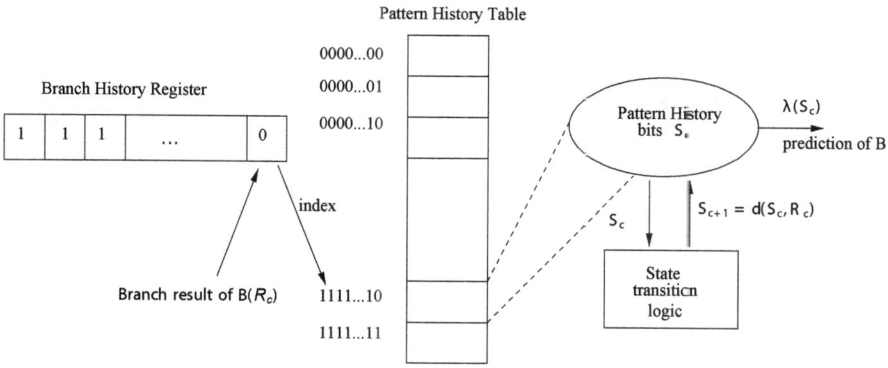

Fig. 3.9 Two level adaptive branch prediction

3.3.2.3 Two-Level Adaptive Predictor

Conditional branches that are taken in a regular recurring pattern are not predicted well by the bimodal predictor. In such cases a two-level adaptive predictor works better. The predictor remembers the last k occurrences of a branch instruction and uses an s-bit prediction function (such as an s-bit predictor) for each of the 2^k history of patterns. The first level in the two level adaptive predictor uses a *branch history register*. This is a shift register of size k, which stores the history of the last k branches—either taken or not taken. The branch history register indexes into a second level called *pattern history table*, which can hold 2^k entries, each s bits wide. An s-bit prediction function is used to make a prediction based on the last s branches. Figure 3.9 illustrates the two level predictor.

When a conditional branch say $\mathscr{B}$ is getting predicted,

- content of the k bit history register is used as address to the pattern history table. Let S_c be the contents of the pattern history table.
- S_c is fed to the s-bit prediction decision function (such as the s-bit predictor), which outputs the predicted value $\lambda(S_c)$.

To take an example, consider the case when k = 2, which means that the last two occurrences of the branch are stored in the branch history register. The register therefore takes 4 values (00, 01, 10, and 11), which indexes into the 0th, 1st, 2nd, and 3rd entry in the pattern history table. The contents of the pattern history table is fed to a bimodal predictor.

Now consider a branch sequence is 0011001100. Further, assume that s is 2 and uses a 2 bit predictor function. The 00 entry would predict 1, the 01 entry would predict 1, the 10 entry would predict 0, and the 11 entry would predict 0.

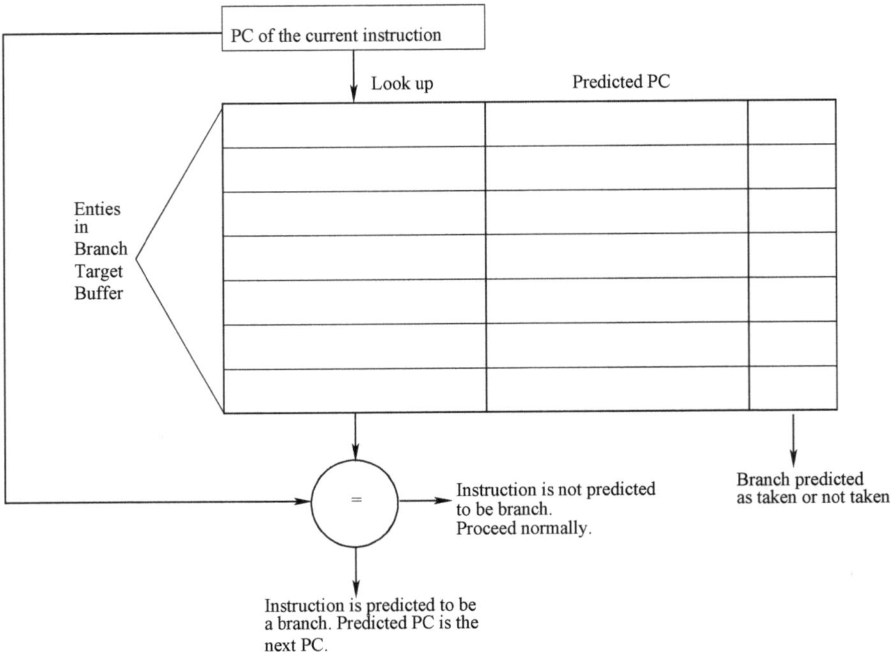

Fig. 3.10 The branch target buffer in a processor

3.3.3 Branch Target Buffers

Branch target buffers (BTB) is a cache in the processor that stores branch targets. Each BTB entry is primarily composed of tag bits, the target address of the branch, status bits indicating if the branch is taken or not taken. Whenever a branch instruction is encountered, a part of the instruction address (PC—program counter) is used as an index to the BTB location.

The BTB has limited size and thus can hold a limited number of branch targets. If a new branch instruction is encountered that is not in the BTB, then an entry in the BTB would be evicted. Figure 3.10 shows the structure of the BTB.

3.4 Conclusion

This chapter laid the foundation for superscalar processors, cache memories, and branch prediction units. Timing attacks discussed from here on would rely significantly on how these components operate when a cipher executes. In the next chapter, we would demonstrate this by showing how the cache memories affect the cipher execution and how they can be exploited to reveal information about the cipher's secret key.

Reference

1. Hennessy JL, Patterson DA (2006) Computer architecture: a quantitative approach, 4th ed. Morgan Kaufmann
2. Shen JP, Lipasti MH (2005) Modern processor design: Fundamentals of superscalar processors. McGraw-Hill
3. Juan T, Navarro JJ, Temam O (1997) Data Caches for Superscalar Processors, in International Conference on Supercomputing, pp 60–67

Chapter 4
Time-Driven Cache Attacks

When a load instruction incurs a cache-miss, a block of memory from the lower level of the memory subsystem is loaded into a cache line. Consequently, the memory access would require considerably more time and power, and has a characteristically different electromagnetic radiation compared to when a cache-hit occurs. Figure 4.1, for instance, shows the power consumption trace for eight memory load instructions on a PowerPC processor. The loads that result in cache misses are easily distinguishable from the cache hits. These indirect manifestations of a memory access can be used by an attacker to gain considerable insight about the application currently being executed. In this chapter, we show how information about the secret key of a cipher can be gleaned from the execution time of a block cipher. We start the chapter with a simple illustration showing how information can be obtained from memory access patterns before discussing attacks on ciphers.

4.1 A Simple Illustration

Consider a program having a look-up table T and a function called `AccessT`, which takes as input two parameters x_1 and x_2. The function accesses the table T twice—at locations $k_1 \oplus x_1$ and $k_2 \oplus x_2$ as shown in Listing 4.1, where k_1 and k_2 are a pair of application specific secret data.

Suppose a malicious user in the system wants to obtain information about the secrets k_1 and k_2. However, she can only invoke `AccessT` and does not have sufficient privileges to observe the internals of `AccessT`. For instance, `AccessT` can be a system call executing in the operating system, while the malicious user has only user level permissions. Since no direct channels to determine k_1 and k_2 are present, the malicious user could resort to using indirect channels. For instance, the user could invoke `AccessT` and observe the power consumption while the function executes. These side-channels would indirectly provide information about the accesses made to T.

Assuming that the cache memory does not contain any part of the look-up table before the first invocation of `AccessT`, the first access at an offset $(x_1 \oplus k_1)$ in the look-up table would result in a cache miss. Consequently, a memory block containing

C. Rebeiro et al., *Timing Channels in Cryptography*,
DOI 10.1007/978-3-319-12370-7_4

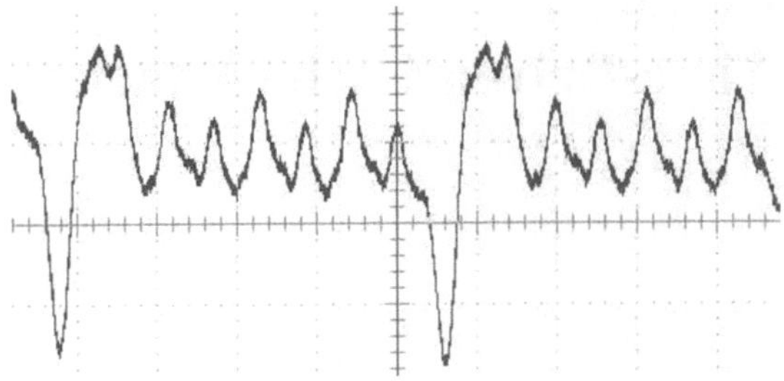

Fig. 4.1 The power consumption trace of eight load instructions in a PowerPC processor. The first and the fifth load resulted in a cache miss, while the others were cache hits

the data $T[x_1 \oplus k_1]$ would get loaded into a cache line. For instance, if the size of the cache line is 32 bytes and each element in the table requires 1 byte then at most $32 = (32/1)$ elements of the look-up table are loaded into the cache.

Listing 4.1 function AccessT reads two locations from a look-up table

```
unsigned char T[256] = {0×33,0×65,···};
unsigned char y;

unsigned char AccessT(unsigned char x1, unsigned char x2){
  unsigned char k1 = load_secret1();
  unsigned char k2 = load_secret2();

  y=T[k1⊕x1]⊕T[k2⊕x2];

  return y;
}
```

During the second access to table T at an offset $(x_2 \oplus k_2)$, a cache hit will arise if (1) data corresponding to the first access still resides in the cache and (2) the access at offsets $(x_1 \oplus k_1)$ and $(x_2 \oplus k_2)$ in the look-up table fall in the same memory block. We call this a *collision*. This cache hit can be inferred by monitoring the power consumption of the device. The cache hit implies that $(x_1 \oplus k_1)$ and $(x_2 \oplus k_2)$ differ by at most 32 (for a cache line of 32 bytes). This indicates that $(x_1 \oplus k_1)$ and $(x_2 \oplus k_2)$ differ by $\log_2 32 = 5$ least significant bits. We represent this relationship as $\langle x_1 \oplus k_1 \rangle = \langle x_2 \oplus k_2 \rangle$, where $\langle z \rangle$ is $z \gg 5$. Since the malicious user chooses x_1 and x_2, a relationship between k_1 and k_2 can be inferred as follows,

$$\langle k_1 \oplus k_2 \rangle = \langle x_1 \oplus x_2 \rangle \ . \tag{4.1}$$

For instance, if x_1 and x_2 are chosen as 0×45 and $0\times c4$, respectively and the memory block size is 32 bytes, then $\langle k_1 \oplus k_2 \rangle$ is 4. This is obtained as follows,

$$\begin{aligned} \langle k_1 \oplus k_2 \rangle &= (x_1 \oplus x_1) \gg \log(32/1) \\ &= (0\times45 \oplus 0\times c4) \gg 5 \\ &= 4 \end{aligned} \tag{4.2}$$

This means that the value of the three most significant bits of $k_1 \oplus k_2$ is 4. The set of candidates are therefore $\{0\times80, 0\times81, 0\times82, \cdots 0\times9f\}$. Only one value in the candidate set is correct, all others are wrong.

In terms of the information gained, since k_1 and k_2 are defined as unsigned char in Listing 4.1, each can take one of 2^8 different values. Together, the uncertainty of k_1 and k_2 is 2^{16} (or 16 bits). After observing the cache hit, the uncertainty of k_1 and k_2 reduces. In the example taken above, with the 32 byte memory block size, the uncertainty of k_1 and k_2 reduces to 2^{13} (or 13 bits).

The amount by which the uncertainty reduces depends on the size of the look-up table, the memory block size which is equivalent to the cache line size, and the alignment of the table in memory. This reduction in uncertainty is discussed in Sects. 4.1.1 and 4.1.2.

A cache miss could also be a source providing information about k_1 and k_2. A cache miss in the second table access would indicate that the first and second memory access to table T are to different memory blocks. Thus , $\langle x_1 \oplus k_1 \rangle \neq \langle x_2 \oplus k_2 \rangle$ and

$$\langle k_1 \oplus k_2 \rangle \neq \langle x_1 \oplus x_2 \rangle \ . \tag{4.3}$$

In Listing 4.1, a cache miss in the second access would eliminate 32 values for $k_1 \oplus k_2$.

A cache miss, however, can also occur if the memory block corresponding to $T[x_1 \oplus k_1]$ gets evicted from the cache before $T[x_2 \oplus k_2]$ is accessed. Thus unlike a cache hit, which guarantees a collision, a cache miss does not always imply that a collision has not occurred. Said differently, Eq. 4.3 may not always hold even if a cache miss has occurred during the second access to table T.

4.1.1 Relation Between Size and Bits Revealed

The size of the memory block, look-up table, and the size of each element in the table dictate the number of bits that get revealed. If N is the number of elements in the look-up table, L the size of each memory block in terms of bytes and B the number of bytes required to store each element in the look-up table, then the number of elements in a memory block is L/B and the number of memory blocks occupied by the table is $(N \times B)/L$.

Power consumption traces of a cache hit determine collisions in a memory block. Thus number of bits revealed is $\log_2(\frac{N \times B}{L})$. Since multiple elements of the table share the same block, there is an ambiguity about which element was actually accessed within the block. This results in multiple candidates of size $\frac{L}{B}$. In Listing 4.1, N is 256, B is 1 (since sizeof(unsigned char) = 1), and assume L to be 32. Thus, the number of bits of $k_1 \oplus k_2$ revealed is 3 and the number of candidates for $k_1 \oplus k_2$ is 32. The number of bits revealed reduces as the size of the look-up table reduces. The ideal case of zero bits revealed occurs when the look-up table fits within a single memory block.

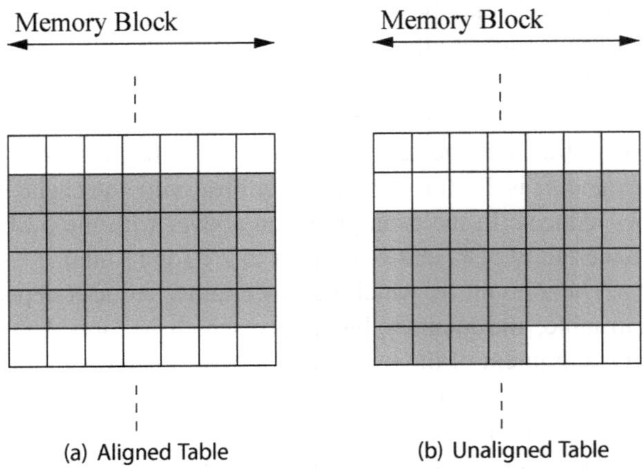

Fig. 4.2 The alignment of a table in memory could affect the bits revealed. Each row in the figure is a memory block, while the *shaded regions* are the look-up table of size l. In the aligned table every block has δ elements, while in an unaligned table the first and last block have fewer elements

4.1.2 Relation Between Alignment of Tables and Bits Revealed

A look-up table is said to be *aligned* if its base address is a multiple of the memory block size. Figure 4.2a shows how an aligned table is placed in memory, while Fig. 4.2b shows an *unaligned* look-up table. Each row in the figures indicates a memory block. The alignment of the table depends considerably on the compiler used to build the program's executable. Some compilers align tables in order to achieve better performance, while others do not align tables thus utilizing memory better. If the tables are dynamically allocated, then the operating system's memory management scheme dictates the table's alignment.

In an aligned table, all memory blocks loaded into the cache contain the same number of elements of the look-up table. Thus the size of the candidate key set is always L/B. This size is independent of which memory block is loaded into the cache. In an unaligned table, the first and the last memory block of the table contain lesser number of elements compared to the other blocks. Consequently, a cache hit in the second access occurring due to a collision in either the first or the last memory block would result in a candidate key set which is smaller than L/B.

For instance, let A be a memory address which is a multiple of the memory block size. Assume that the table T in Listing 4.1 has the base address $A + 31$. Since the table is of size 256 bytes, it occupies 8 memory blocks assuming each block is of 32 bytes. The first memory block contains 1 element of the table, the last block contains 15 elements, while all other memory blocks holding the table contain 16 elements. Consequently, the size of the candidate key set would be 1, 16, or 15 depending on which memory block had the collision.

4.1.3 Initial State of Cache Memory

In the discussion above, we assumed that no part of table T was present in the cache before function `AccessT` was first invoked. In order to ensure this, the attacker could clear the cache of the contents of T before invoking `AccessT`. Some processors have instructions that would allow parts of the cache to be flushed. Intel and AMD x86 processors, for instance, have an instruction called clflush for the purpose. The instruction takes a linear address and invalidates lines in all caches that contain the address. Alternatively the attacker could fill the cache with its own data, thus evicting all other data. This is done by allocating an array as large as the cache itself and then accessing it. The access would load the array elements into the cache, thereby evicting all other data.

4.2 Collisions from Execution Time

The previous section showed how power consumption of a processor can be used to detect cache hits and misses. Power consumption side-channels are generally suited for small embedded processors such as the PowerPC, where tapping into the power lines is easy. In more complex processors, the noise in the power consumption would significantly hinder the determination of a collision. Time-based side-channels are generally preferred in such environments as they can be easily measured, do not require physical colocation with the device, and do not require any special equipment.

In a timing side-channel microarchitectural events such as a cache hit and a miss are distinguished using time measurements. Making these measurements requires a highly precise clock source. Many microprocessors have such precise clocks in the form of a time stamp counter. This is a highly precise low-overhead way to make time measurements. In processors that do not have such counters, virtual time-stamp counters (VTSCs) have been built to measure microarchitectural events. VTSCs have a small loop that increments a counter continuously. Since the loop is small, the counter is incremented at every clock cycle of the processor. The value of the counter, therefore, gives a notion of time. Such ad hoc clocks are feasible because time is only used to distinguish between microarchitectural events, the exact duration of the events are not required. For instance, time side-channels just use the fact that a cache miss takes more time than a cache hit. The exact duration of a cache miss and cache hit is not important. In this section we discuss these two clock sources.

4.2.1 Clocks Using Hardware Time Stamp Counters

Intel machines since the Pentium have a 64-bit time-stamp counter register. The register is made 0 when the processor resets and is monotonically incremented in every clock cycle. Thus, on a 1 GHz machine, the counters could measure time periods as low as a nanosecond. An instruction called rdtsc is used to read the time-stamp counter. When rdtsc is invoked, the high-order 32 bits of the counter is loaded into the edx register, while the low-order 32 bits is loaded into the eax register. The following pseudoassembly code shows how the rdtsc instruction can be used to time a load instruction.

Listing 4.2 *timestamp* (using rdtsc to measure the time required for a load instruction)

```
1 rdtsc                  ; read time stamp
2 mov     time, eax      ; move counter into variable
3 load    ebx, (ebp)     ; a load from memory
4 rdtsc                  ; read time stamp again
5 sub     eax, time      ; find the difference
```

The load instruction in line 3 loads the ebx register with contents pointed to by the register ebp. The load instruction is sandwiched between two rdtsc instructions. The first rdtsc records the counter value in the variable time before the load is carried out, while the second rdtsc instruction records the counter value soon after the load instruction. The difference between the two counter values gives the duration of the load instruction. Since microarchitectural events last only for a few clock cycles, the contents of the eax register is sufficient. The higher bits of the counter stored in the edx register are generally not changed between the two rdtsc calls. The variable time would be higher if there is a cache miss compared to a cache hit.

Several sources of error are possible in the time measurements described above. A common source of error is from hardware interrupts that occur between the two rdtsc instructions. This would result in a measurement that is significantly larger than expected. Errors introduced by hardware interrupts can easily be eliminated by using a threshold. Values above the threshold would most probably be due to a hardware interrupt, therefore not considered. The value of the threshold should be set just above the maximum cycles required for a load.

Besides interrupts, several microarchitectural features could cause errors. These are subtle and more difficult to identify. The errors are considerably smaller, hence cannot be identified by thresholds. One such source of error is the out-of-order execution of instructions that is present in all x86 machines from Pentium Pro onwards. This feature allows instructions to be executed in an order that is different from that specified by the source code. For instance, since the load instruction in Listing 4.3 takes considerable amount of time, it may be performed earlier. This could potentially result in the first rdtsc instruction being performed after the load has completed. If this happened, the time measurement would not take the load into account.

In order to keep the rdtsc instruction from being performed out-of-order, a *serializing* instruction is required. This instruction waits until all preceding instructions complete. One such serializing instruction is the cpuid, which is normally used to identify the processor on which the program is executed. The cpuid instruction can be used to ensure that rdtsc is executed in-order. Listing 4.3 shows the time measurement using an rdtsc instruction that is serialized. Each rdtsc instruction is preceded and followed by a cpuid instruction. The proceeding cpuid instruction ensures that previous instructions are completed, while the cpuid instruction that follows ensures that no instruction is executed before the time-stamp counter is read for the second time. Newer x86 processors have an instruction rdtscp that reads the time-stamp counter and is a serialized instruction by itself.

Listing 4.3 *timestamp* (using rdtsc serialized rdtsc to measure the time required for a load instruction)

```
cpuid              ; ensure preceding instructions complete
rdtsc              ; read time stamp
cpuid              ; ensure preceding instructions complete
mov  time, eax     ; move counter into variable
load ebx, (ebp)    ; a load from memory
cpuid              ; ensure preceding instructions complete
rdtsc              ; read time stamp again
cpuid              ; ensure preceding instructions complete
sub  eax, time     ; find the difference
```

Another source of error occurs in processors whose clock frequency is scaled depending on the work load. A processor that is lightly loaded runs at a lower clock frequency in order to save energy. Its clock frequency is increased automatically as the load of the processor increases in order to boost performance. This scaling of frequency could induce errors in the measured cycle count. It should, therefore, be ensured that the processor is sufficiently warmed up so that it runs at the maximum clock frequency before any measurements are made.

Systems can prevent the user space programs from using rdtsc and rdtscp instructions. However, high-precision timing measurements can still be made by using VTSCs. The next section discusses such counters.

4.2.2 *Clocks with Virtual Time-Stamp Counters*

To build a VTSC, two threads (A and B) are required, each running on a separate core of the processor and sharing a memory address CC, which mimics a time-stamp register. Thread A initializes one of the processor's general purpose register to 0, then increments it in an infinite loop, and copies the result to CC. The shared variable CC should be defined volatile to ensure timely copying. The loop would thus have an add instruction followed by a store. Assuming that no other process modifies CC, it would monotonically increment approximately once every clock cycle just like a hardware time-stamp counter. To use CC, thread B would read it before and after the instructions it wishes to measure.

Listing 4.4 shows an example of the VTSC to time an addition. The function `vtsc_thread` runs in a separate thread and increments the global volatile variable CC. This shared variable is read from the function `measure` before and after the addition is done. As in the case of rdtsc, the instruction cpuid may be added before and after CC is read to ensure serialized execution.

Listing 4.4 *timestamp* (using a virtualized time stamp counter to time an addition)

```

volatile unsigned long long CC;

void vtsc_thread(void *ptr)
{
  while(1) CC++;
}

unsigned int measure(x, y)
{
  t1 = CC;
  z = x + y;
  t2 = CC;

  return (t2 - t1);
}
```

4.2.3 Distinguishing Cache Hit and Miss Events Using Time

Now that we have established how to measure time accurately, we look at how to use time to distinguish microarchitectural events such as a cache hit and a cache miss. We go back to Listing 4.1. The malicious user can invoke `AccessT` with chosen inputs and monitor the execution time required. A longer execution time would likely be due to a cache miss in the second access to the table T, while a shorter execution time would indicate a cache hit. To identify the cache hit, x_1 is fixed and different values of x_2 are chosen in each invocation of `AccessT`. The range of x_2 is 0 to 255. The values of x_2 that results in shorter execution time is likely to be due to a cache hit in the second access to table T.

However, things are not so straightforward. First, we first observe that in a system there are several microarchitectural components working simultaneously, each affecting execution time. This adds noise to the measurements. Since the difference between a cache hit and a miss is often just a few clock cycles, the noise would have a significant effect. The noise is due to several factors such as interrupts, direct memory access (DMA), instruction cache misses, activities in other processor cores, etc.

The noise can be reduced considerably by invoking `AccessT` several times with the same value of x_2 and then finding the average execution time. However, the noise is not uniform. For instance, a burst of network activity would momentarily add noise. This would affect some values of x_2 more than others. To avoid this, we randomize the order in which values of x_2 are chosen in the invocations of `AccessT`.

When a function is invoked repeatedly, the initial invocations take considerably more time. This is because some components in the system need to adjust to the new instructions. For instance, the RAM needs to be loaded with the new code, the instruction cache needs to be filled with the new instructions, and the branch predictors needs to learn about the branches in the program. To avoid the initial overheads, we let the system have a warm up period by ignoring the first couple of invocations of `AccessT`. The warm up period also ensures that the processor frequency is scaled up to its maximum. Later invocations of `AccessT` would then be executed at the maximum frequency. Listing 4.5 shows how this is done.

Listing 4.5 Function FindK monitors execution time of `AccessT` to determine $\langle k_1 \oplus k_2 \rangle$

```
FindK(unsigned char x1, unsigned char x2){
  m= size of table in terms of number of blocks
  s= number of bits needed to address elements in a memory block
  T= {0};
  C= {0};
  x1 = 0x45;
  i= 0;

  while(i <2^n){
    x2 = random() %256;
    flushcache();
    t1 = timestamp();
    AccessT(x1, x2);
    t2 = timestamp();

    i= i+ 1;
    if (i< 200) continue;                  /* the warm up */
    b= (x1 ⊕x2)>>s;
    T[b] = T[b] + (t2- t1);
    C[b] = C[b] +1;
  }

  for (0≤j< m){
    D[j] = (T[j]/C[j]);
  }

  return j for which D is minimum;
}
```

In the implementation

- m stores the number of memory blocks occupied by the look-up table. If the number of elements in the table is N, the size of each memory block is L , and size of each element in the table is B, then $m = (N \times B)/L$. For simplicity, we assume that the table is aligned to a block. In `AccessT`, for instance, there are 256 elements in table T, with each element occupying 1 byte. If we assume a memory block size of 64 bytes, then $m = 4$.
- s contains the number of bits needed to address elements in a memory block. This is equal to $\log_2(L/B)$. In the example configuration above $s = \log_2 64/1 = 6$.
- n determines the number of times the loop is run. As would be seen later in this chapter, as n increases, the probability of detecting the collision also increases.

Table 4.1 Average execution time for AccessT with $\langle x_1 \oplus x_2 \rangle$ in clock cycles on an Intel i7 Haswell Processor running at 3 GHz

$\langle x_1 \oplus x_2 \rangle$	Average execution time (in clock cycles)
0	484
1	483
2	446
3	457

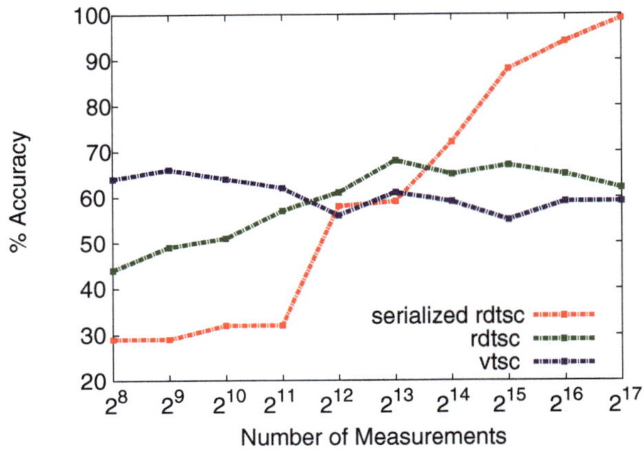

Fig. 4.3 Accuracy with which a cache hit is identified in AccessT with different clock sources. Large number of measurements along with a precise clock source can accurately identify cache hit events in the L1 data cache

- $\mathcal{T}$ is an array of size m , which accumulates the execution time for each block in the table.
- $\mathcal{C}$ is an array of size m , which counts the number of times a block in $\mathcal{T}$ is accessed in the second memory access.
- $\mathcal{D}$ computes the average execution time for each $\mathcal{T}$. The index of the block corresponding to the minimum $\mathcal{D}$ is returned.
- `flushcache` is a function that invalidates the contents of the look-up table in the cache memory.

If $k_1 \oplus k_2$ is 0×94, the two most significant bits is 2. Table 4.1 shows the average execution time for each of value of $\langle x_1 \oplus x_2 \rangle$. When $\langle x_1 \oplus x_2 \rangle = 2$, a cache hit occurs in the second access to the table T. All other values of $\langle x_1 \oplus x_2 \rangle$ results in a cache miss. As a result, the execution time for AccessT when $\langle x_1 \oplus x_2 \rangle = 2$ is slightly faster than the remaining. We, therefore, can infer that $\langle k_1 \oplus k_2 \rangle$ is 2.

The accuracy with which the cache hit is detected depends on several factors. The first is the penalty incurred by a cache miss. A small miss penalty would be more difficult to identify. The miss penalty depends on the system architecture and the level of the cache. Second, the precision with which timing measurements are made also critically affects the accuracy of a cache hit detection. The clock source should be precise enough to measure microvariations in execution time. The third factor affecting accuracy is the number of measurements that are considered while computing the average execution time. A larger number of measurements would help eliminate noise to a greater extent. Figure 4.3 shows the accuracy with which the cache hit in `AccessT` is determined on an Intel i7 Haswell machine. As one would expect, the serialized rdtsc instruction provides the best accuracy when large number of measurements are made. With 2^{17} measurements for instance, cache hits are determined with an accuracy of about 99 %

4.3 Timing Attacks on Block Ciphers Based on Internal Collisions

Now that we have seen how time can be used to distinguish cache hits from misses, we show how it can be used to retrieve the secret key from a block cipher. We discuss this with AES (Advanced Encryption Standard) as an example. The 128-bit version of the AES cipher takes a plaintext block of 128 bits as input and a 128-bit secret key. Encryption starts with a key whitening followed by ten rounds. Each round except the last has four operations: SubBytes, ShiftRows, MixColumns, and AddRoundKeys. The last round does not have the MixColumns operation but has the three other operations. The SubBytes operation performs nonlinear transformations on the input, thereby adding confusion. The ShiftRows and MixColumns diffuse bits of the input. Details of the AES algorithm are present in Sect. 2.2.1.

Several implementations of AES are present. In the most common implementation, five 1024-byte look-up tables called T0, T1, T2, T3, T4 are used. The tables T0 to T3 are used in the first nine rounds. These tables abstract the SubBytes, ShiftRows, and MixColumns operations as shown in Sect. 2.2.1.1. Table T4 is exclusively used in the tenth round and abstracts the SubBytes and MixColumns operations. Listing 4.6 shows the first round of the code. This consists of 16 accesses to the tables.

Listing 4.6 First round of AES implementation

$$y_0 = \mathcal{T}_0[s_0] \oplus \mathcal{T}_1[s_5] \oplus \mathcal{T}_2[s_{10}] \oplus \mathcal{T}_3[s_{15}] \oplus [k_0^{(1)}\ k_1^{(1)}\ k_2^{(1)}\ k_3^{(1)}]^T$$
$$y_1 = \mathcal{T}_0[s_4] \oplus \mathcal{T}_1[s_9] \oplus \mathcal{T}_2[s_{14}] \oplus \mathcal{T}_3[s_3] \oplus [k_4^{(1)}\ k_5^{(1)}\ k_6^{(1)}\ k_7^{(1)}]^T$$
$$y_2 = \mathcal{T}_0[s_8] \oplus \mathcal{T}_1[s_{13}] \oplus \mathcal{T}_2[s_2] \oplus \mathcal{T}_3[s_7] \oplus [k_8^{(1)}\ k_9^{(1)}\ k_{10}^{(1)}\ k_{11}^{(1)}]^T$$
$$y_3 = \mathcal{T}_0[s_{12}] \oplus \mathcal{T}_1[s_1] \oplus \mathcal{T}_2[s_6] \oplus \mathcal{T}_3[s_{11}] \oplus [k_{12}^{(1)}\ k_{13}^{(1)}\ k_{14}^{(1)}\ k_{15}^{(1)}]^T$$

```
return y;
```

Algorithm 4.1: Strategy in a Time-Driven Attack for AES

Output: The value of $\langle k_a \oplus k_b \rangle$ for b ($b > a$)

1 **begin**
2 $d_a \leftarrow$ random value in the range $[0, 255]$.
3 **for** ρ *number of times* **do**
4 $\mathbf{x} \leftarrow (x_1 \| x_2 \| \cdots \| d_a \| \cdots \| x_m)$ where $x_a = d_a$, $x_i \xleftarrow{\mathcal{R}} [0, 255]$, $1 \leq i \leq m$ and $i \neq a$.
5 $t \leftarrow$ Time(Encrypt using $\mathbf{x}$).
6 Compute $\langle d_b \rangle$ from $\mathbf{x}$.
7 $t_{\langle d_b \rangle} \leftarrow t_{\langle d_b \rangle} + t$
8 $c_{\langle d_b \rangle} \leftarrow c_{\langle d_b \rangle} + 1$
9 **end**
10 $t_{avg\langle d_b \rangle} \leftarrow t_{\langle d_b \rangle} / c_{\langle d_b \rangle}$.
11 $t_{avg} \leftarrow$ average encryption time for all encryptions.
12 **return** $\underset{\langle d_a \oplus d_b \rangle}{\text{argmax}}(|t_{avg\langle d_b \rangle} - t_{avg}|)$.
13 **end**

In the first round, s_i has the form $x_i \oplus k_i$, where $1 \leq i \leq 16$. We see a similarity between Listings 4.1 and that of 4.6—both have loads from look-up tables that depend on the secret key. While the former has two key related table loads, the latter has 16; 4 to each look-up table. The time-driven attack in Listing 4.5 can therefore be extended to determine collisions between pairs of table accesses in AES. Algorithm 4.1 gives a high level overview.

To determine a collision between s_a and s_b ($1 \leq a < b \leq 16$) that access the table, the adversary keeps x_a constant, thereby fixing the table access made by s_a. For each of the memory blocks of the table accessible by $s_b = (x_b \oplus k_b)$, she determines the average execution time. Of all the average execution time computed, exactly one corresponds to a collision with s_a. This will have a distinctively different execution time compared to the others. If this value is identified, then we have found the collision. The collision reveals bits of $\langle k_a \oplus k_b \rangle$ as follows:

$$\begin{aligned} \langle s_a \rangle &= \langle s_b \rangle \\ \langle x_a \oplus k_a \rangle &= \langle x_b \oplus k_b \rangle \\ \langle k_a \oplus k_b \rangle &= \langle x_a \oplus x_b \rangle \end{aligned} \tag{4.4}$$

Algorithm 4.1 selects a value of d_a uniformly from the set 0 to 255 (line 2) and a plaintext $\mathbf{x}$ is chosen such that the value of d_a is fixed. The value of $\langle d_b \rangle$ is then computed from $\mathbf{x}$ and an encryption is invoked and timed. Lines 7 and 8 of the algorithm classifies the time. It comprises of a list of average timing corresponding to each line occupied by $\langle d_b \rangle$ (*i.e.*, the top $\log_2 l$ bits of d_b), Line 11 computes the overall average encryption time. The algorithm returns the value of $\langle d_a \oplus d_b \rangle$, which has maximum deviation in time from the average case. An important requirement is that the remaining plaintext bytes (other than d_a and d_b) be sufficiently random in order to build the timing distribution. Algorithm 4.1 can be parallelized. That is for a fixed value of a, $\langle k_a \oplus k_b \rangle$ can be determined simultaneously for multiple values of b. In the parallel version multiple $t_{avg\langle d_b \rangle}$ are measured (one for each b). Consequently, the algorithm would return a set $\langle d_a \oplus d_b \rangle$.

Each look-up table in the AES implementation has 256 elements and each element occupies 4 bytes. On a system with a 64-byte memory block, the table would occupy 16 memory blocks with each memory block containing 16 elements. The timing side-channels essentially identifies a collision in a memory block. It would therefore reveal 4 bits. The keys k_a and k_b are bytes and have a combined entropy of 16 bits. The attack would reduce this entropy to $16 - 4 = 12$ bits.

Each look-up table used in the first round of AES is accessed four times. At most three collisions in each table can, therefore, be obtained. The AES implementation uses four look-up tables in the first round, therefore a maximum of 12 collisions are obtained. Each collision reveals 4 bits. In all 48 bits is revealed. Consequently, the entropy of the 128-bit AES key reduces to 80 bits.

The drawback of the attack is that it is highly sensitive to the initial state of the cache. A cache memory which is not cleaned at the start of encryption would either render the attack ineffective or adversely affect its efficiency. In section 4.4 we describe another timing attack that can reveal more bits about the keys and is not sensitive to the initial cache state.

4.3.1 Max, Min, or Max Deviation

Intuitively, a collision caused by a cache hit should result in a faster execution. Finding the collision using Algorithm 4.1 should have, therefore, identified the minimum average execution time. However in the algorithm, we chose to select the one with the maximum deviation from average. The reason for this is that in certain cases the collision could increase the execution time instead of reducing it.

This being said, the choice of maximum, minimum, or max deviation depends on the system and the cipher implementation. The attacker would try different choices to boost her success in obtaining the secret key. Chapter 6 provides the reason for this.

4.4 Time-Driven Attack Based on Induced Cache Miss

Suppose the function `AccessT` in Listing 4.1 were to be called twice with the same inputs and in quick succession. The first call may cause cache misses to occur due to the look-up table accesses. This would load part of the table into the cache. The second call to `AccessT` , on the other hand, would mostly incur cache hits, as a result is much faster. If a single block of the table is evicted from the cache in between invocations, then the second call to `AccessT` would incur a cache miss if the evicted block is accessed. However, a no cache miss is likely to occur if the evicted block is not accessed. The attack works as follows.

1. For inputs x_1 and x_2 , invoke `AccessT`.
2. Evict a cache line (say c) occupied by the table.
3. Invoke `AccessT` again with the same inputs and this time monitor the execution time.

A longer execution time would possibly be due to a cache miss that occurred. Since the invocations were made in quick succession, it is highly likely that the cache miss is due to the memory block at an offset c in the table being accessed. We thus know that at least one of the two accesses made to the table T were possibly to the memory block c. Therefore,

$$\begin{gathered}\langle c \rangle = \langle x_1 \oplus k_1 \rangle \\ or \\ \langle c \rangle = \langle x_2 \oplus k_2 \rangle\end{gathered} \tag{4.5}$$

Thus,

$$\begin{gathered}\langle k_1 \rangle = \langle x_1 \oplus c \rangle \\ or \\ \langle k_1 \rangle = \langle x_2 \oplus c \rangle\end{gathered} \tag{4.6}$$

We can determine which of these two equations is correct by changing the values of either x_1 or x_2 and checking if the cache miss persists. Listing 4.7 shows how it is done.

Listing 4.7 Function FindK monitors execution time of `AccessT` to determine $\langle k_1 \rangle$

```
FindK2(unsigned char x1, unsigned char x2){
  m= size of table in terms of number of blocks
  s= number of bits needed to address elements in a memory block
  T= {0};
  C= {0};
  i= 0;

  while(i < 2^n){
    x2 = random()%256;
    x1 = random()%256;
    line= random()%256;
    AccessT(x1, x2);              /* First Invocation */
    flushline(line);
    t1 = timestamp();
    AccessT(x1, x2);              /* Second Invocation */
    t2 = timestamp();

    i= i+ 1;
    if (i < 200) continue;                   /* the warm up */

    k1 = (x1 ⊕ line) >> s;
    T[k1] =T[k1] + (t2 - t1);
    C[k1] =C[k1] +1;
  }

  for (0 ≤ j ≤ m){
    D[j] = (T[j]/C[j]);
  }

  return j for which D is minimum;
}
```

Details of the implementation are as follows:

- m stores the number of memory blocks occupied by the look-up table. For simplicity, we assume that the table is aligned to a block. In `AccessT` for instance, there are 256 elements in table $\mathcal{T}$. If we assume a memory block size of 64 bytes, then $m = 4$.
- s contains the number of bits needed to address elements in a memory block. In *AccessT*, each element in the table requires one byte. On a system with a 64-byte memory block $s = \log_2(64/1) = 6$.
- $\mathcal{T}$ is an array of size m, which accumulates the execution time for each block in the table.
- $\mathcal{C}$ is an array of size m, which counts the number of times a block in $\mathcal{T}$ is accessed in the second memory access.
- `flushline` is a function which invalidates the cache line that holds the element $T[line]$.

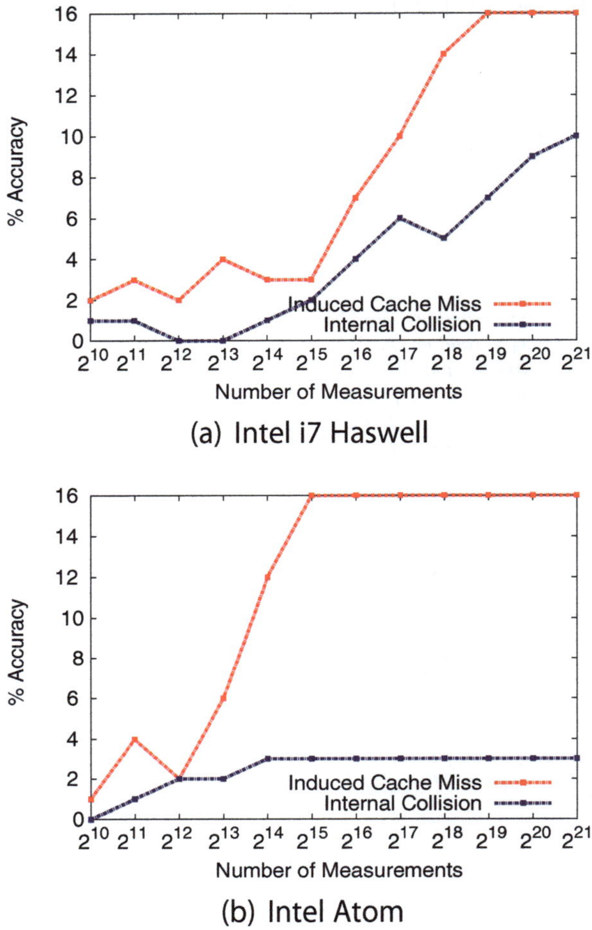

Fig. 4.4 Number of keys successfully determined in attacks on AES versus the number of measurements made. The induced cache miss technique is much more effective. As seen, it finds all keys correctly with significantly lesser number of timing measurements

The function `FuncK2` selects random values for x_1 and x_2 and invokes `AccessT` twice with these inputs. Another random value `line` is generated. This is used in the function `flushline` to evict a line from the cache, thereby potentially inducing a cache miss in the second invocation of `AccessT`. This invocation is timed and the loop is repeated several times. The value of $\langle k_1 \rangle$ is one that has the minimum execution time.

Working of `flushline`**.** The function `flushline` takes an input `line` and evicts T[`line`] from the cache if present. To be able to do this, `flushline` needs to know the cache sets which hold T. This is often not directly possible, since T is not accessible in function `FuncK2`. There are, however, indirect ways of obtaining the base address of T.

Consider that the adversary has an array as large as the cache memory. She first accesses all elements in the array. This would load the entire array into the cache.

She then invokes `AccessT` several times with different values of x_1 and x_2, hoping that the entire table T gets loaded into the cache. The accesses made to T would evict parts of the array from the cache. She then accesses the array again but this time times the accesses. The data in the array that take longer to load are possibly due to those evicted by T. She can thus identify the sets in the cache that hold T.

4.5 Results

Figure 4.4a shows the success in determining bits of the AES key on an Intel i7 Haswell system while Fig. 4.4b shows results on the Intel Atom system. The figures show that the induced cache miss is much better exploiting leakage compared to the using internal collisions.

4.6 Conclusion

Making accurate timing measurement is probably the most important step in a timing attack. The clocks used for making the measurements must be accurate enough to distinguish between microarchitectural events in the microprocessor, such as a cache hit and miss. This chapter described two ways to implement high precision clocks, namely using the in built *rdtsc* instruction and by software timestamp counters called VTSCs. These clock sources are used to analyze the memory access patterns of AES. The cache hits and misses identified from the AES execution is then used to develop two time-driven cache attacks that recover the AES secret key. Though the attacks work considerably well, the number of key bits they reveal is restricted by the size of the cache line. In the next chapter we discuss how cipher properties can be used to obtain more bits of the secret key.

Reference

1. Bertoni G, Zaccaria V, Breveglieri L, Monchiero M, Palermo G (2005) "AES Power Attack Based on Induced Cache Miss and Countermeasure," in ITCC (1). IEEE Computer Society, pp 586–591
2. Lauradoux C (2005) Collision attacks on processors with cache and countermeasures. In: Wolf C, Lucks S, Yau P-W (eds) WEWoRC, ser. LNI, vol 74. GI, pp 76–85
3. Fournier JJA, Tunstall M (2006) Cache based power analysis attacks on AES. In: Batten LM, Safavi-Naini R (eds) ACISP, ser. Lecture notes in computer science, vol 4058. Springer, pp 17–28
4. Bonneau J, Mironov I (2006) Cache-Collision timing attacks against AES. In: Goubin L, Matsui M (eds) CHES, ser. Lecture notes in computer science, vol 4249. Springer, pp 201–215
5. Aciiçmez O, Koç ÇK (2006) Trace-driven cache attacks on AES (Short Paper). In: Ning P, Qing S, Li N (eds) ICICS, ser. Lecture notes in computer science, vol 4307. Springer, pp 112–121

6. Page D (2002) Theoretical use of cache memory as a cryptanalytic side-channel. Departement of Computer Science, University of Bristol, Tech. Rep. http://eprint.iacr.org/2002/169
7. Tsunoo Y, Tsujihara E, Minematsu K, Miyauchi H (2002) Cryptanalysis of block ciphers implemented on computers with cache. In International Symposium on Information Theory and Its Applications, pp 803–806
8. Tsunoo Y, Saito T, Suzaki T, Shigeri M, Miyauchi H (2003) Cryptanalysis of DES Implemented on Computers with Cache. In: Walter CD, Kaya Koç Ç, Paar C (eds) CHES, ser. Lecture notes in computer science, vol 2779. Springer, pp 62–76
9. Aciiçmez O, Schindler W, Koç ÇK (2007) Cache based remote timing attack on the AES. In: Abe M (ed) CT-RSA, ser. Lecture notes in computer science, vol 4377. Springer, pp 271–286

Chapter 5
Advanced Time-Driven Cache Attacks on Block Ciphers

In Chapter 4 we saw how an adversary can use the execution time of the Advanced Encryption Standard (AES) block cipher to obtain information about the secret key. The number of key bits that the adversary determines is however restricted by the size of the cache line. If a cache line holds 2^δ elements of a look-up table used in AES, then at least δ bits of the key are hidden from the adversary. In this chapter, we discuss how properties of the block cipher along with timing side channels can be used to determine more key bits. The chapter begins with a second round time-driven cache attack on AES and then dwells into differential cache attacks on Feistel ciphers.

5.1 Second Round Attack on AES

The attack on AES discussed in Sect. 4.3 is able to determine relations of the form $\langle k_i^{(0)} \oplus k_j^{(0)} \rangle$ using collisions in the first round. The key bytes $k_i^{(0)}$ and $k_j^{(0)}$ $(0 \le i, j < 15)$ are whitening key bytes. On a system with a 64-byte cache line, the entropy of the 128-bit AES key reduces to 80 bits (for details see Sect. 4.3). The second attack described in Section 4.4 performs better. It retrieves 4 bits of each of the 16 key bytes thus reducing the entropy of the AES key to $(128 - (4 \times 16)) = 64$ bits. In order to reduce the entropy further, Aciiçmez, Schindler, and Koç suggest targeting the second round of AES instead of the first [1]. Here we provide the basic second round attack technique. The attack considerably depends on the AES specification. Readers should refer Sect. 2.2.1 wherever needed.

Let a tilde over the symbol represent a guess. For instance, $\tilde{\mathbf{k}}_i^{(0)}$ is a guess for the whitening key byte $\mathbf{k}_i^{(0)}$, where $0 \le i \le 15$. If there is a collision between the states $s_0^{(0)}$ in the first round and $s_0^{(1)}$ in the second round (i.e., the first access to table $\mathcal{T}_0$ in the first and second rounds respectively), then Eq. 5.1 is obtained. In the equation y_0 is the index of the first access to the table $\mathcal{T}_0$ in the second round.

$$\langle s_0^{(0)} \rangle = \langle s_0^{(1)} \rangle \tag{5.1}$$
$$\langle x_0 \oplus k_0^{(0)} \rangle = \langle y_0 \oplus k_0^{(1)} \rangle$$

C. Rebeiro et al., *Timing Channels in Cryptography*,
DOI 10.1007/978-3-319-12370-7_5

Using the AES algorithm (Sect. 2.2.1), y_0 can be written in terms of the plaintext and whitening key as follows.

$$y_0 = 2 \cdot S(x_0 \oplus \tilde{k}_0^{(0)}) \oplus 3 \cdot S(x_5 \oplus \tilde{k}_5^{(0)}) \oplus S(x_{10} \oplus \tilde{k}_{10}^{(0)}) \oplus S(x_{15} \oplus \tilde{k}_{15}^{(0)}) \quad (5.2)$$

Further, from the key expansion algorithm

$$k_0^{(1)} = k_0^{(0)} \oplus S(\tilde{k}_{13}^{(0)}) \oplus 1 \quad (5.3)$$

Substituting Eqs. 5.2 and 5.3 in Eq. 5.1, we obtain

$$\langle x_0 \rangle = \langle 2 \cdot S(x_0 \oplus \tilde{k}_0^{(0)}) \oplus 3 \cdot S(x_5 \oplus \tilde{k}_5^{(0)}) \oplus S(x_{10} \oplus \tilde{k}_{10}^{(0)}) \oplus S(x_{15} \oplus \tilde{k}_{15}^{(0)}) \oplus S(\tilde{k}_{13}^{(0)} \oplus 1 \rangle \quad (5.4)$$

There are five unknown key bytes involved: $\tilde{k}_0^{(0)}$, $\tilde{k}_5^{(0)}$, $\tilde{k}_{10}^{(0)}$, $\tilde{k}_{13}^{(0)}$, and $\tilde{k}_{15}^{(0)}$. These collectively take 2^{40} different values. For each of these values a set is created (thus there are 2^{40} sets). A random plaintext is chosen and the encryption time is found. This encryption time is added into all sets which satisfy Eq. 5.1 for the corresponding plaintext and key guess. The measurement is repeated several times with different plaintexts after which the average time in each of the 2^{40} key sets is found. The set in which the average time is the least corresponds to the correct key. Unlike the first round attacks, all bits of the five keys are obtained this way. In a same way other collisions between the first and second round can be used to determine the remaining key bytes.

The reason for the attack to work is as follows. The adversary chooses plaintexts randomly. Some of these plaintexts cause collisions between states $s_0^{(0)}$ and $s_0^{(1)}$ while others do not. Further, each plaintext will satisfy Eq. 5.1 for the correct key as well as for several wrong keys. Two cases arise,

- The execution time collected in a wrong key set may or may not correspond to collision between $s_0^{(0)}$ and $s_0^{(1)}$.
- The execution time in the set corresponding to the correct key always corresponds to collisions.

Thus, on average, the encryption time added to the correct set will have one cache miss less compared to the encryption time added to a wrong set. The smaller number of cache misses in the correct set will result in a smaller average execution time. This is used to distinguish the correct key.

5.2 Differential Cache Attacks on Feistel Ciphers

We now discuss attacks on Feistel ciphers that have a substitution–permutation (SP)-round function. These ciphers have a Feistel structure, with each round having a substitution followed by a permutation operation. This class of ciphers are of growing interest because they combine useful properties of SPN networks (like fast diffusion) with that of Feistel structures (like identical encryption and decryption). Notable

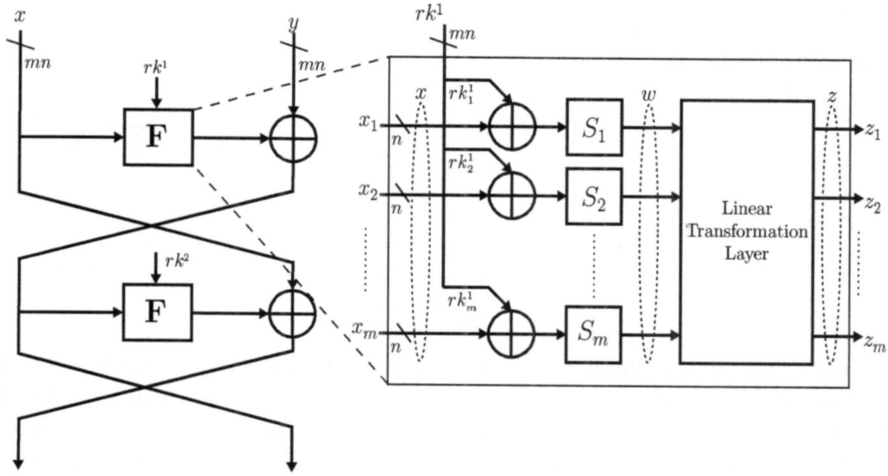

Fig. 5.1 Two rounds of a Feistel structure with SP-round function

block ciphers that have this structure are CAMELLIA [2], CLEFIA [3], and SMS4 [4]. The cipher E2 [5] is a deviant with an SPS round function (i.e., a permutation layer sandwiched between two substitution layers). To understand the working of the attack, we discuss with respect to a two-round Feistel structure having a SP-round function as shown in Fig. 5.1.

Let x and y be the arms of the Feistel structure with each arm having mn bits. The SP function, F, splits the input x into m equal parts (for example $x = (x_1|x_2|\cdots|x_m)$), each of n bits. The function F can be represented as

$$z = \mathsf{P}(\mathsf{S}_1(x_1 \oplus rk_1^1), \mathsf{S}_2(x_2 \oplus rk_2^1), \ldots, \mathsf{S}_m(x_m \oplus rk_m^1)) , \tag{5.5}$$

where S_1 to S_m are the substitution functions (s-boxes), P the permutation function that provides diffusion, and rk_i^1 (for $1 \leq i \leq m$) are n bit round key parts. The s-boxes in the F function are generally implemented as look-up tables in software. The number of tables used depends on the number of different s-boxes present in the cipher. Each table would have 2^n elements and occupy 2^l memory blocks (where $l = n - v$ and 2^v is the number of elements in a cache line).

Suppose the cache is initially clean and y is chosen such that all table accesses in the second round result in collisions, the following equality is obtained.

$$\langle x_i \oplus rk_i^1 \rangle = \langle z_i \oplus y_i \oplus rk_i^2 \rangle \qquad 1 \leq i \leq m , \tag{5.6}$$

where $\langle\,\rangle$ represents the most significant l bits, indicating the memory blocks occupied by the table.

Equation 5.6 can be deduced by monitoring the execution time using Algorithm 4.1. A cache hit in the first round can be determined by mapping $d_a \leftarrow x_i$ and $d_b \leftarrow x_j$ for $1 \leq i < j \leq m$ in the algorithm provided x_i and x_j access the same table. If there are g tables used in the F function, then there are a minimum of

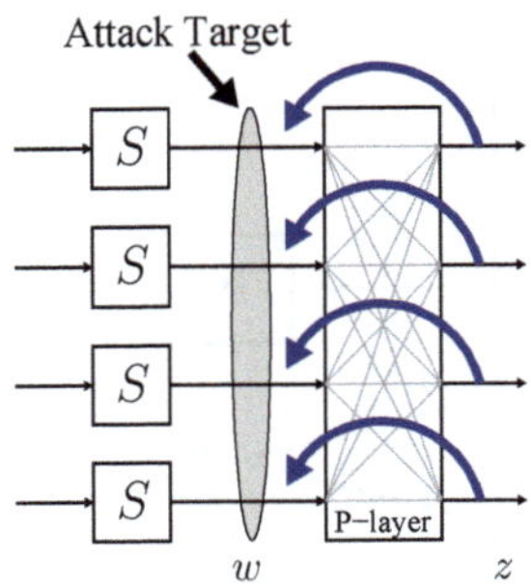

Fig. 5.2 The inverse of the permutation layer is used to obtain differences at the output of the s-box

g misses and a maximum of $m - g$ collisions that can be obtained in the first round (assuming a clean cache at the start of encryption). The state in which x results in the maximum collisions in the first round is known as the *one-round colliding state*.

A cache hit in the second round is obtained by choosing x so that a one-round colliding state is obtained and then mapping y_j $(1 \le j \le m)$ to d_b in Algorithm 4.1, while x_j is mapped to a corresponding d_a. Additionally x has to be kept constant. Thus Eq. 5.6 can be obtained with two invocations of Algorithm 4.1; the first invocation obtains collisions in the first round and the second invocation obtains collisions in the second round. In all 2ρ encryptions are required to be monitored—where ρ is the number of iterations of the loop in Algorithm 4.1.

Now consider another input $x' = (x'_1|x'_2|\cdots x'_m)$ such that $x' \neq x$. Let x' result in the s-box output z', and y' the corresponding input which causes collisions in the second round. These collisions would result in the following equalities:

$$\langle x'_i \oplus k^1_i \rangle = \langle z'_i \oplus y'_i \oplus rk^2_i \rangle \quad \text{where } 1 \le i \le m. \tag{5.7}$$

Adding Eqs. (5.6) and (5.7) we obtain

$$\langle x_i \oplus x'_i \oplus y_i \oplus y'_i \rangle = \langle z_i \oplus z'_i \rangle. \tag{5.8}$$

Thus the differences in the inputs $\Delta x_i = x_i \oplus x'_i$ and $\Delta y_i = y_i \oplus y'_i$ can be used to determine the output difference of the F function in the first round.

For Eq. 5.7, ρ encryptions are sufficient (or 1 invocation of Algorithm 4.1 because the collisions in the first round can be established for free using the previously determined equalities $\langle rk^1_i \oplus rk^1_j \rangle = \langle x_i \oplus x_j \rangle$, $1 \le i < j \le m$. Thus in total, 3ρ encryptions are needed to obtain Eq. 5.8.

In order to derive candidates for the round key rk^1, we use the inverse of the permutation layer (Fig. 5.2), which can be represented as m functions having the form $w_j = LT_j^{-1}(z)$, where $1 \le j \le m$. For the input difference Δx, the collisions in the second round would reveal $\langle \Delta z_i \rangle$ (for all $1 \le i \le m$). These can be used in $LT_j^{-1}(\cdot)$ to obtain few bits of Δw_j (for all $1 \le j \le m$); the output difference of the s-box. The input difference and the partial output difference of the s-box is then used to reduce the key space of rk^1_j.

Expected Number of Candidate Keys For any of the $n \times n$ s-boxes in Fig. 5.1, the total number of output differences for a given input difference is 2^{n-1}. Let o_b be the number of bits of Δw_j that gets revealed by the application of $LT_j^{-1}(\cdot)$. The number of bits o_b can have a minimum value of 0 and a maximum of l. The possible number of output differences of the s-box gets reduced to 2^{n-o_b-1}. This reduces the set of candidate keys for k_j^1 from 2^n to N_b. On average N_b is

$$N_b = 2^{(n-o_b)} \,. \tag{5.9}$$

5.3 Differential Cache Attack on CLEFIA

Now that we have developed the theory behind differential cache attacks, we show how it can be used on the block cipher CLEFIA.

CLEFIA is a 128-bit block cipher and a standard for lightweight encryption. It has a double Feistel structure and uses two F functions named $F0$ and $F1$ (Sect. 2.2.2). Analysis of the CLEFIA algorithm shows that the knowledge of any set of four round keys $(RK4i, RK4i+1, RK4i+2, RK4i+3)$, where $i \bmod 2 = 0$, is sufficient to revert CLEFIA's key expansion process and obtain the secret key. In the differential cache attack on CLEFIA that we describe here, round keys $RK0$, $RK1$, $RK2$, and $RK3$ are determined from which K is obtained.

The attack on CLEFIA comprises three steps. First $RK0$ and $RK1$ are determined, then $WK0 \oplus RK2$ and $WK1 \oplus RK3$, and finally $RK4$ and $RK5$. With these round keys, CLEFIA's key expansion algorithm is used to obtain 57 bits of $(RK2|RK3)$. In all, obtaining the 121 bits of the round keys $RK0$, $RK1$, $RK2$, and $RK3$ requires 2^{14} collisions.

5.3.1 Differential Properties of CLEFIA's F Functions

The attack uses the following observations on the F functions:

- Matrices $M0$ and $M1$ in the F functions do not attain complete diffusion in all bits of the output. If the five most significant bits (*MSBs*) of the input of each byte of the matrices $M0$ and $M1$ are known then few bits of the output can be computed (see Fig. 2.4). In particular three *MSBs* of each byte in $M0$'s output and two *MSBs* of each byte in $M1$'s output are computable. Since $M0$ and $M1$ are self inverting matrices, the inverse of the above statement also holds. That is, given five *MSBs* of each byte of the output, three *MSBs* of the input in $M0$ and two *MSBs* of the input in $M1$ is computable.
- For a pair of inputs, the non-linearity in the s-boxes causes several (60 % in $S0$ and 50 % in $S1$) of the possible input difference–output difference combinations to be invalid. Additionally, for a valid combination, $S0$ has 1.28 choices on average for the inputs to the s-box, while $S1$ has 1.007.

If the inputs to an s-box is $(x_i \oplus k)$ and $(x_i' \oplus k)$, then the ex-or difference is $(x_i \oplus x_i')$. This is known to the adversary. Additionally, on a system with a 32-byte cache line, the cache traces reveal three bit differences per byte of the output of each s-box of $F0$. For the remaining five bits of each output, there are 32 possible input difference–output differences for each s-box resulting in an average of 32 key (k) candidates for each byte. Similarly there are about 64 choices for each key byte in $F1$. We now show how these differential properties of CLEFIA are used in the recovery of the round keys.

5.3.2 *Determining* $RK0$ *and* $RK1$

The equations for the indices to the tables in the first round is given by:

$$\begin{aligned}
I1^0_{s0} &= P_0 \oplus RK0_0 & I1^1_{s0} &= P_2 \oplus RK0_2 \\
I1^2_{s0} &= P_9 \oplus RK1_1 & I1^3_{s0} &= P_{11} \oplus RK1_3 \\
\\
I1^0_{s1} &= P_1 \oplus RK0_1 & I1^1_{s1} &= P_3 \oplus RK0_3 \\
I1^2_{s1} &= P_8 \oplus RK1_0 & I1^3_{s1} &= P_{10} \oplus RK1_2
\end{aligned} \tag{5.10}$$

where $I\alpha^i_{s\beta}$ denotes the index to the $(i+1)$th access to table $s\beta$ in round α.

If we make the assumption that no part of the table is present in cache before the start of encryption, then the first access to each table, i.e., $I1^0_{s0}$ and $I1^0_{s1}$, results in cache misses. Keeping P_0 and P_1 fixed and by varying the other bytes in Algorithm 4.1, collisions can be determined for all table accesses in the first round. Such a state of the cipher is called a *one-round colliding state*.

In the second round, the indices to the tables $S0$ and $S1$ in $F0$ are given by equations in (5.11), where $P_{(0\cdots3)}$ indicates the concatenation of P_0, P_1, P_2, and P_3.

$$\begin{aligned}
I2^0_{s0} &= P_4 \oplus WK0_0 \oplus F0(RK0, P_{(0\cdots3)})_0 \oplus RK2_0 \\
I2^0_{s1} &= P_5 \oplus WK0_1 \oplus F0(RK0, P_{(0\cdots3)})_1 \oplus RK2_1 \\
I2^1_{s0} &= P_6 \oplus WK0_2 \oplus F0(RK0, P_{(0\cdots3)})_2 \oplus RK2_2 \\
I2^1_{s1} &= P_7 \oplus WK0_3 \oplus F0(RK0, P_{(0\cdots3)})_3 \oplus RK2_3
\end{aligned} \tag{5.11}$$

Starting from the one-round colliding state, four collisions are forced in $F0$ of the second round by varying, P_4, P_5, P_6, and P_7. This results in identifying five collisions in table $S0$ (three in the first round and two in the second). The *MSBs* of the indices to the table are all the same, *i.e.*, $\langle I1^0_{s0}\rangle = \langle I1^1_{s0}\rangle = \langle I2^0_{s0}\rangle = \langle I2^1_{s0}\rangle$. We therefore get the following equalities:

$$\langle P_0 \oplus P_4\rangle = \langle F0(RK0, P_{(0\cdots3)})_0 \oplus RK0_0 \oplus WK0_0 \oplus RK2_0\rangle \tag{5.12}$$

$$\langle P_2 \oplus P_6 \rangle = \langle F0(RK0, P_{(0\cdots3)})_2 \oplus RK0_2 \oplus WK0_2 \oplus RK2_2 \rangle$$

Similarly the five cache hits in table $S1$ result in the following equalities:

$$\begin{aligned} \langle P_1 \oplus P_5 \rangle &= \langle F0(RK0, P_{(0\cdots3)})_1 \oplus RK0_1 \oplus WK0_1 \oplus RK2_1 \rangle \\ \langle P_3 \oplus P_7 \rangle &= \langle F0(RK0, P_{(0\cdots3)})_3 \oplus RK0_3 \oplus WK0_3 \oplus RK2_3 \rangle \end{aligned} \quad (5.13)$$

For another plaintext Q, with $Q_0 \neq P_0$ and $Q_1 \neq P_1$, equations similar to (5.12) and (5.13) can be obtained by tracing cache collisions in the first and second rounds. These are shown in Eq. 5.14, where $0 \leq i < 4$.

$$\langle Q_i \oplus Q_{4+i} \rangle = \langle F0(RK0, Q_{(0\cdots3)})_i \oplus RK0_i \oplus WK0_i \oplus RK2_i \rangle \quad (5.14)$$

From Eqs. 5.12, 5.13, and 5.14, and the fact that $\langle P_0 \oplus P_2 \oplus P_4 \oplus P_6 \rangle = \langle Q_0 \oplus Q_2 \oplus Q_4 \oplus Q_6 \rangle$, and $\langle P_1 \oplus P_3 \oplus P_5 \oplus P_7 \rangle = \langle Q_1 \oplus Q_3 \oplus Q_5 \oplus Q_7 \rangle$ the following equations are generated:

$$\begin{aligned} \langle P_0 \oplus P_4 \oplus Q_0 \oplus Q_4 \rangle &= \langle F0(RK0, P_{(0\cdots3)})_0 \oplus F0(RK0, Q_{(0\cdots3)})_0 \rangle \\ \langle P_1 \oplus P_5 \oplus Q_1 \oplus Q_5 \rangle &= \langle F0(RK0, P_{(0\cdots3)})_1 \oplus F0(RK0, Q_{(0\cdots3)})_1 \rangle \\ \langle P_2 \oplus P_6 \oplus Q_2 \oplus Q_6 \rangle &= \langle F0(RK0, P_{(0\cdots3)})_2 \oplus F0(RK0, Q_{(0\cdots3)})_2 \rangle \\ \langle P_3 \oplus P_7 \oplus Q_3 \oplus Q_7 \rangle &= \langle F0(RK0, P_{(0\cdots3)})_3 \oplus F0(RK0, Q_{(0\cdots3)})_3 \end{aligned} \quad (5.15)$$

It is now possible to apply the differential properties of the F functions to derive possible key candidates. Considering just two blocks of plaintexts, P and Q, would result in 32 candidate key values (on average) for each byte of $RK0$. In order to identify a single key with probability greater than 1/2, cache hits in four plaintexts must be considered, and the intersection between all possible candidate key sets must be found.

In a similar way round key $RK1$ can be determined by analyzing cache hits in $F1$. The set of equations that should satisfy $RK1$ is shown below. where $0 \leq i < 4$.

$$\langle P_{8+i} \oplus P_{12+i} \oplus Q_{8+i} \oplus Q_{12+i} \rangle = \langle F1(RK0, P_{(8\cdots11)})_i \oplus F1(RK0, Q_{(8\cdots11)})_i \rangle$$

Due to the matrix $M1$, which only reveals two bits of the difference of outputs in each s-box, six plaintext blocks are required instead of four.

5.3.3 Determining $WK0 \oplus RK2$ and $WK1 \oplus RK3$

A cache hit in the first table accesses in the third round can be found by varying byte P_8 for $S0$ (and P_9 for $S1$) (Fig. 2.4). The cause of this cache hit could be collisions with any of the eight previous accesses to that table. To reduce the number of "causes" that result in cache hits, the plaintext bytes are chosen in a way such that the first two rounds have only one cache miss in each table (i.e., the first accesses).

Such a state of the cipher is called the *two-round colliding state*. The two-round colliding state has 14 cache hits in the first two rounds. Such a state is obtained by first obtaining the one-round colliding state and then varying bytes P_4 to P_7 and P_{12} to P_{15} independently until eight cache hits in the second round are also obtained.

The third round first access cache hit caused by changing P_8 (or P_9) starting from the two-round colliding state has three causes. The first two causes are due to collisions with $S0$ table accesses in $F1$ in round two. The third cause is due to collisions with $S0$ accesses in $F0$; this is of interest and is estimated to occur once every three collisions. The uninteresting cache hits due to the first two reasons are caused by the changing P_8, which in turn changes $Y1$-1 (Fig. 2.4). On obtaining a cache hit in the first table access in the third round, it is required to identify whether the hit is interesting. This is done by changing the value of P_{12} (or P_{13}) and re-doing the encryption. If a cache hit still occurs in round three, then with significant probability it is of interest.

Similar cache hits for the other $F0$ table accesses in round three can be obtained. With these collisions the following equalities are satisfied for a pair of plaintexts P and Q.

$$\begin{aligned}\langle P_i \oplus Q_i \oplus P_{8+i} \oplus Q_{8+i}\rangle =& \langle F0(RK2, WK0 \oplus P_{(4\cdots7)} \oplus Y0\text{-}1^P)_i \\ & \oplus F0(RK2, WK0 \oplus Q_{(4\cdots7)} \oplus Y0\text{-}1^Q)_i\rangle,\end{aligned}$$

where $0 \le i < 4$, and $Y0$-1 is as defined in Fig. 2.4. $Y0$-1 can be computed using the $RK0$ found in the first step of the attack. Differential properties of the $F0$ function and four plaintext blocks can be used to completely determine $RK2 \oplus WK0$.

Finding WK1 ⊕ RK3 In a similar manner $RK3 \oplus WK1$ can be found in less than 2^{12} invocations of Algorithm 4.1 by considering collisions in $F1$ in round three and varying plaintext bytes $P_{(0\ldots3)}$. The difference equations that is to be satisfied is given by the following (where $0 \le i < 4$) :

$$\begin{aligned}\langle P_i \oplus Q_i \oplus P_{8+i} \oplus Q_{8+i}\rangle =& \langle F1(RK3, WK1 \oplus P_{(12\cdots15)} \oplus Y1\text{-}1^P)_i \\ & \oplus F1(RK3, WK1 \oplus Q_{(12\cdots15)} \oplus Y1\text{-}1^Q)_i\rangle\end{aligned}$$

5.3.4 Determining $RK4$ and $RK5$

$RK4$ and $RK5$ can be determined in 2^{13} invocations of Algorithm 4.1 using the same idea as the second step of the attack. To find $RK4$, a two-round colliding state is first obtained from which cache hits in $F0$ of the fourth round is forced by varying the fourth word of the plaintext. $RK4$ can be determined from this using the equations:

$$\begin{aligned}&\langle P_i \oplus Q_i \oplus P_{12+i} \oplus Q_{12+i} \oplus Y1\text{-}1_i^P \oplus Y1\text{-}1_i^Q\rangle \\ &\qquad = \langle F0(RK4, X0\text{-}3^P)_i \oplus F0(RK4, X0\text{-}3^Q)_i\rangle,\end{aligned}$$

where $Y1\text{-}1^P$, $Y1\text{-}1^Q$, $X0\text{-}3^P$, and $X0\text{-}3^Q$ are computed from previously determined round keys and $0 \leq i < 4$. Similarly, $RK5$ is determined by cache hits in $F1$ in the 4th round. The equalities for determining $RK5$ are :

$$\langle P_{4+i} \oplus Q_{4+i} \oplus P_{8+i} \oplus Q_{8+i} \oplus Y0\text{-}1_i^P \oplus Y0\text{-}1_i^Q \rangle$$
$$= \langle F1(RK5, X1\text{-}3^P)_i \oplus F1(RK5, X1\text{-}3^Q)_i \rangle$$

5.3.5 *Determining* $RK2$ *and* $RK3$

In the key expansion algorithm, if $i = 0$ then $T = (RK0|RK1|RK2|RK3)$, and $T = L \oplus (CON_{24}|CON_{25}|CON_{26}|CON_{27})$. Sixty-four bits of the key dependent constant L can be computed using the values of $RK0$ and $RK1$, which were determined in the first step of the attack.

$$(L_0|L_1) = (RK0|RK1) \oplus (CON_{24}|CON_{25}) \tag{5.16}$$

The double swap operation on L places 57 known bits of L in the lower bit positions. This is given by $L'_{(0\cdots56)} = L_{(7\cdots63)}$. Again, in the key expansion algorithm, if $i = 1$, then $T = (RK4|RK5|RK6|RK7)$. This is represented as $T = L' \oplus (CON_{28}|CON_{29}|CON_{30}|CON_{31}) \oplus (WK0|WK1|WK2|WK3)$. Therefore,

$$WK0|WK1_{(0\cdots24)} = L'_{(0\cdots56)} \oplus (CON_{28}|CON_{29(0\cdots24)}) \oplus (RK4|RK5) \tag{5.17}$$

Using $RK4$ and $RK5$ that were determined in the third step of the attack, the whole of $WK0$ and 25 bits of $WK1$ can be determined. Then the result from the second step of the attack is used to obtain 57 bits of $(RK2|RK3)$. Thus 121 out of the 128 bits of $(RK0|RK1|RK2|RK3)$ is retrieved.

5.4 Conclusion

Using properties of block ciphers along with information leakage from timing channels lead to deadly attacks. This chapter showed how properties of AES and CLEFIA can be used to build efficient time-driven cache attacks. The attack on AES for instance requires just 2^{16} time measurements to reveal all bits of the secret key. These attacks must be prevented with as little overheads as possible and with guarantees of security. To do so, we would require an in depth analysis of the attack and the leakage in the timing channels. The next chapter builds a framework to study the information leakage from cache memories. The framework is used to build efficient implementations of block ciphers that are secure against this class of timing attacks.

References

1. Aciiçmez O, Schindler W, Kaya Koç Ç (2007) Cache based remote timing attack on the AES. In: Abe M (ed) CT-RSA. Lecture notes in computer science, vol 4377. Springer, Berlin, pp 271–286
2. Aoki K, Ichikawa T, Kanda M, Matsui M, Moriai S, Nakajima J, Tokita T (2000) Camellia: a 128-bit block cipher suitable for multiple platforms—design and analysis. In: Stinson DR, Tavares SE (eds) Selected areas in cryptography. Lecture notes in computer science, vol 2012. Springer, Berlin, pp 39–56
3. Shirai T, Shibutani K, Akishita T, Moriai S, Iwata T (2007) The 128-bit blockcipher CLEFIA (extended abstract). In: Biryukov A (ed) FSE. Lecture notes in computer science, vol 4593. Springer, Berlin, pp 181–195
4. Diffie W, Liden G (trans) (2008) SMS4 encryption algorithm for wireless networks. Cryptology ePrint Archive, Report 2008/329 (http://eprint.iacr.org/)
5. Kanda M, Moriai S, Aoki K, Ueda H, Takashima Y, Ohta K, Matsumoto T (2000) E2—a new 128-bit block cipher. IEICE Trans Fundam Electron Commun Comput Sci E82(A-1):48–59
6. Fournier JJA, Tunstall M (2006) Cache based power analysis attacks on AES. In: Batten LM, Safavi-Naini R (eds) ACISP, ser. Lecture notes in computer science, vol 4058. Springer, pp 17–28
7. Gallais JF, Kizhvatov I, Tunstall M (2010) Improved trace-driven cache-collision attacks against embedded AES implementations. In: Chung Y, Yung M (eds) WISA, ser. Lecture notes in computer science, vol 6513. pp 243–257
8. Rebeiro C, Poddar R, Datta A, Mukhopadhyay D (2011) An enhanced differential cache attack on CLEFIA for large cache lines. In the Proceedings of the 12th International Conference on Cryptology in India, INDOCRYPT 2011, Chennai, India, pp 58–75, LNCS 7107
9. Poddar R, Datta A, Rebeiro C (2011) A cache trace attack on CAMELLIA. In the Proceedings of First International Conference on Security Aspects in Information Technology, InfoSecHiComNet 2011, Haldia, India, pp 141–156, LNCS 7011
10. Rebeiro C, Mukhopadhyay D (2011) Cryptanalysis of CLEFIA using differential methods with cache trace patterns. In the Proceedings of the Topics in Cryptology—CT-RSA 2011—The Cryptographers' Track at the RSA Conference 2011, San Francisco, CA, USA, pp 89–103, LNCS
11. Rebeiro C, Nguyen PH, Mukhopadhyay D, Pochmann A (2013) Formalizing the effect of feistel cipher structures on differential cache attacks. IEEE Trans Inf Foresics Secu 8(8):1274–1279
12. Nguyen PH, Rebeiro C, Mukhopadhyay D, Huaxiong W (2012) Improved differential cache trace attacks on SMS4. In the Proceedings of the 8th International Conference, Inscrypt 2012, Beijing, China, pp 29–45, LNCS 7763

Chapter 6
A Formal Analysis of Time-Driven Cache Attacks

Chapter 4 introduced two forms of time-driven cache attacks. It showed how the system configuration, number of measurements taken, and the attack algorithm can adversely affect an adversary's success. In this chapter we build a formal framework to analyze time-driven cache attacks on block ciphers. The framework is used to determine the causes of information leakage in modern cache memories. The framework shows that in these memories, there are two causes of information leakage. The first is due to the number of cache misses that occur during the encryption while the second is due to microarchitectural features in the cache such as nonblocking, pipelined, out-of-order, and parallel servicing of cache misses. The framework is used to evaluate commonly used block ciphers and identify the best way to implement them.

6.1 Memory Access Model for a Block Cipher

A block cipher takes as input a plaintext block (typically 64 or 128 bits) and encrypts it with a secret key to produce a ciphertext. Typically block ciphers first subject the plaintext to a key whitening phase where a secret key is added. These are the keys a side-channel adversary tries to recover. The whitening phase is followed by a number of iterations called rounds. Each round has three operations: key mixing, substitution, and permutation. Key mixing modifies the intermediate state of the cipher by adding secret data to it. Permutation permutes bits in the state, while substitution adds randomness to the state. Figure 6.1 shows a typical block cipher structure.

When speed of encryption is important, the bottleneck comes from the substitution operation. To improve speeds, developers implement the substitution operations using look-up tables, denoted $\mathcal{T}$ in Fig. 6.1. In some block cipher implementations, further speed-ups are achieved by huge tables that implement both the substitution and permutation. These tables are known as T-tables. For instance, see the implementation of Advanced Encryption Standard (AES) in Sect. 2.2.1.1 and CLEFIA in Sect. 2.2.2.1 use T-tables.

C. Rebeiro et al., *Timing Channels in Cryptography*,
DOI 10.1007/978-3-319-12370-7_6

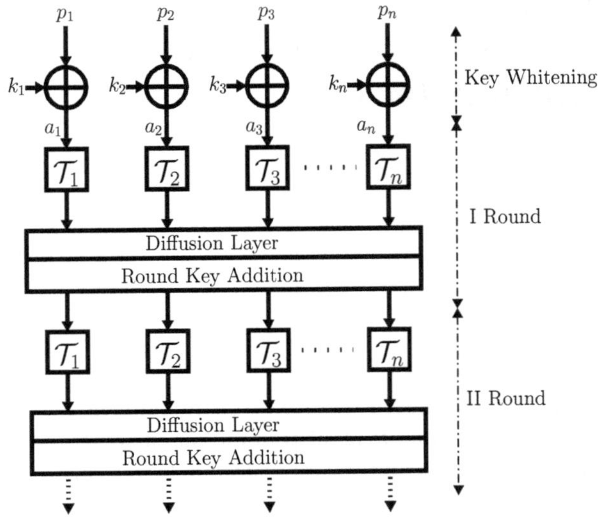

Fig. 6.1 Typical iterative block cipher structure. Each *round* comprises functions that perform key mixing, substitution, and permutation

Most block cipher implementations can be mapped to the structure shown in Fig. 6.1 using five parameters. Table 6.1 shows these parameters while Table 6.2 shows some examples of block cipher implementations mapped to this structure. In later sections we use this model of the block cipher to analyze time-driven cache attacks.

6.2 Cache Misses in a Block Cipher

A block cipher execution makes several memory accesses during its execution due to the load and store instructions. However, when cache attacks are considered, we are only interested in the memory loads or stores of key related data—for instance when the memory location accessed depends on the key. The cache hits or misses that occurs as a result leaks information about the secret key. In block ciphers such as in Fig. 6.1, such key related memory accesses occur due to the loads from the look-up tables. In this section we mathematically analyze the cache misses that occur due to such memory loads.

Suppose the cipher's memory model (Table 6.1) has g look-up tables denoted $\mathcal{T}_1, \mathcal{T}_2, \ldots, \mathcal{T}_g$, each occupying l memory blocks and accessed m times per round in γ rounds. We define the set $\mathbb{S}_i$ on the look-up table $\mathcal{T}_i$ (where $1 \leq i \leq g$) as follows:

Definition 6.1 Let $\mathbb{S}_i$ be a set of sequences defined as $\mathbb{S}_i = \{(p_{i,1}, p_{i,2}, \ldots, p_{i,\gamma}) \mid$ where $p_{i,r}$ is the number of cache misses due to $\mathcal{T}_i$ that occur in the rth round of an encryption and $1 \leq r \leq \gamma$ $\}$.

Table 6.1 Memory access parameters of a block cipher

γ	Number of rounds in the cipher. Each round has a layer of substitution, permutation, and key mixing
g	Number of look-up tables used in the implementation. This includes the look-up tables implemented as T-tables
m	Number of key related look-ups per look-up table per round
l	Number of memory blocks required to hold a table (*table size*). We assume that all look-up tables are of the same size
2^δ	Number of elements in the table sharing a memory block

Table 6.2 Memory access parameters of popular block cipher implementations

	g	γ	m	l	2^δ	T-tables
CLEFIA[a]	2	18	4	4	64	No
CAMELLIA[b]	4	18	2	4	64	No
DES[c]	8	16	1	4	16	Yes
AES[d]	4	9	4	16	16	Yes

DES Data Encryption Standard, *AES* Advanced Encryption Standard

[a] Reference Code 1.0.0, *http://www.sony.net/Products/cryptography/clefia*

[b] PolarSSL 1.1.1, *http://polarssl.org/*

[c] OpenSSL 0.9.8a, *http://www.openssl.org*. The last round is not considered in the model because it uses a different table. However all empirical results reported in this work include the last round as well

[d] In all cases we assume a cache memory with a cache line size of 64 bytes. Therefore, in bytes the look-up table has size $64 \cdot l$ and number of elements $2^\delta \cdot l$

For instance, a possible element in $\mathbb{S}_i$ is $\{4, 3, 4, 1, 0, \ldots, 0\}$. This means that due to table $\mathcal{T}_i$ there were four cache misses in the first round, three in the second, four in the third round, and so on.

The number of cache misses in the m accesses to look-up table $\mathcal{T}_i$ in each round of the cipher are discrete random variables. We denote them by $(\mathsf{P}^m_{i,1}, \mathsf{P}^m_{i,2}, \ldots, \mathsf{P}^m_{i,\gamma})$. Any sequence in $\mathbb{S}_i$ satisfies the following properties:

- If we assume that the contents of the cache is flushed before encryption then at least one cache miss due to $\mathcal{T}_i$ will occur in the first round. Further if we assume no conflict misses during encryption, then a memory block once loaded into the cache would remain in the cache till the encryption completes.
  With these assumptions, the minimum number of cache misses that occur in the first round due to table $\mathcal{T}_i$ is 1. In this case, the first load from the table results in a compulsory miss. All other accesses to $\mathcal{T}_i$ in the round are cache hits. The lower bound for $\mathsf{P}^m_{i,1}$ is therefore 1.
- In rounds other than the first, all accesses to table $\mathcal{T}_i$ may result in cache hits. Thus the minimum number of cache misses in the round is 0. In other words, the lower bound for $\mathsf{P}^m_{i,r}$ $(2 \le r \le \gamma)$ is 0.

- The number of cache misses in the round due to $\mathcal{T}_i$ is limited by its size (i.e., l) and the number of accesses made to the table in the round (m): $\mathsf{P}^m_{i,r} \leq minimum(l, m)$ $(1 \leq r \leq \gamma)$.
- The sum of all elements in the sequence is limited by l—the size of $\mathcal{T}_i$. That is, $\sum_{r=1}^{\gamma} p_{i,r} \leq l$.

The following theorem computes the probability of obtaining $p_{i,r}$ cache misses in the rth round for table $\mathcal{T}_i$.

Theorem 6.1 *Let* $\mathrm{Q}_{i,r}$ *be a discrete random variable denoting the number of cache misses due to table* $\mathcal{T}_i$ *that have occurred before the start of round r (i.e., in rounds* 1 *to* $r-1$*). Three cases arise:*

- *When* $r = 1$ *then* $\mathrm{Q}_{i,1} = 0$ *and* $\Pr[\mathrm{P}^m_{i,1} = 0] = 0$.
- *When* $r > 1$ *then*

$$\Pr[\mathrm{P}^m_{i,r} = 0 | \mathrm{Q}_{i,r} = q_{i,r}] = \left(\frac{q_{i,r}}{l}\right)^m .$$

- *Otherwise,*

$$\Pr[\mathrm{P}^m_{i,r} = p | \mathrm{Q}_{i,r} = q] = \frac{1}{l^m}\binom{l-q}{p}\sum_{j=1}^{p}(-1)^{p-j}\binom{p}{j}\left[(j+q)^m - q^m\right] .$$

(In this above equation we have denoted $p_{i,r}$ *by p and* $q_{i,r}$ *by q for simplicity of notation.)*

where $q = q_{i,r} = p_{i,1} + p_{i,2} \cdots + p_{i,r-1}$ *and* $q_{i,1} = 0$*; and* $p = p_{i,r}$ *can take values from* 0 *to the minimum of m and* $l - q_{i,r}$.

Proof. Since we assume that the cache is flushed before starting the encryption, in the first round ($r = 1$), $\mathsf{P}^m_{i,1} = 0$ is not possible because at least one compulsory cache miss will occur.

Assuming no conflicts, when $r > 1$, no cache misses will occur if the table accesses were made to blocks that were previously accessed, so data would be read from the cache. Each of the m accesses in the round are therefore cache hits and can take $q_{i,r}$ possible values.

The third case determines the probability of obtaining p cache misses in the rth round given that the previous $r-1$ rounds resulted in q cache misses. The summand when $j = p$, counts the number of ways m memory accesses can be made to $(p+q)$ memory blocks of table $\mathcal{T}_i$, ensuring that there is at least 1 and at most p cache misses. Since we require exactly p cache misses (and not at most), we use the inclusion–exclusion principle to exclude the other events. □

Let $s_i = (p_{i,1}, p_{i,2}, \cdots, p_{i,\gamma}) \in \mathbb{S}_i$ be a particular sequence of cache misses. Then,

$$\Pr[\mathsf{S}_i = s_i] = \prod_{r=1}^{\gamma} \Pr[\mathsf{P}^m_{i,r} = p_{i,r} \,|\, \mathsf{Q}_{i,r} = q_{i,r}] , \tag{6.1}$$

where S_i is a random variable denoting the event of obtaining s_i from $\mathbb{S}_i$. This equation gives us the probability of obtaining a particular sequence of cache misses during the encryption from a single look-up table.

If multiple tables are used in the cipher, then the Cartesian product $(\mathbb{S}_1 \times \mathbb{S}_2 \times \cdots \times \mathbb{S}_g)$ is used to represent the cache misses in all tables. Take for example the $((4,3,1,0,0,\dots) \times (5,2,0,1,1,\dots))$. This means that in the first round, table one had four cache misses and table two five. In the second round, table one had three cache misses and table two five, and so on. The number of tables in the implementation is 2 (or $g = 2$). Since accesses to each table in the cipher is independent of the others we can write

$$\mathsf{Pr}[(s_1, s_2, \dots, s_g)] = \prod_{i=1}^{g} \mathsf{Pr}[\mathsf{S}_i = s_i] \,, \tag{6.2}$$

where $(s_1, s_2, \dots, s_g) \in (\mathbb{S}_1 \times \mathbb{S}_2 \times \cdots \times \mathbb{S}_g)$.

The *expected number of cache misses* that occur during an encryption when a single table is used (*i.e.*, $g = 1$), is determined from Eq. 6.1. This is given by,

$$\mu_{avg} = \sum_{s_1 \in \mathbb{S}_1} \mathsf{Pr}[\mathsf{S}_1 = s_1]\Big(\sum_{r=1}^{\gamma} p_{1,r}\Big) \qquad \text{when} \quad g = 1\,. \tag{6.3}$$

When multiple tables are present (*i.e.*, $g > 1$), Eq. 6.2 can be used to find the *expected number of cache misses* as shown below

$$\mu_{avg} = \sum_{s \in \mathbb{S}_1 \times \mathbb{S}_2 \times \cdots \mathbb{S}_g} \mathsf{Pr}[s]\Big[\sum_{r=1}^{\gamma}\Big(\sum_{i=1}^{g} p_{i,r}\Big)\Big] \text{ when } \quad g > 1. \tag{6.4}$$

Equation 6.4 considers the cache misses in each table. In the next section we use Eqs. 6.1 and 6.2 to approximate the difference in execution times between two encryptions.

6.3 Average Execution Time of a Block Cipher

For memory intensive programs, the time to service cache misses largely dominates execution time. Block ciphers implemented with look-up tables fall into this category of programs. Due to this reason execution time of a block cipher can be estimated from the number of misses that have occurred in the cache.

Memory loads when there is a cache miss takes considerably longer than those that have a cache hit. Due to this, modern computer systems have several microarchitectural acceleration techniques incorporated in the cache—such as parallelization pipelining, nonblocking, and out-of-order servicing of cache misses. Parallelization and pipelining allow multiple cache misses to be serviced simultaneously, while out-of-order loading allows memory accesses to be performed in a sequence not strictly specified by the program. Nonblocking memory accesses allow other memory accesses to be done while a cache miss is being serviced (these architectural features are discussed in Sect. 3.2.2).

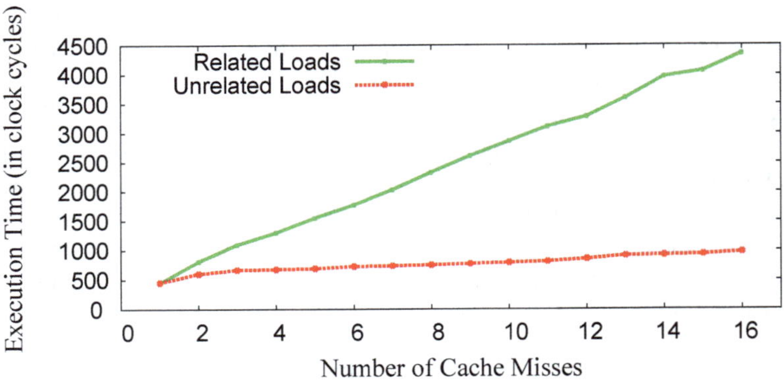

Fig. 6.2 Execution overheads for servicing cache misses on an Intel Dual Core processor (P6100). Related loads cannot be parallelized; therefore, the cache misses cannot be accelerated. Unrelated loads from memory are parallelizable; therefore, the time to service cache misses is less

To show how these microarchitectural techniques affect execution time we consider two programs. The programs make a series of memory loads, each load to a different memory location. In the first program, the address of a location loaded depends on the data read in the previous load (relative loads). In the second program, the memory loads are from independent locations (independent loads). We flush the cache each time to ensure that the required data is not present in the cache. Thus, the execution time for these programs depend mostly on the time required to service cache misses. Figure 6.2 shows the results on an Intel Dual Core (P6100) platform. The x-axis has the number of cache misses. This is equal to the number of loads, since the flushed cache ensures that every load to the table is a cache miss. The y-axis has the execution time for the program. In the case where the loads are related, memory loads cannot be parallelized or pipelined, or performed out of order. These accesses take longer compared to unrelated loads which can be accelerated by these microarchitectural features in the cache.

These two programs can be related to the memory loads made by block ciphers. In a typical block cipher implementation (such as the model in Fig. 6.1), each round has several memory accesses. The accesses within a single round are unrelated to each other. On the other hand, the diffusion layer in the cipher causes the data accessed in a round to be dependent on the previous round accesses. Due to this pattern of memory accesses, the following observations can be made.

1. Cache misses within a single round of the cipher can be accelerated due to the parallelism, pipelining, and out-of-order features in the cache.
2. Cache misses across rounds are done sequentially due to data dependencies.
3. The nonblocking feature in the cache allows hits within a single round to be done while the cache-misses are being serviced thus hiding the hit time.

For the block cipher, the time required to execute cache misses in one round comprising p_r cache misses can be estimated as $(\alpha + \beta(p_r))$. The constant α is the initial latency for the first cache miss and β is the time required to service a cache miss. These constants are CPU-specific and also vary from one core in the system to another. In Fig. 6.2, the initial latency (α) was found to be approximately 180 and β, the slope of the line, was 16.

The total time required for servicing cache misses in the cipher is the sum of the time required for servicing cache misses in each round. The expected time to service cache misses is given as follows.

$$\widehat{\mathfrak{t}}_{avg} = \sum_{s \in \mathbb{S}_1 \times \mathbb{S}_2 \times \cdots \times \mathbb{S}_g} \mathsf{Pr}[s]\Big(\sum_{r=1}^{\gamma} \widehat{\mathfrak{t}}_{r,g}\Big) , \tag{6.5}$$

where $\widehat{\mathfrak{t}}_{r,g}$ is the expected time taken for servicing cache misses from g tables in round r $(1 \le r \le \gamma)$[1]. This is given by

$$\widehat{\mathfrak{t}}_{r,g} = \begin{cases} \Big(\alpha + \beta(\sum_{i=1}^{g} p_{i,r})\Big) & \text{if } \sum_{i=1}^{g} p_{i,r} \neq 0 \\ 0 & \text{otherwise} \end{cases}$$

$p_{i,r}$ is the number of cache misses in round r due to table $\mathcal{T}_i$. Note that $\widehat{\mathfrak{t}}_{r,g}$ is zero when all memory accesses in a round are cache hits. If in a round, even a single memory access results in a cache miss, then there is the large initial latency of α added to $\widehat{\mathfrak{t}}_{r,g}$ for that round. Every subsequent cache miss causes the time to increase by a factor of β.

6.3.1 Estimating the Difference of Means

The time-driven cache adversary uses Algorithm 4.1 to identify collisions in specific memory accesses. When these collisions are detected, the following equation is satisfied.

$$\langle k_a \oplus d_a \rangle = \langle k_b \oplus d_b \rangle . \tag{6.6}$$

The collision results in an average execution time which is visibly different from the average case. If $\widehat{\mathfrak{t}}_{collision}$ is the execution time when a collision is present, then the difference between the two expectations acts as a distinguisher. This is called the DOM or *difference of means* and is given by the following equation.

$$\widehat{\mathfrak{d}} = \widehat{\mathfrak{t}}_{collision} - \widehat{\mathfrak{t}}_{avg} . \tag{6.7}$$

When Eq. 6.6 is satisfied, every encryption would have a collision during the memory load $T[k_b \oplus d_b]$. We refer to this as the *ever-present* collision.

[1] It may be noted that s has the form $((p_{1,1}, p_{1,2}, \ldots, p_{1,\gamma}), (p_{2,1}, p_{2,2}, \ldots, p_{2,\gamma}), \ldots \ldots, (p_{w,1}, p_{w,2}, \ldots, p_{w,\gamma}))$

For an *ever-present* collision in round r_c, the probability of obtaining cache misses (Theorem 6.1) is affected. Out of the m accesses that are made to the look-up table in that round, one always results in a cache hit due to the collision. Thus, the maximum number of cache misses in round r_c (i.e., upper bound for the random variable P^m_{i,r_c}, defined in Sect. 6.2) is $minimum(l, m-1)$ instead of $minimum(l, m)$, where l is the size of the table and m the number of accesses to the table in round r_c.

The following corollary shows the probability of cache misses in a round, given an ever-present collision.

Corollary 6.1 *The probability of having p_{i,r_c} cache misses in a round comprising of m memory accesses to table $\mathcal{T}_i$, given that one of the m accesses always results in a collision, is given by* $\Pr[\mathrm{P}^m_{i,r_c} = p_{i,r_c} | \mathrm{Q}_{i,r_c} = q_{i,r_c}, collision] = \Pr[\mathrm{P}^{m-1}_{i,r_c} = p_{i,r_c} | \mathrm{Q}_{i,r_c} = q_{i,r_c}]$ *when* $m > 1$, *while* $\Pr[\mathrm{P}^1_{i,r_c} = 1 | \mathrm{Q}_{i,r_c} = q_{i,r_c}, collision] = 0$ *when* $m = 1$.

Proof. There is a collision in round r_c which is ever present. Therefore, when $m = 1$, the probability of obtaining a cache miss is 0. Further, since we are only considering cache misses and not cache hits, the ever-present collision when $m > 1$, will not influence the probability. Therefore, obtaining p_{i,r_c} cache misses from m accesses given the ever-present collision is as probable as having p_{i,r_c} cache misses from $m-1$ random table accesses.

Since the cache miss probabilities in round r_c are affected by the ever-present collision, the probability of cache misses of subsequent rounds is also affected. Thus the effect of the ever-present collision spreads to the remaining rounds of the cipher as well. Further, the forced collision only affects the cache misses in the concerned table. For all other tables Theorem 6.1 still holds.

The estimation of $\widehat{\mathrm{t}_{collision}}$ can be done in exactly the same way as $\widehat{\mathrm{t}_{avg}}$. However, Corollary 6.1 has to be used for the round r_c. All other rounds would still follow Theorem 6.1.

6.4 DOM as a Security Metric

The time-driven adversary uses the difference between $\mathrm{t}_{collision}$ and t_{avg} to determine a collision. This difference is termed as the DOM (d). If $|d|$ is large, the adversary will find it easy to distinguish between $\mathrm{t}_{collision}$ and t_{avg} thus would require lesser measurements to identify the collision. As |d| decreases, it becomes increasingly difficult to distinguish between the two timings. The adversary may need more number of measurements, and sophisticated analysis techniques to make out the difference. To take an analogy, lets assume a scientist wants to distinguish between two colors. If the colors are far apart in the electromagnetic spectrum (for instance blue and red), then it would be easy to distinguish. He could probably do it with the naked eye. On the other hand, if the colors were almost similar (two shades of yellow for instance), then he would require more sophisticated tools, such as a spectrometer to make the difference.

We consider the absolute value of the DOM ($|\mathrm{d}|$) as a metric to compare the security of two cipher implementations. Larger the value of $|\mathrm{d}|$, the more easier the attack. Conversely, if $|\mathrm{d}| = 0$, then the collision cannot be identified, there by the attack will not be successful.

For a block cipher, the value of d is influenced by two factors.

F1 : The number of cache misses that occur during the execution of the cipher.
F2 : The influence of microarchitectural features in the cache such as pipelining, parallelization, and out-of-order servicing of cache misses. These components in the cache result in acceleration of memory accesses within a round.

The magnitude by which F1 and F2 affects the value of d depends on the implementation of the cipher, that is the parameters g, γ, m, and l (defined in Table 6.1). We demonstrate the effect of the cipher implementation by using Eq. 6.7. For the analysis we assume the values of (α, β) as $(150, 16)$.

Figure 6.3a, b, and c plot the variation in the DOM, $\widehat{\mathrm{d}} = \widehat{\mathrm{t}}_{collision} - \widehat{\mathrm{t}}_{avg}$, with size of table ($l$) for different γ, g, and m respectively. When $l = 1$, there is no leakage because in every encryption the first memory access loads the entire table into the cache and all subsequent memory accesses would result in cache hits, thus $\widehat{\mathrm{d}} = 0$. For small values of $l > 1$, $\widehat{\mathrm{d}}$ is large and gradually reduces as l increases. For a particular value of l (other than 1), $\widehat{\mathrm{d}} = 0$. This is the ideal table size for the cipher as it implies that the distributions cannot be distinguished. We denote this value of l by l_{ideal}. Implementations of the cipher having $l < l_{ideal}$ are called *small-table implementations* while implementations having $l > l_{ideal}$ are called *large-table implementations*.

In small-table implementations, the size of the look-up table is small compared to the number of accesses made to it. With significant probability the entire table gets loaded into cache in every encryption. Thus every encryption will have the maximum permissible cache misses (which is the number of memory blocks in the table l). Even with a collision satisfied, l cache misses still occur. Thus F1 has no contribution in the leakage and it is only F2 which causes the leakage. The ever-present cache hit in round r_c ($1 \leq r_c \leq \gamma$) restricts acceleration and results in $\widehat{\mathrm{t}}_{collision} > \widehat{\mathrm{t}}_{avg}$. As a result $\widehat{\mathrm{d}}$ is positive. This means that a cache hit due to the collision actually slowed down the encryption. This is counter intuitive. Figure 6.4 shows an example of how this could happen.

In large-table implementations, the probability that the entire table gets loaded into cache is almost zero. When a collision is ever present, due to Eq. 6.6 being satisfied, there is a reduction in the number of cache misses, which in turn results in faster encryptions. Subsequently $\widehat{\mathrm{t}}_{collision}$ is less than $\widehat{\mathrm{t}}_{avg}$ and $\widehat{\mathrm{d}}$ is negative. In this case both F1 and F2 contribute to the information leakage. F2 affects $\widehat{\mathrm{d}}$ in the positive direction while F1 affects $\widehat{\mathrm{d}}$ in the negative direction. However, the impact of the reduced miss is higher compared to the acceleration obtained, thus F1 dominates F2.

When $l = l_{ideal}$, the effects of F1 and F2 cancel each other resulting in $\widehat{\mathrm{d}} = 0$, therefore protected against this form of nonprofiled time-driven cache attacks. This phenomenon was practically visualized on several modern microprocessors including Intel Atom, Dual Core, Core 2 Duo, i3, and Pentium 4. However, on the older Pentium 3 processors, the microarchitectural effects due to F2 were not observed.

Fig. 6.3 The effect of different cipher models on the difference of means (d). All graphs are plotted with the size of the look-up tables on the x-axis. The value of d reduces as the look-up table size increases

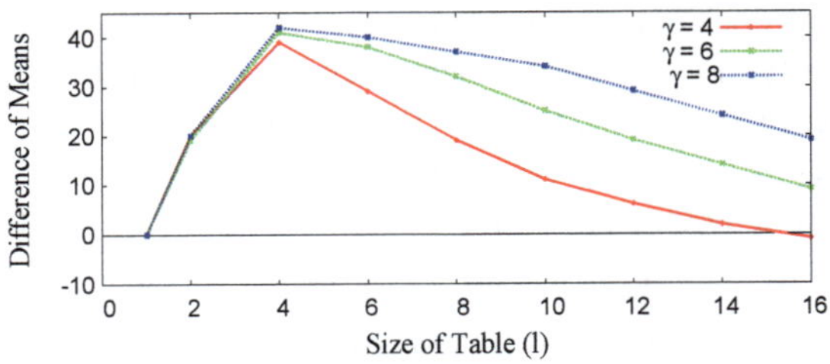

a Cipher models with different numbers of rounds and $m = 4$, $g = 1$. The value of d increases as the number of rounds in the cipher increase.

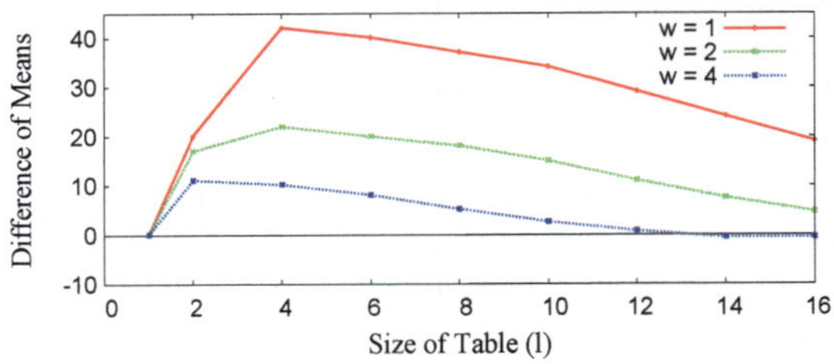

b Cipher models with different number of tables and $m = 4$, $\gamma = 8$. The value of d reduces as the number of tables used in a cipher implementation increases.

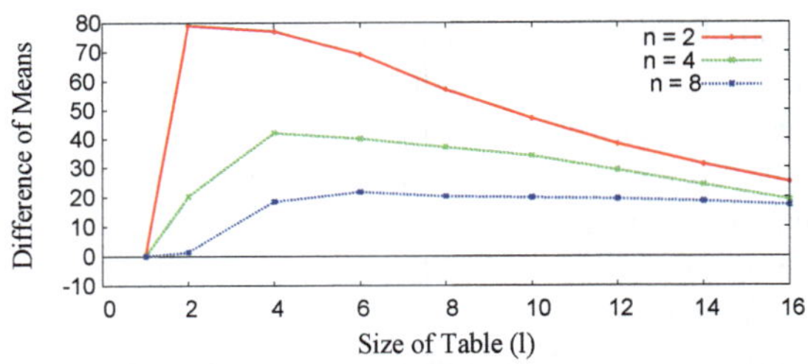

c Cipher models with different number of table accesses per round and $g = 1$, $\gamma = 8$. The value of d reduces as the number of table accesses per round increase.

6.5 Application of the Model

The model for predicting the DOM can be used to compare the security of cipher implementations and also select the best way to implement a cipher. We compare four standard block cipher implementations of AES, CLEFIA, data encryption standard (DES), and CAMELLIA. In the second part of the section we evaluate various implementation options for CLEFIA and CAMELLIA with the aim of finding implementations more resistant against time-driven cache attacks.

$\widetilde{\mathcal{E}}_K(P)$					$\widetilde{\mathcal{E}}_K(P')$			
M	H	H	H	round 1	M	H	M	H
H	M	H	H	round 2	H	M	H	H
M	H	H	H	round 3	M	H	H	H
M	H	H	H	round 4	H	H	H	H
H	H	H	H	round 5	H	H	H	H
H	H	H	H	round 6	H	H	H	H
H	H	H	H	round 7	H	H	H	H
H	H	H	H	round 8	H	H	H	H

Consider a cipher implementation $\widetilde{\mathcal{E}}_K$ with a cipher model $w = 1$, $\gamma = 8$, $l = 4$, and $n = 4$. Consider encryptions of two different plaintexts P and P$'$: $\widetilde{\mathcal{E}}_K(P)$ and $\widetilde{\mathcal{E}}_K(P')$. For each encryption, the figure shows the result of memory accesses to the look-up table. A memory access that resulted in a cache hit is represented by **H** while a memory access that resulted in a cache miss is represented by **M**. It may be noted that both encryptions result in equal number of cache misses (i.e. 4). However the distribution of the misses differ. In $\widetilde{\mathcal{E}}_K(P)$ there is at most a single cache miss in each round, while in $\widetilde{\mathcal{E}}_K(P')$ there exists a round with 2 cache misses. Due to the MAF in the cache, the pair of cache misses in round 1 during the encryption of $\widetilde{\mathcal{E}}_K(P')$ gets serviced faster compared to any pair of the cache misses that occur during the execution of $\widetilde{\mathcal{E}}_K(P)$. This results in a faster execution for $\widetilde{\mathcal{E}}_K(P')$ compared to $\widetilde{\mathcal{E}}_K(P)$ even though both encryptions have the same number of cache misses.

Fig. 6.4 Example of collisions in a cipher that uses a small look-up table

The verification of the results are done using the Intel Xeon E5606 platform. Refer to Table A.1 in the appendix for details on the system's cache memory configuration. However, the cache configuration alone is not sufficient to apply the model developed. Additional parameters of α and β are required as well. The values of α and β found are also depicted in the table for each platform.

6.5.1 Comparing Cipher Implementations

Figure 6.5 shows the DOM obtained for the four cipher implementations considered in Table 6.2. Each implementation was analyzed theoretically to obtain the modeled DOM for the platforms and then the empirical DOM was obtained after 2^{20} measurements.

The histogram shows that the modeled DOM matches the empirical results closely. The OpenSSL implementation of AES is the most secure against time-driven cache attacks compared to the other cipher implementations analyzed as it has the smallest absolute value of DOM. The security of the AES implementation is due to the large number and size of tables, which are implemented as T-tables (Refer Sect. 2.2.1.1). As seen in Fig. 6.3, large number and size of tables result in smaller DOM thus more security. Further, the tables in OpenSSL's AES implementation are T-tables, which reduces the diffusion to just 16 ex-ors per round intertwined with the table accesses.

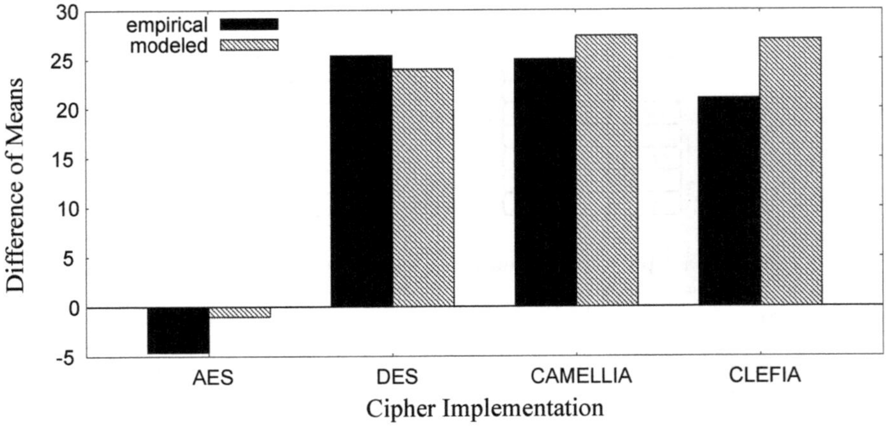

Fig. 6.5 Difference of means obtained from the model and empirically for the cipher implementations in Table 6.2 on Intel Xeon 5606

The T-tables allow cache misses from one round to overlap with another, there by resulting in smaller α (approximately 10). For $\alpha = 10$, the value of l_{ideal} is close to 16, which is the size of the tables in OpenSSL AES on the Intel Xeon processors. This *close-to-ideal* size of the tables further contributes to the small DOM.

6.5.2 Choosing the Right Implementation

In this part of the section we use the DOM model to judge the security of a set of implementation options for CLEFIA and CAMELLIA, with the hope of obtaining more secure implementations for these ciphers compared to their implementations in Table 6.2. Tables 6.3 and 6.4 show the the DOM obtained from the mathematical model for both platforms evaluated (E5606 and E5345). This DOM can be used to

Table 6.3 Implementation options for CLEFIA

No.	l[a]	γ	m	w	Table size (in bytes)	Leaked bits	DOM $\hat{d}$
CL1[b]	4	18	4	2	256	2	27
CL2	8	18	4	2	512	3	25
CL3[c]	16	18	1	8	1024	4	−1
CL4	4	18	2	4	256	2	27
CL5	8	18	2	4	512	3	23
CL6	16	18	2	4	1024	4	10

DOM difference of means
[a] In all cases we assume a cache with a cache line of 64 bytes
[b] This corresponds to the reference implementation (Table 6.2)
[c] This corresponds to CLEFIA implemented with *T*-tables

Table 6.4 Implementation options for CAMELLIA

No.	l[a]	γ	m	w	Table size (in bytes)	Leaked bits	DOM $\hat{d}$
CA1[b]	4	18	2	4	256	2	36
CA2	8	18	2	4	512	3	31
CA3	16	18	2	4	1024	4	12
CA4	4	18	8	1	256	2	29
CA5	8	18	8	1	512	3	32
CA6	16	18	8	1	1024	4	31

DOM difference of means

[a] In all cases we assume a cache with a cache line of 64 bytes

[b] This corresponds to the reference implementation (Table 6.2)

relatively rank the implementations in the order of their security against nonprofiled time-driven cache attacks. Stronger implementations are those with a DOM closer to zero. For CLEFIA (Table 6.3), we can thus conclude that CL3 is the most secure against and CL6 is the next best. The remaining implementations are insecure. For CAMELLIA (Table 6.4), the implementation CA3 turns out to be most secure with respect to the other implementations investigated.

To verify the correctness of the predictions, we subject the identified implementations to time-driven cache attacks using Algorithm 4.1. In all cases the first pair of accesses to the same table were targeted. We gradually increase the number of

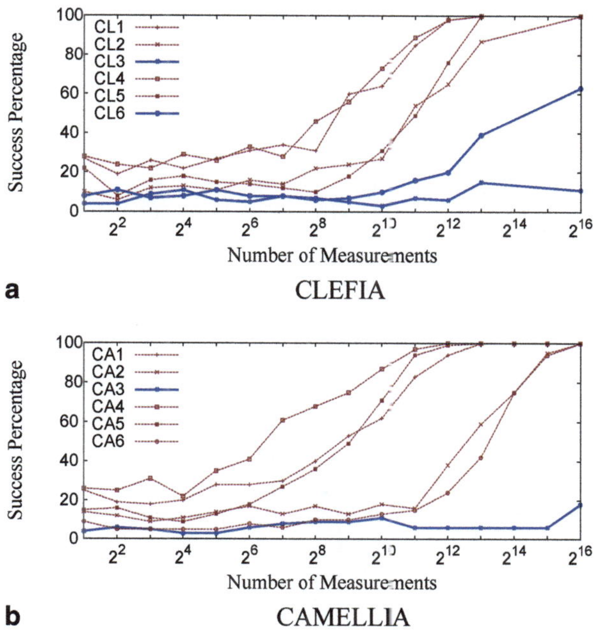

Fig. 6.6 Success of an attack for various implementations of CLEFIA (Table 6.3) and CAMELLIA (Table 6.4) on Intel Xeon 5606

timing measurements made and determine the success rate with which the bits of the ex-or of the key bytes are determined. Figure 6.6 shows this success percentage for CLEFIA and CAMELLIA on both platforms after performing 100 tests. As predicted, CL3 and CL6 are the most secure implementation for CLEFIA, since the attacker achieves minimum success in obtaining the key compared to the other implementations. Similarly, the prediction CA3 for CAMELLIA is the most secure as was predicted in Table 6.4.

6.6 Conclusion

The DOM is used as a distinguisher in time-driven cache attacks. The number of cache misses that occur during the execution affects the DOM negatively, while microarchitectural components in the cache such as out-of-order, parallel, and pipelined servicing of cache misses affect the DOM positively. Generally block ciphers implemented with small look-up tables have a positive DOM while those with large look-up tables have a negative DOM. It is also possible to implement the block cipher with a DOM of 0 or close to 0. Such implementations would be more resistant against time-driven cache attacks without the use of any explicit countermeasures. These implementations provide security with no performance overheads.

Chapter 7
Profiled Time-Driven Cache Attacks on Block Ciphers

In 2005, D. J. Bernstein developed a timing attack capable of retrieving the Advanced Encryption Standard (AES) secret key [4]. Unlike the previous attacks discussed so far, Bernstein's attack has two phases: a profiling phase followed by an attack phase. During the *profiling phase*, the attacker learns the characteristics of the system by building a timing profile called *template* using a key which is known to her. The template captures all the timing characteristics when AES is executed. With this template, any other secret key used in the AES implementation on that system can be attacked. During the *attack phase*, another timing profile is built for the secret key. A statistical comparison of this timing profile with the template reveals the secret key. This chapter provides details of the attack and analyzes the information leaked.

7.1 Bernstein's Cache Timing Attack

7.1.1 Building a Timing Profile

The AES input has 16 bytes as seen in Sect. 2.2.1. To build the timing profiles we first start by choosing random inputs (say $\mathbf{x} = (x_0 \parallel \cdots \parallel x_i \parallel \cdots \parallel x_{15}), 0 \le i \le 15$), invoking AES and measuring the execution time. This is repeated several times with different inputs. We define 16 arrays denoted $\mathcal{A}_i$ $(0 \le i \le 15)$ each of size 256, which stores the average execution time. The element $\mathcal{A}_i[j]$ $(0 \le j \le 255)$ in an array stores the average execution time when the ith byte of the plaintext was j. We then find the overall average execution time (denoted t_{avg}) and compute deviations as follows: $\mathcal{D}_i[j] = \mathcal{A}_i[j] - t_{avg}$, where $0 \le i \le 15$ and $0 \le j \le 255$. The ith deviation array ($\mathcal{D}_i$) is called the *timing profile* for the ith input byte. We thus have 16 deviation arrays; one for each input byte. Algorithm 7.1 shows how the templates are built for AES. The array $\mathcal{T}$ accumulates the encryption time, while $\mathcal{C}$ counts the number of times a particular input is obtained. Figure 7.1 shows the timing profile for the first plaintext byte of AES. It plots the value of x_0 on the x-axis

C. Rebeiro et al., *Timing Channels in Cryptography*,
DOI 10.1007/978-3-319-12370-7_7

Algorithm 7.1: Timing Profile

Output: $\mathcal{D}$
begin
  $\mathcal{C}_i[j] \leftarrow 0$ for $0 \leq i \leq 15$ and $0 \leq j \leq 255$
  $\mathcal{T}_i[j] \leftarrow 0$ for $0 \leq i \leq 15$ and $0 \leq j \leq 255$
  for *at-least* 2^{16} *times* **do**
    $\mathbf{x} \leftarrow$ Random plaintext
    $t_1 \leftarrow$ start time
    $AES_{\mathbf{k}}(\mathbf{x})$
    $t_2 \leftarrow$ end time
    $\mathcal{C}_i[x_i] \leftarrow \mathcal{C}_i[x_i] + 1$ for $0 \leq i \leq 15$
    $\mathcal{T}_i[x_i] \leftarrow \mathcal{T}_i[x_i] + (t_2 - t_1)$ for $0 \leq i \leq 15$
  end
  $\mathcal{A}_i[j] \leftarrow \mathcal{T}_i[j] / \mathcal{C}_i[j]$ for $0 \leq i \leq 15$ and $0 \leq j \leq 255$
  $t_{avg} \leftarrow \sum_{i,j} T_i[j] / \sum_{i,j} C_i[j]$
  $\mathcal{D}_i[j] \leftarrow \mathcal{A}_i[j] - t_{avg}$
  return $\mathcal{D}$
end

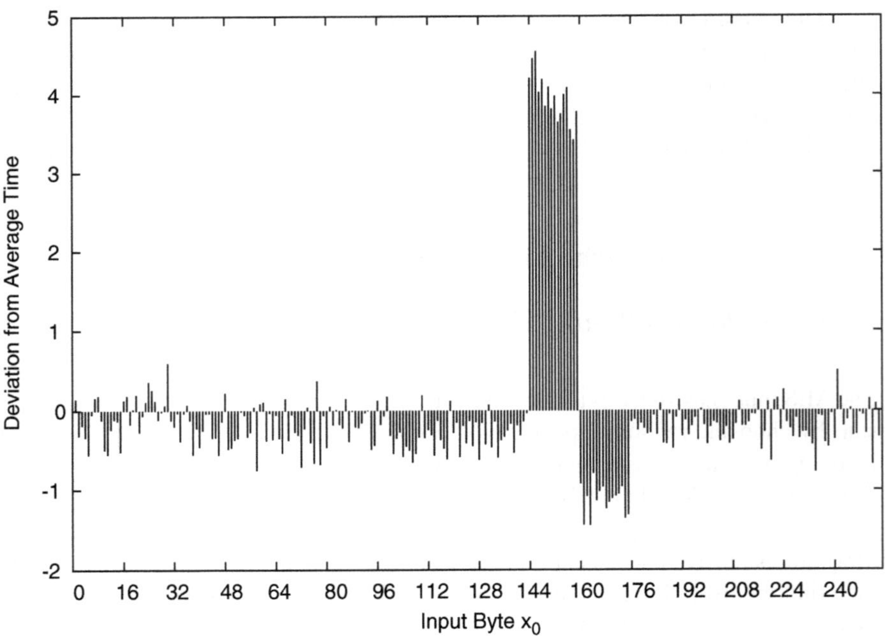

Fig. 7.1 Timing profile for $\mathcal{D}_0$ vs. x_0 for AES on an Intel Core 2 Duo (E7500)

and $\mathcal{D}_0[x_0]$ on the y-axis. To attack a key byte, two such timing profiles are required—one for a known key and the other for the unknown key.

7.1.2 Extracting Keys from Timing Profiles

There are 16 timing profiles created ($\mathcal{D}_0, \mathcal{D}_1, \ldots, \mathcal{D}_{15}$)—one for each input byte. Each vertical line in a timing profile provides an average execution time deviation $\mathcal{D}_i[j]$, when the ith input byte is fixed at j and all other bytes vary randomly. As seen in the AES specification (Algorithm 2.1), the ith input corresponds to the ith look-up table access in the first round. This access is at an index $s_i^{(0)} = x_i \ominus k_i^{(0)}$ in the look-up table. Thus the time deviation measures the deviation when $s_i^{(0)}$ is fixed and all other intermediate values in the cipher vary randomly. With respect to the look-up tables in the implementation, the table access made by $s_i^{(0)}$ is the only fixed operation in the entire encryption. All other table look-up operations are at random locations. When the average of several million encryptions are considered, the influence of all the varying table accesses in the execution time cancel out leaving behind only the effect of the constant table access. Thus, $\mathcal{D}_i[s_i^{(0)}]$ is a characteristic deviation of execution time of $s_i^{(0)}$.

Now consider two timing profiles for the ith byte ($0 \le i \le 15$): one from a known key (denoted $\hat{k}_i^{(0)}$) and another from an unknown key (denoted $\tilde{k}_i^{(0)}$). The deviation $\mathcal{D}_i[s_i^{(0)}]$ is an invariant and a characteristic of the element in the table accessed (i.e., at index $s_i^{(0)}$). There are two ways to access this element in the table using the i-th input byte. Using the known key $\hat{k}_i^{(0)}$ the access can be obtained from $\hat{x}_i \oplus \hat{k}_i^{(0)}$ for some $\hat{x}_i$ ($0 \le \hat{x}_i \le 255$). Second, using $\tilde{k}_i^{(0)}$ the access can be obtained from $\tilde{x}_i \oplus \tilde{k}_i^{(0)}$ for some $\tilde{x}_i$ ($0 \le \tilde{x}_i \le 255$). Thus with reference to the deviation of time, the following relation would be met

$$\mathcal{D}_i[\tilde{x}_i \oplus \tilde{k}_i^{(0)}] = \mathcal{D}_i[\hat{x}_i \oplus \hat{k}_i^{(0)}] \,. \tag{7.1}$$

With respect to the timing profiles, the invariant has shifted from $\hat{x}_i$ in the timing profile of the known key to $\tilde{x}_i$ in the timing profile of the unknown key. In the attack we determine this shift and then compute the unknown key as follows: $\tilde{k}_i^{(0)} = \tilde{x}_i \oplus \hat{x}_i \oplus \tilde{k}_i^{(0)}$.

To determine the shift, the unknown key byte is guessed (say $kguess$) and for every possible value of x ($0 \le x \le 255$), a correlation coefficient is computed as follows:

$$CC_{kguess} = \sum_{x=0}^{255} \mathcal{D}_i[x \oplus \hat{k}_i^{(0)}] \times \mathcal{D}_i[x \oplus kguess] \tag{7.2}$$

This equation essentially computes the correlation between the timing profile of the known key and a shifted version of the timing profile of the unknown key. The shift being $kguess$. There are thus 256 values of CC_{kguess} (one for each $kguess$). The key guess corresponding to the maximum value of CC_{kguess} is most likely the unknown secret byte of the key ($\tilde{k}_i^{(0)}$).

7.2 Causes of Information Leakage

The vector $\mathcal{D}_i$, which represents a timing profile, has 256 different entries. The only difference between each entry in the vector is that x_i $(0 \leq i \leq 15)$ changes. Each x_i represents a unique element in a look-up table that is accessed in the first round during the cipher's execution. The element is accessed at the index $x_i \oplus k_i^{(0)}$. Since the look-up table has 256 entries, the vector $\mathcal{D}_i$ represents execution time deviation corresponding to each element in the table.

Timing variations in the profiles can be attributed due to two sources: intrablock and interblock. *Interblock* sources cause timing variations between memory blocks. In other words if two values of x_i result in table accesses in different memory blocks, the execution time is likely to be different. However, if the two values of x_i fall in the same memory block then they would have the same execution time. Thus interblock sources can only distinguish between accesses in the look-up table that fall in different memory blocks.

Intrablock sources result in execution time of the cipher that is different for each element in the table. Thus each value of x_i is likely to have a different execution time even if the corresponding access at the index $x_i \oplus k_i^{(0)}$ are to adjacent elements in the look-up table. Thus, intrablock sources can distinguish between every access made to the look-up table.

During the final postprocessing phase, shifts between the template and the attack profile are determined. The shift between the two profiles are used to determine $k_i^{(0)}$. The shifts could take any value from 0 to 255. Consequently $k_i^{(0)}$ can take values in this range. However, shifts can only be identified if there are variations in the timing profile. With inter-block sources, there is no timing variation within a memory block. As a result, shifts in a memory block cannot be identified. As a result, attacker would be able to identify only the most significant bits of $k_i^{(0)}$. With intrablock sources, each value of x_i is likely to have different execution time. Therefore, this source can be used to identify all bits of $k_i^{(0)}$.

In the remainder of this section we discuss the various sources that affect the average execution time. We discuss how an attack on OpenSSL's implementation of AES[1], is primarily due to *interblock* sources, while an attack on an implementation of CLEFIA[2], is due to *intrablock* sources. As a result, the bits of the key recovered from the AES attack are restricted to the higher bits. On the other hand, all bits of CLEFIA's key are recovered. This results in better success rates for the CLEFIA attack.

Interblock Sources: A primary interblock source is conflict misses. These cache misses occur due to memory accesses which evict recently accessed data, thus resulting in misses. Conflict misses are ideally sporadic and occur at random cache

[1] OpenSSL version 0.9.8a, http://www.openssl.org details present in Sect. 2.2.1.

[2] CLEFIA Reference code, http://www.sony.net/Products/cryptography/clefia/download/data/clefa_ref.c, details present in Sect. 2.2.1.

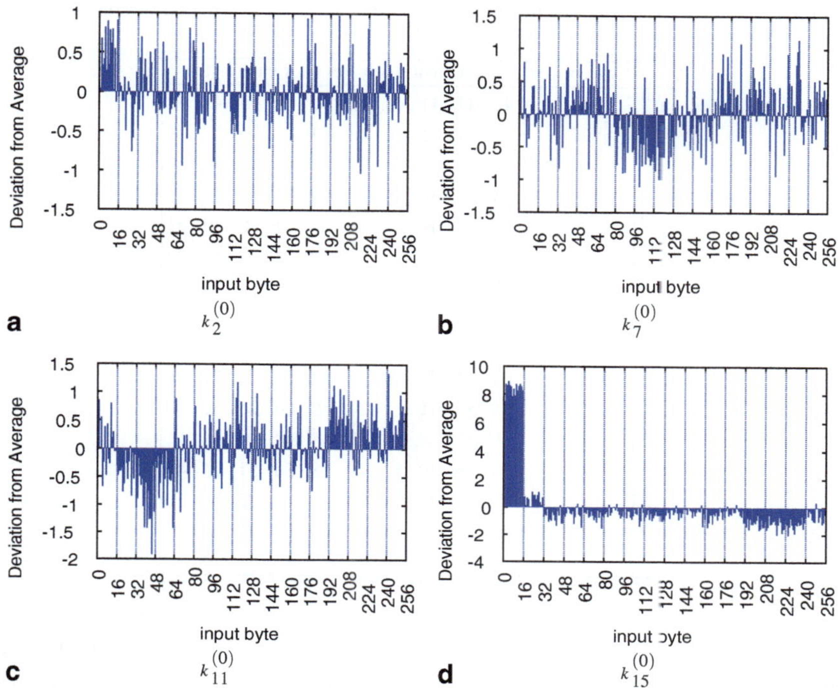

Fig. 7.2 AES timing profiles for four key bytes after 2^{27} encryptions on an Intel Core 2 Duo (E7500) system

sets. Such conflict misses would cause random variations in the timing profiles, thus not posing a threat. But as pointed out by Neve, Seifert, and Wang [1; 2], conflict misses can also be periodic. Periodic conflict misses cause cache misses in the same cache set at regular intervals of time. This periodic form of conflict misses affect the average execution time, because they evict the same bytes from the cache in every encryption. Additionally, a conflict miss requires servicing of a cache miss, which takes considerably long time. Thus conflict misses have a large impact on the information leakage.

Large look-up tables are more likely to be affected by conflict misses, since they occupy a larger region of the cache memory. Thus the AES implementation, which uses tables four times larger than the CLEFIA implementation, is more likely to be affected by conflicts. This can be seen in the timing profiles for an AES attack (some of the profiles are shown in Fig. 7.2). Each table in the AES implementation has 256 values spread over 16 contiguous memory blocks (depicted by the vertical grid in Fig. 7.2), with each block holding 16 elements. Each point in the timing profile has the characteristic time for the access to a unique element in the table. A conflict miss will equally affect the characteristic time for all elements in a block. As a result of this, the correlation in the final phase of the attack produces the same result for shifts that are smaller than a memory block. Since each shift represents a key candidate,

Table 7.1 Ten most likely keys obtained from the AES attack after 2^{27} encryptions on an Intel Core 2 Duo (E7500) system

key	Correct key	Ten most likely keys for each byte									
$k_0^{(0)}$	**11**	4e	47	41	4a	46	4c	48	45	4f	44
$k_1^{(0)}$	**22**	05	**22**	c2	*2f*	ca	33	e1	06	*23*	c9
$k_2^{(0)}$	**33**	**33**	*38*	*3b*	*3a*	*34*	*37*	*39*	0c	*3f*	a7
$k_3^{(0)}$	**44**	83	89	8a	81	*41*	8b	84	*46*	*4b*	*4a*
$k_4^{(0)}$	**55**	d1	de	d9	a8	d0	d3	aa	a5	a0	a1
$k_5^{(0)}$	**66**	8f	52	c3	7a	2b	50	1a	23	f6	4a
$k_6^{(0)}$	**77**	79	*73*	*78*	*74*	**77**	*7e*	*7f*	*75*	8d	8e
$k_7^{(0)}$	**88**	*8e*	*87*	*8f*	*80*	*8a*	*86*	*89*	*8d*	*8b*	**88**
$k_8^{(0)}$	**99**	**99**	39	83	*90*	ba	1e	7a	af	70	13
$k_9^{(0)}$	**aa**	b4	e2	7b	e8	b1	c8	53	7a	79	bb
$k_{10}^{(0)}$	**bb**	65	57	5f	*b2*	24	*b6*	60	25	5e	80
$k_{11}^{(0)}$	**cc**	*c6*	*c2*	*ce*	*ca*	*cb*	**cc**	*c1*	*c0*	14	*cf*
$k_{12}^{(0)}$	**dd**	53	5b	50	52	49	58	5d	51	*d1*	48
$k_{13}^{(0)}$	**ee**	7c	*e0*	4e	98	94	*eb*	*e5*	d7	b3	3b
$k_{14}^{(0)}$	**ff**	ea	*fd*	*fb*	3a	e1	a4	e9	03	*f1*	**ff**
$k_{15}^{(0)}$	**00**	*05*	*01*	*06*	*02*	*04*	*08*	*03*	*0a*	*00*	*0c*

the key space is partitioned based on the size of the memory block, and all keys in the same partition are ranked together. Table 7.1 shows the top ten candidate keys for each byte for an AES attack. The third column is ordered from the most likely key to the least likely of the ten keys. The correct key is represented in bold red, while the keys which fall in the same cache line are emphasized in blue. As an example consider the key $k_6^{(0)}$, whose correct value is 77. The keys which map to the same memory block are 70 to $7f$. These have ranks which are close together. In most cases the attack on AES fails to distinguish between these keys. This results in a low success rate.

Another intersource of leakage is the hardware prefetcher. Most superscalar systems have a hardware prefetcher which anticipates future memory accesses and prefetches them into the cache in order to reduce the miss rate. The prefetcher works by detecting patterns in memory accesses. In cipher implementations, the interesting memory accesses (i.e., to the look-up tables) are expected to be at random locations, thus following no deterministic pattern. However, prefetchers are still capable of prefetching parts of the look-up table albeit inefficiently. This can nevertheless result in information leakage. A detailed analysis of this leakage source is done in Sect. 7.4.

Intrablock Sources: Microarchitectural components in the processor and memory such as the cache banks may cause a small variation in execution time of the cipher. Bernstein gives an example of how recently occurring stores to certain memory locations affect load time. Another example is loads from memory, which cause conflicts in cache banks. The effects of these microarchitectural components in the timing profile is more profound when the size of the look-up tables used in the implementation is small. In such cases conflict misses are less likely and every encryption is likely

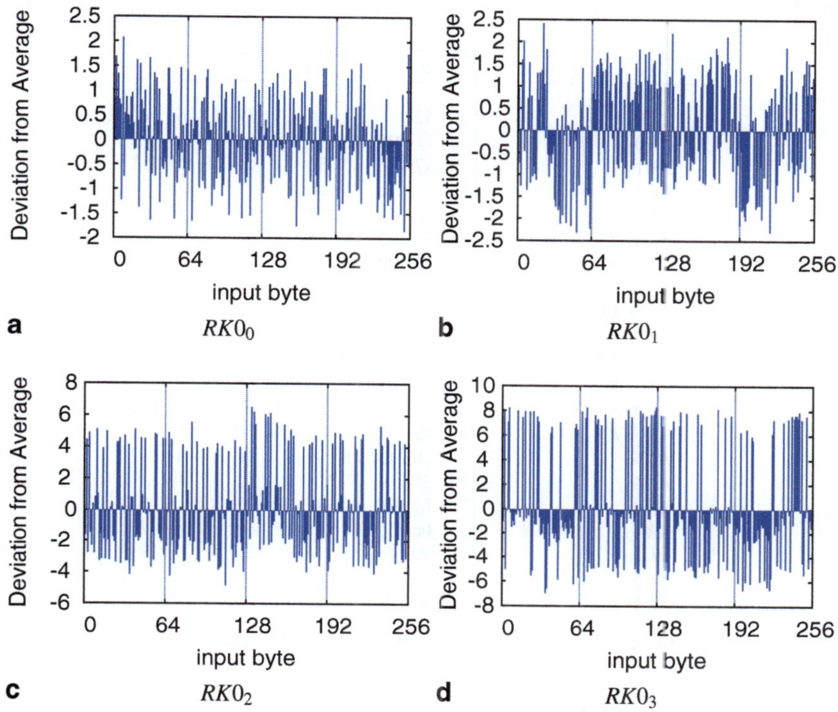

Fig. 7.3 CLEFIA timing profiles for four key bytes after 2^{27} encryptions on an Intel Core 2 Duo (E7500) system

to have the same number of cache misses. Minor variations in execution time due to the microarchitectural components become important. Some of these variations are at the word level and not the block level. These can distinguish keys within a block. Sources which cause such variations are called intrablock sources.

CLEFIA uses two small tables of 256 bytes (each occupying 4 memory blocks and each block holding 64 elements). Every encryption is likely to load the entire table into the cache. Also the effect of conflict misses is less due to the smaller tables. Thus leakage due to microarchitectural features (including the prefetcher) becomes important. This can be seen from the timing profiles for CLEFIA (Fig. 7.3), where (unlike the AES profiles in Fig. 7.2), no distinct patterns are present that can identify a memory block (each memory block is represented by the vertical grid lines). As a consequence, the keys obtained from the correlation phase are not necessarily ordered based on the memory blocks. This can be seen in Table 7.2 which shows the ten most likely key candidates for an attack on CLEFIA ordered from the most likely to the least likely of the ten. The correct key is represented in bold red and is always ranked first for this particular trial, while keys which fall in the same block are emphasized in blue. The smaller impact of interblock leakage sources is evident from the results.

Table 7.2 Ten most likely keys obtained from the CLEFIA attack after 2^{27} encryptions on an Intel Core 2 Duo (E7500) system

key	Correct key	Ten most likely keys for each byte									
$RK0_0$	**f4**	**f4**	*e2*	*c1*	*eb*	52	18	*e1*	*d7*	14	44
$RK0_1$	**d0**	**d0**	52	*f0*	*df*	46	51	*d8*	44	*f2*	*d7*
$RK0_3$	**6a**	**6a**	*5f*	94	92	e8	*48*	*6c*	*75*	a9	*b6*
$RK1_0$	**ca**	**ca**	a7	5b	40	54	52	bf	58	51	53
$RK1_1$	**7b**	**7b**	*46*	db	d1	c6	c4	*52*	*56*	8f	*79*
$RK1_2$	**91**	**91**	13	5a	*8c*	f2	14	64	*a8*	f6	36
$RK1_3$	**60**	**60**	ab	07	*68*	c5	ec	9c	*78*	e9	16
$RK2_0 \oplus WK0_0$	**fe**	**fe**	*f8*	00	06	*ec*	14	11	1c	*f6*	1b
$RK2_1 \oplus WK0_1$	**57**	**57**	*51*	*62*	a7	*5a*	f7	*64*	24	e1	9f
$RK2_2 \oplus WK0_2$	**3c**	**3c**	ea	c5	eb	*3d*	8c	be	92	*11*	ec
$RK2_3 \oplus WK0_3$	**80**	**80**	51	02	58	57	3c	d8	*89*	10	74
$RK3_0 \oplus WK1_0$	**6b**	**6b**	*7b*	*42*	90	*6f*	a3	*56*	d6	3d	a9
$RK3_1 \oplus WK1_1$	**40**	**40**	*4a*	b1	88	fd	92	16	2b	05	13
$RK3_2 \oplus WK1_2$	**16**	**16**	*05*	94	fd	45	6b	b9	*15*	f8	6e
$RK3_3 \oplus WK1_3$	**36**	**36**	f2	42	a8	ad	86	80	c5	*1b*	*34*
$RK4_0$	**7e**	**7e**	e0	fe	e8	01	11	ff	07	1c	12
$RK4_1$	**32**	**32**	*2f*	*34*	*26*	*38*	*31*	*35*	*3f*	*3e*	*29*
$RK4_2$	**50**	**50**	*5d*	00	a0	81	f0	*65*	82	b0	03
$RK4_3$	**e1**	**e1**	0e	37	*dc*	63	*cc*	*c8*	*e5*	89	77
$RK5_0$	**eb**	**eb**	9b	*da*	85	1e	*f8*	3e	*fe*	4c	99
$RK5_1$	**11**	**11**	*24*	e9	ef	*33*	93	cd	*0e*	d2	*17*
$RK5_2$	**47**	**47**	37	92	f8	99	8c	bb	34	b2	*52*
$RK5_3$	**35**	**35**	b7	*38*	7f	e7	5f	*31*	e8	8b	ed

Execution Time of Initial Rounds: The timing profile logs the variation in encryption time as a result of an access to a look-up table in the cipher's implementation. Besides this *specific* memory operation, all other operations done during the execution of the cipher can be considered as noise. The influence this operation has on the execution time of the cipher can determine how efficiently a key is recovered. A large influence implies that the execution time for the *specific* memory operation significantly affects the total encryption time, which can be therefore easily distinguished. A small influence on the other hand implies that execution time for the *specific* memory operation cannot be easily distinguished from the total encryption time.

The *specific* memory operation is in the first round for the AES attack and in the first three rounds for the CLEFIA attack. Thus the influence of the memory operations in these rounds on the encryption time will determine how efficiently the key can be recovered. Figure 7.4 shows the percentage influence in the encryption time for the first nine rounds of CLEFIA and AES.

Due to the small tables used in the CLEFIA implementation, the first three rounds of the cipher load the entire table into the cache. Subsequently, cache misses are most likely in the first three rounds. These rounds will have both hits and misses, while all remaining rounds will generally have only cache hits. As a result, over 95 % of the variations in the encryption time are due to the first three rounds. Since these are the rounds targeted during the attack, the variations due to memory operations from these rounds significantly influence the encryption time and are easily captured.

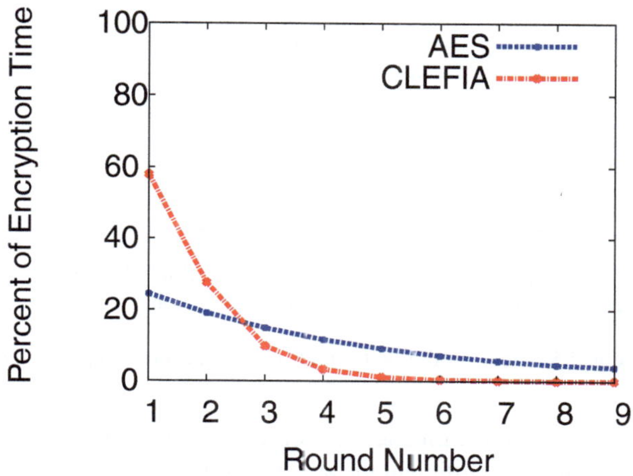

Fig. 7.4 Percentage of encryption time by each round of AES and CLEFIA (only first nine rounds shown)

AES on the other hand has large tables therefore cache misses are spread over a larger number of rounds. As a result, the targeted round (the first round), makes up only 25 % of the total encryption time. Thus time variations due to memory accesses made in this round do not influence significantly the encryption time, thus difficult to capture. This makes recovery of an AES key more difficult than that of a CLEFIA key and more number of AES encryptions need to be monitored in order to recover more bits of the key. In the next section we provide a generalized profiled time-driven cache attack, in order to boost the success rate of the attacks without increasing the number of encryptions required.

7.3 Quantifying Information Leakage in a Timing Profile

Bernstein's attack uses correlation to determine the shift between the timing profile for the known key and that of the unknown key. Each timing profile has 256 points corresponding to $x_i \oplus k_i$, where x_i varies from 0 to 255. Variations in the values at these points are characteristic features of the timing profile These features are matched during the correlation. The efficiency of the attack is determined by how well features of one profile match with the other for the correct shift. An efficient attack would give a correlation close to 1 when the right shift is applied, while a correlation close to 0 when the shift is wrong.

The attack would fail if the correlation values obtained are the same irrespective of the shifts in the timing profile. For instance, consider a constant time implementation of a cipher. The average deviations in execution time would be zero irrespective of the inputs. The timing profiles for the known key and unknown key would be just a set of zeros. The correlation computed between these profiles is the same irrespective of the shift applied. A constant time implementation would therefore be able to resist the attack since the keys are not distinguishable. Thus the unique features in the

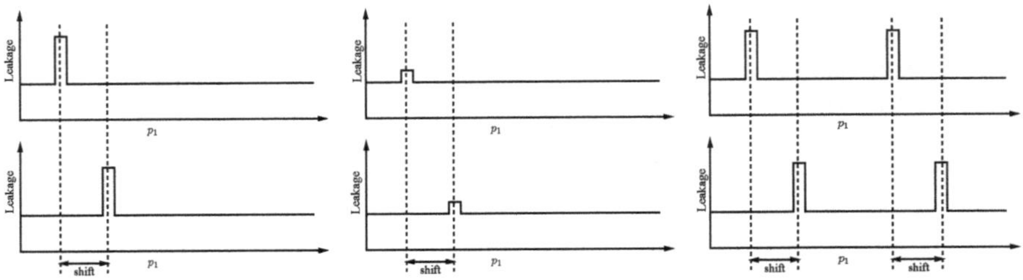

a Single, Large Deviating Distribution
b Single, Small Deviating Distribution
c Multiple Deviating Distributions

Fig. 7.5 Cache profiles with single or multiple deviating distributions

timing profile influences the efficiency of the attack. More the unique features in the profile, more efficient the attack. For instance, the timing profile in Fig. 7.5a is likely to result in more efficient attacks than that of Fig. 7.5b, and less efficient attacks compared to Figure 7.5c. As an other example, consider the attacks on AES and CLEFIA. The AES timing profile, which is more influenced by interblock leakage sources, has less unique features compared to the CLEFIA timing profile. As a result, AES attacks are less efficient compared to CLEFIA attacks.

In order to quantify the unique features in a timing profile, the metric in Eq. 7.3 can be used.

$$\mathcal{L} = \sum_{\substack{\forall pairs\, of \\ j\, and\, j'}} |\mathcal{D}[j] - \mathcal{D}[j']| \tag{7.3}$$

The index of the summation, j, ranges from 0 to 255. There are two parts in Eq. 7.3: (a) the difference term and (b) the summation over all pairs of distributions in the cache profile. The difference term quantifies how much one point in the profile differs from another. When two points on the timing profile are the same, they have a difference zero and cannot be distinguished. If the two points have different values, the difference is positive, and they can be distinguished. The distinguishability becomes easier as the difference between the points increase. The summation in Eq. 7.3 quantifies cases where the profile has more variations in the distributions leading to more information leakage. In the next section we show how the information leaked through hardware prefetching can be analyzed.

7.4 Information Leakage due to Hardware Prefetching

One way to prevent Bernstein's attack is to have the cipher execute at constant time. There are various strategies that have been adopted to provide constant time implementations. For example, the use of specialized instructions, such as Intel's AES-NI [3], eliminate memory-based look-up tables resulting in constant time execution of the cipher. However, these instructions are specific for the AES block

Algorithm 7.2: *SP* : Sequential Prefetching Algorithm

```
Input: The access to memory block at address t_i
begin
    if (t_i not in cache) or (t_i was prefetched and this is the first access to t_i) then
        if t_{i+1} not present in the cache then
            prefetch t_{i+1}
        end
    end
end
```

cipher, leaving a large number of other block ciphers still vulnerable. An alternate direction is to build platforms which inherently protect against cache attacks. This approach is more general and would be usable with a variety of ciphers. A step towards this is to be able to compare the vulnerability of components in a computing platform. In this section we analyze the information leakage due to sequential prefetching algorithm; a prefetching algorithm in the cache memory.

The misses in cache memory have considerable overheads in the performance of a program. In order to reduce these overheads many modern processors have hardware that automatically prefetches data into the cache. The algorithms work by predicting the memory locations accessed by a program. A commonly used prediction algorithm called *sequential prefetching*, monitors memory accesses made and prefetches subsequent blocks. Sequential prefetching is presented in Algorithm 7.2. If t_i is a memory address and t_{i+1} its subsequent address location, an access to t_i would automatically prefetch t_{i+1}.

Each point in the timing profile ($\mathcal{D}_i[j]$) corresponds to the deviation in average execution time when a plaintext byte x_i is kept constant. Owing to this constant, a specific block in the look-up table (corresponding to the index $x_i \oplus k_i$) is accessed in every encryption that is considered in the average. Since all other inputs to the cipher are random, and the randomness diffuses to every round, the other accesses to the look-up tables are at random locations. With respect to the number of cache misses that occur due to the look-up table, probability of a cache miss in the memory block corresponding to the index $x_i \oplus k_i$ is one assuming that there are no conflict misses.

The probability of cache misses in all other memory blocks in the look-up table is less than one, depending on wheather the block was accessed during the encryption. In a cache memory that does not support prefetching, all these memory blocks have an equal probability of being loaded into the cache. However, when prefetching is present, the uniformity in the probability of a cache miss across all memory blocks of the table is not true. Depending on the prefetching algorithm used, the probability of obtaining a cache miss may vary from one memory block of the table to another. We explain the behavior with respect to the sequential prefetching algorithm. We initially assume that the table is isolated, which means that no memory access outside the table can prefetch a memory block of the table. Later we will relax this assumption.

With respect to the sequential prefetcher in Algorithm 7.2, the memory blocks of the table can be split into three categories (depicted in Fig. 7.6).

Fig. 7.6 Prefetchable blocks of a table used in a block cipher

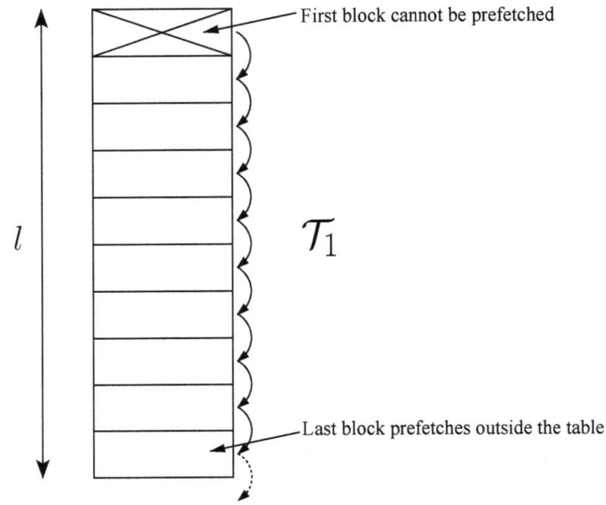

- The first memory block of the table can never get prefetched because it would require a memory access to a block preceding the first block. This is outside the table boundary, therefore by our assumption, cannot prefetch any part of the table.
- All blocks of the table except the last, prefetch a memory block inside the table (a valid prefetching by our assumption).
- An access to the last block of the table prefetches a memory block outside the table boundary.

Since the number of cache misses that occur during the encryption have a significant impact on the execution time, this difference in memory blocks results in nonconstant time executions of the cipher and a timing profile which leaks information. In the following discussion we understand the impact of this on the timing profile.

Assuming that the table occupies l memory blocks, and no conflict misses arise, one of three scenarios occur depending on which memory block is accessed by the fixed $x_i \oplus k_i$.

- If the first memory block in the table is accessed by $x_i \oplus k_i$, then the second block of the table is prefetched. Assuming a clean cache at the start of encryption, there are $l - 2$ memory blocks remaining, which can result in cache misses. All these $l - 2$ memory blocks are also prefetchable.
- If the last memory block in the table is accessed by $x_i \oplus k_i$, the prefetching is done outside the table and will not affect the table accesses. In this case there are $l - 1$ cache misses that can still occur due to the look-up table and all these memory blocks can be prefetched except for the first.
- If any other memory block other than the first and last is fixed by $x_i \oplus k_i$, then $l - 2$ cache misses can occur and $l - 3$ of these blocks can be prefetched.

These observations are summarized in Table 7.3 along with the average timing expected. When the first memory block is the invariant, the fastest execution is achieved

Table 7.3 Cache misses in a table of size l used in a block cipher with sequential prefetching enabled

Memory block of table fixed	Number of cache misses possible	Number of sequential prefetching possible	Remarks on average time
First	$l-2$	$l-2$	t_1
Any except first and last	$l-2$	$l-3$	t_2
Last	$l-1$	$l-2$	t_3

Comparison of the average timings : $t_1 < t_2 < t_3$

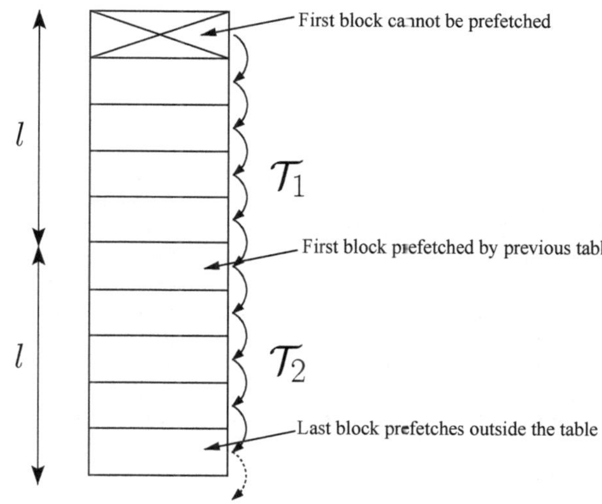

Fig. 7.7 Prefetchable blocks of two adjacent tables used in a block cipher

as it has the least number of cache misses possible and all memory blocks can be prefetched. When the last memory block is the invariant, the slowest encryptions occur because it has the most number of cache misses. Thus, prefetching causes nonconstant encryption time to occur, which is captured in the timing profile.

Relaxing the Assumptions: The previous analysis was made under the assumption that the table is isolated and other memory accesses made by the program do not affect the prefetching in the table. In reality, however, this assumption would not always hold. To show the effect we consider a block cipher implemented with two tables $\mathcal{T}_1$ and $\mathcal{T}_2$ placed side-by-side in memory (as shown in Fig. 7.7). If $x_i \oplus k_i$ accesses $\mathcal{T}_1$, then the last memory block prefetches the first memory block of $\mathcal{T}_2$. Conversely, if $x_i \oplus k_i$ accesses $\mathcal{T}_2$, then the first memory block can be prefetched by the last memory block of $\mathcal{T}_1$. This can cause a reduction in the leakage as there are fewer variations in the encryption time.

7.5 Conclusion

If an adversary is capable of learning the execution behavior of a cipher then she can mount profiled attacks. These are typically the most powerful form of side-channel attacks. The profiled time-driven cache attacks discussed in this chapter is one such profiled attack. Leakage in profiled time-driven cache attacks are due to several causes such as conflict misses, hardware prefetching, and differential timing characteristics of memory loads. The leakage is also significantly affected by the cipher implementation. This chapter analyzed information leakage in hardware prefetching and showed how the information leaked in these attacks can be quantified.

References

1. Neve M, Seifert J-P, Wang Z (2006) A refined look at Bernstein's AES side-channel analysis. In: Lin FC, Lee D-T, Lin B-S, Shieh S, Jajodia S (eds) ASIACCS. ACM, New York, p 369
2. Neve M, Seifert J-P, Wang Z (2006) Cache time-behavior analysis on AES, http://citeseerx.ist.psu.edu/viewdoc/summary?doi=10.1.1.88.4091. Accessed Dec 2014
3. Gueron S (2010) Intel Advanced Encryption Standard (AES) instructions set (Rev : 3.0)
4. Bernstein DJ (2005) Cache-timing Attacks on AES. Tech. Rep.
5. Rebeiro C, Mukhopadhyay C (2012) Boosting profiled cache timing attacks with apriori analysis. IEEE Trans Inf Forensics Security 7(6):1900–1905
6. Bhattacharya S, Rebeiro C, Mukhopadhyay D (2012) Hardware prefetchers leak: a revisit of SVF for cache-timing attacks. In the Workshop Proceedings of the 45-th IEEE/ACM International Symposium on Microarchitecture, MICRO 2012 Workshops, Vancouver, BC, Canada, pp 17–23, IEEE

Chapter 8
Access-Driven Cache Attacks on Block Ciphers

Cache memories are a shared resource in a system. If a process running in a system fetches a block of data into the cache, the data will remain in the cache memory unless evicted. The eviction can be done by any process in the system and can result in a covert timing channel. The channel was first discovered by Wray in [1] and Hu in [2]. Attacks on cryptographic algorithms using the shared feature in cache memories were suggested independently by Percival in [3] and Osvik, Shamir, and Tromer in 2005 [4]. These attacks are known as *access-driven* attacks and use time as a side channel instead of power or electromagnetic traces. We briefly describe selected works in this category of cache attacks.

8.1 Access-Driven Attacks on Block Ciphers

Consider the T-table implementation for AES (Sect. 2.2.1.1). For the first round, the table $\mathcal{T}_j$ is accessed at index $s_{i+j}^{(0)} = (x_{i+j} \oplus k_{i+j}^{(0)})$ for $0 \leq j \leq 3$ and $i \in \{0, 4, 8, 12\}$, where x_{i+j} is an input byte and $k_{i+j}^{(0)}$ the corresponding whitening key byte. Suppose the key byte $k_{i+j}^{(0)}$ is guessed (the guess represented by $\tilde{k}_{i+j}^{(0)}$), then a corresponding index in table $\mathcal{T}_j$ can be computed as $\tilde{s}_{i+j}^{(0)} = x_{i+j} \oplus \tilde{k}_{i+j}^{(0)}$. If the guessed key is correct (i.e., $k_{i+j}^{(0)} = \tilde{k}_{i+j}^{(0)}$) then the computed index is also correct (i.e., $s_{i+j}^{(0)} = \tilde{s}_{i+j}^{(0)}$) and vice versa.

In [4–6], Osvik, Tromer, and Shamir showed two methods that an adversary monitoring the covert timing channels in cache memories can use to identify when the following equality $\langle s_{i+j}^{(0)} \rangle = \langle \tilde{s}_{i+j}^{(0)} \rangle$ holds. The methods *evict+time* and *prime+probe* require that the adversary knows (or learns) the cache sets occupied by the table $\mathcal{T}_j$. We summarize the methods below:

Evict + Time: For every possible value of the table index $(\langle \tilde{s}_{i+j}^{(0)} \rangle)$ perform the following operations:

1. Trigger an encryption (Algorithm 2.1) for a random input $\mathbf{x}$.
2. For a guess of $\tilde{s}_{i+j}^{(0)}$, determine the memory address of $\mathcal{T}_j[\tilde{s}_{i+j}^{(0)}]$ and compute the cache set it gets mapped into. Denote this cache set as M.

C. Rebeiro et al., *Timing Channels in Cryptography*,
DOI 10.1007/978-3-319-12370-7_8

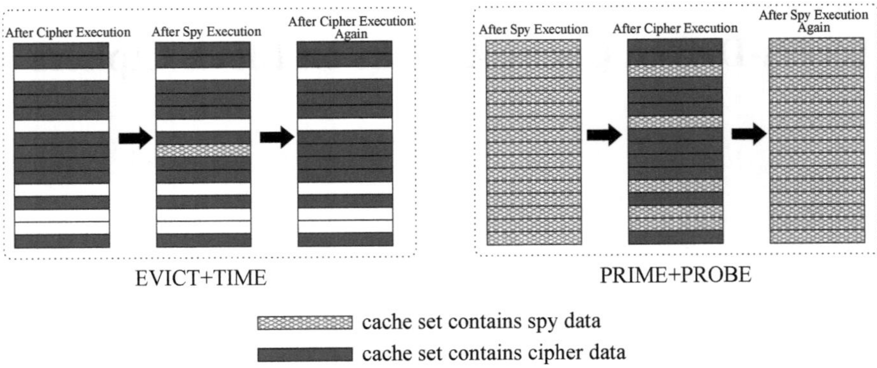

Fig. 8.1 Cache sets corresponding to table $\mathcal{T}_j$ for evict+time and prime+probe access driven attacks

3. Perform operations so that all data present in cache set M due to the first encryption is evicted.
4. Trigger a second encryption with $\mathbf{x}$ again and time it.

Figure 8.1 shows the various stages of the cache sets used by the table $\mathcal{T}_j$. There are two possible outcomes after the second encryption:

- If $\langle s_{i+j}^{(0)} \rangle = \langle \tilde{s}_{i+j}^{(0)} \rangle$ implying $(\langle k_{i+j}^{(0)} \rangle = \langle \tilde{k}_{i+j}^{(0)} \rangle)$, then the cache set M is always accessed by both encryptions (in steps 1 and 4). Particularly, the second encryption would *always* (with probability $= 1$) result in a cache miss for the access $\mathcal{T}_j[s_{i+j}^{(0)}]$. We call such a miss as an *ever-present* cache miss.
- If $\langle s_{i+j}^{(0)} \rangle \neq \langle \tilde{s}_{i+j}^{(0)} \rangle$ implying $(\langle k_{i+j}^{(0)} \rangle \neq \langle \tilde{k}_{i+j}^{(0)} \rangle)$, then for a value of $\mathbf{x}$, the cache set M is not accessed at $\mathcal{T}_j[\tilde{s}_{i+j}^{(0)}]$. However, the set *may* be accessed by the other memory accesses to $\mathcal{T}_j$. Thus for the second encryption (in step 4), the probability of a cache miss is ≤ 1 for the access $\mathcal{T}_j[s_{i+j}^{(0)}]$.

If several values of $\mathbf{x}$ are considered, the ever-present cache miss at $\mathcal{T}_j[s_{i+j}^{(0)}]$ when $\langle s_{i+j}^{(0)} \rangle = \langle \tilde{s}_{i+j}^{(0)} \rangle$, would cause a slight increase in the expected encryption time compared to when $\langle s_{i+j}^{(0)} \rangle \neq \langle \tilde{s}_{i+j}^{(0)} \rangle$ where the cache miss at $\mathcal{T}_j[s_{i+j}^{(0)}]$ is not always present. This difference in time can be detected by an adversary to determine the correct value of $\langle s_{i+j}^{(0)} \rangle$ there by obtaining $\langle k_{i+j}^{(0)} \rangle$ (from the known values of the plaintext).

In [7], Bogdanov, Eisenbarth, and Paar use the Evict+Time strategy to mount a differential cache-collision attack on AES. The main idea of the attack is to choose pairs of plaintexts such that they cause *wide collisions*. Wide collisions in AES are explained using Fig. 8.2. Plaintexts P_1 and P_2 are chosen randomly as inputs to AES in such a way that

- Values on the main diagonals $A = \{a_i\}$ and $E = \{e_i\}$, $0 \leq i \leq 4$ are chosen randomly and independently of each other such that $\exists i, a_i \neq e_i$.
- The remaining bytes of P_1 and P_2 are chosen randomly and are made pairwise equal.

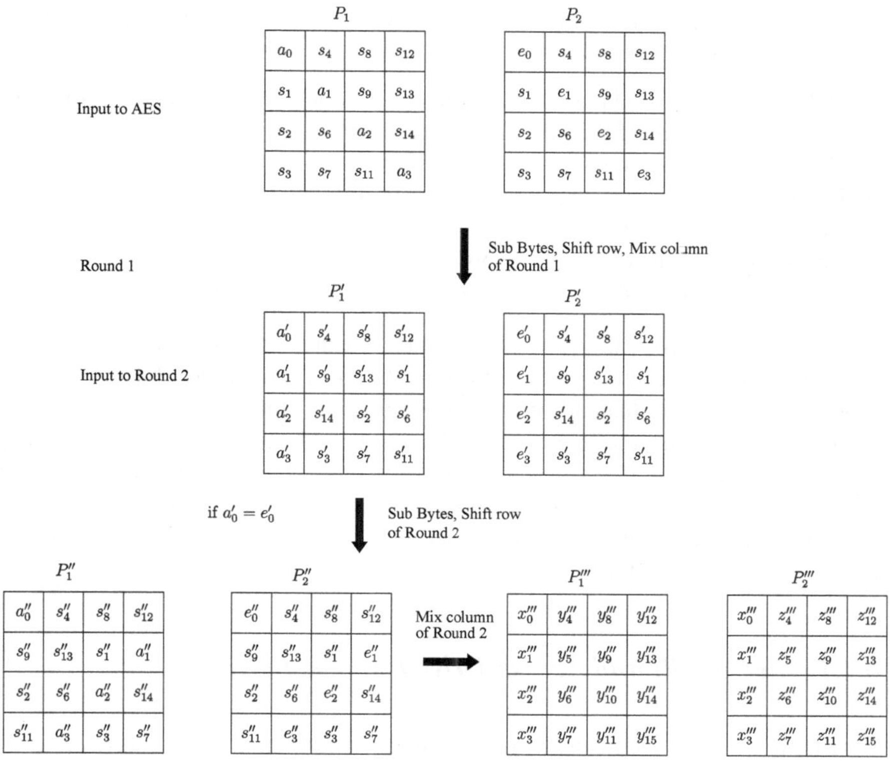

Fig. 8.2 Wide collisions in AES

From Fig. 8.2, P_1' and P_2' are the outputs of the first round. It is observed that in both of them, only the bytes in the first column varies. The other bytes are the same. If we trace the inputs to the third round, the values of first column of P_1''' and P_2''' are pairwise equal which leads to five AES s-box operations (one in the second round and four in the third), which possess either pair wisely equal or pair wisely distinct values.

Wide collision can only be observed when $a_0' = e_0'$. The evict+time technique is used to identify plaintext pairs with wide collisions. An existence of a wide collision would cause on average the second encryption to be faster by a margin of five compared to when there is no wide collision. The wide collisions are then used to construct a set of four nonlinear equations, which when solved reveal bytes of the key.

Prime + Probe: In the evict+time method, the time for the memory access to $\mathcal{T}_j[s_{i+j}^{(0)}]$ (where $0 \leq j \leq 3$ and $i \in \{0, 4, 8, 12\}$) gets reflected in the total execution time of the cipher, thus leading to the attack. However, this technique delivers a low success due to the presence of additional memory accesses and other code that executes during the encryption. Further, there is considerable noise from sources such as instruction scheduling, conditional branches, and cache contention thus resulting in a low signal-to-noise ratio (SNR). In the prime+probe method, smaller codes are timed thus leading to a attacks that are more successful. The steps involved in the prime+probe is as follows:

1. Define an array A as large as the cache memory and read a value of A for every memory block (thus filling the entire cache with A).
2. Trigger an encryption with a random input $\mathbf{x}$.
3. For a guess of $\langle\tilde{s}^{(0)}_{i+j}\rangle$, determine the cache set that $\mathcal{T}_j[\langle\tilde{s}^{(0)}_{i+j}\rangle]$ gets mapped into. Denote this cache set as M.
4. Access A at indices which get mapped to the cache set M and time the individual accesses. Figure 8.1 shows the various stages of the cache sets used by the table $\mathcal{T}_j$.

If $\langle\tilde{s}^{(0)}_{i+j}\rangle$ is correct (i.e., $\langle\tilde{s}^{(0)}_{i+j}\rangle = \langle s^{(0)}_{i+j}\rangle$), then $A's$ data present in the cache set M would be evicted with probability $= 1$ during the encryption in step 2. However, this probability can be less than one if $\langle\tilde{s}^{(0)}_{i+j}\rangle$ is incorrect. Further, an eviction of $A's$ data at M would result in a cache miss in the fourth step; identified by a longer memory access time. However, if $A's$ data at M is not evicted, then the last step would have a cache hit therefore a shorter access time. The correct $\langle\tilde{s}^{(0)}_{i+j}\rangle$ (thus $\langle\tilde{k}^{(0)}_{i}\rangle$, again assuming that the plaintext is known) can therefore be identified by repeating the four steps with several inputs.

8.1.1 Second Round Access-Driven Attack on AES

For a table with $l \cdot 2^{\delta}$ elements, each element has an entropy $n = \delta + \log_2 l$, where 2^{δ} is the number of elements in the table sharing a memory block and l is the number of memory blocks occupied by the table. The first round attack can reveal at most $\log_2 l$ bits of each key byte. In order to reveal the entire key byte, the second round of AES needs to be targeted [4–6].

Consider the first table access in the second round. From Algorithm 2.1, Fig. 2.3, and the key scheduling algorithm of AES, this is

$$s^{(1)}_0 = 2 \cdot S[x_0 \oplus k^{(0)}_0] \oplus 3 \cdot S[x_5 \oplus k^{(0)}_5] \oplus S[x_{10} \oplus k^{(0)}_{10}] \oplus S[x_{15} \oplus k^{(0)}_{15}]$$
$$\oplus k^{(0)}_0 \oplus S(k^{(0)}_{13}) \oplus 1 . \qquad (8.1)$$

The value of $\langle s^{(1)}_0\rangle$ is affected by keys $k^{(0)}_0$, $k^{(0)}_5$, $k^{(0)}_{10}$, $k^{(0)}_{13}$, and $k^{(0)}_{15}$ each occupying one byte. The first round attack reveals $\log_2 l$ bits of each of these key bytes leaving δ bits unknown, Thus there is a space of $5 \cdot 2^{\delta}$ keys that need to be searched. The evict+time or prime+probe methods can be used along with Eq. 8.1 to identify the correct key from this key space.

8.1.2 A Last Round Access-Driven Attack on AES

As described in Sect. 2.2.1.1, the last round of some implementations of AES use a different table (called $\mathcal{T}_4$). This table is exclusively used in the last round and can lead to ciphertext only attacks. A byte of the ciphertext can be expressed as $y_i = k_i^{(10)} \oplus S(s_i^{(9)})$ where $0 \leq i \leq 15$ (here we have ignored the last round ShiftRows operation as it does not affect the attack complexity). Thus $k_i^{(10)} = y_i \oplus S(s_i^{(9)})$. Evict+time or prime+probe methods can be used to determine $\langle S(s_i^{(9)}) \rangle$, and therefore $\langle k_i^{(10)} \rangle$ can be determined.

Neve and Seifert in [8] developed two ways to determine all bits of $k_i^{(10)}$. The first method called the *nonelimination method*, is based on the fact that for a given value of y_i, the input to the SubBytes in the last round (i.e., $s_i^{(9)}$) is fixed. This implies that the cache set accessed for a given y_i is always the same. Now consider that $y_i = t_1$ corresponds to $s_i^{(9)} = o_1$, where $0 \leq t_1, o_1 \leq 255$. Similarly $y_i = t_2$ corresponds to $s_i^{(9)} = o_2$, where $0 \leq t_2, o_2 \leq 255$. Thus,

$$
\begin{aligned}
t_1 &= k_i^{(10)} \oplus S(o_1) \\
t_2 &= k_i^{(10)} \oplus S(o_2)
\end{aligned} \tag{8.2}
$$

The access attack reveals $\langle o_1 \rangle$ and $\langle o_2 \rangle$. This leaves 2^δ possible key options for $k_i^{(10)}$. This key space can then be reduced by using the relation $t_1 \oplus t_2 = S(o_1) \oplus S(o_2)$ obtained from Eq. 8.2. Further reduction of key space is possible by considering different pairs of values for y_i. The authors in [8] estimate that 186 pairs are required to uniquely identify the correct key.

The second method is based on eliminating incorrect values of $s_i^{(9)}$. It uses the fact that the table $\mathcal{T}_4$ is accessed only 16 times and not all memory blocks of the table are likely to be accessed during the encryption. The memory blocks not accessed can be detected by the evict+time or the prime+probe methods. It is certain that $S(s_i^{(9)})$ is not one of the values in the unaccessed blocks. Thus if n blocks of $\mathcal{T}_4$ are not accessed during the encryption, then $n \cdot 2^\delta$ values of $s_i^{(9)}$ can be eliminated. By using different inputs to the cipher, more values of $s_i^{(9)}$ can be eliminated until a unique value is obtained. This corresponds to a unique value of $k_i^{(10)}$.

8.2 Asynchronous Access-Driven Attacks

The access-driven attacks discussed so far were synchronous, wherein it is assumed that the adversary can trigger an encryption. A less restrictive form of the attack is the asynchronous mode, where a nonprivileged adversary running concurrently in an independent user space, needs only to monitor the cache activity to determine cache sets accessed by the executing cipher. No triggering of encryption is required. Instead, the only requirements is that the adversary executes in the same processor and concurrently with the cipher. Asynchronous access driven attacks were pioneered by Percival in [3], who demonstrated an attack on RSA. Osvik et al. and Neve then explored the feasibility of asynchronous attacks on AES [4–6, 8], though the practicality of the work remains unclear. Aciiçmez et al. [9] were the first to present a practical asynchronous access-driven using instruction caches on OpenSSL's DSA, while [10] demonstrated the first practical asynchronous attack on AES.

In an asynchronous attack, the adversary runs a spy process which periodically reads or writes data into cache lines and monitors the time it takes. An eviction of the spy's data by another process (such as an executing cipher) from the cache will result in a longer memory access time due to the cache miss that occurs. This gives clues about the memory access patterns of the cipher process. Such leakages are also possible when the spy and cipher are executing in different security zones. For instance, in cloud computing virtualized environments [11], several users share the same hardware, albeit in different security zones.

A system vulnerable to such asynchronous attacks utilizes at least one of the following features in the computer system: symmetrical multithreading [3; 6] or preemptive operating system (OS) scheduler [8; 10].

Symmetrical multithreading allows multiple processes to execute simultaneously on a single processor. The hardware resources including the cache memories are shared between the processes. The cache memory acts as a covert channel transmitting information about the memory access patterns of the cipher.

Preemptive OS Scheduling divides CPU time into equally spaced intervals called *slices*. At the beginning of a slice, a process gets allocated to the CPU and executes until it voluntarily relinquishes the CPU or the time slice completes. A new process may then get allocated to the CPU by a process known as *context switching*. In [8], Neve and Seifert suggest that spy processes can exploit such schedulers to obtain covert information about a cipher's execution, though no explicit details about the construction were given. In [12], Tsafrir, Etsion, and Feitelson present a practical malicious code that can exploit context switching in OS schedulers. The intuition is that the malicious code starts executing at the beginning of a slice, but yields the processor before the slice completes. Another process is then scheduled for the remaining time interval in the slice. The authors show several applications of the malicious code such as denial of services, bypassing profiling and administrative polices, etc. Bangerter, Gullasch, and Krenn use such a malicious code to develop a fine grained access-driven cache timing attack in [10]. In the next section we show how the OS scheduler can be used monopolize the system. The find grained attack on AES is then described.

8.3 Secretly Monopolizing the CPU scheduler

Computer systems are governed by two clocks—*hardware clocks* drive the instruction cycle while the *operating system clocks* control the system activity, measure time passage, provide timing services, and maintain control. Unlike the hardware clock, whose resolution is determined by the processor frequency (and hence cannot be tuned), the system clock frequency is determined by the OS at boot-up.

Clock resolution in the OS clock has a strong influence on scheduling. In [12], it is shown how these clocks can be tricked to allow a malicious code to secretly consume a large percentage of the CPU cycles without any accounting. The phenomenon is called "cheat" attack and exploits two paradigms of some OS schedulers.

- CPU usage is accounted by periodic sampling interrupts.
- Processes that use less CPU have higher priority.

A periodic hardware clock interrupt termed as *tick* is used for measuring CPU usage. This accounting information is mostly used by priority-based schedulers for calculating priority of processes. When a tick occurs, the process executing is billed for the entire interval since the previous tick. The schedulers rely on this periodic sampling interrupts of a low-precision clock to account for the CPU usage. Due to this coarse granularity of the periodic sampling the billing may not be accurate, thus leading to the inaccurate information being used by the scheduler. This inaccuracy is exploited in the *cheat* attack procedure.

The exact time for which the process is scheduled to run is the quantum of a process. The effective quantum can have widely varying values and may or may not be greater than a clock tick, though CPU usage is accounted only at the clock ticks. Thus when a clock tick occurs the running process gets billed in integral multiples of the periodic sampling interval. If a quantum is shorter than the tick duration, then the process has higher probability of not getting billed. So for processes having intervals of quanta shorter than a clock tick, many smaller quanta can consume arbitrarily small CPU cycles remaining unbilled, while if a short quanta includes a clock tick then it gets overbilled by the entire tick duration though actually consuming a small fraction of the periodic interval. Similarly, the situation is true for bigger quanta. Since the probability that a quantum includes a tick is proportional to its duration, on average the overbilling and underbilling effect tends to cancel out to provide reasonably accurate billing.

If a process maliciously manages to exploit this coarse granularity phenomenon for accounting the CPU usage then it never gets billed. The cheat process thus performs the following:

- Schedules to run just after the tick occurs.
- Finishes its work within the fraction of the tick duration.

These are repeated over several ticks. Thus the cheat process makes sure that it is never scheduled when the periodic tick occurs. This work demonstrates the working principles of a cheat program in which a process at user level is not only capable of consuming any percentage of the CPU but also the accounting of this extra CPU usage does not get reflected in the monitoring tools since other processes gets billed instead of the consuming process.

As shown in Fig. 8.3, when a tick occurs, the OS performs billing and allows the pending process having highest priority to execute in the CPU. Thus the cheat process can easily take control over the CPU just after a clock tick occurs. In order to avoid billing, if the cheat process is capable of preempting itself using a fine-grained timer of high frequency before the next tick occurs, then it successfully prevents itself

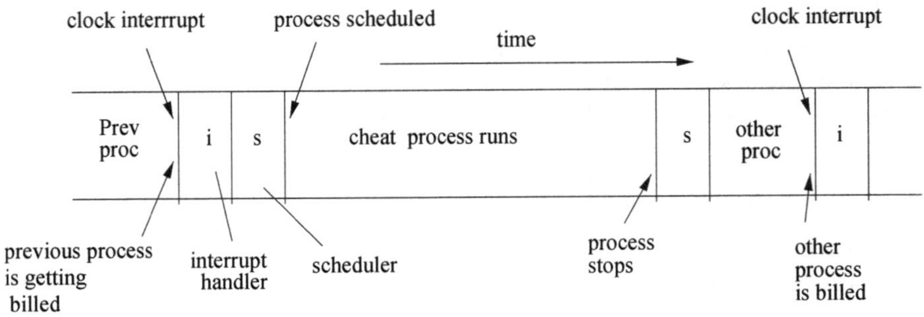

Fig. 8.3 Philosophy of the cheat program

from getting billed. Thus the OS is forced to believe that the malicious process is consistently sleeping. This illusion increases the priority of the malicious code from the OS perspective. The policy for the recent schedulers is such that it always assign a higher priority to processes that have lower CPU usage and thus their requests of consuming CPU cycles are serviced first. Thus for a malicious process which is accounted as sleeping, it is easy to be fired just after a periodic clock interrupt occurs.

In order to prevent itself from getting billed the process constructs a fine-grained timing measurement mechanism by reading the values of a timestamp counter. The timestamp counter is read using the "rdtsc" assembly call, which returns the value of the hardware timestamp counter. This assembly call can be made from the user level and returns the timestamp of the system in clock cycles.

Listing 8.1 illustrates a C code implementation of the cheat program. The subroutine *get_cycles* enables the user to read the timestamp counter value using a *rdtsc* assembly call. The assembly call is explained in details in Sect. 4.2.1. The interval between two consecutive ticks is determined using subroutine *cycles_per_tick*. The interval is observed for 1000 such measurements and is averaged to obtain a precise value of the number of clock cycles per tick. A zero sleep is inserted in the function which wakes up the process at the next clock interrupt.

These subroutines are actually used by the *cheat_attack* which successfully steals a fraction of CPU cycles without getting billed. The cheat process continuously polls to see that whether it has consumed more that the fraction it actually wants to consume. The attack code continuously iterates over the *while* loop and checks whether the desired fraction is over. When the counter crosses the limit, the process blocks itself till the next tick by calling nanosleep function with zero time. While it polls to check for the desired interval, additionally it executes some short work side-by-side. Thus when the desired fraction of the periodic interrupt has been used by the *cheat_attack* it preempts itself and allows other processes to run.

Listing 8.1 Simple C implementation of cheat program

```
#define _POSIX_C_SOURCE 199309
#include <stdio.h>
#include <stdlib.h>
#include <sys/time.h>
struct timespec zero = {0,0};  // sleep for zero ns before
                               // the next tick
typedef unsigned int cycle_t; // unsigned 32 bit integer to store
                               // the clock cycles

// returns the value of the timestamp
// counter in number of clock cycles
inline cycle_t get_cycles()
{
  cycle_t ret;
  asm volatile("rdtsc" : "=A" (ret));
  return ret;
}

//calculating the number of clock cycles per clock tick
cycle_t cycles_per_tick()
{
  int i;
  nanosleep(&zero,0);       // sync with tick
  cycle_t start = get_cycles();
  for(i=0 ; i<1000 ; i++)     // measures cycles for 1000 ticks
    nanosleep(&zero,0);
  return (get_cycles() - start)/1000;    // measure average cycles
                                          // between clock ticks
}


//Subroutine for the Attack
void cheat_attack( double fraction )
{
  cycle_t work, tick_start, now;
  work = fraction * cycles_per_tick();     // clock cycles to be
                                            // consumed within a
                                              tick duration
  nanosleep(&zero,0);                 // sync with tick
  tick_start = get_cycles();          // timestamp counter at the
                                        // start of tick
  // poll continuously to check
        // whether the desired fraction is consumed
  while( 1 ) {
    now = get_cycles();
    if( now - tick_start >= work ) {
      nanosleep(&zero,0); // avoid bill
      tick_start = get_cycles();
    }
    // do some short work here...
    system("gcc test.c -o t1"); // compiling a test process
    system("./t1"); // calling a test process to run

  }

}


int main()
{
  cheat_attack(0.95);       // calling the cheat attack with the
                                   // desired fraction
}
```

But the above implementation has certain drawbacks as in the cheat process polls continuously to check whether the desired interval has expired. This continuous polling leads to a significant overhead for the cheat process. The following subsection provides a brief discussion on a server–client setup which shows a significant improvement upon the previous strategy.

8.3.1 Cheat Server

The cheat server is a machine with which the client machine (in which the cheat code executes) communicates, such that the cheat process blocks itself on receiving a predefined message from the server. This communication with the cheat server can be viewed as a substitute to the finer grained timing measurements required to preempt the cheat process successfully.

The communication between the cheat process and the network server is carried out through a predefined protocol named as cheat protocol. The protocol is explained briefly as follows: Initially the cheater process allows the cheat client to establish a connection with the remote cheat server. After the client machine establishes the connection, the client sends a packet to the cheat server requesting the server to send it a message after a fine-grained time interval which is essentially less than the interval to next periodic interrupt of the client. This request is sent via a User Datagram Protocol (UDP) data packet which contains the desired interval in nanoseconds. The server on receiving a request, waits for the interval to expire with a precise granularity and communicates with the client as soon as the timer exceeds the interval. Thus, there can be two possibilities: either the server is having an OS of higher tick rate or the server busy waits on a cycle counter for the request time interval to signal the client precisely. The client on the other hand allows the target application to run in the local processor while it polls the network for the message to arrive. As soon as it receives the message, in order to prevent itself from getting billed the client goes to nanosleep with zero offset. This essentially prevents the target application from getting billed, on addition to this due to zero-sleep, the application gets rescheduled on the immediate next tick. The cheat client again performs the same sequence of communication on every tick.

The immediate advantages of this cheat server above the previous implementation is that since the cheat server is dedicated to provide the reminder to go for sleep, the client system thus can sleep wait until it receives a message while the target application runs on the local host unmodified. Since the cheat code avoids the polling, the loss in throughput is prevented to a great extent.

But the communication of the cheat client with the server can be detected if the network traffic is monitored. In addition to this there is a serious drawback regarding the delay caused by the network latency. The success of the procedure truly relies on the performance of the server. The next subsection provides an alternative to this procedure and is free from the drawbacks mentioned above.

8.3.2 *Binary Instrumentation*

Binary instrumentation is an easier alternative for implementation of the cheating code and by incorporating it, any benign code can be converted into a cheater code. The cheat analysis subroutine is listed in Listing 8.2. Any arbitrary program can include this subroutine and if this routine is invoked a number of times then essentially it turns into a cheat program.

Listing 8.2 Cheat analysis routine

```
void cheat_analysis()
{
  cycle_t c = get_cycles();

  if(c - tick_start >= work)
  {
    nanosleep(&zero, 0);

    tick_start = get_cycles();
  }
}
```

Due to the two features—accounting the CPU usage and system timer services being done at periodic ticks in general purpose systems any arbitrary user level code can systematically sleep at the periodic ticks monopolizing the CPU.

The following section provides an attack strategy which exploits this fine grained scheduling to practice a *denial of service* attack. The attack falls in the category of Access driven attacks and is demonstrated using timing as side channel on AES block cipher.

8.4 Fine Grained Access-Driven Attack on AES

The previous access-driven cache-timing attacks on AES [4-6, 8] in Sect. 8.1 and 8.2, used the fact that the spy determines all the cache sets accessed by AES only after the encryption is complete. Due to the large granularity of measurements in the attacks, it was not possible for the spy to determine intermediate accesses done by AES. In [10], Bangerter, Gullasch, and Krenn show a way of exploiting the OS scheduler to obtain information about every single cache access performed by AES.

In order to practically mount the attack, the completely fair scheduler (CFS), which is typically used in the Linux kernels, is exploited. About 100 spy threads are launched in the system. When a spy thread gets scheduled into the processor, it first makes the timing measurements of cache accesses, and then waits until most of the time slice is complete before blocking itself. The cipher then executes in the small time interval that remains in the time slice. Typically this attack features the *denial of service* to the cipher process, exploiting the underlying scheduler policy.

The cipher gets scheduled for a very small interval of time before being preempted by the scheduler.

This interval is just sufficient for the cipher to make one memory access. The next spy thread then gets scheduled into the processor and the single access made by the cipher is determined by measuring the memory access time to access its data. In this way, the cipher is executed very slowly, and each memory access it makes can be tracked by the spy.

Bangerter et al. [10] demonstrated the first practical asynchronous attack on AES. In this work the OS scheduler vulnerabilities are exploited to perform a fine grained access driven cache attack on practical implementation of AES. A malicious code is executed with multithreading capabilities exploiting the CFS policies in order to obtain cache access patterns of running encryption. Thus this attack does not require any direct synchronization with the victim process though exploits the timing as the source of information to decide whether an intermediate cache access is a hit or miss.

8.5 Attack Procedure

8.5.1 *Completely Fair Scheduler*

Any general purpose multitasking OS provides an illusion to the user to run several parallel processes at a time. The scheduler is responsible for the multiplexing between processes and threads. Recent Linux systems are equipped with CFS, which justifies its name by asymptotically behaving like an ideal system. Virtual run times (τ_is) are associated with every running process P_i and to ensure complete fairness of the scheme, the virtual run times of all executables would increase simultaneously. But in real systems this phenomenon of simultaneous increment in the virtual run time is not possible since at any point of time, a single process takes hold of the CPU and thus run time of that particular running process gets modified.

For real systems, the scheduler maintains a time line of the respective virtual run times of processes that are ready to run. Consider a queue of processes ordered in increasing sequence from left to right. The leftmost entry in the sequence is the process that is least favored since it has been waiting for greatest interval while the process on the rightmost in the ordered sequence is the most favored process because it has been serviced recently. The scheduler defines *unfairness* as the difference in the virtual run times of the rightmost and the leftmost entries in the sequence as $\Delta\tau = \tau_{right} - \tau_{left}$ and the scheduler ensures fairness by bounding the unfairness value ($\Delta\tau$) at any point of time to be lesser than a predefined maximum unfairness ($\Delta\tau_{max}$). The policy of CFS is such that the scheduler preempts the rightmost process as soon as its virtual run time exceeds the maximum unfairness value and in turn activates the leftmost process in the time line. This process is repeated to provide all running processes a fair share of CPU.

Similar to the running processes, if a process blocks itself at virtual run time τ_{block}, the virtual run time of the process on unblocking ($\tau_{unblock}$) itself must be initialized to a value such that it is serviced as soon as possible. So initialization of $\tau_{unblock}$ is made such that the unfairness amount $\Delta\tau$ exceed $\Delta\tau_{max}$. Thus the scheduler immediately preempts the running process and allows the unblocked process to perform the events for which it was waiting. The *sleeper fairness* criteria of the scheduler says that if the process sleeps for a significant amount of time such that $\tau_{block} < \tau_{right} - \Delta\tau_{max}$, then the blocking process on being unblocked is initialized as $\tau_{unblock} \leq \tau_{right} - \Delta\tau_{max}$ and the process eventually occupies the leftmost position in the time line. For the generic case, the scheduler policy maintains that the virtual run time for an unblocking process is initialized to a value ($\tau_{unblock}$) which is greater than or equal to the virtual run time when it blocked itself (τ_{block}). Thus, $\tau_{unblock} = max(\tau_{block}, \tau_{right} - \Delta\tau_{max})$.

8.5.2 Denial of Service Exploiting Completely Fair Scheduler

Bangerter et al. in [10] exploits the sleeper fairness of the CFS to surface an attack on the encryption process. The malicious attacker observes the timing measurements of cache accesses via a multithreaded spy process $\mathcal{S}$. The spy process $\mathcal{S}$ launches large number of threads and the blocking or unblocking of the subsequent threads are scheduled sequentially in a way that the victim process $\mathcal{V}$ gets scheduled between two subsequent thread scheduling. The interval for which the victim process runs is sufficient for making one cache access before it is pre-empted due to the unblocking of the subsequent thread. Thus each cache accesses performed by the victim is monitored by the multiple spy threads at fine grained timing interval. The attack principle is detailed as follows:

1. The spy process activates a large number of threads, where initially the virtual run times of all threads are initialized to a significantly low value by blocking itself for sufficient time. After this initialization all the launched threads performs the timing measurements in a round-robin fashion.
2. At any point of time, thread i determines the memory accesses performed by victim process by observing the access times.
3. Spy thread i before blocking itself computes t_{sleep} and t_{wakeup} which are virtual run times in the time line for blocking of the ith thread and the unblocking time for the $(i + 1)^{th}$ thread. Thread i requests a timer to wakeup thread $i + 1$ at t_{wakeup} before entering into the busy-wait loop until t_{sleep} is reached.
4. The ith thread blocks itself at t_{sleep} and at this point of time there are no spy threads that are ready to run. The $(i + 1)^{th}$ thread is only activated when t_{wakeup} is reached.
5. The scheduler activates the victim process at t_{sleep}. The victim process can at most run for the interval $t_{wakeup} - t_{sleep}$ since at t_{wakeup} the timer expires and unblocks thread $i + 1$. Since the spy threads are activated one after another in order, thread $i + 1$ will be serviced immediately at t_{wakeup} since it is the leftmost thread in the time line.
6. Thus, unblocking of the $(i + 1)^{th}$ thread causes a preemption of victim V and the above scenario can be repeated successively for a large number of spy threads in cache access driven attacks exploiting denial of service for the victim process $\mathcal{V}$.

In this work, $t_{wakeup} - t_{sleep}$ is set as 1500 machine cycles, where only 200 cycles are actually being used by the victim to perform the encryption.

8.5.3 Using Timing as Side Channel for Cache Access Attack

Listing 8.3 is a C code implementation for the timing measurements of cache accesses. The spy process can use this implementation to learn the information about the table look-ups only up to the cache line granularity. The reason behind this is that the requested data is loaded from the main memory to the cache memory in data blocks, equal to the size of cache line. So if any data is requested to be loaded from the main memory location, the entire cache line containing the main memory location gets loaded into the cache memory. The spy process measures the time for each look-up table accesses and determines whether any particular access is a hit or miss by looking at the time taken to access the respective locations.

Listing 8.3 Subroutine used by spy process checking the parts of look-up table that have been accessed by other processes

```
#define CACHELINESIZE 64
#define THRESHOLD 200
unsigned measureflush(void *table, size_t tablesize, uint8_t *
    bitmap) {
  size_t i;
  uint32_t t1,t2;
  unsigned bit, n_hits = 0;
  for(i=0; i< tablesize/CACHELINESIZE; i++) {
    _asm_(" xor %%eax, %%eax    \n"
          " cpuid               \n"
          " rdtsc               \n"
          " mov %%eax, %%edi    \n"
          " mov (%%esi), %%ebx  \n"
          " xor %%eax, %%eax    \n"
          " cpuid               \n"
          " rdtsc               \n"
          " clflush (%%esi)     \n" :
          "=a" (t2),
          "=D" (t1) :
          "S" ((const char*) table + CACHELINESIZE * i) :
          "ebx", "ecx", "edx", "cc");
      bit = (t2 - t1 < THRESHOLD) ? 1 : 0;
      n_hits += bit;
      bitmap[i/8] &= ~(1 << (i%8));
      bitmap[i/8] |= bit << (i%8);
  }
  return n_hits;
}
```

The timing measurements for the respective table locations are treated as following:

1. The *rdtsc* as appears in Sect. 4.2.1 instruction is used to observe the timestamp counter values before and after the look-up accesses.
2. The event cache hits and misses can be distinguished through a thresholding scheme where if a timing measurement is less than a threshold then the spy process decides that it is cache hit, otherwise if the measurement exceeds the threshold then the spy concludes the access to be a miss. Rigorous testing on a desired platform will reveal the actual threshold for a system. To the best of our knowledge, there are no well defined techniques to estimate the threshold accurately.
3. A bitmap is maintained so as to keep track at which location cache hits and misses occurred.

The steps are performed on the entire look-up table for AES for each and every cache lines. Since data from the main memory is accessed in cache block, access attacks can only reveal cache accesses in the granularity of cache lines. The timing information is thus used by the attacker to construct a bitmap containing the information of cache hits and misses.

8.5.4 Denoising by Neural Network

Normally the timing measurements as obtained from the explained procedure are highly affected by noise from other processes running concurrently on the system. The events of cache hit and misses for the victim process get hugely affected by the cache accesses from other running processes in the system. This may result in faulty conclusions regarding the key space. Another drawback of this procedure is that the fine-grained timing measurements are not perfect, thus single timing measurements may lead to misleading results.

In this work, Bangerter et al. [10] proposes a denoising technique using the artificial neural network (ANN) which is used efficiently to improve the success of the attack. The inputs to the ANN is bitmap which represents cache hits and misses for the respective cache access across time. The ANN filters out the noise and outputs the probabilities for each memory accesses being actually performed by the victim process.

8.6 Conclusion

The chapter discussed strategies for access-driven timing attacks, which are capable of identifying individual cache accesses made by the cipher. Several attacks on block cipher were discussed. The attacks are immensely benefited by multithreaded CPU environments. Further, by exploiting CPU scheduler policies, every memory access made by the cipher can be determined.

References

1. Wray JC (1991) An analysis of covert timing channels in research in security and privacy. In: Proceedings of the 1991 IEEE Computer Society Symposium, May 1991, pp 2–7
2. Hu W-M (1992) Lattice scheduling and covert channels. In: Proceedings of the 1992 IEEE symposium on security and privacy. SP '92. IEEE Computer Society, Washington DC, pp 52–61
3. Percival C (2005) Cache missing for fun and profit. In: Proceedings of BSDCan 2005
4. Osvik DA, Shamir A, Tromer E (2005) Cache attacks and countermeasures: the case of AES. Cryptology ePrint Archive, Report 2005/271
5. Tromer E, Osvik DA, Shamir A (2010) Efficient cache attacks on AES, and countermeasures. J Cryptol 23(2):37–71
6. Osvik DA, Shamir A, Tromer E (2006) Cache attacks and countermeasures: the case of AES. In: Pointcheval D (ed) CT-RSA. Lecture notes in computer science, vol 3860. Springer, Berlin, pp 1–20
7. Bogdanov A, Eisenbarth T, Paar C, Wienecke M (2010) Differential cache-collision timing attacks on AES with applications to embedded CPUs. In: Pieprzyk J (ed) CT-RSA. Lecture notes in computer science, vol 5985. Springer, Berlin, pp 235–251
8. Neve M, Seifert J-P (2006) Advances on access-driven cache attacks on AES. In: Biham E, Youssef AM (eds) Selected areas in cryptography. Lecture notes in computer science, vol 4356. Springer, Berlin, pp 147–162
9. Aciiçmez O, Schindler W (2008) A vulnerability in RSA implementations due to instruction cache analysis and its demonstration on OpenSSL. In: Malkin T (ed) CT-RSA. Lecture notes in computer science, vol 4964. Springer, Berlin, pp 256–273
10. Gullasch D, Bangerter E, Krenn S (2011) Cache games—bringing access-based cache attacks on AES to practice. In: IEEE symposium on security and privacy. IEEE Computer Society, Washington DC, pp 490–505
11. Ristenpart T, Tromer E, Shacham H, Savage S (2009) Hey, you, get off of my cloud: exploring information leakage in third-party compute clouds. In: Al-Shaer E, Jha S, Keromytis AD (eds) ACM conference on computer and communications security. ACM, New York, pp 199–212
12. Tsafrir D, Etsion Y, Feitelson DG (2007) Secretly monopolizing the CPU without superuser privileges. In: Proceedings of 16th USENIX security symposium. SS'07. USENIX Association, Berkeley, pp 17:1–17:18. http://dl.acm.org/citation.cfm?id=1362903.1362920. Accessed March 2013

Chapter 9
Branch Prediction Attacks

So far we have looked at timing channels that arise due to cache memories in the system. Cache memories leak information about the memory accesses made by a cipher. In this chapter, we look at information leakage due to branch instructions. If a cipher implementation uses a conditional branch that depends on the secret key, then information about the key can leak through the processor's branch predictor. A misprediction causes the execution to take considerably more time compared with predictions that are correct. This variation in execution time is exploited by attackers to determine bits of the secret key. Several timing attacks based on branch prediction have been proposed. Unlike cache attacks, which are mostly effective on block ciphers, branch prediction attacks are applicable only on public key ciphers such as the Rivest, Shamir, and Adleman (RSA) algorithm. This chapter begins with a review of the RSA implementation before discussing various branch prediction attacks.

9.1 Implementation of RSA

RSA performs modular exponentiation to encrypt or decrypt messages. The encryption of message (M) with the public key e is done as follows: $C = M^e \bmod N$. The corresponding decryption is done using the private key d as follows: $M = C^d \bmod N$. For details about RSA, please refer to Sect. 2.5. In this section, we review how the exponentiation is implemented.

9.1.1 Square and Multiply Exponentiation Algorithm

The square and multiply algorithm performs a squaring at each step, while the multiplication operation is performed only for the exponent bits that are set to one. Algorithm 9.1 represents the details. The function returns $S \leftarrow M^d \bmod N$, where the private key d is $(d_0||d_1||d_2||\cdots||d_i||\cdots||d_{n-1})$. Here, d_0 is the most significant bit and d_{n-1} is the least significant bit of d.

C. Rebeiro et al., *Timing Channels in Cryptography*,
DOI 10.1007/978-3-319-12370-7_9

Algorithm 9.1: Binary version of square andmultiply exponentiation algorithm

```
Input: M
Output: S
begin
    S ← M
    for i from 1 to n − 1 do
        S ← S ∗ S  mod N
        if d_i = 1 then
            S ← S ∗ M  mod N
        end
    end
    return S
end
```

This algorithm is unbalanced because multiplication is performed only when $d_i = 1$. Common side channels such as power and timing are capable of identifying this conditional execution. Simple power attacks (SPA) and timing attacks exploit this conditional instruction execution retrieving the bits in d that are 1.

9.1.2 Balanced Montgomery Powering Ladder Implementation

A modification to the square and multiply algorithm protects against obvious side-channel leakages. This is called the Montgomery ladder algorithm. Algorithm 9.2 shows this modification.

The Montgomery ladder performs the entire exponentiation by alternatively modifying the values of two variables that depend on the exponent bits. Algorithm 9.2 has both "if" and "else" statements and every iteration has either of the corresponding code executed. Unlike the square and multiply algorithm, here the number of multiplications performed is always a constant and independent of the value of the key. This balanced mode of operation eliminates obvious side-channel leakage.

9.1.3 Montgomery Multiplication

The integer division required for the modular multiplication and squaring in Algorithms 9.1 and 9.2 is the most time consuming. Montgomery multiplication provides a means to avoid this. For two multiplicands a and b, the algorithm computes $(a * b) \bmod N$. There are four steps in the multiplication.

1. First, an integer R is found such that $gcd(R, N) = 1$. If the RSA modulus N is a k-bit number, then R is generally taken to be 2^k. The extended Euclidean algorithm is used to determine the inverse of R and N, denoted R^{-1} and N^{-1} respectively. Thus $R * R^{-1} = 1$ and $N * N^{-1} = 1$. This has to be done only once before any encryption can start.

Algorithm 9.2: Montgomery Ladder Algorithm

```
begin
    R0 ← 1
    R1 ← M
    for i from 0 to n − 1 do
        if d_i = 0 then
            R1 ← (R0 ∗ R1) mod N
            R0 ← (R0 ∗ R0) mod N
        end
        else if d_i = 1 then
            R0 ← (R0 ∗ R1) mod N
            R1 ← (R1 ∗ R1) mod N
        end
    end
    return R0
end
```

2. R is used to convert the multiplicands into their Montgomery domain as follows:

$$A = (a * R) \bmod N \qquad (9.1)$$
$$B = (b * R) \bmod N.$$

3. The next step is to perform the Montgomery multiplication. This is shown in Algorithm 9.3. This involves three multiplications and some less expensive operations such as addition and right shifts (assuming R is a power of 2, division by R is performed by shifting right). The multiplication $S * N^{-1}$ is not too expensive either, because $\bmod R$ implies that only the least significant R bits need to be considered.
4. The inverse Montgomery transformation is then performed to convert the result to an ordinary integer,

$$z = S * R^{-1} \bmod N. \qquad (9.2)$$

From the side-channel perspective, there is an extra reduction step in 4th line of Algorithm 9.3. Due to the conditional branch statement, variations in the execution time will occur. These variations are the source of side-channel exploits. The next section demonstrates this leakage through the processor's branch predictor.

9.2 Timing Branch Mispredictions

The branch prediction unit in the processor is responsible for predicting whether a branch in a program is taken or not-taken. If a branch is predicted taken, instructions from the destination address are fetched. On the other hand, a not-taken prediction causes the CPU to continue fetching instructions in a sequential fashion. A prediction

Algorithm 9.3: Montgomery Multiplication Algorithm

```
begin
    S ← A ∗ B
    S ← (S + (S ∗ N^{-1} mod R) ∗ N)/R
    if S > N then
        S ← S − N
    end
    return S
end
```

that goes wrong is called a *misprediction*. It causes the pipeline to be flushed and instructions fetched from the correct destination. This incurs a significant overhead in superscalar processors. The difference in execution time between a correct and a wrong branch prediction is used to break ciphers such as the RSA. In this section, we demonstrate the timing difference caused by mispredictions.

We take the example of a Montgomery multiplication implemented to return the square of its input. As seen in Algorithm 9.3, the Montgomery multiplication has an extra reduction step which is executed only if $S > N$. We determine the execution time difference caused by the mispredictions that occur due to this condition in the algorithm.

The experiment performs the operation $a^2 \bmod N$ for an input a using the Montgomery multiplication algorithm. Listing 9.1 shows the pseudocode. The modulus N is a 1024-bit integer similar to the modulus used in RSA. Since this involves numbers much bigger than what normal CPU registers can hold, GNU Multiple Precision Arithmetic Library is used to perform the big number arithmetic. R is chosen to be 2^{1025}, which is greater than N and the greatest common divisor (GCD) of R and N is 1. Around 10,000 randomly chosen inputs were considered within the range 0 to $N - 1$. Since there is a predominant effect of noise, we determine the average execution time for each input a from 1000 runs.

Listing 9.1 Pseudocode for modular squaring using Montgomery multiplication

```
1  int sqmult(int a, int R, int N)
2  {
3      A = (a * R) % N;
4
5      S = A * A;
6      S = (S + (((S * N^{-1}) % R) * N)) / R;
7      if (S > N)
8        S = S - N;
9      return S;
10
11 }
```

The conditional branch instruction occurs due to the *if* statement in line 8 of the listing. This causes two different execution paths:

- Branch *not-taken* path occurs when $S > N$. This causes the execution of the subtraction $S \leftarrow S - N$. In this case, there is no deviation from the sequential execution of the program.
- Branch *taken* path occurs when $S \leq N$. No subtraction is done and the execution jumps to the return statement following the "if" block. This is the branch target statement.

A part of the assembly code corresponding to Listing 9.1 is illustrated in Listing 9.2 for better understanding. The function call to the GMP function `__gmpz_cmp` makes the comparison between *S* and *N*. On its return, the `eax` register contains a positive, negative, or zero value corresponding to `A > N`, `A < N`, and `A = N`, respectively. The instruction `testl` compares this value with zero, while `jle` branches to the label `L4` if `eax` is less than or equal to zero, thereby not performing the subtraction. If `eax` is greater than zero, then the subtraction (using function `__gmpz_sub`) is performed.

Listing 9.2 Partial Assembly Code for the Squaring Function (sqmult)

```
    movq  %rdx, %rsi
    movq  %rax, %rdi
    call  __gmpz_cmp
    testl %eax, %eax
    jle .L4
    movl  $N, %edx
    movl  $S, %esi
    movl  $S, %edi
    call  __gmpz_sub
.L4:
    leave
    .cfi_def_cfa 7, 8
    ret
    .cfi_endproc
```

Since the inputs *a* were randomly chosen, some cause the branch to be taken while others do not. We classify the execution time into two sets depending on whether the branch was taken or not-taken. Figure 9.1 shows the average execution time for the function corresponding to inputs where the branch was not taken and Fig. 9.2 shows the average execution time where the branch was taken.

In both graphs, we see a separation in the execution time. The regions of the graphs with average execution time between 11,000 and 11,500 correspond to inputs where the branch prediction was correct and the regions above 12,000 correspond to the mispredictions. A clear separation is observed in both graphs. This experiment will be helpful for understanding the attack algorithms exploiting the branch mispredictions. The following sections discuss the details of these attacks.

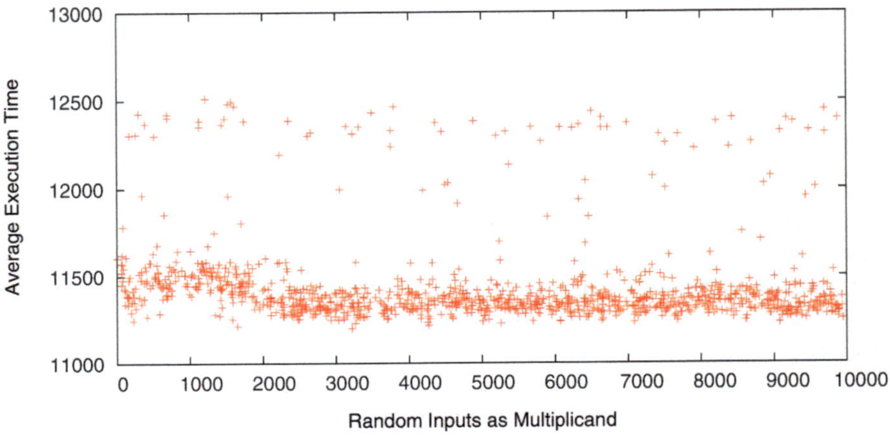

Fig. 9.1 Execution time from not-taken branches in Montgomery multiplication (on Intel i5). The *lower band* (around 11,500 clock cycles) indicates runs where the branches were correctly predicted not-taken. The *upper band* (around 12,500 clock cycles) indicates mispredictions

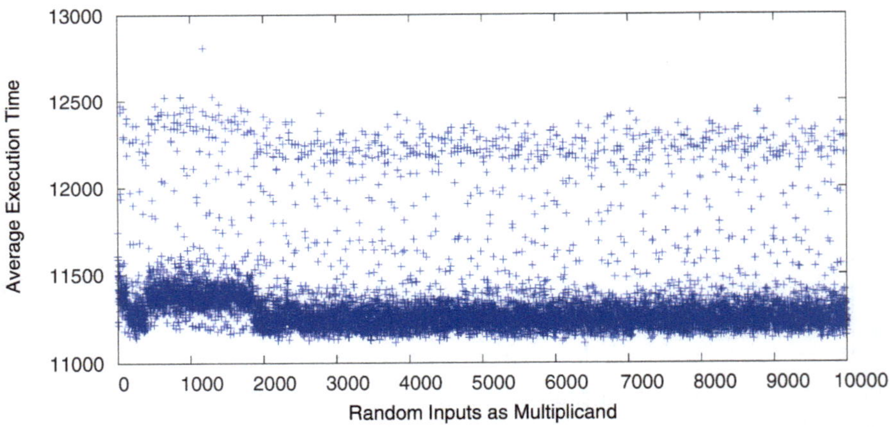

Fig. 9.2 Execution time from taken branches in Montgomery multiplication (on Intel i5). The *lower band* (around 11,000–11,500 clock cycles) indicates runs where the branches were correctly predicted taken. The *upper band* (around 12,500 clock cycles) indicates mispredictions

9.3 Attacking the Square and Multiply Exponentiation Algorithm

In this section, we show how timing mispredictions can be used to recover the secret exponent in the square and multiply algorithm (Algorithm 9.1) if the Montgomery multiplication algorithm is used. We assume that the adversary can invoke the function implementing the algorithm with different values of M and monitor the execution time. She however cannot determine the intermediate and any other internal values of the function.

The most significant bit of the exponent (i.e., d_0) in Algorithm 9.1 is set to one. The attacker determines all other bits iteratively. That is, bit d_i is determined only after d_{i-1} is found. The process to determine bit d_i is as follows:

Assumptions: The attacker knows the branch prediction algorithm used in the attack system. She also knows the initial state of the branch predictor.

Offline Phase: The attacker selects a large set of plaintexts $\mathcal{M}$ and then performs the following operations:

- Assumes that $d_i = 1$ and forms an exponent as follows: $d^{(1)} = d_0||d_1||d_2|| \cdots ||d_{i-1}||1$. The guess of $d_i = 1$ causes the multiplication in line 6 of Algorithm 9.1 to be performed in the ith iteration.
- Now, for each plaintext in $\mathcal{M}$, the attacker simulates Algorithm 9.1 to determine if there is a misprediction in the squaring in the $(i+1)^{th}$ iteration. Such an analysis is possible because the attacker knows how the branch predictor works.
- Based on whether there is a correct or wrong prediction in the Montgomery multiplier (MM), the set of plaintexts $\mathcal{M}$ can be partitioned into two sets: $\mathcal{M}_1$ and $\mathcal{M}_2$. These sets are defined as follows:
  - $\mathcal{M}_1 = \{m | m$ causes a misprediction during MM of $(i+1)^{th}$ squaring if $d_i = 1\}$
  - $\mathcal{M}_2 = \{m | m$ does not cause a misprediction during MM of $(i+1)^{th}$ squaring if $d_i = 1\}$
- The attacker now assumes $d_i = 0$ and forms an exponent as follows: $d^{(0)} = d_0||d_1||d_2|| \cdots ||d_{i-1}||0$. The above steps are repeated to partition the set of plaintexts $\mathcal{M}$ into two sets, $\mathcal{M}_3$ and $\mathcal{M}_4$. These sets are defined as follows:
  - $\mathcal{M}_3 = \{m | m$ causes a misprediction during MM of $(i+1)^{th}$ squaring if $d_i = 0\}$
  - $\mathcal{M}_4 = \{m | m$ does not cause a misprediction during MM of $(i+1)^{th}$ squaring if $d_i = 0\}$

Now equipped with four sets of plaintexts $\mathcal{M}_1$, $\mathcal{M}_2$, $\mathcal{M}_3$, and $\mathcal{M}_4$, the attacker is all set for the online phase.

Online Phase: Corresponding to each set $\mathcal{M}_i$ $(1 \leq i \leq 4)$, the attacker determines four average execution times for Algorithm 9.1. Since the ith bit of the exponent is either 0 or 1, two of the sets will show a separation in timing. For instance, consider the correct value of d_i is 1. This means,

- All plaintexts in $\mathcal{M}_1$ have a misprediction during MM of $(i+1)^{th}$ squaring. Thus, these plaintexts would correspond to a longer time.
- All plaintexts in $\mathcal{M}_2$ have no misprediction during MM of $(i+1)^{th}$ squaring. Thus, these plaintexts would correspond to a shorter time.
- In both $\mathcal{M}_3$ and $\mathcal{M}_4$, there are some plaintexts that would have a misprediction during MM of $(i+1)^{th}$ squaring, while other plaintexts in these sets will not have the misprediction.

When the average time is computed, there would be a separation between the time for $\mathcal{M}_1$ and $\mathcal{M}_2$ but no (or lesser) separation between $\mathcal{M}_3$ and $\mathcal{M}_4$. Algorithm 9.4 shows the details of the attack.

Algorithm 9.4: Determining bit d_i of the exponent

Output: M, M_1, M_2, M_3, M_4
Output: d_i
begin
  $C_j \leftarrow 0$ for $1 \le j \le 4$
  $T_j \leftarrow 0$ for $1 \le j \le 4$
  for *at-least* 2^{16} *times* **do**
    $\mathbf{x} \leftarrow$ choose random plaintext from $\mathbf{M}$
    $t_1 \leftarrow$ start time
    $ScalarMultiply(\mathbf{x})$
    $t_2 \leftarrow$ end time
    for $k \in \{1,2,3,4\}$ **do**
      if $\mathbf{x} \in M_k$ **then**
        $C_k \leftarrow C_k + 1$
        $T_k \leftarrow T_k + (t_2 - t_1)$
      end
    end
  end
  $A_j \leftarrow T_j / C_j$ for $1 \le j \le 4$
  if $(A_1 - A_2) > (A_3 - A_4)$ **then**
    return 1
  end
  else
    return 0
  end
end

In the algorithm, a random plaintext $\mathbf{x}$ is chosen from $\mathcal{M}$ and the scalar multiplication invoked and timed. The measured time is added to all sets where $\mathbf{x}$ is present. Since $\mathcal{M} = \mathcal{M}_1 \cup \mathcal{M}_2$ and $\mathcal{M} = \mathcal{M}_3 \cup \mathcal{M}_4$, $\mathbf{x}$ would be present in exactly two of the sets.

The average time $\mathcal{A}_i$ is found for each set and then the separation is determined. If there is more separation between $\mathcal{A}_1$ and $\mathcal{A}_2$, then d_i is 1, otherwise d_i is 0.

Evaluation: There are two limitations of this attack technique. First, the attacker needs to know the branch predictor algorithm present in the system. Second, she has to know the initial state of the branch predictor. Both these requirements are needed to partition $\mathcal{M}$ into sets $\mathcal{M}_1$, $\mathcal{M}_2$, $\mathcal{M}_3$, and $\mathcal{M}_4$.

Attackers may be able to tackle the initial state requirement of the predictor, for instance, by performing dummy operations that force the predictor to a particular state—such as branch taken. The first requirement however is more difficult to fulfill especially for commercially available processors. Branch prediction algorithms in modern commercial processors are a closely guarded secret and most manufactures do not reveal their internals. Thus, attackers would only be able to make smart guesses about the algorithm or use approximations.

In the experiments we consider an Intel i5 machine. We do not know the branch prediction internals in the i5 but consider the 2-bit bimodal algorithm (see

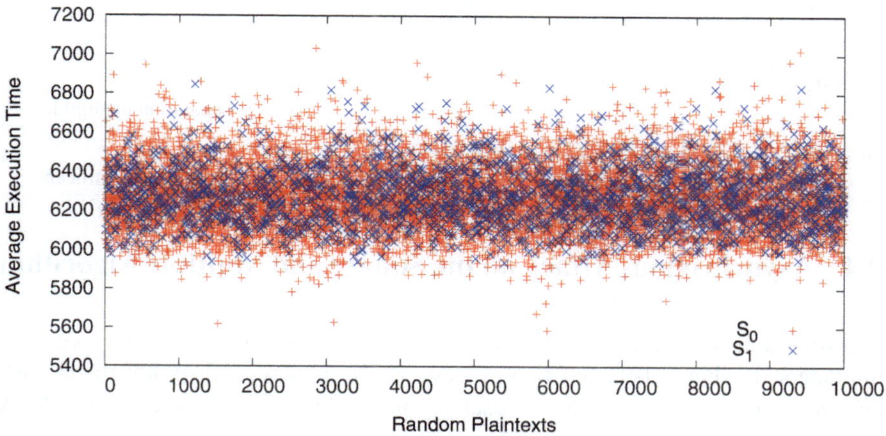

Fig. 9.3 Timing variation due to plaintexts for sets $\mathcal{M}_3$ and $\mathcal{M}_4$ when $d_i = 1$—the wrong guess

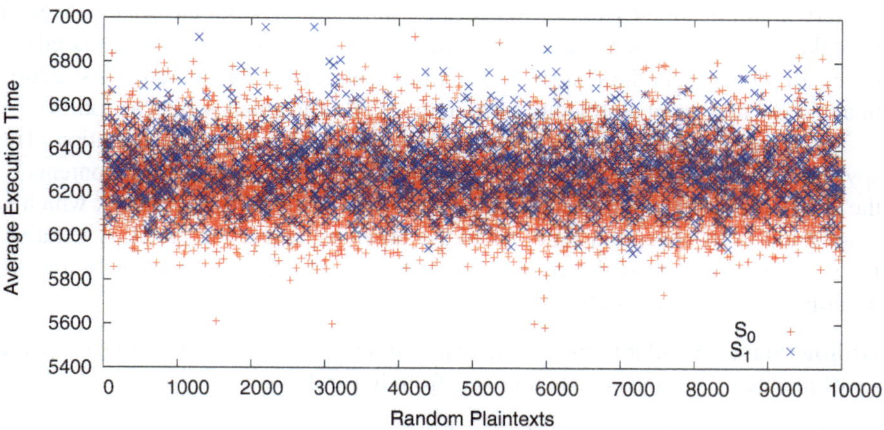

Fig. 9.4 Taming variation for random plaintexts for sets $\mathcal{M}_1$ and $\mathcal{M}_2$ when $d_i = 1$—the correct guess

Sect. 3.3.2.2), which we assume is a close match. Figure 9.3 shows the average execution time for the exponentiation (Algorithm 9.1) for plaintexts $\mathcal{M}_3$ and $\mathcal{M}_4$ and Fig. 9.4 shows the average execution time for plaintexts in $\mathcal{M}_1$ and $\mathcal{M}_2$.

Table 9.1 tabulates the difference in the average separation between the sets. It can be observed that the difference in timing for the wrong assumption is small whereas the difference in timing for the plaintext partitions for the correct guess of d_i is significantly large. Thus, in spite of the assumed predictor, the attack was able to correctly identify d_i as 1. We expect that as better models for the predictor emerge, this attack's success would improve considerably. Alternatively, the attacker can use other attack strategies. Some of these strategies are discussed in the following sections.

Table 9.1 Separation in timing for assumptions $d_i = 0$ or $d_i = 1$

Assumption	Separation (in clock cycles)	
$d_i = 0$	$(\mathcal{A}_3 - \mathcal{A}_4) = 6278 - 6271 = 7$	*Wrong guess* Average separation negligible
$d_i = 1$	$(\mathcal{A}_1 - \mathcal{A}_2) = 6312 - 6257 = 55$	*Correct guess* Average separation significant

9.4 Asynchronous Attack on the Square and Multiply Algorithm

In a simultaneous multithreading (SMT) environment, a single processor core is simultaneously shared between two processes. The processor does not distinguish between processes from different users. This could lead to attacks on the RSA exponentiation using the square and multiply algorithm (Algorithm 9.1).

The asynchronous attack we describe now does not require knowledge of the branch predictor algorithm or its initial state. It uses the fact that when the Montgomery multiplication (Sect. 9.1.3) executes, information about the branch taken or not-taken is stored in the branch target buffer (BTB). The BTB is a cache shared between all processes in the system. It is of limited size, therefore a branch instruction in a process may evict a previous entry in the BTB. The attack works as follows:

The adversary runs a spy process simultaneously with the exponentiation. The spy process uses dummy branch instructions to continuously evict the contents of the BTB. This induces mispredictions in the square and multiply process, which is manifested in its execution time. As in the previous attack, the attack is iterative. Bit d_i in the exponent is determined only after bits d_0–d_{i-1} are known. Below we describe how the bit d_i is obtained.

Offline Phase: Similar to the offline phase in Sect. 9.3, the attacker selects a large set of plaintexts $\mathcal{M}$ and creates four sets from it as follows:

- $\mathcal{M}_1 = \{m | m \in \mathcal{M}$ causes the branch to be taken during MM of $(i+1)^{th}$ squaring if $d_i = 1\}$
- $\mathcal{M}_2 = \{m | m \in \mathcal{M}$ causes the branch not to be taken during MM of $(i+1)^{th}$ squaring if $d_i = 1\}$
- $\mathcal{M}_3 = \{m | m \in \mathcal{M}$ causes the branch to be taken during MM of $(i+1)^{th}$ squaring if $d_i = 0\}$
- $\mathcal{M}_4 = \{m | m \in \mathcal{M}$ causes the branch not to be taken during MM of $(i+1)^{th}$ squaring if $d_i = 0\}$

As in the attack in Sect. 9.3, $\mathcal{M}_1$ and $\mathcal{M}_2$ partition $\mathcal{M}$. Similarly, $\mathcal{M}_3$ and $\mathcal{M}_4$ partition $\mathcal{M}$. However, there is a subtle difference between the two offline phases. The attack in Sect. 9.3 required the partitions to be based on correct and wrong branch predictions. However, in the asynchronous attack described here, the partitions are based on whether the branch is taken or not taken in the $(i+1)^{th}$ iteration.

Online Phase: In the online phase, the attacker executes a spy process simultaneously with the cipher in a multithreaded environment and times the execution of the exponentiation. The spy is capable of evicting the contents of the BTB. Since the BTB stores information about the branch targets of the Montgomery multiplication used during the exponentiation, evicting the BTB would evict these branch target entries.

The aim of the spy is to evict the BTB entries just before the $(i+1)^{th}$ iteration. In the $(i+1)^{th}$ iteration, when the CPU executes the conditional reduction branch instruction in the Montgomery squaring, it is unable to find the branch target address in the BTB. This may result in a compulsory misprediction if the branch is taken.

Depending on which set the plaintext belongs to and the value of d_i, the following will happen. Let us say, $d_i = 1$, then

- All plaintexts in $\mathcal{M}_1$ take the branch in the $(i+1)^{th}$ squaring. Therefore, these plaintexts would cause a compulsory misprediction. Thus the execution time is longer.
- All plaintexts in $\mathcal{M}_2$ do not take the branch in the $(i+1)^{th}$ squaring. Therefore, these plaintexts would not have a misprediction (since we assume the default is branch not taken). Thus, the execution time is shorter.
- In both $\mathcal{M}_3$ and $\mathcal{M}_4$, there are some plaintexts that take the branch while others do not. Thus, the forced misprediction does not occur all the time.

When the average time is computed, there would be a separation between the time for $\mathcal{M}_1$ and $\mathcal{M}_2$ but no (or lesser) separation between $\mathcal{M}_3$ and $\mathcal{M}_4$.

If $d_i = 0$, the separation in the average execution time is expected to occur between $\mathcal{M}_3$ and $\mathcal{M}_4$ instead of $\mathcal{M}_1$ and $\mathcal{M}_2$. The attack process is exactly similar to Algorithm 9.4.

Evaluation: The critical factor in the entire attack is the spy process that evicts the BTB. The spy needs to flush the BTB just before the squaring in the $(i+1)^{th}$ iteration. Since the exact instant when the $(i+1)^{th}$ iteration is executed is not known, the spy has to periodically flush the BTB and hope that a flush happens before the target iteration.

The resolution of clearing a BTB has a major impact in determining the attack's success. Either the BTB can be evicted completely or partially. Each has its own advantages.

- *Total eviction.* In this method, the entire BTB is cleared by the adversary. This is the easiest eviction strategy because the knowledge of the target destination address in the Montgomery multiplication is not required. However in practice, clearing the entire BTB between two consecutive squaring operations is unrealistic since it requires more time than performing the operations between two consecutive squarings. To improve the performance of the attack, the adversary can evict the BTB partially.
- *Partial eviction.* The adversary continuously clears only a part of the BTB that stores the BTB set of the target address of the branch taken. In the extreme case, the adversary only clears the particular set of the BTB that stores the target address

of the target branch. In these cases, the adversary needs to know how the branch targets get mapped in the BTB.

In this attack, it is assumed that the adversary does not have any means to control a particular misprediction event since the synchronization between the spy and the attacker was deliberately avoided. In the next section, a synchronized version of the explained attack is presented.

9.5 Synchronous Attack on the Square and Multiply Algorithm

The main idea of the synchronous attack that reveals d_i are the following observations:

1. In the ith iteration of Algorithm 9.1, if $d_i = 1$, then the branch is not taken and the multiplication in line 5 is done. If $d_i = 0$, the branch is taken and the multiplication is skipped.
2. If the processor is forced to predict branch not taken, $d_i = 0$ will cause a misprediction, thus resulting in a pipeline flush and an increase in execution time. On the other hand, $d_i = 1$ will not cause any misprediction, therefore no increase in execution time.
3. The adversary forces the branch not taken state in the processor by evicting the branch target in the BTB just before the ith iteration.

The attack works as follows:

- The attacker chooses an M and determines the execution time for Algorithm 9.1, say T_1.
- The execution is run again with the same M and the execution time is measured. Let T_2 be the time. The only difference compared to the first execution is that just before the ith iteration, the adversary clears a single target location of the BTB corresponding to the branch instruction.
- if $(|T_2 - T_1|)$ is greater than some threshold τ, then $d_i = 0$ else $d_i = 1$.

In all the attacks discussed so far, the attacker measures the execution time of the misprediction while running an implementation of the square and multiply algorithm. In the following section, a new attack strategy is explored which is asynchronous in nature and measures the execution time of the branches of a spy process.

9.6 Trace Driven Attack Targeting the BTB

This attack is similar to that of the prime and probe attack on cache memories (Sect. 8.1). Just like the prime and probe attack, the adversary executes a spy program. The spy has a series of branch instructions in a loop, which are executed and timed. The execution time for the branch instructions would initially be large due to several mispredictions that occur. The execution time would reduce gradually as the BTB gets filled. When a cipher executes simultaneously with the spy, some of the BTB entries of the spy get evicted. This forces branch mispredictions in the spy causing an increase in the execution time. The increase in time leaks information about the cipher's secret key. The steps in the attack are enumerated below:

- The adversary initiates the attack by allowing the spy process to execute a number of branch statements that gradually fill the BTB with branch target addresses from the spy process.
- After a while, the cipher is allowed to execute.
- When the exponentiation code in the cipher encounters a branch, the CPU will not find the target address of the branch in the BTB. The prediction then defaults to *not-taken*.
- If $d_i = 0$, the conditional branch in the square and multiply algorithm (Algorithm 9.1) results to *taken* and one of the entries in the BTB corresponding to the spy process is evicted in order to accommodate the new target destination. If $d_i = 1$, then there is no misprediction, thus no entries in the BTB get evicted.
- After the cipher completes execution, the spy reexecutes its branches and measures the time required. If the branches take a longer time than usual, it is attributed to the mispredictions in the BTB due to the cipher's evictions. Thus, d_i is taken 0. If there is no change in the execution time of the branches, it is attributed to no mispredictions and d_i is taken 1.

9.7 Conclusion

Branches in programs can lead to information leakage due to the branch prediction units present in modern processors. Eliminating branches from programs is not always possible, thus alternate approaches to counter branch prediction based timing attacks need to be determined. The next chapter provides an overview of the countermeasures available for timing attacks. Many of these countermeasures can be applied or tuned to prevent branch prediction attacks.

References

1. Aciiçmez O, Koç ÇK, Seifert JP (2006) On the power of simple branch prediction analysis. IACR Cryptology ePrint Archive, vol 2006, p 351
2. Aciiçmez O, Gueron S, Seifert JP (2007) New branch prediction vulnerabilities in OpenSSL and necessary software countermeasures. In: Galbraith SD (ed) IMA international conference on security lecture notes in computer science, vol 4887. Springer, pp 185–203
3. Aciiçmez O, Koç ÇK, Seifert JP (2007) Predicting secret keys via branch prediction. In: Abe M (ed) CT-RSA, ser. lecture notes in computer science, vol 4377. Springer, pp 225–242

Chapter 10
Countermeasures for Timing Attacks

The criteria for a timing attack to be successful are the following:

1. The run time behavior of the crypto application should vary depending on the secret key. For instance, the instructions executed or memory locations accessed should depend on the secret. Even the data operated on could also cause run time behavior to vary.
2. These variations in the run time behavior must be manifested through the time required to execute operations.
3. The adversary must have sufficient capabilities to monitor time variations. For instance, the adversary should have access to the system executing the crypto application.
4. A clock source precise enough to capture variations in the run time behavior of the application should be available.
5. A synchronization signal to determine when the application starts and/or completes its processing.
6. In addition to these criteria, the success obtained in the attack largely depends on the number of time measurements made.

An attack would be successful only if all these criteria are met. It is therefore not surprising that countermeasures try to make one or more of these criterias difficult to fulfill. Completely preventing a criterion can be done, for example, by eliminating the cache memory from processors. However, this leads to significant performance degradation, which far supersedes the security gains obtained. Thus, the aim of practical countermeasures is to obtain the right balance between security and performance. The upcoming sections discuss the countermeasures.

10.1 Application Level Countermeasures

Countermeasures in the crypto application are generally easy to apply and provide sufficient levels of security against timing attacks. However, they suffer from significant shortcomings. First, most of the application-level countermeasures are heavy

C. Rebeiro et al., *Timing Channels in Cryptography*,
DOI 10.1007/978-3-319-12370-7_10

and inefficient. Further, all applications require to be patched with the same countermeasures. A system may thus have several applications which are modified with these bulky countermeasures. This adversely affects performance and energy requirements of the system. Neverthe-less, these countermeasures provide a quick fix to prevent timing attacks. We therefore discuss some of these countermeasures here.

10.1.1 Countermeasures Involving Look-Up Tables

In block cipher implementations, key dependent look-up table operations are performed. While these operations improve speed, it leads to vulnerabilities through timing channels. One method to prevent the attack is designing new ciphers which have no s-boxes, for example [1]. For the general class of ciphers that use s-boxes, implementations without look-up tables would have a drastic affect on performance. An alternate implementation without look-up tables is using bitslicing [2, 3]. This allows multiple encryptions to be done concurrently in a single processor. Bitslicing yields high speed encryption, which however is restricted to certain nonfeedback modes of operation, and therefore cannot be universally applied.

An alternate approach is to implement the cipher with appropriately chosen look-up tables or restrict their usage. This approach tries to achieve security with increased performance by either restricting the size of the look-up tables or the accesses made to it.

10.1.1.1 Look-Up Tables in Selected Rounds

Cache attacks on block ciphers discovered so far, rely on collisions in the first few or the last round of the cipher. For instance in Advanced Encryption Standard (AES), cache attacks target collisions in the look-up tables either in the first, second, or the last round. It is only in these rounds that collisions can be exploited by an adversary.

Preventing usage of look-up tables in these "sensitive" rounds would therefore prevent the attack. For instance, a hybrid implementation can be built for AES, which uses look-up tables in all but the first, second, and final round. In these sensitive rounds alternate implementations can be used for the s-box, for example, by logical equations or by using small look-up tables as discussed in the next section. The execution time for these implementations of s-boxes must be independent of the secret key. This approach provides security because there are no timing variations in run-time execution due to the sensitive rounds. For instance, compared to an AES implementation with look-up tables, the overhead of an implementation with two rounds protected is 10.8 compared to 66.7 for an implementation that does not use look-up tables. The rounds implemented without look-up tables shield the implementation against cache attacks, while the look-up tables used in the intermediate rounds boost performance.

10.1.1.2 Small Look-Up Tables

Consider an implementation of a cipher with a look-up table of size $l \cdot 2^{\delta}$, where l is the number of memory blocks occupied by the table and 2^{δ} the number of elements in the table that share the same memory block. The number of bits required to access an element in the table is $\delta + \log_2 l$; $\log_2 l$ bits to select the memory block and δ bits to address the element within the block. Most cache attacks are not able to distinguish between elements that lie in the same block. Assuming that the table is aligned to a memory block, these cache attacks can only retrieve $\log_2 l$ bits leaving the remaining δ bits uncertain. The uncertainty can be increased by packing more elements into a memory block thereby reducing l.

In the extreme, if the entire look-up table fits in a single cache line, then the adversary cannot ascertain any information from the memory access patterns. In such cases $\log_2 l = 0$, that is the adversary retrieves 0 bits. The problem however is that, s-boxes used in most ciphers are considerably larger than the memory block. Therefore compressing them into a single block is not easy. In [4] and [5], techniques that compress s-boxes of standard ciphers like CLEFIA into a single memory block were proposed. One technique that is especially suited for s-boxes based on the multiplicative inverse in a finite field uses composite field isomorphisms [6]. This allows the use of smaller tables which stores subfield operations. For instance the AES s-box, which typically requires atleast 256 bytes, can be compressed into a single memory block of 32 or 4 bytes.

10.1.1.3 Cache Warming

Loading the contents of the look-up tables prior to encryption could reduce the risk of cache attacks. As a result, the accesses to the look-up table during the cipher execution would be cache hits. However, this alone is not sufficient to thwart attacks. Conflict misses that occur during the execution may evict the look-up table entries from the cache, potentially resulting in cache misses that leak information about the secret key. The system should therefore ensure that elements in the look-up table are not evicted during the encryption. This is not easy to satisfy.

10.1.1.4 Choosing Look-Up Tables

The number and size of the look-up tables play an important role in an attack's success. Choosing appropriate number and size of look-up tables can therefore be used to mitigate time-driven cache attacks. For a specific system, Chap. 6 shows how appropriate look-up tables can be chosen without compromising on the performance.

10.1.2 Data-Oblivious Memory Access Pattern

This allows the pattern of memory accesses to be performed in an order that is oblivious to the data passing through the algorithm. To a certain extent, modern microprocessors provide such data oblivious memory accesses by reordering instructions. For example, four memory accesses to different locations say, A, B, C, and D, would get executed in any of the four ways (say *BCDA* or *DCAB*). This reordering would increase the difficulty of certain time-driven and access-driven attacks. However, the reordering is restricted to memory accesses which do not have data dependencies. For block ciphers, such independent memory accesses are present within a round but not across rounds, therefore only partially fulfills data-oblivious requirements.

A naïve method to attain complete data-oblivious memory accesses is to read elements from every memory block of the table, in a fixed order, and use just the one needed. Another option is to add noise to the memory access pattern by adding spurious accesses, for example, by performing a dummy encryption in parallel with the real one. A third option is to mask the sensitive table accesses with random masks that are stripped away at the end of the encryption. Alternatively, the table can be permuted regularly to thwart attacks using statistical analysis.

Generic program transformations are also present for hiding memory accesses [7]. However, there are huge overheads in performance and memory requirements. More practical proposals have been developed using shuffling [8] and permutations [9].

10.1.3 Constant and Random Time Implementations

A combination of the countermeasures discussed so far can be used to deliver constant time implementations [10, 11]. To obtain such implementations requires cache warming to ensure that all memory accesses result in cache hits. A timing probe which measures the current execution time for the cipher, and a compensation loop which is designed to increase the encryption time until the worst case execution time is obtained. Additionally, to prevent evictions of the table occurring during an interrupt, an interrupt detector is required to rewarm the cache.

As opposed to constant time implementations, adding a random delay to each execution time can be used to increase the attack difficulty. The countermeasure was first proposed for DPA in [12]. Tunstall and Benoit in [13] and later Coron and Kizhvatov in [14, 15] provide improvements by modifying the distribution of the delay.

10.2 Countermeasures Applied in the Hardware

Subtle changes in the hardware can make attacks much more difficult with little overhead in the performance. In this section we survey some of the hardware modifications that have been suggested to counter cache attacks.

10.2.1 Noncached Memory Accesses

Certain memory pages can be flagged as noncacheable. This would prevent memory accesses from loading data into the cache. Consequently, every memory access would read data from the RAM and every access would thereby be a cache miss preventing all cache attacks.

10.2.2 Specialized Cache Designs

Specialized cache memory designs have been proposed for thwarting cache attacks. They work on the fact that information leakage is due to sharing of cache resources, thus leading to *cache interference*. These solutions provide means of preventing access-driven attacks. Their effectiveness in blocking time-driven attacks has not yet been analyzed. In [16], Percival suggests eliminating cache interference by modifying the cache eviction algorithms. The modified eviction algorithms would minimize the extent to which one thread can evict data from another thread.

In [17], Page proposed to partition cache memory, which is a direct-mapped *partitioned cache* that can dynamically be partitioned into protected regions by the use of specialized cache management instructions. By modifying the instruction set architecture and tagging memory accesses with partition identifiers, each memory access is hashed into a dedicated partition. Such cache management instructions are however only available in the operating system. While this technique prevents cache interference from multiple processes, the cache memory is underutilized due to rigid partitions. For example, a process may use very few cache lines of its partition, but the unused cache lines are not available to another process.

In [18], Wang and Lee provide an improvement on the work by Page using a construct called *partition-locked cache* (PLCache), where the cache lines of interest are locked in cache, thereby creating a private partition. These locked cache lines cannot be evicted by other cache accesses not belonging to the private partition. In the hardware, each cache line requires additional tags comprising a flag to indicate if the line is locked and an identifier to indicate the owner of the cache line. The underutilization of Page's partitioned cache still persists because the locked lines cannot be used by other processes, even after the owner no longer requires them.

Wang and Lee also propose a *random-permutation cache* (RPCache) in [18], where, as the name suggests, it randomizes the cache interference, so that the difficulty of the attack increases. The design is based on the fact that information is leaked only when cache interference is present between two different processes. The architecture requires an additional hardware called the *permutation table*, which maps the set bits in the effective address to obtain new set bits. These are then used to index the cache set array. Changing the contents of the permutation table will invalidate the respective lines in the cache. This causes additional cache misses and a randomization in the cache interference. In [19], Wang and Lee use an underlying direct-mapped cache and dynamically reprogrammable cache mapping algorithms to achieve randomization. From the security perspective, this technique is shown to be as effective in preventing attacks as RPCache, but with less overheads on the performance. In [20], Kong et al. show that partition locked and random permutation caches, although effective in reducing performance overhead, are still vulnerable to advanced cache attack techniques. They then go on to propose modifications to Wang and Lee's proposals to improve the security [21, 22].

In [23], Domnitser et al. provide a low cost solution to prevent access-driven attacks based on the fact that the cipher evicts one or more lines of the spy data from the cache. The solution, which requires small modifications of the replacement policies in cache memories, restricts an application from holding more than a predetermined number of lines in each set of a set-associative cache. With such a cache memory, the spy can never hold all cache lines in the set, therefore the probability that the cipher evicts spy data is reduced. By controlling the number of lines that the spy can hold, trade-off between performance and security can be achieved.

10.2.3 Specialized Instructions

A few recent microprocessors support specialized instructions for ciphers. For example, Intel supports the AES-NI for their new processors [24], which allows AES encryption or decryption using a combination of six instructions. These instructions have dedicated hardware support [25] and therefore do not require to use look-up tables. Dedicated hardware has the added advantage of boosting performance. With Intel's AES-NI, speed of encryptions can be increased more than an order of magnitude. The drawback however is that, such dedicated hardware are only applicable for AES-based crypto systems. There are several other ciphers that are used in many applications, which cannot gain from these instructions. These ciphers are still vulnerable to attacks. An alternate direction is to develop improved instruction sets, which would be friendly for general cipher algorithms. For example, bit-permutation instructions were suggested in [26, 27] as a means to speedup a typical block cipher. For public-key ciphers, several instruction enhancements have been suggested such as [28–30]. In [31], a new instruction to be added to the microprocessor ISA called PERMS, was suggested. This instruction allows any arbitrary permutation of the bits in an n-bit word, and can be used to accelerate diffusion layer implementations in block ciphers, thereby providing high-speed encryptions.

10.2.4 Hardware Prefetching

Prefetching data automatically into cache memory is present in modern microprocessors to reduce the memory latencies by anticipating accesses. In [32], automatic hardware prefetching is suggested as a means to counter cache attacks, since it would confuse the adversary while distinguishing between a cache hit and a miss.

While prefetching was in fact demonstrated to make access-driven attacks slightly more difficult [33]. It has the opposite effect in profiled time-driven cache attacks. In profiled time-driven attacks such as the one by Bernstein in 2005, prefetching can be a source of information leakage rather than a means to mitigate the attack.

10.2.5 Fuzzying Clocks

Current timing attacks require distinguishing between events by their execution time. This requires highly accurate timing measurements to be made. A popular method for making these measurements is by the rdtsc (*read time stamp counter*) instruction, which is present in most modern day processors. This instruction allows the processor's time stamp counter (TSC) to be read.

Since the TSC is incremented in every clock cycle, rdtsc can provide nanoscale accuracy. Denying the adversary access to the rdtsc instructions robs her of a simple and highly precise clock source. However, the main drawback preventing user access to the TSC is that several benign applications (such as Linux kernels, multimedia games, and certain cryptographic algorithms [34]) rely on rdtsc. These applications will cease to function if the instruction is disabled. A less stringent way is to therefore *fuzz* the timestamp returned by the rdtsc instruction. This is possible because unlike timing attacks, benign applications do not require highly accurate timing measurements. Time fuzzing can be either done by reducing the accuracy of the measurement (such as by masking least significant bits of the timestamp [35]) or by reducing the precision of the measurement (such as by injecting noise into the timestamp [36]).

If `t1` is the current value of the TSC, then masking returns $\lfloor \texttt{t1/E} \rfloor$ (`E` is a time duration called *epoch*). Consequently, `t` has the form $\lfloor \frac{\texttt{t2}}{\texttt{E}} \rfloor - \lfloor \frac{\texttt{t1}}{\texttt{E}} \rfloor$. The size of the epoch is crucial. It should be large enough to protect against timing attacks, yet small enough to not affect benign applications. In [37], Vattikonda et al. propose a way to bypass such masking schemes. The technique uses the fact that timing attacks do not need absolute timing measurements, rather it is only required to distinguish between two or more events. In the proposal, the adversary first loops continuously to synchronize with the start of the epoch. This is denoted by $\texttt{t1e} = \lfloor \frac{\texttt{t1}}{\texttt{E}} \rfloor$. Then the operation to be timed is called and an rdtsc instruction is invoked in a tight loop till the end of the epoch (detected by the change in the rdtsc value). A counter c counts the number of rdtsc invocations made. The value of c in the end of the loop can be used to distinguish between events executed by the operation, as a smaller value of $t2e = \lfloor \frac{t2}{E} \rfloor$ will lead to a larger value of c.

```
c = 0;
/* Synchronize to the start of epoch */
t1e = rdtsc;
while(t1e != rdtsc);
function() /* the operation that needs to be timed */
/* Continuously scan TSC until end of epoch */
t2e = rdtsc;
while(t2e != rdtsc) c = c + 1;
return c and (t2e - t1e)
```

In the noise injection technique, a random offset between $[0, E)$ is added to the timestamp value returned by rdtsc. There are however two possible problems. The first is likely to occur when multiple rdtsc instructions are invoked. The first invocation of rdtsc returns a value of the form $t1 + r1$, while the second invocation return has the form $t2 + r2$, where t1 and t2 are the timestamp values and r1 and r2 are the random numbers ($0 \leq r_1, r_2 < E$). If the two invocations are done in quick succession then $t1 = t2$, while it is possible that $r1 > r2$. This gives the impression of time moving backward. The second drawback of the noise injection scheme is that the randomness is limited by the size of E. If sufficient number time measurements are made and average taken, the effect of the injected noise can be eliminated.

In [34], Martin, Demme, and Sethumadhavan propose a scheme called *Timewarp*. The scheme prevents the attack in [37] by restricting the TSC reading to the end of an epoch. Additionally, the rdtsc instruction is delayed by a random time (less than an epoch) before the result is returned. Noise is further added to the TSC value to increase the difficulty of the attack.

Even without using the rdtsc instructions, it is possible to make highly accurate timing measurements. For example, on multiprocessor systems, a counter can be run in a tight loop in a different thread, which can be used to make sufficiently precise timing measurements. These are called *virtual time-stamp counters* (VTSC). To prevent use of such counters, [34] proposes to add hardware in the processor, which detects these loops and injects random delays into them.

10.3 Countermeasures in the Operating System

While hardware countermeasures are expensive and application layer countermeasures are heavy, the search is still on for the countermeasure that is a panacea. The operating system may hold the key—especially in attacks that use a spy process to snoop at a victim's execution behavior.

In the operating system for instance, the scheduler can be tweaked to ensure processes from different security domains do not share the same hardware resources. Fine-grained attacks, which exploit scheduler policies may be mitigated by changes to the scheduling algorithm. For instance, by ensuring that a minimum threshold is forced before a context switch happens between two processes.

10.4 Conclusion

Countering timing attacks can be done at either the application layer, hardware, or in the operating system. In each category there are several options available. While application layer countermeasures can easily be applied, they have huge overheads that affect system performance. On the other hand, countermeasures applied at the hardware level are efficient but are expensive requiring system redesign. Comparatively, countermeasures in operating systems and virtual managers are less explored.

Engineers would need to make learned decisions on what countermeasures to incorporate. These decisions should be made early in the design phase. Trade-offs between the performance, cost, and security need to be considered. However, this is easier said than done. A countermeasure which works well with for one system and algorithm may not be suitable in another environment. While performance and costs can be easily gauged, the same cannot be done with security. For this reason, metrics need to be developed that could quantify the security of the system. For time-driven cache attacks, this book provided a means to do this. Similar frameworks need to be developed for other leakage models and attacks. The hope is for a unified framework that could assess the vulnerability of a system to all forms of timing attacks.

References

1. Klimov A, Shamir A (2002) A new class of invertible mappings. In: Çetin Kaya Koç Jr BSK Paar C (eds) CHES. Lecture notes in computer science, vol 2523. Springer, Berlin, pp 470–483
2. Biham E (1997) A fast new DES implementation in software. In: Biham E (ed) FSE. Lecture notes in computer science, vol 1267. Springer, Berlin, pp 260–272
3. Rebeiro C, Selvakumar AD, Devi ASL (2006) Bitslice implementation of AES. In: Pointcheval D, Mu Y, Chen K (eds) CANS. Lecture notes in computer science, vol 4301. Springer, Berlin, pp 203–212
4. Rebeiro C, Mukhopadhyay D (2011) Cryptanalysis of CLEFIA using differential methods with Cache Trace Patterns. In: Kiayias A (ed) CT-RSA. Lecture notes in computer Science, vol 6558. Springer, Berlin, pp 89–103
5. Rebeiro C, Poddar R, Datta A, Mukhopadhyay D (2011) An enhanced differential cache attack on CLEFIA for large cache lines. In: Bernstein DJ, Chatterjee S (eds) INDOCRYPT. Lecture notes in computer science, vol 7107. Springer, Berlin, pp 58–75
6. Paar C (1994) Efficient VLSI architectures for bit-parallel computation in Galois fields. Ph.D. dissertation, Institute for Experimental Mathematics, Universität Essen, Germany, June 1994
7. Goldreich O, Ostrovsky R (1996) Software protection and simulation on oblivious RAMs. J ACM 43(3):431–473

8. Zhuang X, Zhang T, Lee H-HS, Pande S (2004a) Hardware assisted control flow obfuscation for embedded processors. In: Irwin MJ, Zhao W, Lavagno L, Mahlke SA (eds) CASES. ACM, New York, pp 292–302
9. Zhuang X, Zhang T, Pande S (2004b) HIDE: an Infrastructure for efficiently protecting information leakage on the address bus. In: Mukherjee S, McKinley KS (eds) ASPLOS. ACM, New York, pp 72–84
10. Bernstein DJ (2005) Cache-timing attacks on AES. Technical report, 2005
11. Canteaut A, Lauradoux C, Seznec A (2006) Understanding cache attacks. INRIA, Research Report RR-5881, 2006. http://hal.inria.fr/inria-00071387/en/
12. Clavier C, Coron J-S, Dabbous N (2000) Differential power analysis in the presence of hardware countermeasures. In: Çetin Kaya Koç Paar C (eds) CHES. Lecture notes in computer science, vol 1965. Springer, Berlin, pp 252–263
13. Tunstall M, Benoît O (2007) Efficient use of random delays in embedded software. In: Sauveron D, Markantonakis C, Bilas A, Quisquater J-J (eds) WISTP. Lecture notes in computer science, vol 4462. Springer, Berlin, pp 27–38
14. Coron J-S, Kizhvatov I (2009) An efficient method for random delay generation in embedded software. In: Clavier C, Gaj K (eds) CHES. Lecture notes in computer science, vol 5747. Springer, Berlin, pp 156–170
15. Coron J-S, Kizhvatov I (2010) Analysis and improvement of the random delay countermeasure of CHES 2009. In: Mangard S, Standaert F-X (eds) CHES. Lecture notes in computer science, vol 6225. Springer, Berlin, pp 95–109
16. Percival C (2005) Cache missing for fun and profit. In: Proceedings of BSDCan 2005, Ottawa
17. Page D (2005) Partitioned cache architecture as a side-channel defence mechanism. IACR cryptology eprint archive, vol 2005, p 280
18. Wang Z, Lee RB (2007) New cache designs for thwarting software cache-based side channel attacks. In: Tullsen DM, Calder B (eds) ISCA. ACM, New York, pp 494–505
19. Wang Z, Lee RB (2008) A novel cache architecture with enhanced performance and security. In: MICRO. IEEE Computer Society, pp 83–93
20. Kong J, Aciiçmez O, Seifert J-P, Zhou H (2008) Deconstructing new cache designs for thwarting software cache-based side channel attacks. In: Jaeger T (ed) CSAW. ACM, New York, pp 25–34
21. Kong J, Aciicmez O, Seifert J-P, Zhou H (2012) Architecting against software cache-based side channel attacks. IEEE Transactions on Computers, vol 99, no. PrePrints
22. Kong J, Aciiçmez O, Seifert J-P, Zhou H (2009) Hardware-software integrated approaches to defend against software cache-based side channel attacks. In: HPCA. IEEE Computer Society, pp 393–404
23. Domnitser L, Jaleel A, Loew J, Abu-Ghazaleh NB, Ponomarev D (2012) Non-monopolizable caches: low-complexity mitigation of cache side-channel attacks. TACO 8(4):35
24. Gueron S (2010) Intel Advanced Encryption Standard (AES) instructions set (Rev : 3.0)
25. Mathew S, Sheikh F, Agarwal A, Kounavis M, Hsu S, Kaul H, Anders M, Krishnamurthy R (2010) 53Gbps native $GF(2^4)^2$ composite-field AES-encrypt/decrypt accelerator for content-protection in 45nm high-performance microprocessors. In: VLSI Circuits (VLSIC), 2010 IEEE Symposium on. 16–18 June 2010, pp 169–170
26. Shi ZJ, Yang X, Lee RB (2008) Alternative application-specific processor architectures for fast arbitrary bit permutations. IJES 3(4):219–228
27. Lee RB, Shi Z, Yin YL, Rivest RL, Robshaw MJB (2004) On permutation operations in cipher design. In: ITCC (2). IEEE Computer Society, pp 569–577
28. Großschadl J, Kamendje G-A (2003) Instruction set extension for fast elliptic curve cryptography over binary finite fields $GF(2^m)$. In: ASAP. IEEE Computer Society, pp 455–468
29. Großschadl J, Savas E (2004) Instruction set extensions for fast arithmetic in finite fields $GF(p)$ and $GF(2^m)$. In: Joye M, Quisquater J-J (eds) CHES. Lecture notes in computer science, vol 3156. Springer, Berlin, pp 133–147
30. Vejda T, Page D, Grossschadl J (2007) Instruction set extensions for pairing-based cryptography. In: Takagi T, Okamoto T, Okamoto E, Okamoto T (eds) Pairing. Lecture notes in computer science, vol 4575. Springer, Berlin, pp 208–224

31. Khurana S, Kolay S, Rebeiro C, Mukhopadhyay D (2013) Light weight cipher implementations on embedded processors. In: Design and technology of integrated systems (DTIS). IEEE Computer Society
32. Lauradoux C (2005) Collision attacks on processors with cache and countermeasures. In: Wolf C, Lucks S, Yau P-W (eds) WEWoRC. LNI, vol 74. GI, Bonn, pp 76–85
33. Demme J, Martin R, Waksman A, Sethumadhavan S (2012) Side-channel vulnerability factor: a metric for measuring information leakage. In: ISCA. IEEE, pp 106–117
34. Martin R, Demme J, Sethumadhavan S (2012) TimeWarp: rethinking timekeeping and performance monitoring mechanisms to mitigate side-channel attacks. In: ISCA. IEEE, pp 118–129
35. Osvik DA, Shamir A, Tromer E (2005) Cache attacks and Countermeasures: the case of AES. Cryptology ePrint Archive, Report 2005/271
36. Jayasinghe D, Fernando J, Herath R, Ragel R (2010) Remote cache timing attack on advanced encryption standard and countermeasures. Information and Automation for Sustainability (ICIAFs), 2010 5th International Conference on. December 2010, pp 177–182
37. Vattikonda BC, Das S, Shacham H (2011) Eliminating fine grained timers in Xen. In: Cachin C, Ristenpart T (eds) CCSW. ACM, New York, pp 41–46

Appendix A: CPUs Used for Experiments

The attacks and models developed in this book were tested on several systems. We selected a wide range of systems based on the technology used and the target application. The technology ranged from 22 to 90 nm and covered servers, desktops, laptops, and notebooks. Table A.1 has the details of the machines, the technology, clock, microarchitecture, and their cache configurations.

Table A.1 Cache configurations for test platforms

	Cache	Size	Line	Associativity
AMD Opteron (275)	*90 nm, 2.2 GHz, Italy*			
	L1-data	64 KB	64 byte	2-way
	L1-instruction	64 KB	64 byte	2-way
	L2-unified	1 MB	64 byte	16-way
Intel Xeon (E5345)	*65 nm, 2.33 GHz, Core*			
($\alpha = 255, \beta = 20$)	L1-data	32 KB	64 byte	8-way
	L1-instruction	32 KB	64 byte	8-way
	L2-unified	4 MB	64 byte	16-way
Intel Core 2 Duo (E7500)	*45 nm, 2.93 GHz, Wolfdale-3M*			
	L1-data	32 KB	64 byte	8-way
	L1-instruction	32 KB	64 byte	8-way
	L2-unified	3 MB	64 byte	12-way
Intel Atom (N455)	*45 nm, 1.66 GHz, Pineview*			
	L1-data	24 KB	64 byte	6-way
	L1-instruction	32 KB	64 byte	8-way
	L2-unified	512 KB	64 byte	8-way

C. Rebeiro et al., *Timing Channels in Cryptography*,
DOI 10.1007/978-3-319-12370-7

Table A.1 (continued)

	Cache	Size	Line	Associativity
Intel i3 (550)	*32 nm, 3.2 GHz, Clarkdale*			
	L1-data	32 KB	64 byte	8-way
	L1-instruction	32 KB	64 byte	4-way
	L2-unified	256 KB	64 byte	8-way
	L3-unified	4 MB	64 byte	16-way
Intel Dual Core (P6100)	*32 nm, 2 GHz, Arrandale*			
($\alpha = 180, \beta = 16$)	L1-data	32 KB	64 byte	8-way
	L1-instruction	32 KB	64 byte	4-way
	L2-unified	256 KB	64 byte	8-way
	L3-unified	3 MB	64 byte	12-way
Intel Xeon (E5606)	*32 nm, 2.13 GHz, Westmere*			
($\alpha = 193, \beta = 7.5$)	L1-data	32 KB	64 byte	8-way
	L1-instruction	32 KB	64 byte	4-way
	L2-unified	256 KB	64 byte	8-way
	L3-unified	8 MB	64 byte	16-way
AMD Opteron (6272)	*32 nm, 1.4 GHz, Interlagos*			
	L1-data	16 KB	64 byte	4-way
	L1-instruction	64 KB	64 byte	2-way
	L2-unified	2 MB	64 byte	16-way
	L3-unified	6 MB	64 byte	64-way
Intel i7 (4770)	*22 nm, 3.4 GHz, Haswell*			
	L1-data	32 KB	64 byte	8-way
	L1-instruction	32 KB	64 byte	8-way
	L2-unified	256 KB	64 byte	8-way
	L3-unified	8 MB	64 byte	16-way

MIX
Papier aus verantwortungsvollen Quellen
Paper from responsible sources
FSC® C105338

If you have any concerns about our products,
you can contact us on
ProductSafety@springernature.com

In case Publisher is established outside the EU,
the EU authorized representative is:
Springer Nature Customer Service Center GmbH
Europaplatz 3, 69115 Heidelberg, Germany

Printed by Libri Plureos GmbH
in Hamburg, Germany

Cashew Nut Shell Liquid

Parambath Anilkumar
Editor

Cashew Nut Shell Liquid

A Goldfield for Functional Materials

Springer

Editor
Parambath Anilkumar
Department of Pathology & Laboratory
Medicine
University of British Columbia
Vancouver, BC
Canada

ISBN 978-3-319-47454-0 ISBN 978-3-319-47455-7 (eBook)
DOI 10.1007/978-3-319-47455-7

Library of Congress Control Number: 2016954715

Printed on acid-free paper

This Springer imprint is published by Springer Nature
The registered company is Springer International Publishing AG
The registered company address is: Gewerbestrasse 11, 6330 Cham, Switzerland

Preface

The building blocks currently used for organic functional materials are mostly synthesized from starting materials derived from fossil fuel resources, but the overdependence of these sources is associated with chemical pollution and resource depletion. Therefore, there is growing scientific and technological attempts to utilize renewable resources as a feasible alternative for fossil fuel resources. Renewable resource-based organic synthesis has distinct advantages for the generation of new building blocks since (a) they are obtainable from renewable resources and (b) they are cheap and widely available. Amongst the various renewable resource raw materials, the cashew nut shell liquid (CNSL), which is an industrial waste and pollutant from the cashew nut processing industry, has gained a renewed interest in recent times owing to its wide availability and easy isolation. CNSL is a unique natural source for unsaturated long-chain phenols. Traditionally, this renewable material has wide applications in the form of brake linings, surface coatings, paints, and varnishes. Cardanol is a phenol obtained from anacardic acid, the main component of technical CNSL, by thermal decarboxylation during vacuum distillation. Cardanol has a unique structure which allows a variety of functional groups to be introduced to the molecule for derivatization. This book is an effort to summarize the research on the utilization of CNSL as a starting material for the synthesis of advanced functional materials.

This book is organized into a collection of 10 chapters. Chapter 1 describes the downsides of the overdependence of fossil fuel resources and the importance of finding sustainable crop-based alternatives. Chapter 2 is focused on the origin and chemical compositions of the different types of CNSL and utilization of its major component cardanol. This chapter details the various methods involved in the isolation of cardanol from CNSL and describes some of the basic chemistries involved in cardanol utilization. Chapter 3 reviews the chemistries involved in the synthesis of heterocyclic compounds from CNSL-based starting materials. This chapter also describes some applications of porphyrins synthesized from cardanol. Chapter 4 outlines various monomeric and polymeric surfactants derived from cardanol and its applications. Chapter 5 discusses the liquid crystalline properties of molecules and polymers derived from cardanol. The chapter also describes the interesting nonlinear optical properties exhibited by cardanol-based materials. Chapter 6 describes the utilization of cardanol-derived amphiphilic molecular assemblies as a template for conducting polymer nanostructures and also explains the influence the amphiphiles on the physical properties of the template polymer nanostructures. Chapter 7 is focused on self-assembled gel formation associated with cardanol derivatives. Also, indicated are some of the potential applications of cardanol molecular gels. Chapter 8 discusses the use of CNSL-derived molecules in medicinal applications such as cancer therapy, drug-delivery, and imaging applications. Chapter 9 reviews the research on the synthesis of step-growth polymers using cardanol-derived difunctional monomers. This chapter also highlights the performance of these classes of polymers with special stress on processability and thermal properties. Chapter 10 gives an industrial perspective on the utilization of CNSL-derived molecules and polymers. This chapter outlines a broad spectrum of high-performance industrial products and formulations beads on CNSL.

I highly appreciate all contributing authors for helping me to bring the CNSL research work in the form of a book. I would like to express my thanks to all reviewers for giving suggestions to improve the quality of the manuscripts. I also take this opportunity to thank my Ph.D. mentor Dr. M. Jayakannan, who introduced me to the field of CNSL research. I extend my thanks to Dr. Mayra Castro of Springer Nature for inviting me to edit a book on this topic.

Vancouver, BC, Canada Parambath Anilkumar

Contents

Chapter 1
Emergence of Sustainable Approaches for Functional Materials: Cashew Nut Shell Liquid and Other Relevant Crop-Based Renewable Resources

Srinivas Abbina and Parambath Anilkumar

Abstract Development of renewable resources as an alternative to fossil fuel-based feedstock for the production of different materials is vital due to various concerns associated with petroleum-based resources. Cashew nut shell liquid, an industrial waste from cashew nut (*Anacardium occidentale*) processing industry, is widely exploited as a renewable resource for developing several sustainable materials. This chapter presents few projections on the development of biorenewable resources, utilization and implementation; especially, it highlights the recent progress in the production of cardanol-derived products.

1.1 The Current Status of the Depleting Petroleum-Based Feedstock

Asphalt, although it was not processed from petroleum resources at that time or at least it was not known as a one of the petroleum by-products, today's well-known petroleum by-product, was first used around ca. 4000 BC on the banks of the Euphrates River, Iraq [1]. It is believed that, over million years ago, petroleum resources or fossil fuel feedstock was formed, thousands of feet below the sea bed, by gradual decomposition of buried dead plants and animals along with

S. Abbina (✉) · P. Anilkumar (✉)
Department of Pathology and Laboratory Medicine,
Life Sciences Centre, University of British Columbia, 2350 Health Sciences Mall,
Vancouver, BC V6T 1Z3, Canada
e-mail: sabbina@mail.ubc.ca

P. Anilkumar
e-mail: p.anilkumar@live.in

P. Anilkumar (ed.), *Cashew Nut Shell Liquid*,
DOI 10.1007/978-3-319-47455-7_1

the other sediments (rocks, mud and sand) under high temperature and pressure. Petroleum and its by-products have quite known throughout the history; however, prior to the nineteenth century, they have known to very few people and limited to for special use in very few countries such as Iraq, Egypt, China, Persia, Rome and Azerbaijan. The modern era of the petroleum industry was dawn in 1815 with the first time production of crude oil from brine wells in Pennsylvania, USA [1]. Following this, various developments in the field of petroleum industry around North America, Europe and Asia have been lead to successful growth of oil production and processing industry [2]. Later, early in nineteenth century, inception of automobile industry sharply increased the great demand for petroleum-based feedstock. In continuation to this, around the period of Second World War, drastic developments in the area of fossil oil processing/refinery approaches gave a great boost to the petroleum-based industry and made the petroleum-based feedstock a high-end value and versatile feedstock for variety of applications [3].

The petroleum-based feedstock/fossil fuel resources will not be replenished over the time, and they are limited. The most common fossil fuel resources are coal, petroleum and natural gas. The advantages of these resources are relatively inexpensive, easy extraction, inexpensive transportation and superior performance of the by-products; they became very reliable energy resource by the mid-twentieth century. This feedstock offers a wide variety of compounds in response to

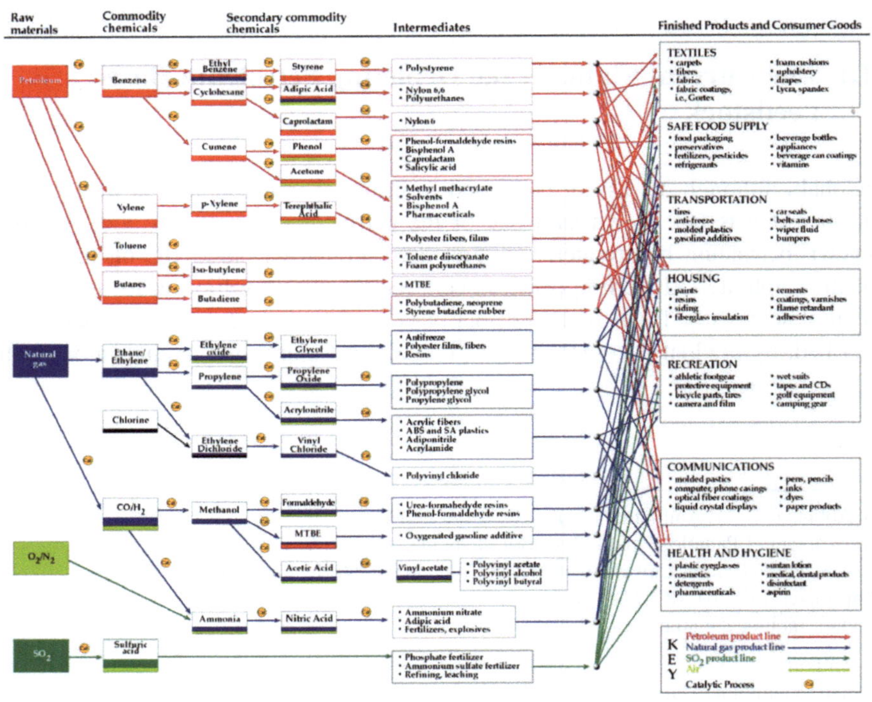

Fig. 1.1 Availability of products from the petroleum feedstock (adapted from Ref. [4] http://www.osti.gov/scitech/servlets/purl/15008859)

consumer demand. Figure 1.1 depicts the availability of products from the petroleum feedstock [4]. The domestic economies of the most nations in the world are still largely dependent on these fossil fuel resources including coal, oil and natural gas, as these are vital for functioning of the modern society. According to the international energy network review 2015, in 2013, approximately 80 % of the world's energy consumption was coming from petroleum-based feedstock, 4.6 % from nuclear energy and remaining from renewable energy resources and others [4].

Rapid increase in global population and industrialization has led to a marked increase in consumption and production of petroleum-originated products. The utilization of these crude oil resources is ranging from energy production to synthesis of various polymeric materials/chemicals that are indispensable in variety of applications including structural and packing materials, commodity materials, appliance, electronics, biomedical tools, lubricants, asphalt, paints, wax [5]. For instance, synthetic plastics alone contribute to ~7–8 % worldwide consumption of fossil fuel resources [6]. In another survey, it is estimated that around 1 % of each barrel oil is consumed for plastics production. According to a recent report in 2010, a single person in the USA in average consumed approximately 140 kgs of plastics per year [5–7]. If the consumption of petroleum-based feedstock is continued at the current pace, they will be depleted within the next few centuries. In addition to the issues with synthesis of conventional polymers, usage and disposal of synthetic polymers are also a huge environmental challenge. Most of these fossil fuel-derived products are not degradable; they accumulate in landfills and in natural habitats and create physical problems for wildlife from ingestion of microplastics. Over the last few decades, burning of fossil fuel resources like coal, oil and natural gas was also one of the major contributors to raise the CO_2 levels in the earth atmosphere, which is a key contributor to greenhouse gas emission and associated effects of global warming [8].

1.2 The Emergence of Renewable Resource Alternatives

Despite our current widespread dependence on fossil fuels or petroleum feedstock, there is a need for the transition from a non-sustainable fossil fuel economy to a sustainable economy due to the growing environmental concerns and inevitable depletion of the fossil-based resources. The continued utilization and burning of fossil fuel resources negatively impact our ecosystems and climate with an overloaded carbon footprint [9–14]. Renewable resources or renewable feedstock is defined as "a raw material of feedstock should be renewable rather depleting over a short period of time by natural processes or replenishment" [15]. In contrast to fossil resources, renewable resources are a good option to mitigate the climate change as they are clean, sustainable, partially and or fully degradable, inexpensive and abundantly available. Altogether, it enables us to protect the environment from toxic pollutants and hazardous chemicals, which in turn keeps people healthier. Most of the companies, for instance Dupont and Dow Chemicals, have

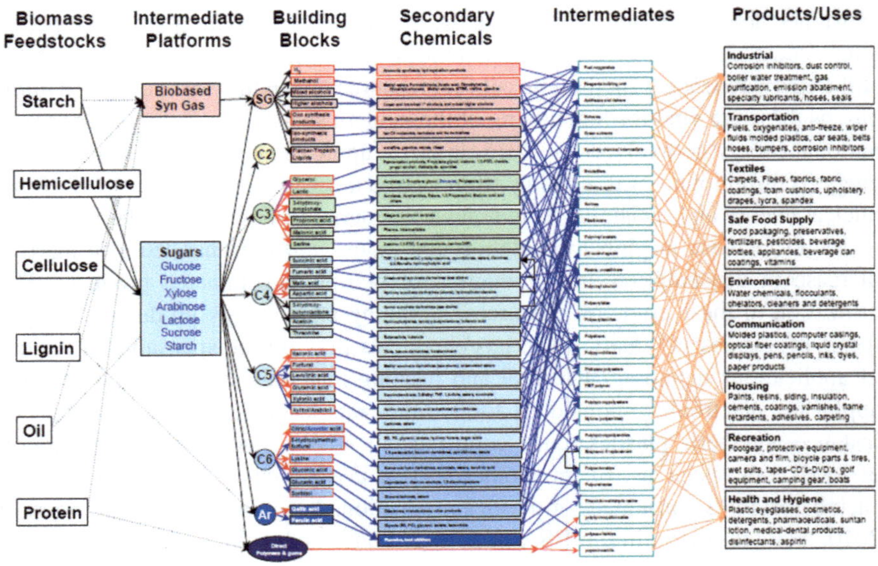

Fig. 1.2 A succinct example of products derived from renewable resources [4] http://www.osti.gov/scitech/servlets/purl/15008859

largely focused their attention on developing products from renewable resources to replace the non-renewable feedstock [16].

Renewable resources that are used in this prospective are classified as two types: plant resources and animal resources (Fig. 1.2). Similar to non-renewable resources, vast number of materials can be derived from plant-based renewable resources (Fig. 1.2). The plant resources are highlighted in this chapter.

1.3 The Advantages of Renewable Resource Alternatives

As the name defines the renewable resources are limitless or naturally replenished resources over a short period of time in contrast to the scarce fossil fuel-based feedstock. Over the past two decades, utilization of renewable resources for energy production as well as for the development of various materials has been steadily increasing. These developments create new opportunities in different sectors including manufacturing, product development, construction, operations and maintenance, transportation and logistics and consulting services. Moreover, the renewable resources-based technologies and/or industries are often more labor-intensive relative to highly capital-intensive and mechanized fossil fuel technologies, so it helps in creating new job market [17–20]. The maintenance and operational cost of renewable-based manufacturing units are quite low too [21, 22]. In addition to this, the utilization of agricultural renewable raw materials for newer technologies directly benefits rural communities and farmers.

Despite the significant progress in utilization of renewable resources around the world, large numbers of countries are still largely dependent on imported fossil fuel resources. Most often, these imports are vulnerable to political instabilities, war and trade disputes. These factors significantly affect the economy of a country, whereas the renewable dependent economy can stabilize the fuel prices and minimize the ups and downs of the economy of the countries. [23]. Thus, investigation of new renewable resources as an alternative to traditional fossil fuel resources has become an attractive topic for both academic and industrial research due to their less environmental impact, degradability, high abundance and cost-effectiveness.

1.4 Crop-Based Renewable Resources

Over the last two decades or so, there is a growing effort among scientific communities to develop process and technologies based on crop-based raw materials. This approach has distinct advantages over the conventional approaches since the crop-based building blocks are obtainable from renewable resources, economical and widely available. The most widely explored renewable resources/raw materials include polysaccharides, vegetable/plant oils, wood, sugars, proteins, natural rubbers. In this prospective, we mainly discussed the recent advancements in the field cashew nut shell liquid (CNSL).

1.4.1 Polysaccharides

Polysaccharides, as a part of natural polymers, are one of the most abundant and inexpensive biorenewable energy resources on earth and have a prominent role in energy and other commodity supplies. The most heavily used polysaccharides are cellulose, starch and chitin, which have almost similar repetitive units [24]. These widely available resources show tendency to interact with different molecules by both chemically and physically. Their tunable mechanical properties, crystallinity, solubility, viscosity, gelation, chelating efficiency and hydrophobicity/hydrophilicity make them ideal candidates for numerous applications [24–27]. Different strategies have been well exploited to tailor the structural parameters of these widely abundant biorenewable resources. The most common strategies are (1) chemical modification of surface functional groups (hydroxyl, carboxylic acids, amines and carbonyl), end groups and branch residues, (2) enzymatic modifications (transglycosylations, debranching and oxidation) and (3) physical modifications, for instance thermal, ultrasonic and alkali treatments [28, 29]. Although the utility of these three major representatives, with or without modifications, has known for quite long time for food, clothing, paper, etc., recently, these resources have been extensively explored as biorenewable macromolecular materials for applications including films and coatings, adhesive and packing materials, cosmetic and textile industries, polyelectrolytes, emulsifying agents, as well as antimicrobial agents [30–35].

1.4.2 Vegetable Oils

In recent days, plant/vegetable oils became viable alternative to petroleum-based resources for fine chemical industry and material production. The annual global production of vegetable oils has been sharply increasing from 90.5 million metric tons (Mt) in 2000 to 179.5 million Mt in 2015 [36]. In general, vegetable oils are extracted from different plants and named with their source of origin.

The naturally occurring vegetable oils consist of triglycerides (tri-esters of glycerol with long-chain fatty acids), differing in the composition of chain length and position and number of double bonds [37]. The major structural component of triglycerides is fatty acids, which account for 95 % of the total weight, and each plant oil is unique with the type of fatty acid and unsaturation [38]. The physical and chemical properties of the vegetable oils are largely dependent on the nature of the side chains. The double bonds in the side chain are the great source for deriving plant oil-based materials. Taking the advantage of unsaturation in the side chain, different polymerization techniques have been well explored, for instance thermal, cationic, free radical polymerization methods; more recently new approaches such as acyclic metathesis polymerization and ring-opening metathesis polymerization have been employed to generate vegetable oil-based polymeric materials [39, 40]. In recent decades, various plant oils (palm trees, soybeans, rapeseeds, cotton, sunflower, palm kernel, olives and coconuts) widely used for the production of polymer composites, coatings, adhesives, plastics and other biomedical applications include surgical sealants, glues, pharmaceutical patches, wound healing devices as well as drug carriers scaffolds for tissue engineering [38, 41 and references therein].

1.4.3 Biopolyesters

In the search of alternatives to petroleum-based conventional polymers, biodegradable polyesters can be considered as prominent choice as they are safe for the environment and degradable. A number of biodegradable polyesters have been developed in the last 40 years. Most of these polymers derived from renewable monomers have microbial as well as plant origin [42]. Among these environmental benign and biocompatible polyesters, poly(hydroxyalkanoate)s, poly(lactide) s, aliphatic poly(amide)s and poly(ester-amide)s have been extensively explored [42–44]. These polyesters can be synthesized by three prominent approaches: polycondensation, ring-opening polymerization and copolymerization [43]. The polyesters have attracted great attention due to their easily tunable material properties which allow them to find numerous applications in different fields as medical appliances, packaging and recycling materials, textiles, resins, fibers, films, thermoplastics and molding compounds [42–46].

1.4.4 Cashew Nut Shell Liquid (CNSL)

The important sources of crop-based raw materials are wood waste, soybean, tung, linseed, rapeseed, castor oil and cashew nut shell liquid (CNSL). Currently, this feedstock is being widely used as raw materials or after appropriate modification in different industries, for instance surfactants, lubricants, paints, household materials, wastewater purification, textiles and application in coatings and resins [47–49]. Among these renewable raw materials CNSL, which is an industrial waste and pollutant from cashew nut (*Anacardium occidentale*) processing industry, gained particular interest owing to its wide availability and easy isolation in good yields.

CNSL, a greenish-yellow to brownish viscous liquid, is generated by thermal treatment of cashew nut shell, and its approximate annual industrial production is 450,000 metric tons [50]. Cashew nut trees, belong to the family of Anacardiaceae, originally native to northeastern Brazil, are currently widely spread over different tropical regions of the world such as India, Brazil, Bangladesh, Tanzania, Kenya, Mozambique and Southeast Asia [51]. The cashew nut grows outside of the cashew apple, which is an edible fruit. The cashew nut shell is leathery and contains a reddish brown viscous liquid which amounts approximately 67 % of the nut weight [52–54]. CNSL is one of the very limitedly available aromatic biofeedstocks. It is a rich source of variety of phenolic compounds with unsaturated aliphatic chains, including anacardic acid, cardol, cardanol and 2-methylcardol [55]. The phenolic residues and highly hydrophobic long alkyl chains make CNSL unique for wide range of applications. Addition to its inherent antimicrobial and antitermite properties, it is found industrial applications, for instance brake lining formulations [56], flame-resistant materials [57], surfactants [58], plasticizers [59], lubricants [60], resins [61], CNSL nanomaterials [62], soft materials [63], doping agents [64] and other medical applications [65].

As CNSL being non-edible oil, it does not show any negative effect on food production and supply in contrast to some other food crop-based biofeedstock. Traditionally, different chemical methods including hot oil process, solvent extraction, mechanical extraction, vacuum distillation, supercritical fluid bath are well established for extracting the CNSL from the crude cashew nut shells. The final compositions of CNSL vary based on the extraction used. Solvent extraction generates CNSL with anacardic acid in high proportions (solvent extracted or natural CNSL), whereas roasting approach produces CNSL with major portions of cardanol (technical CNSL) [51]. The compositions of two main CNSLs are given in Table 1.1.

Cardanol is an aromatic biomonomer and is derived from CNSL; it is sustainable, inexpensive and largely available as agricultural waste in major parts of the world. Typically, when CNSL is treated with high temperature (180–220 °C) under vacuum (3–4 mm torr), anacardic acid decarboxylated into cardanol and separated as a pale yellow liquid. It contained a mixture of cardanol compounds with varying the degree of unsaturation in meta-positioned side chain (Fig. 1.3).

Table 1.1 Compositions of two different CNSL

Compound	Natural CNSL (%)	Technical CNSL (%)
Cardanol	10	60–65
Cardol	11	15–20
2-Methyl cardol	2.04	2
Anacardic acid	60–65	Trace amounts
Polymeric material		10

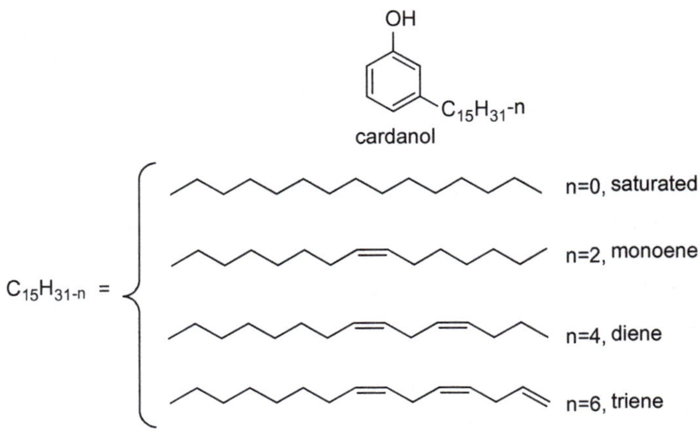

Fig. 1.3 Different components of cardanol

The structural features of the cardanol are very peculiar; it is a phenol-based monomer with meta-substituted unsaturated alkyl side chain (C15) and has different possible functionalization sites (circled in Fig. 1.4). The phenolic hydroxyl group directs the next incoming groups into ortho and/or para positions of the aromatic ring, providing more synthetic flexibility. The unsaturated/saturated hydrophobic alkyl chain with odd-numbered carbons induces both amphiphilicity and lipid nature to the system. The phenolic hydroxyl group and *cis* double bonds in the side chains are further useful for functionalization with different monomers.. Considering the difficulty of synthesizing phenols with a long unsaturated carbon chain at *meta*-position and its low price and abundance made cardanol as a precious starting material for the synthesis of various derivatives [52–54].

Fig. 1.4 A schematic view of active sites of cardanol (*circled*)

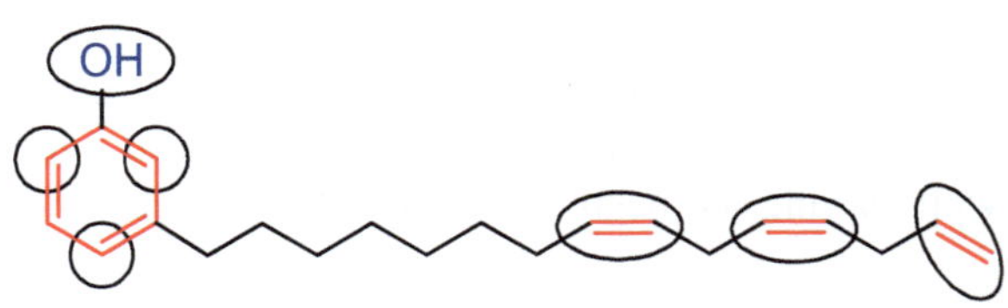

1.5 Cardanol-Based Functional Materials

The versatility of cardanol as a starting material is derived from the virtue of its unique chemical structure. Cardanol is greatly exploited, particularly in the countries where cashew nut trees are widely grown, as a renewable feedstock for generating different functional materials (Fig. 1.5). Several functional materials are derived from cardanol by the reactions of active sites of cardanol (aromatic ring, hydroxyl group as well as the unsaturation in side chains) with different molecules. The phenolic hydroxyl group is a great source for esterification, alkylation, etherification, propoxylation, polymerization and phosphatation (Fig. 1.6). The unsaturation in the side chains can be used for hydrogenation, epoxidation, metathesis and hydrosilylation reactions. Different substitution (bromination, nitration, sulfation and amination), condensation and hydrogenation reactions can be performed on aromatic ring (Fig. 1.6). These features are greatly exploited to design and develop a diverse range of materials including paints and varnishes, epoxy resins, rubbers, polyurethanes, phenolic resins, acrylics using cardanol as a biomonomer [56–65].

A comprehensive list of the materials that are derived from cardanol is listed in Fig. 1.6. We, briefly, in this chapter, highlighted the highly commercial valued cardanol-based polymers. Different types of resins were synthesized from condensation of formaldehyde with cardanols [66]. The mechanical properties of these resins can be easily manipulated by utilizing the unsaturation in the side chains [67]. Moreover, these materials have high toughness, elasticity, superfine adhesive qualities. The copolymers of cardanol with phenols and glycidyl ethers are used to produce epoxy-type resins. These resins are perhaps best of the modified phenolic epoxy resins as alternative to traditional synthetic epoxy resins. They offer good

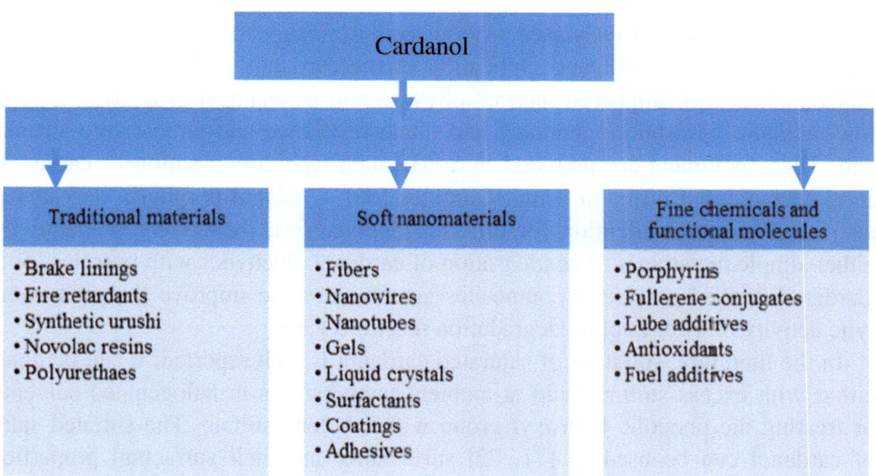

Fig. 1.5 A comprehensive list of materials derived from cardanol feedstock (Reprinted with permission from Ref. [54] (copyright 2013), The Royal Society of Chemistry)

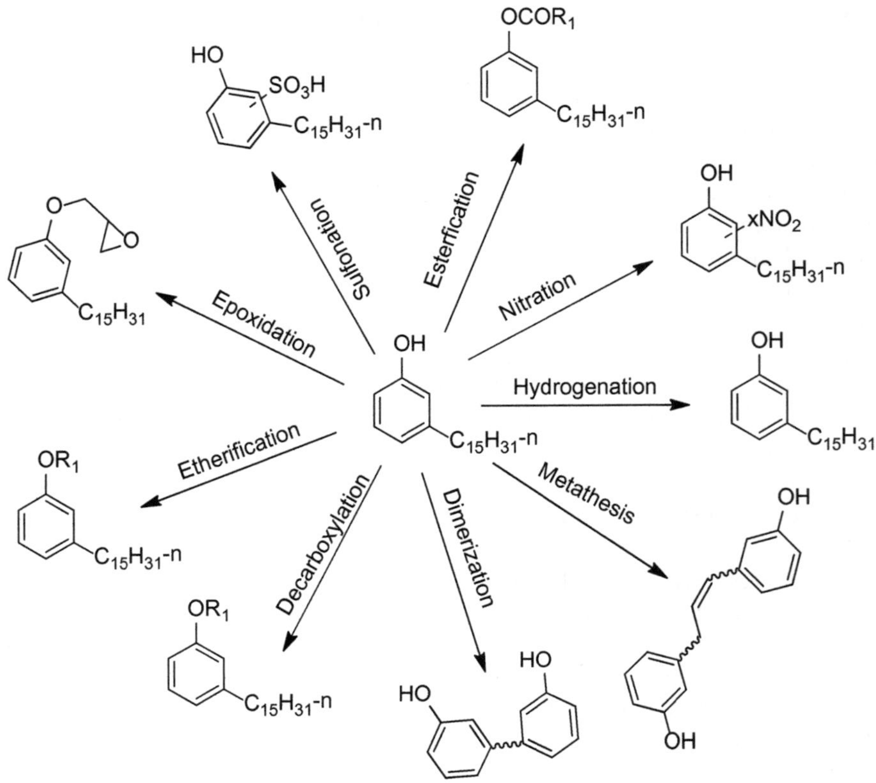

Fig. 1.6 A set of possible reactions of cardanol

flexibility and thermal resistance without compromising the oil bleeding properties [56]. Another significant application of the cardanol is development of the benzoxazines with improved thermal physical and mechanical properties such as low moisture absorbtion, chemical and flame resistance and almost zero shrinkage. These polymers are prepared by a Mannich-type condensation of cardanol, formaldehyde and different primary amines [68]. Cardanol-porphyrin derivatives also received good attention for different applications; these were prepared by either simple metathesis or condensation of cardanol aldehydes with pyrroles [69]. Cardanol-derived porphyrin composites are also used to improve the photocatalytic activity of TiO_2 in photodegradation reactions [70].

In the literature, sulfation of saturated cardanol is well reported; it can be done either with excess sulfuric acid at ambient temperatures in halogenated solvents or treating the phenolic hydroxyl group with ethylene sulfate. The sulfated salts of cardanol can be used as [71, 72] surfactants and their surfactant properties are apparently similar to the properties of commercial available detergents, for instance dodecylbenzene sulfonate [73].

The industrial scale production of cardanol sulfates is very economical, and the synthetic process of cardanol sulfate was not complicated unlike in the case of traditional detergents. Cardanol-based surfactants are one of the potential alternatives to petroleum-product-derived household surfactants [74, 75]. Sulfonic acid derivates of cardanol are also found as detergents, and few studies showed that these molecules can also act as doping agents for conducting polymers [76]. Halogenated cardanol products are found to have pesticidal and water repellency properties [77]. The unsaturated C_{15} hydrocarbon side chain of the cardanol is well known for its crosslinking properties. Self-polymerization of cardanol-azo benzene forms highly transparent liquid crystalline material films. These materials showed both thermotropic and lyotropic properties [78].

Another important type of functionalization of cardanol is phosphorylation of hydroxyl group of phenol. These phosphorylated cardanols are well known as an effective plasticizer for different rubbers [79]. The cardanol polymers with formaldehyde produced rubbery-type gel materials, which can be used in cement hardening [56]. Thermoplastic polyurethanes were also developed from cardanol, which possess very good thermal and mechanical properties [59, 80]. These environmental benign cardanol-derived polyemic coatings has found applications in paint, enamel and varnish industry [51]. Most of the time these polymeric coatings are formed by either autooxidation or crosslinking of oily materials. Cardanol can undergo easily autooxidation at ambient conditions due to the advantage of double bonds in the side chain and forms robust, transparent and scratch and wrinkle-free polymeric films [79]. Cardanol-based amphiphilic compounds were found to exhibit remarkable self-assembly properties in various types of nanostructures. For example, cardanyl glycolipids molecules formed nanotubes, twisted ribbons, tapes, etc., and these structures might have possible applications in nanofabrication, inclusion chemistry and catalysis [63].

1.6 Future Prospectives

Before the discovery of inexpensive crude oil feedstock, our society was dependent on the natural renewable resources for various basic needs. Later, the steep growth in the world's population forced to move toward less expensive fossil fuel feedstock. Since petroleum-based feedstock is the primary resource for energy production and manufacturing, consumption of this feedstock has also drastically increased. This continual usage along with growing industrial activity creates several challenges: depletion of fossil fuel resources, climate change, random fluctuations in crude oil prices and political instability. These concerns turned the scientific community's attention toward green and sustainable approaches in research. Utilization of renewable resources became a viable option to address the challenges with non-renewable resources and to create a sustainable society. Various renewable resources have been well explored to replace the petroleum feedstock-based products, and a great number of attempts have also been placed to modify the existing synthetic routes in greener route by adapting twelve principles of Green Chemistry. Adaption of this environmental benign research discipline is lead to both environmental and economical benefits. In 2013, renewable energy policy network estimated that the renewable resources supply around 23 % of the total energy consumption throughout the world and it has been steadily increased [81]. A recent Pike research estimate concluded that green chemicals will save approximately $65.5 billion by 2020 [82]. Plant-based biomass feedstock is one of the most studied resources in both academia and industry and offers a vast range of possibilities to the chemical, energy and material sectors. Cardanol, a major constituent of CNSL which is an industrial waste product in cashew nut industry, cheap and renewable resource, is also well exploited to generate novel functional materials for polymer and coating applications. It showed a great promise as a one of the alternatives to the non-renewable resources and generated high-end value and eco-friendly materials. The unique structural features of cardanol allow it to participate in different synthetic methodologies spanning from esterification, sulfation, nitration, epoxidation and polymerization. Moreover, cardanol-derived amphiphiles, lipids and self-assembled nanostructures are quite promising as next-generation materials. Some of the cardanol-based products such as resins, fibers, lining agents have already found in market. Different industries are synthesizing cardanol-derived value-added materials in large scale. For instance, Cardolite, Plamer, Satya Cashew Chemicals Ltd., Sanoor Cashew and Adarsh Industrial Chemicals have been developing and manufacturing numerous compounds using CNSL as a starting material [52]. The intriguing features of cardanol-derived nanostructures might also provide opportunities to evaluate them as biotherapeutics and other drug delivery applications. In addition to this, the recent progress in olefin metathesis with cardanol monomers opened new opportunities for generating different hybrid materials.

Development and implementation of renewable materials are a key component of sustainable development due to the facts: (1) less environmental damage, (2) being replenish, (3) stable economy and (4) possible decentralized energy. Despite the considerable advances in the development of sustainable materials from renewable resources, the successful examples are still scarce. Future efforts should be focused on improvement of current process, minimizing the operational expenses, finding new resources, industrial implementation, reducing the dependence on food-related biocrops. The sustainable or eco-friendly economy cannot be adequately fulfilled merely by the basic use of renewable resources. They must meet the necessary barriers to surpass the ultimate economic competition with traditionally available inexpensive and superior fossil fuel-based products (Fig. 1.7). Finally, governments should also encourage renewable-based initiatives by decreasing subsidization benefits for fossil fuel-based technologies and offer more incentives for renewable energy industries. And also, administrative and planning communities should try to create more awareness among the public about sustainable society.

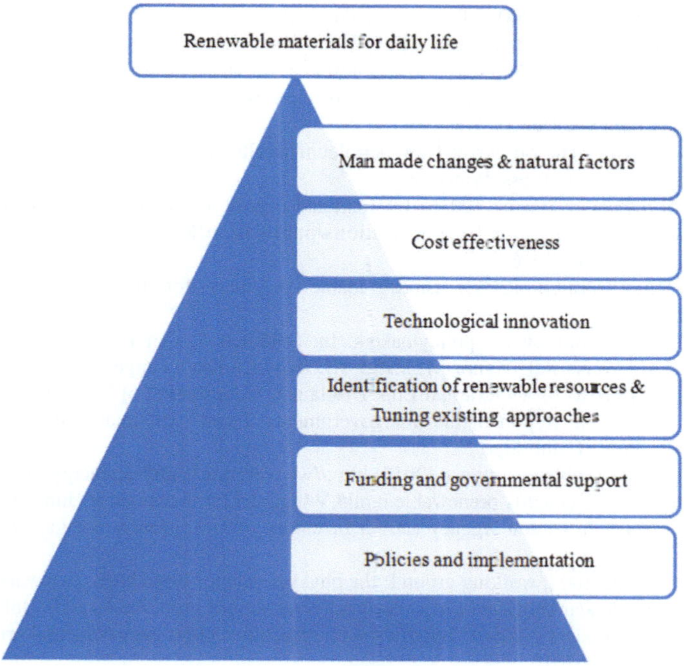

Fig. 1.7 A list of hurdles to renewable resources to compete with mature technologies

1.7 Conclusion

Renewable resources are becoming potential alternative to fossil fuel resources in order to replace the various petroleum-derived, non-degradable, less sustainable materials. The recent studies showed that cardanol is one of the promising renewable resources due to its diverse structural features; it is less expensive and widely abundant. A wide array of cardanol-derived products showed promising properties to compete with petroleum-based products; moreover, several of them are commercially available and few more in the development process.

References

1. A Timeline of highlights from the histories of ASTM committee D02 and the petroleum Industry (1996) ASTM International, West Conshohocken. http://www.astm.org/COMMIT/D02/to1899_index. Accessed 02 Feb 2016
2. "Petroleum." The columbia encyclopedia, 6th edn. 2016. Retrieved July 18, 2016 from Encyclopedia.com: http://www.encyclopedia.com/doc/1E1-petroleu.html
3. Malhotra R (ed) (2013) Fossil energy: selected entries from the encyclopedia of sustainability science and technology. Springer, New York

4. Werpy T, Petersen G (2004) Top Value Added Chemicals from Biomass: Volume I-Results of Screening for Potential Candidates from Sugars and Synthesis Gas. United States. doi:10.2172/15008859. http://www.osti.gov/scitech/servlets/purl/15008859
5. Danner H, Braun R (1999) Biotechnology for the production of commodity chemicals from biomass. Chem Soc Rev 28:395–405
6. Ross ML (2012) The oil curse: How petroleum wealth shapes the development of nations. Princeton University Press, Princeton
7. Key World Energy Stastics (2015) International energy agency (IEA), Paris Cedex.https://www.iea.org/publications/freepublications/publication/KeyWorld_Statistics_2015.pdf Accessed on 02 Feb 2016
8. Gross RA, Kalra B (2002) Biodegradable polymers for the environment. Science 297:803–807
9. IPCC (2007) Summary for policymakers. In: Solomon S, Qin D, Manning M, Chen Z, Marquis M, Averyt KB, Tignor M, Miller HL (eds) Climate change 2007: the physical science basis. Link to EPA's External Link Disclaimer Contribution of Working Group I to the Fourth Assessment Report of the Intergovernmental Panel on Climate Change. Cambridge University Press, Cambridge
10. Benefits of renewable energy use (2012) http://www.ucsusa.org/cleanenergy-choices/renewable-energy/public-benefits-orenewable.html#.V48gjNIrLIU. Accessed 10 June 2015
11. Environmental Protection Agency (2012) Inventory of U.S. greenhouse gas emissions and sinks: 1990–2010
12. Climate change 2007: working group I: the physical science basis https://www.ipcc.ch/publications_and_data/ar4/wg1/en/spmsspm-direct-observations.html. Accessed 08 July 2016
13. CFR backgrounders (2016) http://www.cfr.org/china/chinas-environmental-crisis/p12608. Accessed 05 May 2016
14. Natural resources Canada (2009) http://www.nrcan.gc.ca/energy/renewable-electricity/7293 Accessed on 30 May 2016
15. Anastas P, Eghbalia N (2010) Green chemistry: principles and practice. Chem Soc Rev 39:301–312
16. Chemical research brief (2016) http://www.sasb.org/wp-content/uploads/2015/03/RT0101_Chemicals_Brief.pdf. Accessed on 30 May 2016
17. Mohanty AK, Misra M, Drzal LT (2002) Sustainable bio-composites from renewable resources: opportunities and challenges in the green materials world. J Polym Environ 10:19–26
18. Stevens ES (2002) Green plastics. Princeton University Press, Princeton
19. Thompson RC, Moore CJ, Saal FS, Swan SH (2009) Our plastic age. Philos Trans R Soc B 364:2153–2166
20. Gkatsou S, Kounenou M, Papanagiotou P et al (2014) The impact of green energy on employment: a preliminary analysis. Int J Bus Soc Sci 5:29–41
21. Renewable energy jobs: status, prospects & policies, biofuels and grid-connected electricity generate (2013) http://www.irena.org/documentdownloads/publications/renewableenergy-jobs.pdf. Accessed on 20 June 216
22. Pimentel D, Herz M, Glickstein M et al (2002) Renewable energy: current and potential issues. Bioscience 52:1111–1123
23. Alsahlawi MA (2010) The future prospect of world oil supply: depletion of resources or price trends. OPEC Energy Rev 34:73–81
24. Gandini A (2008) Polymers from renewable resources: a challenge for the future of macromolecular materials. Macromolecules 41:9491–9504
25. Gandini A, Lacerda TM, Carvalho AJF et al (2016) Progress of polymers from renewable resources: furans, vegetable oils, and polysaccharides. Chem Rev 116:1637–1669
26. Ciesielski W, Lii CY, Yen MT et al (2003) Interactions of starch with salts of metals from the transition groups. Carbohydr Polym 51:47–56
27. Belgacem MN, Gandini A (eds) (2008) Monomers, polymers and composites from renewable resources. Elsevier, Amsterdam

28. Yalpani M (1985) A survey of recent advances in selective chemical and enzymic polysaccharide modifications. Tetrahedron 41:2957–3020
29. Cumpstey I (2013) Chemical modification of polysaccharides. ISRN Organic Chemistry, vol 2013, Article ID 417672, 27 pages, 2013. doi:10.1155/2013/417672
30. Hansen NML, Plackett D (2008) Sustainable films and coatings from hemicelluloses: a review. Biomacromlecules 9:1493–1505
31. Tatarka PD, Cunningham RL (1998) Properties of protective loose-fill foams. J Appl Polym Sci 67:1157–1176
32. Shinoj S, Visvanathan R, Panigrahi S et al (2011) Oil palm fiber (OPF) and its composites: A review. Ind Crops Prod 33:7–22
33. Boddohi S, Moore N, Johnson AP et al (2009) Polysaccharide-based polyelectrolyte complex nanoparticles from chitosan, heparin, and hyaluronan. Biomacromolecules 10:1402–1409
34. Bravin B, Peressini D, Sensidoni A (2004) Influence of emulsifier type and content on functional properties of polysaccharide lipid-based edible films. J Agric Food Chem 52:6448–6455
35. Sánchez-Gonzáleza L, Cháfera M, Hernández M et al (2011) Antimicrobial activity of polysaccharide films containing essential oils. Food Control 22:1302–1310
36. Global production of vegetable oils from 2000/01 to 2015/16 (in million metric tons) (2016) http://www.statista.com/statistics/263978/global-vegetable-oil-production-since-2000-2001/. Accessed on 0808.2016
37. Lligadas G, Ronda JC, Galià M et al (2013) Renewable polymeric materials from vegetable oils: a perspective. Mater Today 16:337–343
38. Xia Y, Larock RC (2010) Vegetable oil-based polymeric materials: synthesis, properties, and applications. Green Chem 12:1893–1909
39. Valverde M, Andjelkovic D, Kundu PP (2007) Conjugated low-saturation soybean oil thermosets: free-radical copolymerization with dicyclopentadiene and divinylbenzene. J Appl Polym Sci 107:423–430
40. Andjelkovic DD, Valverde M, Henna P et al (2005) Novel thermosets prepared by cationic copolymerization of various vegetable oils—synthesis and their structure–property relationships. Polymer 46:9674–9685
41. Quirino RL, Garrison TF, Kessler MR (2014) Matrices from vegetable oils, cashew nut shell liquid, and other relevant systems for biocomposite applications. Green Chem 16:1700–1716
42. Rydz J, Sikorska W, Kyulavska M et al (2015) Polyester-based (bio)degradable polymers as environmentally friendly materials for sustainable development. Int J Mol Sci 16:564–596
43. Nair LS, Laurencin CT (2007) Biodegradable polymers as biomaterials. Prog Polym Sci 32:762–798
44. Rodriguez-Galan A, Franco L, Puiggali J (2011) Degradable poly(ester amide)s for biomedical applications. Polymers 3:65–99
45. Migneco F, Huang YC, Birla RK (2009) Poly(glycerol-dodecanoate), a biodegradable polyester for medical devices and tissue engineering scaffolds. Biomaterials 30:6479–6484
46. Seyednejad H, Ghassemi AH, van Nostrum CF et al (2011) Functional aliphatic polyesters for biomedical and pharmaceutical applications. J Control Release 152:168–176
47. Bentsen NC, Felby C (2012) Biomass for energy in the european union–a review of bioenergy resource assessments. Biotechnol Biofuels 5:25–35
48. Ben-Iwo J, Manovic V, Longhurst P (2016) Biomass resources and biofuels potential for the production of transportation fuels in Nigeria. Renew Sustain Energy Rev 63:172–192
49. Pimentel D, Hepperly P, Hanson J (2005) Environmental, energetic, and economic comparisons of organic and conventional farming systems. Bioscience 55:573–582
50. Velmurugan A, Loganathan M (2011) Performance and emission characteristics of a DI diesel engine fuelled with cashew nut shell liquid (CNSL)-diesel blends. World Acad Sci Eng Technol 58:889–900
51. Banchhor M, Baid R (2007) CNSL (Cashew Nut Shell Liquid)—a versatile renewable natural resource. Plant Arch 7:497–501

52. Voirin C, Caillol S, Nilakshi V et al (2014) Functionalization of cardanol: towards biobased polymers and additives. Polym Chem 5:3142–3162
53. Lubi MC, Thachil ET (2000) Cashew nut shell liquid (CNSL)—a versatile monomer for polymer synthesis. Des Monomers Polym 3:123–153
54. Balachandran VS, Jadhav SR, Vemula PK et al (2013) Recent advances in cardanol chemistry in a nutshell: from a nut to nanomaterials. Chem Soc Rev 42:427–438
55. Paramashivappa R, Kumar PP, Vithayathil PJ et al (2001) Novel method for isolation of major phenolic constituents from cashew (*anacardium occidentale* l.) nut shell liquid. J Agric Food Chem 49:2548–2551
56. Devi A, Srivastava D (2007) Studies on blends of cardanol-based epoxidized novolac type phenolic resin and carboxyl-terminated polybutadiene (CTPB) I. Mater Sci Eng A 458:336–347
57. Ravichandran S, Bouldin RM, Kumar J et al (2011) A renewable waste material for the synthesis of a novel non-halogenated flame retardant polymer. J Clean Prod 19:454–458
58. Scorzza C, Nieves J, Vejar F, Bullon J (2010) Synthesis and physicochemical characterization of anionic surfactants derived from cashew nut shell oil. J Surfactants Deterg 13:27–31
59. Anilkumar P, Jayakannan M (2007) Fluorescent tagged probing agent and structure-directing amphiphilic molecular design for polyaniline nanomaterials via self-assembly process. J Phy Chem C111:3591–3600
60. Rios MAS, Sales FAM, Mazzetto SE (2009) Study of antioxidant properties of 5-*n*-pentadecyl-2-*tert*-amylphenol. Energy Fuels 23:2517–2522
61. Pillai CKS, Prasad VS, Sudha JD (1990) Polymeric resins from renewable resources. II. Synthesis and characterization of flame-retardant prepolymers from cardanol. J Appl Polym Sci 41:2487–2501
62. John G, Vemula PK (2006) Design and development of soft nanomaterials from biobased amphiphiles. Soft Matter 2:909–914
63. John G, Masuda M, Okada Y et al (2001) Nanotube formation from renewable resources via coiled nanofibers. Adv Mater 13:715–719
64. Radhakrishan S, Rao CRK, Vijayan M (2009) Electrochemical synthesis and studies of polypyrroles doped by renewable dopant cardanol azophenylsulfonic acid derived from cashew nutshells. J Appl Polym Sci 114:3125–3131
65. Toyomizu M, Sugiyama S, Jin RL et al (1993) Glucosidase and aldose reductase inhibitors: constituents of cashew, *Anacardium occidentale*, nut shell liquids. Phytother Res 7:252–254
66. Sultania M, Rai JSP, Srivastava D (2011) Process modeling, optimization and analysis of esterification reaction of cashew nut shell liquid (CNSL)-derived epoxy resin using response surface methodology. J Hazard Mater 185:1198–1204
67. Mwaikambo LY, Ansell MP (2001) Cure characteristics of alkali catalysed cashew nut shell liquid-formaldehyde resin. J Mater Sci 36:3693–3698
68. Pathak SK, Rao BS (2006) Structural effect of phenalkamines on adhesive viscoelastic and thermal properties of epoxy networks. J Appl Polym Sci 102:4741–4748
69. Mele G, Sole RD, Vasapollo G et al (2004) Polycrystalline TiO_2 impregnated with cardanol-based porphyrins for the photocatalytic degradation of 4-nitrophenol. Green Chem 6:604–608
70. Vasapollo G, Giuseppe Mele G, Sole RD et al (2011) Use of novel cardanol-porphyrin hybrids and their TiO_2-based composites for the photodegradation of 4-nitrophenol in water. Molecules 16:5769–5784
71. Kattimuttathu S, Foerst G, Schubert R, Bartsch E (2012) Synthesis and micellization properties of new anionic reactive surfactants based on hydrogenated cardanol. J Surfactants Deterg 15:207–215
72. Bruce IE, Mehta L, Porter MJ et al (2009) Anionic surfactants synthesized from replenishable phenolic lipids. J Surfactants Deterg 12:337–344
73. Souza FG Jr, Soares BG, Siddaramaiah et al (2006) Influence of plasticizers (DOP and CNSL) on mechanical and electrical properties of SBS/polyaniline blends. Polymer 47:7548–7553

74. John HPT, Mahesh P (2007) Phenolic structure and colour in Mannich reaction products. J Chem Res 1:34–37
75. Scorzza C, Nieves F, Vejar NF, Bullon J (2010) Synthesis and physicochemical characterization of anionic surfactants derived from cashew nut shell oil. J Surfactants Deterg 13:27–31
76. Paul RK, Pillai CKS (2001) Melt/solution processable polyaniline with functionalized phosphate ester dopants and its thermoplastic blends. J Appl Polym Sci 80:1354–1367
77. Lepage ES, Delelis AT (1980) Protecting wood against dry wood termite with cashew nut shell oil. Forest Prod J 36:35–36
78. Saminathan MS, Pillai CKS (2000) Synthesis of novel liquid crystalline polymers with cross-linked network structures. Polymer 41:3103–3108
79. Menon ARR, Pillai CKS, Nando GB (1998) Modification of natural rubber with phosphatic plasticizers: a comparison of phosphorylated cashew nut shell liquid prepolymer with 2-ethyl hexyl diphenyl phosphate. Eur Polym J 34:923–929
80. Mythili CV, Retna AM, Gopalakrishnan S (2004) Synthesis, mechanical, thermal and chemical properties of polyurethanes based on cardanol. J Mater Sci 27:235–241
81. Renewables global status report 2015 [http://www.ren21.net/wp-content/uploads/2015/07/REN12-GSR2015_Onlinebook_low1.pdf]. Accessed on 10 July 2016
82. Green Chemicals Will Save Industry $65.5 Billion by 2020 https://www.navigantresearch.Com/newsroom/green-chemicals-will-save-industry.-65-5-billion-by-2020. Accessed on 16 July 2016

Chapter 2
Cashew Nutshell Liquid (CNSL): From an Agro-industrial Waste to a Sustainable Alternative to Petrochemical Resources

Diego Lomonaco, Giuseppe Mele and Selma E. Mazzetto

Abstract The global concern about the problems caused by the growing consumption of petrochemical compounds, such as global warming and the diminishing of fossil fuels reserves, has given rise to the green chemistry field—based on the design of products and processes that minimize the use and generation of hazardous substances—in both academy and industry. In face of the increasing demand for chemicals, sustainable and innovative technologies are necessary in order to overcome current difficulties and also avoid the evolution of new ones. The advent of agro-industry waste as a resource of raw materials has attracted the attention of researchers and opened a window of opportunity for the breakthrough of alternative products to the oil industry. In this sense, cashew nutshell liquid (CNSL), a by-product from cashew industrial processing and one of the richest natural sources of phenolic compounds, emerges as a promising and renewable feedstock for the development of a wide range of functional products. This chapter is focused on the discussion of the origin and chemical composition of the different types of CNSL, and also on the challenges involved in the methods of isolation and main applications of its major component, cardanol, in the context of a sustainable development.

2.1 Introduction

The cashew tree, scientifically named as *Anacardium occidentale* L., a member of the Anacardiaceae family, is considered an indigenous element from the tropical region of South America. Appointed by Lindley in 1830, this family includes

D. Lomonaco (✉) · S.E. Mazzetto
Department of Organic and Inorganic Chemistry, Federal University of Ceará, Campus Pici, Fortaleza 60440-900, Brazil
e-mail: lomonaco@ufc.br

G. Mele
Department of Engineering for Innovation, University of Salento, Via Arnesano, 73100 Lecce, Italy

P. Anilkumar (ed.), *Cashew Nut Shell Liquid*,
DOI 10.1007/978-3-319-47455-7_2

Fig. 2.1 Cashew pedicel, fruit, and mesocarp filled with CNSL

more than 700 species, distributed in 82 genera that are majorly found in tropical zones, although few of them extend to temperate zones. Members of this family are cultivated all around the world due to its edible fruits, medicinal compounds, wood properties, and scenic effects. Besides cashew, other agricultural products from Anacardiaceae are: mangos (*Mangifera indica* L.), pistachios (*Pistacia vera* L.), and marula (*Sclerocarya birrea* (A. Rich.) Hochst.) [35, 40].

The colorful cashew apple (Fig. 2.1), an overdeveloped pedicel, is edible and commonly mistaken as the fruit of the cashew tree, but it is in fact a pseudocarp (accessory fruit), scientifically known as flower stack, which is used in the production of sweet beverages and foods. On the other hand, the true fruit is a reniform (kidney-shaped) seed (also considered a nut) that grows in one of the ends of the apple, and is covered by a hard shell (pericarp). The cashew nut (kernel) is the primary commercial product of cashew plantation as it is very appreciated as a snack and is also used in the Asiatic cuisines, especially Thai and Chinese.

The pericarp that involves the nut, the cashew nutshell, which has an alveolar mesocarp, is filled with a dark and caustic oily substance called cashew nutshell liquid (CNSL), obtained as a by-product during the industrial processing of the cashew nut.

This oil is composed mainly by phenolic lipids and, for this particular reason, has drawn the attention of scientist and researchers all over the globe, since it has been considered as a viable alternative to petroleum products [16, 24].

2.2 Origin and Production

The origin of cashew is based on evidences that strongly indicate Brazil, or at least tropical South America, as the center of origination of the species currently known as *Anacardium occidentale*. The warm and moist climate conditions found

Fig. 2.2 An illustration made by André Thévet in his book "The singularities of Antarctic France" (Reproduced from ref. [55])

especially in the coastal regions of Northeast Brazil, with air humidity about 70 %, are ideal for cashew growing. The name cashew is in fact derived from the indigenous dialect word "*acaju*" (pronounced [kɐˈʒu]) [16].

A substantial indication about the Brazilian origin of cashew can be found in an old reference about this plant made by the French naturalist monk André Thévet (1502–1590). An illustration, describing few natives harvesting what seems to be a cashew tree (Fig. 2.2), found in his book entitled "The singularities of Antarctic France," written in 1557 during the French expeditions to south of the equator regions of Brazil between 1555 and 1567 [56].

Although native from Brazil, the plant is already disseminated in many countries around the world, especially in India, Nigeria, Viet Nam, Cote d'Ivoire, Tanzania, Philippines, Indonesia, and Guinea-Bissau. This widespread of cashew is because the Portuguese colonization of Brazil, who, in late the 1500, introduced the cashew in India, then in the African colonies, and Asia, where it quickly adapted due to the similar climate conditions.

From a commercial perspective, the worldwide production of cashew is growing every year, along with the total area of plantation.

According to the Food and Agriculture Organization of the United Nations [17], considering the period of 20 years, from 1993 to 2013, the global annual production of cashew nut (with shell) increased more than 400 %, rising from 1.053.232 tonnes to 4.439.960 tonnes per year. In its turn, considering the same period, the total area harvested increased from 2.089.710 ha to 5.457.009 ha per year, a growing around 260 % (Graphic 2.1).

The top 5 producers of cashews, in this interval, were, in decreasing order, Viet Nam, Nigeria, India, Côte D'Ivoire, and Brazil, responsible for the production of 43.099.972 tonnes, around 75 % of the global production of 57.297.951 tonnes [17] (Graphic 2.2).

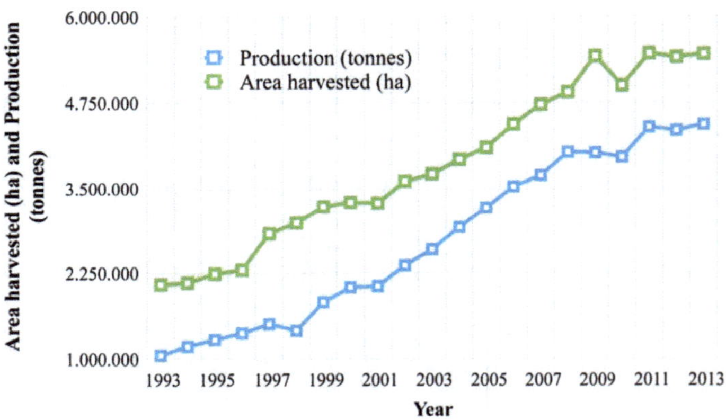

Graphic 2.1 Global production and area harvested of cashew nut

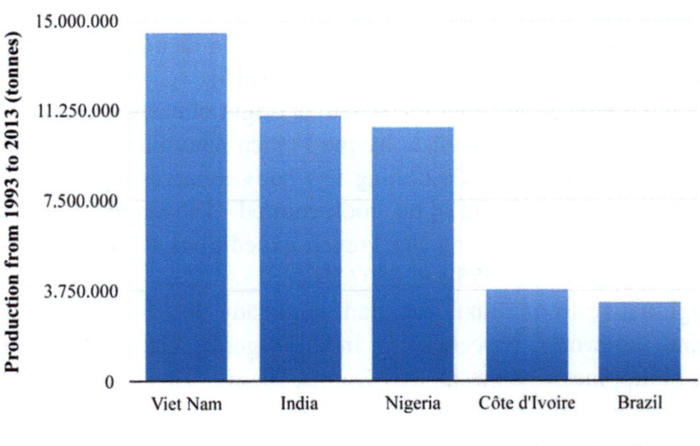

Graphic 2.2 Total cashew nut production (1993–2013) of the top 5 world producers

Taking into consideration the numbers presented above and the fact that CNSL comprises around 25 % of the cashew nut total weight, one can estimate that, only in 2013, about 1.000.000 tonnes of CNSL were obtained as by-product from the cashew nut industrial processing.

As mentioned before, being composed mainly by phenolic lipids, the dispose of such a large volume of CNSL becomes a real concern in light of the ecological issues it may raise. In this way, the other alternative is to make use of this agro-industrial by-product as a raw material in the synthesis of different chemical products, which will not only prevent the disposal-related problems, but also turn CNSL into a new commodity.

According to the Green Chemistry principles, the use of renewable feedstocks is an elegant approach to overcome some of the issues that the world is facing nowadays in order to achieve a more sustainable society.

Another interesting aspect important to point out is that the utilization of CNSL as a replenishable raw material will not compete for land or agricultural resources, not even food production, since the actual purpose in the cashew plantation is the obtainment of the valuable cashew kernel [23].

2.3 CNSL: Extraction and Chemical Composition

Cashew nutshell liquid is a reddish brown dark oil, with a peculiar odor, considered as one of the main natural sources of non-isoprenoid phenolic lipids. The main chemical compounds present in CNSL are: anacardic acid (1), cardanol (2), cardol (3), and 2-methylcardol (4).

These phenolic compounds have a very peculiar chemical structure, bearing an alkyl side chain comprised of 15 carbon atoms located in *meta*-position in comparison with their hydroxyl group(s). Another interesting characteristic is that these alkyl side chains may present different degrees of unsaturations, varying from none up to 3 double bonds, with cis (or Z) configuration, located in the carbons 8, 11, and/or 14 [35] (Fig. 2.3).

Depending on the method of obtainment, CNSL may present different chemical composition and, therefore, be classified into two main types: solvent-extracted CNSL (natural CNSL) and technical CNSL (tCNSL).

Natural CNSL is obtained by the utilization of some solvent extraction technique (commonly Soxhlet, supercritical carbon dioxide, or subcritical water) in order to obtain its constituents under mild conditions, without promoting any chemical modification. In this way, natural CNSL represents the original composition found in nature, which is basically composed by anacardic acids (60–70 %), cardols (10–20 %), cardanols (3–10 %), 2-methylcardols (2–5 %), and other minor constituents [19, 44].

On the other hand, technical CNSL is the by-product of the cashew nut industrial processing. Since the primary goal in the cashew industry is the obtainment of the valuable kernel, CNSL comes as a secondary product.

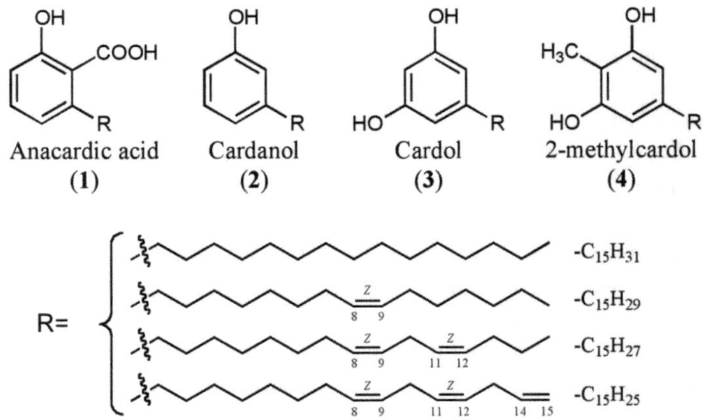

Fig. 2.3 Chemical structures of the main components of CNSL

Fig. 2.4 Decarboxylation of anacardic acid into cardanol

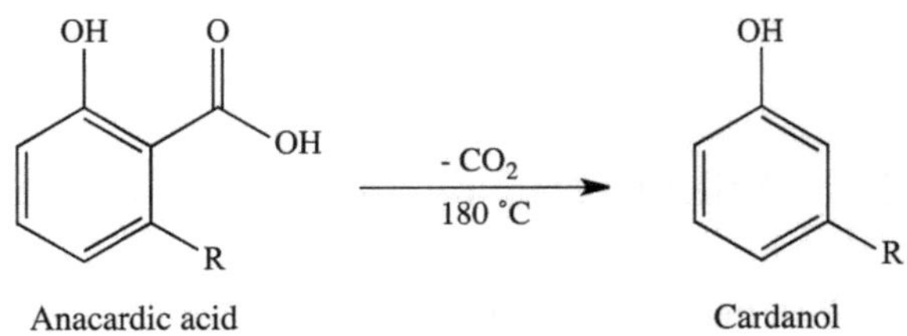

Initially, technical CNSL was employed as a source of phenolic compounds for the synthesis of phenol/formaldehyde polymers. Nowadays, with the advances in the chemistry of these phenolic lipids, tCNSL appears as an economically feasible source of phenolic components. In the industry, CNSL is extracted by an automated process that employs high temperatures in order to open the shell and recover the cashew kernel. In this process, named "hot-oil process," cashew nutshells are immersed in the CNSL and heated to 180–190 °C.

Under these temperatures, anacardic acid present in natural CNSL suffers a decarboxylation reaction converting itself into cardanol, generating technical CNSL, which is chemically differently from natural CNSL, composed mainly by cardanol (60–70 %), cardol (10–20 %), 2-methylcardol (2–5 %), polymeric materials (5–10 %), and other minor constituents (Fig. 2.4).

In Table 2.1 are described different compositions of natural and technical CNSL reported in the literature.

At first, these differences can be attributed to the fact that CNSL is a natural product, and therefore, its composition may vary according to the geographical coordinates, climate, and soil conditions.

Other important factors that may contribute for the differences in composition are the extraction techniques and the methods of analysis. On the basis of the extraction mode (solvent used, temperature, time, etc.) chemical compounds will

Table 2.1 Chemical composition of natural and technical CNSL

Compound	Natural CNSL			Technical CNSL		
	Tyman [57] (%)	Oliveira et al. [42] (%)	Paramashivappa et al. [43] (%)	Tyman [57] (%)	Andrade et al. [2] (%)	Kumar et al. [26] (%)
Anacardic acid (1)	71.65	62.90	63.00	–	–	–
Cardanol (2)	5.10	6.99	10.50	67.80	40.26	67.00
Cardol (3)	22.30	23.98	22.50	18.20	29.95	22.00
2-Methylcardol (4)	1.10	–	–	3.30	–	–

present distinct affinities for the mobile phase, thereby altering their concentration in the final product [63]. Although gas chromatography (GC) is an useful method for the study of phenolic compounds, HPLC is a much more efficient and sensitive technique [59].

Another interesting difference between the chemical composition of natural and technical CNSL is the existence of oligomeric and polymeric materials.

Tyman et al. [59], by using HPLC analysis under gradient elution with THF and acetonitrile, estimated quantitatively the presence of polymeric material in three diverse tCNSL samples.

They have found that the concentration and complexity of polymeric materials raise with the age of the sample, increasing from 7.38 %, in the case of a fresh sample, to as far as 21.63 %, for a five-year-old sample.

The eluted polymeric material was recovered and evaluated by TLC, which indicated an increase in complexity and polarity with the level of polymerization.

The authors also suggested that the majority of theses materials is composed by dimeric and trimeric oligomers with a relatively saturated alkyl side chain.

Since the recovery of the almond (kernel) in perfect conditions as an edible product is the main purpose in the industrial environment, the continuous heating of this process leads to the degradation of the CNSL obtained as a by-product, promoting the polymerization of its unsaturated compounds as well.

The mechanism of the polymerization reaction through the double bonds in the alkenyl side chain of cardanol has been studied in different conditions.

Manjula et al. [34] investigated the kinetics and mechanism of the oligomerization of cardanol using acid catalysts, while Rodrigues et al. [49] studied the influence of the temperature on the oligomerization of cardanol.

Since the harsh conditions in which cashew nutshells are submitted during industrial processing involve temperatures as high as 190 °C, and that in this condition the oxidation of organic compounds may lower the pH, these two situations may present a synergistic effect and therefore can be considered as the main factors that initiate the oligomerization of the double bounds present in the alkenyl side chain of the phenolic compounds of CNSL.

Upon consideration of data obtained by rheological analyses, ^{1}H nuclear magnetic resonance (NMR), and infrared spectroscopy (FT-IR), it was suggested that, in thermal oligomerization, the dimeric structure is the main oligomer that was

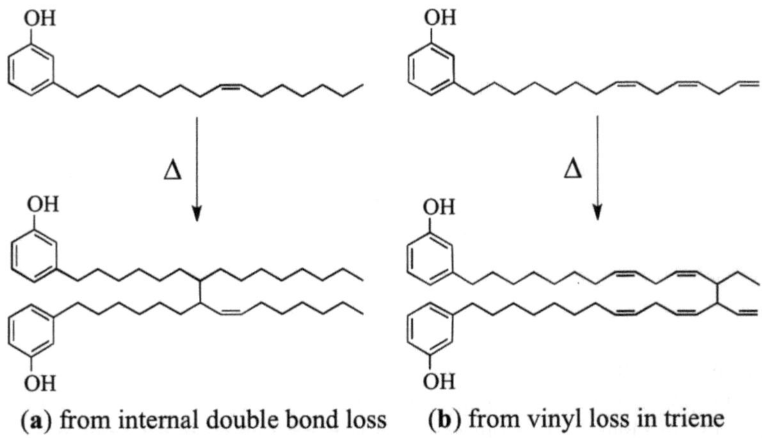

Fig. 2.5 Possible oligomerization reactions of cardanol under high temperatures

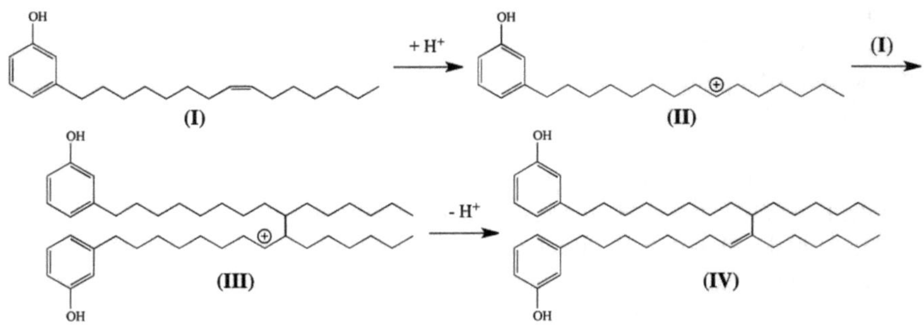

Fig. 2.6 Possible oligomerization reactions of cardanol under acidic conditions

formed by two possible reactions: (a) from internal double bond loss and (b) from vinyl loss in triene, as described in Fig. 2.5 [49].

Another possible route is described under acidic conditions. A suggested mechanism involves protonation of the side chain of cardanol, followed by a cationic chain growth reaction, as depicted in Fig. 2.6 [34].

2.4 Cardanol as a Sustainable Feedstock

With the continuous exploration for new renewable and biodegradable chemical resources, the large quantities of cashew nutshell liquid, especially technical CNSL, which possess elevated percentual of cardanol, have need for the development of purification/isolation processes of its main constituents in expectation to a authentic acceptation by the industrial market structure.

This should represent the first step into the promotion of a sustainable development, supported by a technological autonomy, due to the potential achievement of new high added value products derived from CNSL.

Based on the above assumptions, cardanol turns into one of the most important and promising chemical constituents of CNSL.

Once isolated in high purity, this green start material could be employed in many different industrial sectors, with emphasis in those that the prices of the final products are substantially elevated, like fine chemicals and pharmaceuticals.

Although cardanol is a phenol, compared to the main commercial mono-hydroxyl phenolic compounds used as starting materials, e.g., *p*-cresol, it presents some peculiarities in its physicochemical properties, due mainly by the long aliphatic side chain containing up to three double bonds (*Z* configuration), which enables a great number of functionalizations and hydrophobicity, low volatilization, no aggressive odor, and a higher boiling point, important characteristics for safe operations.

Furthermore, are also available the traditional reactive sites of phenolic chemical structures: the nucleophilic aromatic carbon atoms located at the *ortho*- and *para*-positions and the hydroxyl group (Fig. 2.7).

In an academic point of view, cardanol is a compound already known by many groups but, considering its potential, it still has a lot of space to grow. Many useful applications are found every year as it begins to substitute traditional phenolic raw materials in the synthesis of different classes of products, or new one is developed taking advantage of its unique properties.

Considering a bibliographic research involving the period of the last 30 years (1985–2015), made by a simple search using Scopus and SciFinder, the term "cardanol" exhibited 575 and 848 articles as results, respectively.

A more careful look at these results (Graphic 2.3) shows a progressive accession in the number of published articles, which indicates an increasing interest in this chemical compound and its applications more recently. Nevertheless, these numbers can be considered very inexpressive in view of the true potential that cardanol has as feedstock.

Taking into account all these particularities, it is not difficult to understand that the isolation of the constituents of cashew nutshell liquid is one of the fundamental drawbacks to reach its summit as an attractive starting material in a industrial sector searching for more sustainable alternatives.

In spite of the few different methods reported in the literature dealing with the obtainment of cardanol (and other constituents) from CNSL, they continue to collide into many adversities, e.g., the elevated price of reagents and solvents

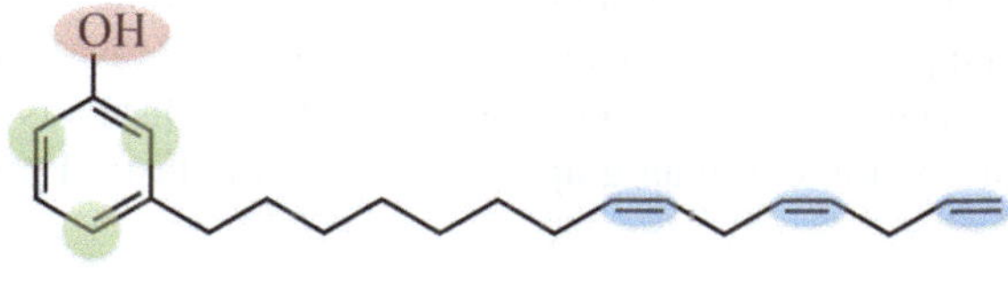

Fig. 2.7 Main reactive sites of cardanol molecule

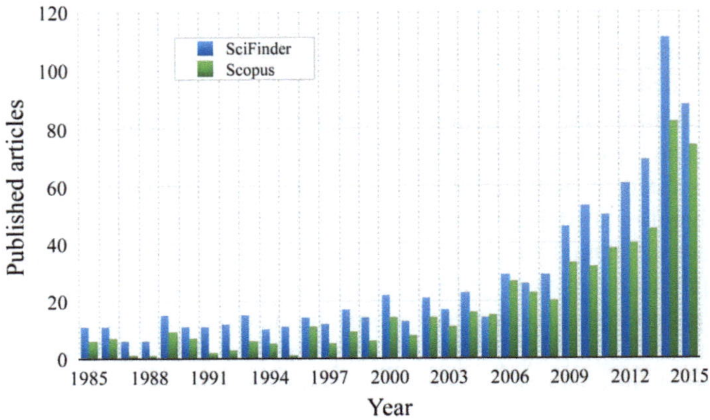

Graphic 2.3 Number of published articles, according to Scopus and SciFinder

involved in the process, low yields, contamination of the final product, and the cost-effectiveness of energy employed, invalidating remarkably its fulfillment in large scales.

Historically, the distillation process under reduced pressure was the first method to isolate cardanol from technical CNSL.

A patent by M. T. Harvey in 1937 [21] described that at a pressure of about 10 mmHg cardanol could be obtained in a temperature at about 225 °C, with very little amount of other compounds.

An interesting fact about this document is that there is made the first mention of the name "cardanol" to indicate the chemical compound originated by the decarboxylation of anacardic acid from natural CNSL under high temperatures, once previous publications used mistakenly the term "anacardol" [62], which identified a different mono-phenol isolated from the kernel of *Semecarpus anacardium*, to make reference to cardanol.

Currently, many articles still mention the use of distillation process in order to collect cardanol from commercially available or decarboxylated natural CNSL. Recently, the utilization of distilled cardanol as an additive for biofuels was reported. In that paper, the authors obtained cardanol at 230 °C at 0.2 mmHg and added in concentrations up to 20 % in diesel [50].

In another work, the synthesis of new oximes based on cardanol described a one-pot procedure for decarboxylation followed by a distillation process of natural CNSL. The flask containing CNSL was heated slowly to 180 °C, under 5 mm Hg, and after 20 min of the evolution of a carbon dioxide mist, the system was heated to 220–230 °C, temperatures at which a light yellow liquid was collected (70 % v/v) and posteriorly characterized by FT-IR and 1H NMR as high-purity cardanol [22].

The isolation of cardanol, with elevated purity, is only achieved through the use of column chromatography, which, in a industrial perspective, is inviable, due not only to the costs involved (silica gel, or other adsorbent) and amounts of solvent employed, but especially for being a time-consuming methodology.

Laboratory-scale processes for obtainment of pure cardanol are still based on chromatographic methods.

One of the first mentions of this process was made by Murthy et al. [41]. In this work, the authors described a procedure for the purification of 1.0 g of cardanol from fresh distilled CNSL (containing approximately 5 % cardol) using a 200-cm long glass column, filled with 50 g of silica gel (100–200 mesh). Different solvents were used (petroleum ether, benzene, ethyl acetate) as the polarity of eluent was increased. The authors also described the process of argentation of silica gel to separate the three different olefinic constituents of cardanol (mono-, di- and tri-unsaturated).

In 1986, Sood et al. [52] improved significantly the chromatographic technique to separate CNSL constituents. Prior to the advent of flash chromatography, the authors detailed the use of TLC type silica gel (much thinner than that used for tradition column chromatography) and pressurized air, in order to enhance the efficiency of the process.

Another progress made was the proportion of CNSL/adsorbent used, in the range of 1:5–1:6, in a column measuring 50 cm long. The experiments reported the use of 25.6 g of technical CNSL for 125 g of silica gel and a total volume of around 5.2 L of different solvents (light petroleum, diethyl ether and methanol), resulting in the separation of pure cardanol (17.68 g) and cardol (2.84 g), besides other fractions with lower efficiency.

By employing a two-step reversed-phase (RP) column chromatography, Yuliana et al. [63] successfully isolated isomers of the three major constituents of solvent-extracted CNSL in high purity. Using a mass ratio of CNSL to silica gel of 1:20 and methanol as mobile phase, the authors reported 81.8 % recovery of 3-*n*-pentadecylphenol (saturated cardanol) with 99.2 % purity, verified by HPLC analysis.

Tyman et al. [58] proposed a different methodology in order to isolate cardanol from technical CNSL.

In a work from 1992, a solvent extraction purification system was described as an improvement on the previous methods available in the scientific literature. A non-aqueous immiscible bi-phase solvent system composed of a solution of technical CNSL in light petroleum and one of many diols (ethane, propane, butane and pentane) was used to separate cardanol, which was recovered from the petroleum phase in yields above 90 % and purity, analyzed by HPLC, over 99 % [58].

Later, also through a phase extraction process, Kumar et al. [26] presented a methodology based on the acidity difference between cardanol and cardol.

Basically, a methanolic solution of technical CNSL mixed with ammonium hydroxide was extracted with portions of hexane, from which pure cardanol was recovered with great efficiency. An important point here, the authors mention, is that the scale used (up to 1 kg of tCNSL is reported) makes this an industrially feasible process for the obtainment of large quantities of cardanol.

Innumerous advances have been made in order to isolate cardanol from CNSL with great yields and high purity. Indeed, the need for improvements in the technical and economical performances of this agro-business demands a new and adequate development of practical techniques for the isolation of cardanol and, overall, its adequate chemical employment in the sense to aggregate value to a versatile by-product that continues to be overlooked inside and outside the academic field.

2.5 Synthetic and Technological Advances in the Cardanol Chemistry

In the last decade, our group has been dedicate its research activities to the development of new products and processes based on CNSL chemistry that justify its uses as ecofriendly technological materials, guided by a policy of aggregate value to its constituents, working in consonance with the green chemistry principles and the existing environmental legislation, in search of a sustainable industrial development.

The utilization of CNSL and its constituents has been reviewed in the literature in the last 20 years [19, 35, 39, 60, 53, 6, 7, 61] .

Although the main applications of cardanol were based on the production of resins and polymers [61], it did not take long for the appearance of new advanced materials, taking advantage of the peculiar chemical structure of cardanol [35, 39, 60, 6].

As presented in the beginning of this chapter, Brazil, Vietnam, Nigeria, Cote d'Ivory, and India are the largest producers and exporters of cashew kernel in the world.

At the same time, Africa, Asia, and South America are endemic regions for the globally known dengue fever. It is estimated that, globally, each year up to 50 million dengue cases occur, from which 500,000 cases are diagnosed as the more seriously related illness, dengue hemorrhagic fever (DHF) [13].

The main vector for dissemination of dengue is a mosquito, also responsible for the spreading of yellow fever and malaria, called *Aedes aegypti*.

Despite the number of cases and the efforts of many countries to control infestations, dengue is considered a neglected disease, which has no effective drug or vaccine.

The method for controlling mosquito proliferation continues to be prevention of the mosquito breeding through the use of larvicides, since in larval stage is easier, compared to adult stage, to control *Aedes aegypti* reproduction.

The main problems associated with the commonly used larvicides are the emergence of insect species showing resistance and the toxicity associated with nontarget organisms.

In face of these arguments, the utilization of CNSL and its components for controlling mosquito infestation could be of great interest and economically feasible in developing countries that cultivate cashew and are also endemic areas.

The biological assays of cardol and cardanol against *Aedes aegypti* larvae presented interesting values of LC_{50} (the concentration of compound to kill half of the sample population) of 14.20 ± 0.62 and 32.90 ± 0.25 mg/mL, which are considered promising results once substances with LC_{50} lower than 100 mg/mL are potential larvicidal candidates [30].

Exploring the biocomposites field, many studies involving the interaction of cardanol–formaldehyde thermoset resins reinforced with different types of natural fibers were made.

Fibers of jute (*Corcorus capsularis*) [8], green-dwarf coir (*Cocos nuciferas* L.) [9], and sisal (*Agave sisalana*) [10] were used in their natural form and chemically modified by alkaline treatment, to increase the thermal and mechanical properties of these biocomposites. In other works, the utilization of bamboo fibers (*Bambusa vulgaris*) [25] and sponge gourd fibers (*Luffa cylindrica*) [51] was also evaluated as reinforcements for cardanol resol-type resins. In all these studies, the chemical modification of fibers surface was important to improve the adhesion between fiber and matrix in the biocomposites and resulted in materials more homogeneous, resistant to thermal decomposition and with increased mechanical resistance.

Polybenzoxazines, a polymer originated by a ring-opening polymerization (ROP) of benzoxazine monomers, are bicyclic benzoheterocycles containing at least one oxazine ring, which are generated by the Mannich-like reaction of a phenol, formaldehyde, and an amine. This class of phenolic resin has attracted much attention due to its enhanced properties in comparison with traditional phenol formaldehyde resins, e.g., molecular design flexibility, no need of catalysts, excellent dimensional stability, low water absorption, and flammability.

Since the most used phenolic raw material for the synthesis of polybenzoxazines is still bisphenol-A (BPA), a pioneer work from Calò et al. [14] reported the utilization of cardanol as a renewable feedstock for the preparation of this type of resin and its application as a polymeric matrix in the production of bio-based composites.

Later on, another paper evaluated the synthetic feasibility of cardanol and other main constituents of technical CNSL as precursors of new mono- and

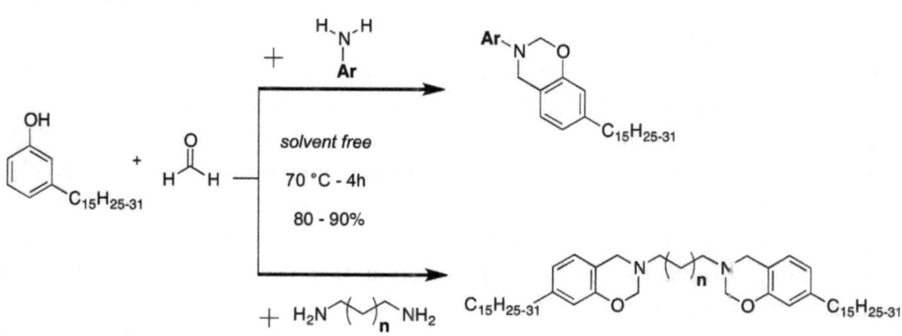

Fig. 2.8 Cardanol-based mono- and bis-benzoxazine synthetic scheme

bis-benzoxazines structures via cyclization reactions with formaldehyde and different aromatic amine derivatives under ecofriendly conditions (Fig. 2.8) [3].

Hindered phenolic compounds are well known for their antioxidant activity.

The mechanism responsible for this activity is based on the donation of a proton (an H atom, from the hydroxyl group) that reduces the free-radical oxidation reaction of organic molecules, promoted by peroxy species (ROO·).

This antioxidant effect can be enhanced by the presence of bulky groups linked to the aromatic ring of the phenolic structure. Butylated hydroxytoluene, commonly called BHT, is the most widely used phenolic antioxidant in the world and is synthesized by the alkylation of p-cresol.

As cardanol already has a 15 carbon atoms long side chain, which is considered a large group although located in *meta*-position, the investigation of its antioxidant capacity was evaluated by computational methods and electrochemical analyses (square wave voltammetry and cyclic voltammetry) and compared to BHT [32]. The results obtained showed that saturated cardanol presented a lower oxidation potential compared to BHT.

A further analysis [33] concluded that unsaturated cardanol isomer has even lower values, and experiments of square wave voltammetry demonstrated that the unsaturated compound presented higher current values than saturated compounds indicating that a higher number of unsaturated molecules were participating in the oxidation process.

In order to increase the steric hindrance over the phenolic hydroxyl of cardanol, Rios et al. synthesized two *ortho*-substituted derivatives of cardanol, adding to the aromatic ring a *tert*-butyl group [48] and a *tert*-amyl group [47, 46].

Both compounds, when added to mineral lubricant oils, reduced the formation of oxidation products and also proved to be more thermally stable than BHT.

Another important class of antioxidants used in the industry is the "secondary type antioxidants," which prevent oxidation by reacting with hydroperoxides formed by thermal oxidation reactions.

Organophosphates and organothiophosphates are the most common compounds in this class. They are usually combined with hindered phenols and applied as a synergetic formulation to retard the initiation of oxidative processes.

Cardanol was used as starting material in the synthesis of many useful substances that showed excellent activity as secondary type antioxidants.

In the work of Façanha et al., the antioxidant activity of a diethylphosphate derivative of hydrogenated cardanol in naphthenic mineral oils (NH10 and NH20) was investigated through thermogravimetric analysis (TGA).

The authors reported improvement in their thermal-oxidative stability by the increase in the temperature of initial degradation of the additivated samples when compared to the pure oils [18].

Another paper described the synthesis of a diethylthiophosphate derivative of cardanol and similar results of antioxidant activity [31].

Other organophosphorus compounds derived from cardanol were also synthesized and evaluated as antioxidants in mineral oils [29, 36], polymers [27], biodiesel [28], and others [45, 47, 46].

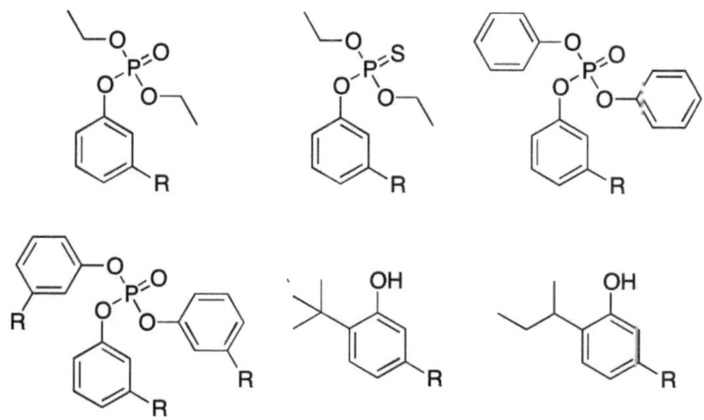

Fig. 2.9 Chemical structure of cardanol-based antioxidants

In Fig. 2.9 is described the chemical structure of some cardanol-based antioxidants.

Porphyrins and phthalocyanines are interesting classes of macrocyclic compounds. Due to their extended π-system and metal ion binding capacity, these compounds are commonly used in optical/electronic devices, as active layers for gas sensors, homogeneous/heterogeneous catalysis, organic dyes for photoinduced electron-transfer processes, liquid crystals, and OLEDs.

In homogeneous catalysis, the utilization of porphyrins and phthalocyanines is limited, since their active sites are strongly hindered due to the low solubility of these molecules in most commonly used organic solvents.

To overcome this situation, cardanol was used as raw material in the synthesis of many of these macroheterocyclic compounds. The presence of long alkyl chains of cardanol has shown a great influence in the physical properties of these derivatives, such as a low melting points and high solubility in organic solvents [4, 5] (Fig. 2.10).

A composite material obtained by impregnation of a Cu-substituted cardanol-based porphyrins onto TiO_2 (anatase phase) increased the photoactivity in the degradation of 4-nitrophenol in aqueous suspension. In this context was confirmed the importance of lipophilicity, photostability of the sensitizer, and the involvement of the Cu(II) in the process [38] (Fig. 2.10).

Cardanol-based porphyrins were also applied in the preparation of a superparamagnetic fluorescence nanosystem composite material, by the interaction of side chains derived from cardanol with oleic acid, which was assembled onto the Fe_3O_4 surface modified [15].

Sandwich-type phthalocyanines containing a rare earth metal core (Pr, Nd, Eu–Lu) and macrocycles peripherally substituted by pentadecylphenoxy groups were synthesized using a cardanol-based phthalonitrile precursor and the respective lanthanide acetate [54] (Fig. 2.10).

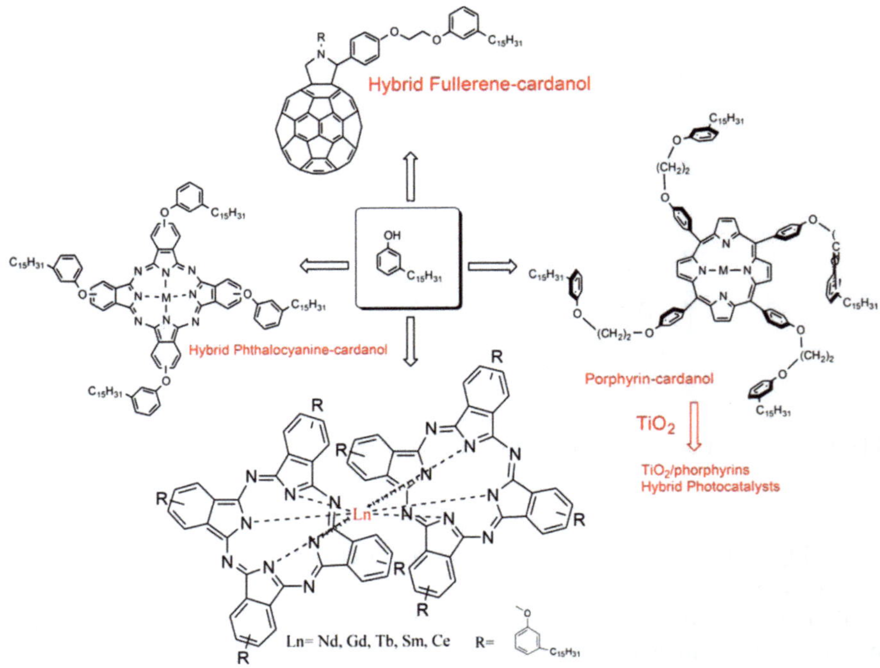

Fig. 2.10 Chemical structure of macroheterocyclic cardanol-based compounds

Transition metal-catalyzed reactions have been applied successfully to the transformation of cardanol-based precursors.

As shown in Fig. 2.11, palladium-catalyzed cyclocarbonylation of allylic cardanol and cardol derivatives produced selectively heterocycles having different ring sizes [1].

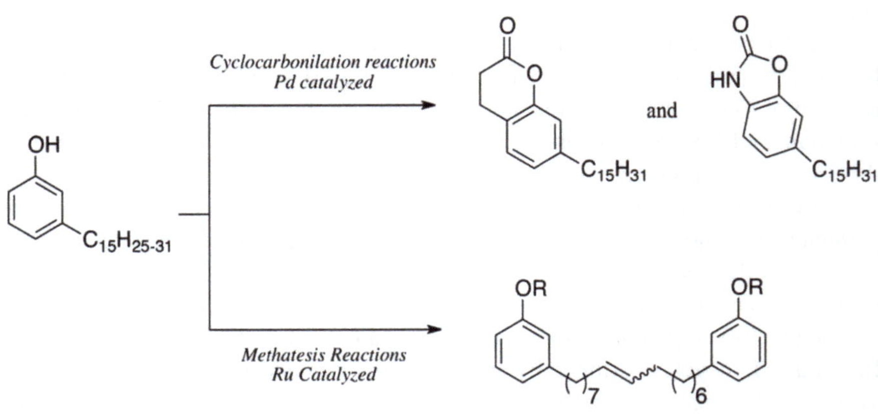

Fig. 2.11 Transition metal-catalyzed reactions involving CNSL-based precursors

Olefin metathesis reaction on cardanol is an important class of reactions that allows for the synthesis of new olefins.

The presence olefin group in the long chain attached to the *meta*-position of the phenolic ring allowed the preparation of new cardanol-based compounds via Ru-catalyzed metathesis reactions [20, 37, 60]. Cardanol was subjected to transfer hydrogenation catalyzed by $RuCl_3$ using isopropanol as a reductant.

Green nanocarriers based on the combination of natural renewable materials like cardanol and cholesterol have been used to encapsulate lipophilic or hydrophilic compounds such as the chlorogenic acids extracted from different rowanberries like *Sorbus Americana* and *Vaccinium* sp. [12, 11].

2.6 Final Considerations

Among the green chemistry concepts developed by P. Anastas and J. Warner in 1998, the 7th principle, which addresses the "use of renewable feedstock," can be highlighted when considering the use of CNSL and its constituents as starting material in the development of new chemical compounds.

As modern society faces the continued growth demand for chemicals, sustainable and innovative technologies are necessary in order to overcome current difficulties and also avoid the evolution of new ones. The advent of agro-industry waste as a resource of raw materials has attracted the attention of researchers and opened a window of opportunity for the breakthrough of alternative products to the oil industry.

The increasing number of research papers dedicated to this theme, and also the increasing production of cashew nuts turns cardanol into a promising renewable feedstock for the development of a wide range of functional products.

However, for this forthcoming industry to develop efficiently and consistently, are mandatory the investments in innovation, especially the development of new products and processes focused on the strengths and peculiarities of these phenolic lipids, not only seeking alternatives for the commercially known products, but also envisioning the arising of new ones with breakthrough properties and applications.

In any case, it is still valid to point out challenges involved in CNSL development, in particular when it comes to the efficiently isolation of cardanol, free from impurities, in high yields and industrial scales. Another important aspect to be considered is the need of a closer relationship between universities and research centers with the industrial segment, which is the driving force for technological transformations.

Acknowledgements The authors would like to thank the Brazilian agencies CNPq (Universal 449060/2014-4, PVE 401359/2014-0 and 402566/2013-0) and Capes for the financial support.

References

1. Amorati R, Attanasi OA, El Ali B et al (2002) Synthesis of new cardanol and cardol derivatives by allylation and regioselective cyclocarbonylation reactions. Synthesis 18:2749–2755
2. Andrade TJAS, Araújo BQ, Citó AMGL et al (2011) Antioxidant properties and chemical composition of technical Cashew Nut Shell Liquid (tCNSL). Food Chem 126:1044–1048
3. Attanasi OA, Behalo MS, Favi G et al (2012) Solvent free synthesis of novel mono- and bis-benzoxazines from cashew nut shell liquid components. Curr Org Chem 16:2613–2621
4. Attanasi OA, Ciccarella G, Filippone P et al (2003) Novel phthalocyanines containing cardanol derivatives. J Porphyr Phthalocyanines 07:52–57
5. Attanasi OA, Del Sole R, Filippone P et al (2004) Synthesis of novel lipophilic porphyrin-cardanol derivatives. J Porphyr Phthalocyanines 08:1276–1284
6. Balachandran VS, Jadhav SR, Vemula PK et al (2013) Recent advances in cardanol chemistry in a nutshell: from a nut to nanomaterials. Chem Soc Rev 42:427–438
7. Balgude D, Sabnis AS (2014) CNSL: an environment friendly alternative for the modern coating industry. J Coating Tech Res 11:169–183
8. Barreto ACH, Esmeraldo MA, Rosa DS et al (2010) Cardanol biocomposites reinforced with jute fiber: microstructure, biodegradability, and mechanical properties. Polym Compos 31:1928–1937
9. Barreto ACH, Júnior AEC, Freitas JEB et al (2013) Biocomposites from dwarf-green Brazilian coconut impregnated with cashew nut shell liquid resin. J Compos Mater 47:459–466
10. Barreto ACH, Rosa DS, Fechine PBA et al (2011) Properties of sisal fibers treated by alkali solution and their application into cardanol-based biocomposites. Compos Part A Appl Sci 42:492–500
11. Bloise E, Becerra-Herrera M, Mele G et al (2014) Sustainable preparation of cardanol-based nanocarriers with embedded natural phenolic compounds. ACS Sustain Chem Eng 2:1299–1304
12. Bloise E, Carbone L, Colafemmina G et al (2012) First example of a lipophilic porphyrin-cardanol hybrid embedded in a cardanol-based micellar nanodispersion. Molecules 17:12252–12261
13. Callaway E (2007) Dengue fever climbs the social ladder. Nature 448:734–735
14. Calò E, Maffezzoli A, Mele G et al (2007) Synthesis of a novel cardanol-based benzoxazine monomer and environmentally sustainable production of polymers and bio-composites. Green Chem 9:754–759
15. Clemente CS, Ribeiro VGP, Sousa JEA et al (2013) Porphyrin synthesized from cashew nut shell liquid as part of a novel superparamagnetic fluorescence nanosystem. J Nanopart Res 15:1–10
16. Duarte O, Paull RE (2015) Exotic fruits and nuts of the new world. CABI Publishing, Boston
17. FAOSTAT (2015) Food and Agriculture Organization of the United Nations. http://faostat.fao.org. Accessed 20 Dez 2015
18. Façanha MAR, Mazzetto SE, Carioca JOB et al (2007) Evaluation of antioxidant properties of a phosphorated cardanol compound on mineral oils (NH10 and NH20). Fuel 86:2416–2421
19. Gedam PH, Sampathkumaran PS (1986) Cashew Nut Shell Liquid: extraction, chemistry and applications. Prog Org Coat 14:115–157
20. Guo Y-C, Mele G, Martina F et al (2006) An efficient route to biscardanol derivatives and cardanol-based porphyrins via olefin metathesis. J Organomet Chem 691:5383–5390
21. Harvey M (1937) Process of destructively distilling cashew nut shell liquid. US Patent 2,098,824, 9 Nov 1937
22. Hoang AS, Tran TH, Nguyen HN et al (2015) Synthesis of oxime from a renewable resource for metal extraction. Korean J Chem Eng 32:1598–1605

23. Horváth IT, Anastasia PT (2007) Innovations and green chemistry. Chem Rev 107:2169–2173
24. Janick J, Paull RE (eds) (2008) Encyclopedia of fruit and nuts. CABI Publishing, Boston
25. Júnior AEC, Barreto ACH, Rosa DS et al (2015) Thermal and mechanical properties of biocomposites based on a cashew nut shell liquid matrix reinforced with bamboo fibers. J Comp Mater 49:2203–2215
26. Kumar PP, Paramashivappa R, Vithayathil PJ et al (2002) Process for isolation of cardanol from technical Cashew (Anacardiumoccidentale L.) Nut Shell Liquid. J Agric Food Chem 50:4705–4708
27. Lomonaco D, Cangane FY, Mazzetto SE (2011) Thiophosphate esters of cashew nutshell liquid derivatives as new antioxidants for poly(methyl methacrylate). J Therm Anal Calorim 104:1177–1183
28. Lomonaco D, Maia FJN, Clemente CS et al (2012) Thermal studies of new biodiesel antioxidants synthesized from a natural occurring phenolic lipid. Fuel 97:552–559
29. Lomonaco D, Maia FJN, Mazzetto SE (2013) Thermal evaluation of cashew nutshell liquid as new bioadditives for poly(methyl methacrylate). J Therm Anal Calorim 111:619–626
30. Lomonaco D, Santiago GMP, Ferreira YS et al (2009) Study of technical CNSL and its main components as new green larvicides. Green Chem 11:31–33
31. Lopes AAS, Carneiro EA, Rios MAS et al (2008) Study of antioxidant property of a thiosphorated compound derived from cashew nut shell liquid in hydrogenated naphthenics oils. Braz J Chem Eng 25:119–127
32. Maia FJN, Clemente CS, Oliveira TMBF et al (2012) Electrochemical and computational studies of phenolic antioxidants from cashew nut shell liquid. Electrochim Acta 79:67–73
33. Maia FJN, Ribeiro FWP, Rangel JHG et al (2015) Evaluation of antioxidant action by electrochemical and accelerated oxidation experiments of phenolic compounds derived from cashew nut shell liquid. Ind Crop Prod 67:281–286
34. Manjula S, Kumar VC, Pillai CKS (1992) Kinetics and mechanism of oligomerization of cardanol using acid catalysts. J Appl Polym Sci 45:309–315
35. Mazzetto SE, Lomonaco D, Mele G (2009) Óleo da castanha de caju: oportunidades e desafios no contexto do desenvolvimento e sustentabilidade industrial. Quím Nova 32:732–741
36. Mazzetto SE, Oliveira LDM, Lomonaco D et al (2012) Antiwear and antioxidant studies of cardanol phosphate ester additives. Braz J Chem Eng 29:519–524
37. Mele G, Li J, Vasapollo G (2008) Fine chemicals from cardanol via cross metathesis reaction. Chim Oggi 26:72–74
38. Mele G, Del Sole R, Vasapollo G et al (2004) Polycrystalline TiO2 impregnated with cardanol-based porphyrins for the photocatalytic degradation of 4-nitrophenol. Green Chem 6:604–608
39. Mele G, Vasapollo G (2008) Fine chemicals and new hybrid materials from cardanol. Mini-Rev Org Chem 5:243–253
40. Mitchell JD, Mori SA (1987) The cashew and its relatives (Anacardium: Anacardiaceae). Mem New York Bot Gard 42:1–76
41. Murthy BG, Samban MA, Aggarwal JS (1968) Identification of some naturally occurring alkyl-substituted phenols in cashew-nut shell liquid by chromatographic techniques. J Chromatogr 32:519–528
42. Oliveira MSC, Morais SM, Magalhães DV (2011) Antioxidant, larvicidal and anti acetylcholinesterase activities of cashew nut shell liquid constituents. Acta Tropic 117:165–170
43. Paramashivappa R, Kumar PP, Vithayathil PJ (2001) Novel method for isolation of major phenolic constituents from Cashew (Anacardium occidentale L.) Nut Shell Liquid. J Agric Food Chem 49:2548–2551
44. Patel RN, Bandyopadhyay S, Ganesh A (2006) Extraction of cashew (Anacardium occidentale) nut shell liquid using supercritical carbon dioxide. Bioresour Technol 97:847–853
45. Rios MAS, Mazzetto SE (2012) Thermal behavior of phosphorus derivatives of hydrogenated cardanol. Fuel Process Technol 96:1–8

46. Rios MAS, Nascimento TL, Santiago SN et al (2009) Cashew Nut Shell Liquid: a versatile raw material utilized for syntheses of phosphorus compounds. Energy Fuel 23:5432–5437
47. Rios MAS, Sales FAM, Mazzetto SE (2009) Study of antioxidant properties of 5-n-pentadecyl-2-tert-amylphenol. Energy Fuel 23:2517–2522
48. Rios MAS, Santiago SN, Lopes AAS et al (2010) Antioxidative activity of 5-n-pentadecyl-2-tert-butylphenol stabilizers in mineral lubricant oil. Energy Fuel 24:3285–3291
49. Rodrigues FHA, Souza JRR, França FCF et al (2006) Thermal oligomerisation of cardanol. e-Polymers 6:1027–1040
50. Sanjeeva SK, Pinto MP, Narayanan MM et al (2014) Distilled technical cashew nut shell liquid (DT-CNSL) as an effective biofuel and additive to stabilize triglyceride biofuels in diesel. Renew Energy 71:81–88
51. Silva AL, Silva LRR, Camargo IA et al (2015) Cardanol-based thermoset plastic reinforced by sponge gourd fibers (Luffa cylindrica). Polímeros (in press)
52. Sood SK, Tyman JHP, Durrani A et al (1986) Practical liquid chromatographic separation of the phenols in technical cashew nutshell liquid from Anacardium occidentale. Lipids 21:241–246
53. Stasiuk M, Kozubek A (2010) Biological activity of phenolic lipids. Cell Mol Life Sci 67:841–860
54. Słota R, Dyrda G, Hofer M et al (2012) Novel lipophilic lanthanide bis-phthalocyanines functionalized by pentadecylphenoxy groups: synthesis, characterization and UV-photostability. Molecules 17:10738–10753
55. Thevet A (1558) The singularities of France Antarctica, otherwise called America, and of several Lands and Isles discovered of our times
56. Thévet A (1997) Le Brésil d'André Thévet: Les singularités de la France antarctique (1557). Edition Chandeigne, Paris
57. Tyman JHP (1996) Synthetic and natural phenols. Elsevier, Amsterdam
58. Tyman JHP, Bruce IE, Payne P (1992) Specific isolation of phenolic lipids by phase separations. Nat Prod Lett 1:117–120
59. Tyman JHP, Tychopoulos V, Colenutt BA (1981) Long-chain phenols. J Chromatogr A 213:287–300
60. Vasapollo G, Mele G, Del Sole R (2011) Cardanol-based materials as natural precursors for olefin metathesis. Molecules 16:6871–6882
61. Voirin C, Caillol S, Sadavarte NV et al (2014) Functionalization of cardanol: towards biobased polymers and additives. Polym Chem 5:3142–3162
62. Wasserman D, Dawson C (1948) Cashew Nut Shell Liquid. III. The cardol component of Indian Cashew Nut Shell Liquid with reference to the liquid's vesicant activity. J Am Chem Soc 70:3675–3679
63. Yuliana M, Tran-Thi NY, Ju Y-H (2012) Effect of extraction methods on characteristic and composition of Indonesian cashew nut shell liquid. Ind Crop Prod 35:230–236

Chapter 3
Cardanol-Based Heterocycles: Synthesis and Applications

Giuseppe Mele, Diego Lomonaco and Selma E. Mazzetto

Abstract This chapter provides an overview of the progresses of research activities related with the synthesis of functional heterocyclic systems utilizing mainly cardanol—or structural analogous (i.e., anacardic acid, cardol, methylcardol) deriving from the cashew nut shell liquid—as starting material, a renewable feedstock having significant perspectives in the field of the green chemistry and design of innovative environmentally friendly chemical processes.

3.1 Introduction

The use of renewable feedstocks rather than depleting natural resources is one of the principles of the green chemistry. In this perspective, the importance of cardanol, a low-cost largely available renewable material obtained from the distillation of CNSL (cashew nut shell liquid) well-known waste of the cashew agro-industry, is increased last years.

This chapter is dedicated to honour the career of Professor Orazio A. Attanasi and Professor Giuseppe Vasapollo. They made significant contributions to researches dealing with heterocycles and technologically sustainable chemical processes related with the chemistry of cardanol and related compounds.

G. Mele (✉)
Department of Engineering for Innovation, University of Salento, Via Arnesano, 73100 Lecce, Italy
e-mail: giuseppe.mele@unisalento.it

D. Lomonaco · S.E. Mazzetto
Department of Organic and Inorganic Chemistry, Federal University of Ceará, Campus Pici, 60440-900 Fortaleza, Brazil

P. Anilkumar (ed.), *Cashew Nut Shell Liquid*,
DOI 10.1007/978-3-319-47455-7_3

The previous chapter (Lomonaco et al.) has been focused mainly on the origin and chemical composition of the different types of CNSL, methods of isolation, and main applications of its major components.

In this chapter, the synthetic strategies for the preparation cardanol-based heterocycles with different ring sizes and their use as precursors of functional organic, soft, and composite materials have been reported.

In particular, will be described how the molecular features of cardanol, such as the –OH group bonded to the aromatic ring, the aromatic ring itself, and the variously (un)saturated long chain attached to the meta position of the phenolic ring, allow many chemical transformations producing a large variety of fine chemicals and multifunctional heterocyclic hybrid materials.

3.2 Cardanol-Based Oxiranes (Epoxides)

Oxiranes are strained cyclic ethers with three-atom rings which can confer to them high reactivity; for this reason, they can be conveniently produced on a large scale for many applications. Generally, many epoxides are produced by treating alkenes with peroxide-containing reagents, which donate a single oxygen atom.

Cardanol, as well as its derivatives such as cardanol acetate (**CA**) and cardanol glycidyl ether (**CGE**) prepared through the substitution of the hydroxyl group with acetate or glycidyl ether groups as shown in the Scheme 3.1, can be exhaustively epoxidated by reacting with 3-chloroperbenzoic acid (3-CPBA) as shown in Scheme 3.2.

The mono-epoxidized cardanol was used as reactive diluent for the curing reactions of diglycidyl ether of bisphenol-A (DGEBA) in the presence of other curing agents such as anhydrides and triethyl amine as catalyst [41].

Enzymatic epoxidation and polymerization of cardanol was also obtained curing of epoxide containing polycardanol [27].

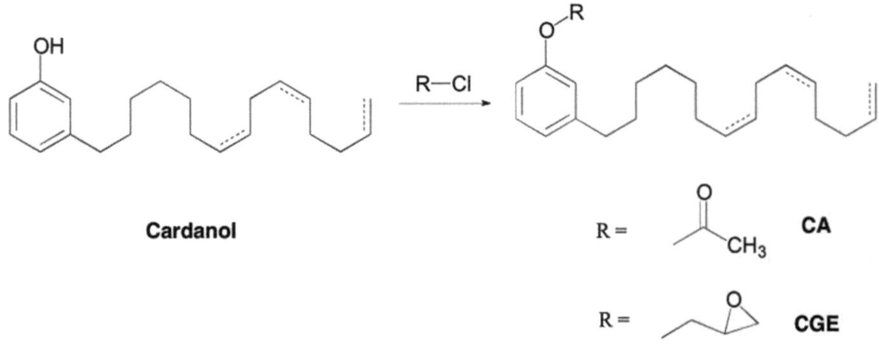

Scheme 3.1 Synthesis of cardanol acetate (**CA**), cardanol glycidyl ether (**CGE**) from cardanol

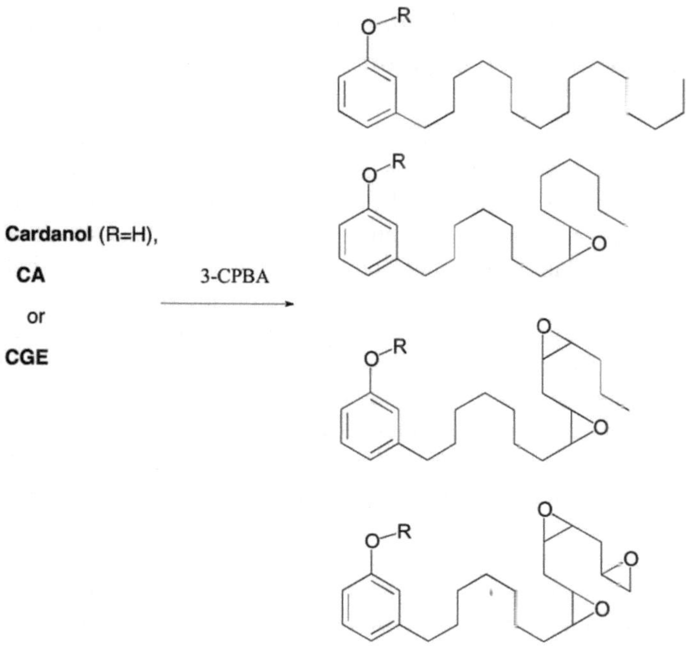

Scheme 3.2 Epoxydation of cardanol, **CA** or **CGE**

CA and **CGE** have been successfully used as plasticizer for poly(vinyl chloride)-based polymers [16, 23]. Self-curability of the coating films has been achieved via anionic polymerization [52].

CGE was also cured by electron-beam radiation in the presence of cationic photoinitiators such as triarylsulfonium hexafluorophosphate (P-type) and triarylsulfonium hexafluoroantimonate (Sb-type), and the properties of the cured product were influenced on the type and concentration of cationic photoinitiators and the electron-beam absorption dose [17].

Cathodic electrodeposition of epoxidated of cardanol–formaldehyde novolacs with epichlorohydrin, and subsequently modified with secondary amine, produced coating binders as modified polyepoxide used as primer coat for the protection of metallic substrates against corrosion [29].

New classes of biobased materials have been prepared from the reaction between cardanol-derived epoxy compounds and diamines [20, 21, 26].

Cardanol-based epoxides have been used as flexibilizers for novolac epoxy resins [2, 22]. Epoxidized cardanol was opportunely cured with two different amines, isophorone diamine and Jeffamine T403, and the materials exhibited good properties but not enough to replace the bisphenol-A diglycidyl ether in epoxy networks. Two kinds of sucrose epoxy derivatives, sorbitol and isosorbide, were used as blends to enhance the properties of the epoxy cardanol-derived materials [19].

3.3 Benzoxazines from Cardanol and Structural Analog Components Deriving from CNSL

Benzoxazines are fused benzoheterocycles containing N and O heteroatoms synthesized through Mannich-like condensation of a phenol, formaldehyde, and amine. Of course, the syntheses of benzoxazines starting from renewable organic raw materials deserve particular attention of both academic and industrial research.

Cardanol and related compounds deriving from the cashew nut shell liquid—through a condensation reaction involving the aromatic ring—are suitable precursors of benzoxazines. CNSL and its components (mainly cardanol, cardol, and methylcardol) can be considered as suitable starting materials for the synthesis of various substituted benzoxazines through sustainable processes.

The high percentage of renewable and the moderate use of other auxiliary chemicals under organic solvent-free and mild reaction conditions are the peculiarities of these reactions, having an overall reduced impact of harmful chemicals in the environment.

Several research groups have reported the synthesis of various benzoxazines (i.e., cardanol/ammonia, cardanol/aniline, cardanol/furfurylamine, cardanol/allylamine, cardanol/ethanolamine, cardanol/aromatic, or aliphatic diamine).

A large variety of cardanol-based mono-benzoxazines have been prepared through Mannich-like condensation involving cardanol, formaldehyde, and ammonia or primary (aromatic or aliphatic) amines as shown in the Scheme 3.3 [10, 14, 61, 62].

Cardanol-based bis-benzoxazines were prepared by reacting with cardanol, formaldehyde, and primary aliphatic (linear/branched) or aromatic diamines as shown in the Fig. 3.1 [10, 28, 32, 33, 49, 50].

Variously substituted bis-benzoxazines whose structures are reported in Scheme 3.4 have been prepared starting from the "resorcinol-type" components of cashew nut shell liquid that are cardol and methylcardol.

Tris- and tetra- benzoxazines shown in Fig. 3.2 have been prepared by using tris and tetra p-aminophenylmethane reacted, respectively, with cardanol and paraformaldehyde [53].

Cardanol-based polymers and polymeric materials are both well documented in the literature and still on focus of continuing research.

Among the most important fields of application of benzoxazines is polymer chemistry and technology, as they are used as monomers of polybenzoxazine resins of unusual but attractive and useful properties with excellent dimensional stability, low water absorption, and flammability.

Effect of structure on the thermal behavior of cardanol-based mono- and bis-benzoxazines was studied [34, 35]. Cardanol-based benzoxazines can be used as reactive diluents or toughening agents of conventional benzoxazines [15, 30].

The mono-functional benzoxazine monomer prepared by condensation reaction of aniline and formaldehyde with pure cardanol [43] or blended with other benzoxazines (no cardanol-based) has been used as precursor of various biocomposites.

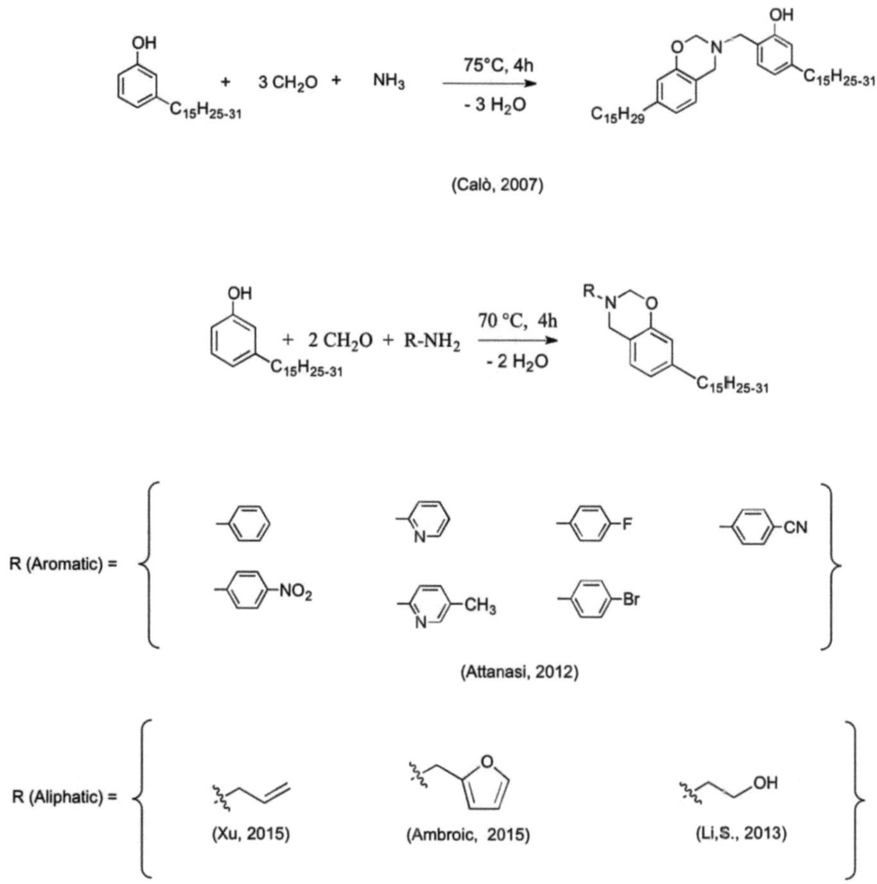

Scheme 3.3 General features of variously substituted 3,4-dihydro-2H-1,3-benzoxazines based on cardanol

Interestingly, the epoxidation of aniline-cardanol-based benzoxazine permitted the achievement of a higher cross-linking of the resulting thermoset polybenzoxazines [3].

Copolymer system based on the cardanol benzoxazine and epoxy [45] as well as benzoxazine and hydroxy benzoxazoline [44] monomers has been developed. Low-temperature curable mixtures were obtained combining opportunely the monomers content in the copolymers.

Biocomposites were also prepared by the combination of mono-functional cardanol-based benzoxazines with natural fibers, such as jute fibers [14, 43]. Hybrid organic–inorganic materials have been prepared by solgel technique combining a benzoxazine monomer obtained by condensation reaction of ethanolamine, formaldehyde, and cardanol with SiO_2 [31].

Graphene or graphene oxide were used in combination with benzoxazines based on cardanol to produce new composites [61] some of which were used for capacitor applications [47].

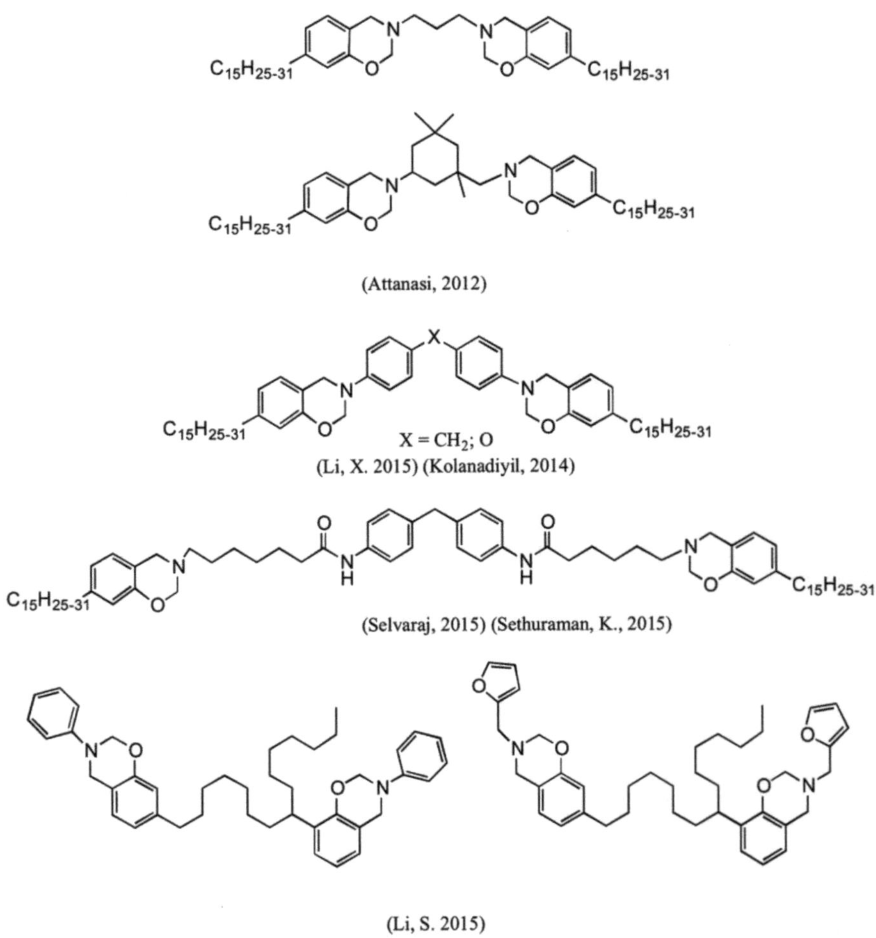

Fig. 3.1 Molecular structure of variously substituted of bis-benzoxazines-based cardanol

Microencapsulated cardanol-derived benzoxazines were used also for self-healing applications, [51] and 1,3-bis (benzoxazine) prepared from cashew nut shell liquid and diaminodiphenyl methane has been used for the preparation of composites with wood flour [1].

3.4 Glycolipids Based on Cardanol

Recent advances in cardanol chemistry are concerning a new class of amphiphilic molecules produced by the combination of cardanol with sugars in their heterocyclic form. In Fig. 3.3 are reported the structures of the glycolipids prepared by

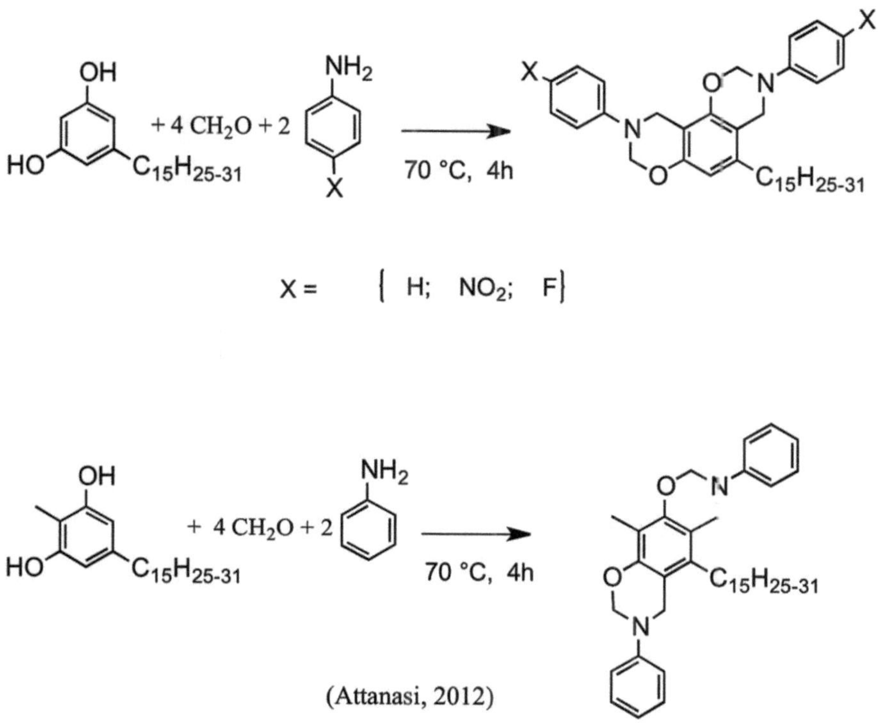

Scheme 3.4 Molecular structure of bis-benzoxazines-based on cardol and methylcardol

attaching glucopyranose to cardanol [11] as well as of the glucose-triazole-hydrogenated cardanol [57] which are interesting for their solgel properties [46].

Closed-chain sugars are also a class of fascinating polar and non-ionic heterocyclic which can be part of an amphiphile moiety [55].

A fluorescent molecular probe has been recently used in order to study the thermotropic gelation-inducing changes in micropolarity and microviscosity of hydrogel derived from glucose-triazole-hydrogenated cardanol conjugate. [56].

3.5 Cyclocarbonylation Derivatives from Cardanol or Structural Analog

Lactones are structural elements often found in natural products, which have medicinal value.

Seven-membered ring lactones can be prepared selectively by a typical palladium-catalyzed reaction of carbon monoxide involving the –OH group and an opportunely bonded allyl group in ortho-position to the –OH group [58].

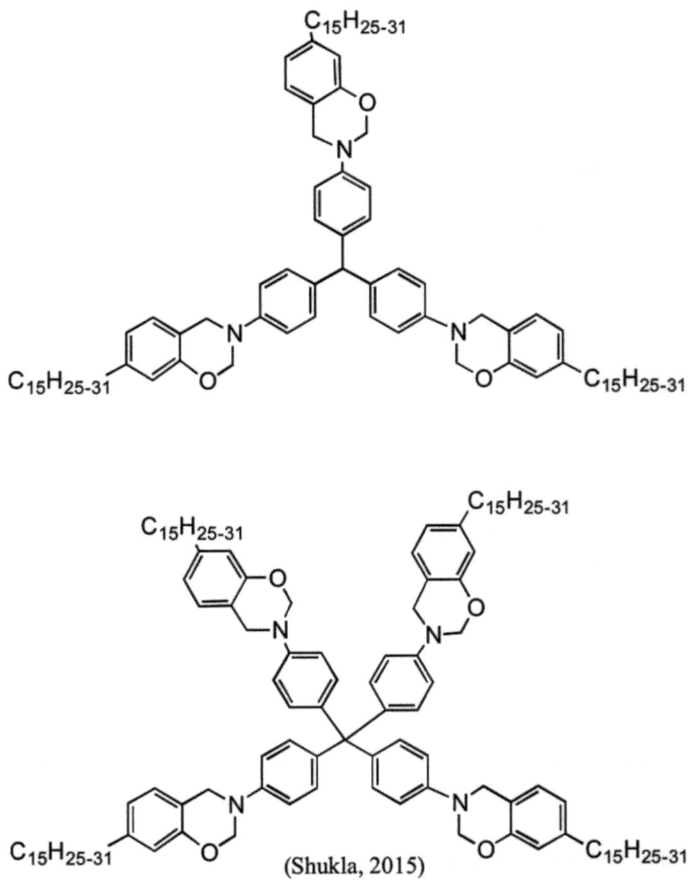

Fig. 3.2 Molecular structure of tris- and tetra-benzoxazines based on cardanol

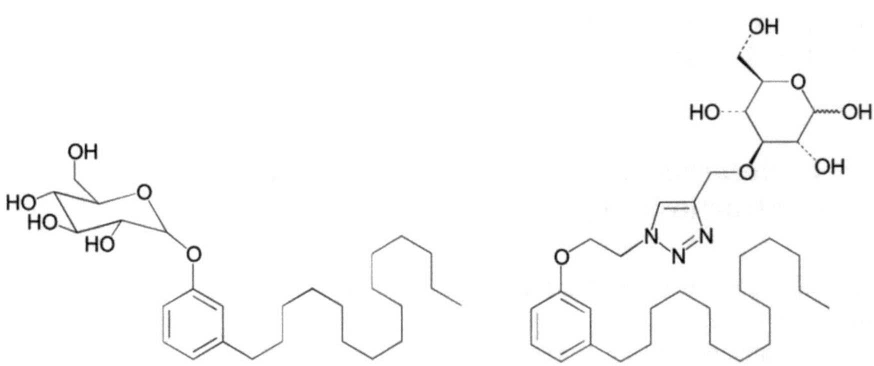

Fig. 3.3 Molecular structure of an amphiphilic glycolipids based on cardanol

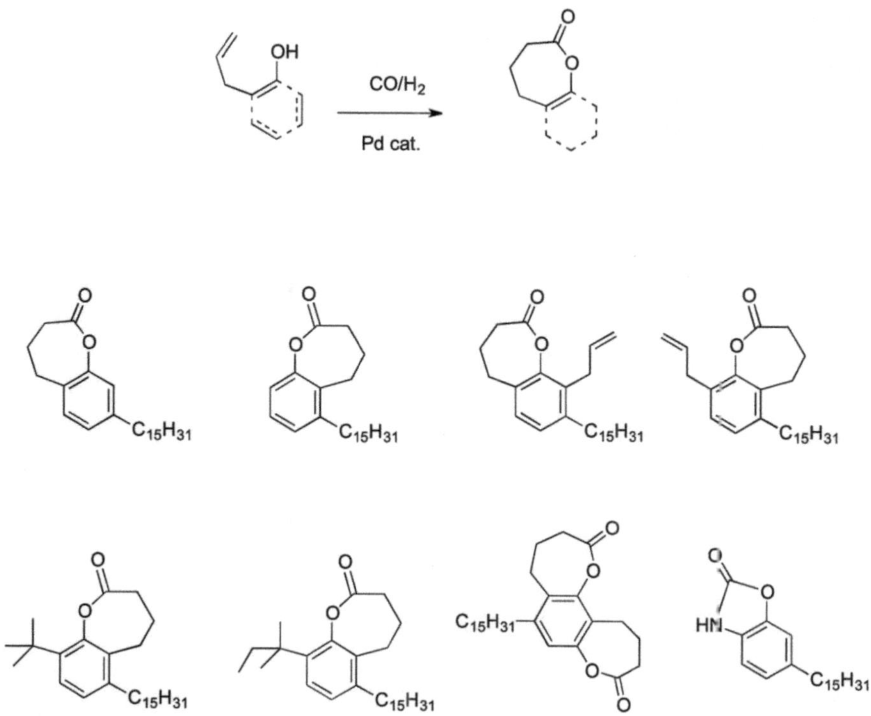

Fig. 3.4 Cyclocarbonilation derivatives obtained by using cardanol (or cardol) as precursor

This reaction has been successfully applied for the syntheses of the ring mono- and bis-cardanol-based lactones as well as at the five-membered ring benzoxazol-2-one as shown in Fig. 3.4 [4, 8, 37].

The unsaturated lactone (Scheme 3.5), 8-hydroxy-3-tridecyl-1*H*-isochromen-1-one, has been synthesized by isomerization of anacardic acid having heterogeneous alkyl side chains (a mixture of mono-, di-, and tri-unsaturated anacardic acid). Hydrogenation of 8-hydroxy-3-tridecyl-1*H*-isochromen-1-one produced a saturated lactone, 8-hydroxy-3-tridecyl-3,4-dihydroisochromen-1-one [39].

3.6 Porphyrins, Phthalocyanines, and Fulleropyrrolidines Based on Cardanol

Porphyrins, phthalocyanines, and fullerenes are compounds with great potential for serving as components for molecular materials that possess unique electronic, magnetic, and photophysical properties. In particular, porphyrins and phthalocyanines are perfectly suited for their integration in light energy conversion systems.

Scheme 3.5 Conversion of anacardic acid to 8-hydroxy-3-tridecyl-3,4-dihydroisochromen-1-one and 8-hydroxy-3-tridecyl-1H-isochromen-1-one

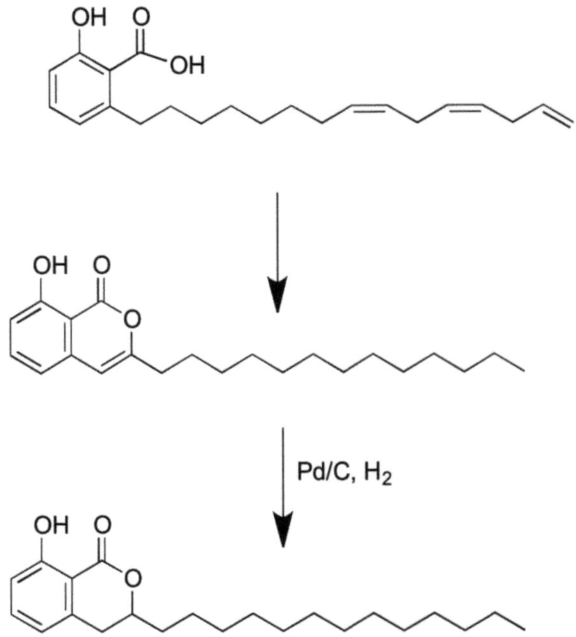

These colored macrocycles exhibit very attractive physical properties, particularly very high extinction coefficients in the visible and near-IR regions, where the maximum of the solar photon flux occurs, that is, necessary for efficient photon harvesting, besides a rich redox chemistry, as well as photoinduced electron transfer and semiconducting capabilities.

Cardanol, also in its hydrogenated form, was used as starting material to prepare the aldehydic precursors used for lipophilic cardanol-based metal-free and metallo-meso-tetraarylporphyrins [6, 7, 42] and fulleropyrrolidines [6, 7] some of which are reported in Scheme 3.6.

These compounds exhibited good solubility in common organic solvents as well as ease of manipulation and suitability for further transformation due to the presence of cardanol chains.

Porphyrins having different symmetry reported in Fig. 3.5 have been prepared by statistic synthesis of 4-[2-cardanil ethoxy]-benzaldehyde with benzaldehyde or other dipyrromethane [59, 60].

Porphyrins bearing cardanol as substituent containing unsaturated chains can be easily epoxidized by reaction with 3-chloroperbenzoic acid to produce typical structures such for example the tetra-epoxidized porphyrins shown in Fig. 3.6.

Selected structures of heterocyclic macromolecules obtained by the transformation of the unsaturated chains of cardanol by metathesis reactions [9, 24, 25, 38, 59, 60] are reported in Figs. 3.7 and 3.8.

Metallophthalocyanines containing a peripheral substituent 3-*n*-pentadecylphenol have been synthesized and characterized. The preparation of the

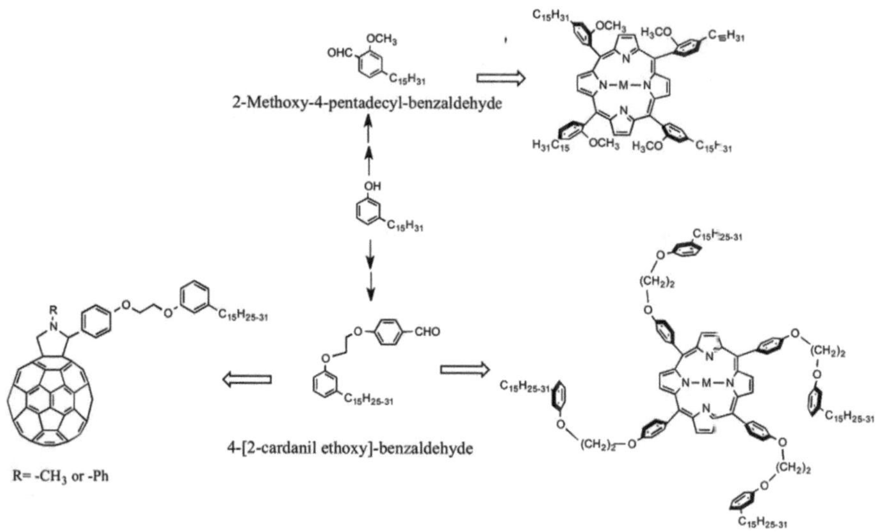

Scheme 3.6 Aldehydic precursors and structures of porphyrins and fulleropyrrolydines based on cardanol

phthalonitrile derivatives, precursors of the phthalocyanines, was accomplished by a simple base-catalyzed nucleophilic aromatic nitro displacement of 4-nitrophthalonitrile with an alcohol or phenol. Thus, as shown in Scheme 3.7, 4-(3-pentadecyl-phenoxy)-phthalonitrile precursors have been synthesized by reacting 3-*n*-pentadecylphenol derivatives with 1,2-dicyano-4-nitrobenzene that was tetramerized in ethanol in the presence of the corresponding metal salt MCl_2 (M = Cu, Zn, Pd, Ni) and 1,8-diazabicyclo [5.4.0] undec-7-ene giving the corresponding metallophthalocyanines [5].

Sandwich-type phthalocyanines containing a rare earth metal core (Pr, Nd, Eu–Lu) and macrocycles peripherally substituted by pentadecylphenoxy groups were synthesized using a cardanol-based phthalonitrile precursor and the respective lanthanide acetate. These products were all found to be thick and viscous substances at room temperature, showing liquid crystalline behavior with a distinct increase in fluidity at ca. 40 °C. The complexes were readily soluble in chloroalkyl solvents and dissolved fairly well in DMF with some tendency to form aggregates. Besides, they were strongly hydrophobic and revealed a peculiar affinity for lipophilic media [54].

3.7 Applications of Porphyrins Based on Cardanol

The introduction of cardanol into various macrocyclic frameworks such as porphyrins, phthalocyanines, or fullerenes has produced the enhancement of their solubility due to the presence of long alkyl chains. The improved processability permits the application of cardanol in different contexts.

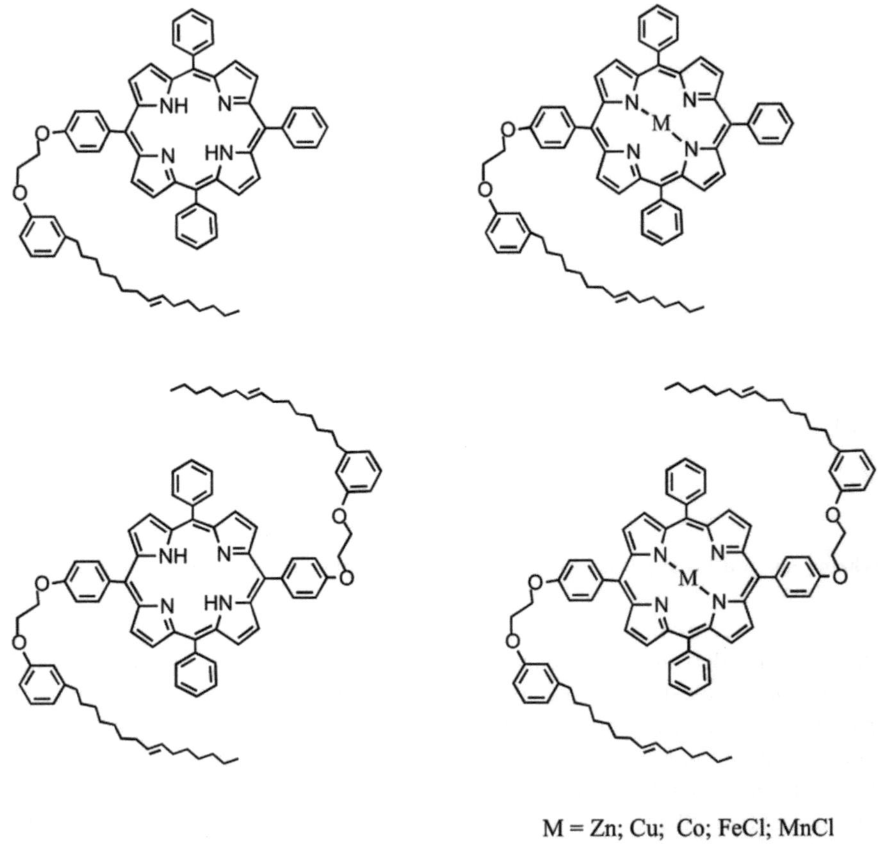

Fig. 3.5 meso-AB3 and trans-A2B2 tetraarylporphyrins bearing cardanol

Lipophilic cardanol-based porphyrins have been used as sensitizers of polycrystalline TiO_2 powders used in the photodegradation of 4-nitrophenol (4-NP) in aqueous suspensions [36, 59, 60]. It was demonstrated that a composite material obtained by impregnation of a Cu-substituted cardanol-based porphyrins onto TiO_2 (anatase phase) increased the photoactivity in the degradation of 4-NP, confirming the importance of lipophilicity, photostability of the sensitizer, and the beneficial effect of Cu(II) species coordinated in the core of the porphyrins [36].

Amphiphilic porphyrin-cardanol derivatives in Langmuir and Langmuir–Blodgett films have been applied for sensing [48] as well as a fluorescent marker for diesel fuel [42].

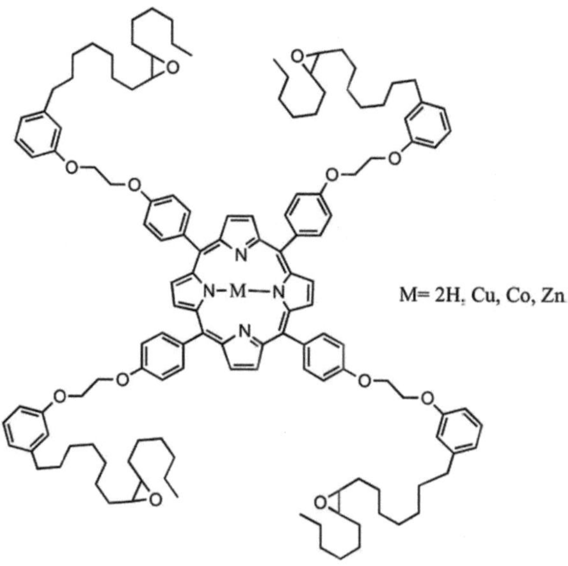

Fig. 3.6 Structure of the tetra-epoxydized porphyrin

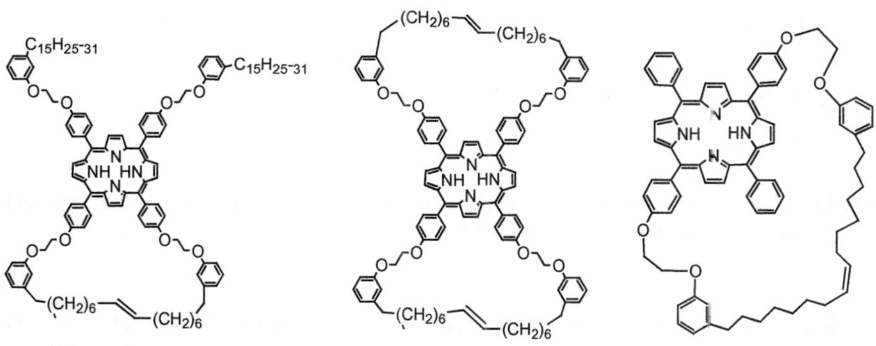

Fig. 3.7 Porphyrin derivatives obtained by intra-molecular metathesis reactions involving the unsaturated chains of cardanol

The Cu-porphyrin derived from cardanol has been used to modify the surface of a gold electrode successively and was used as an effective electrochemical sensor for nitric oxide [40].

The development of soft nanomaterials by the controlled self-assembly of molecules derived from renewable sources has become a major area of interest for scientists across the world.

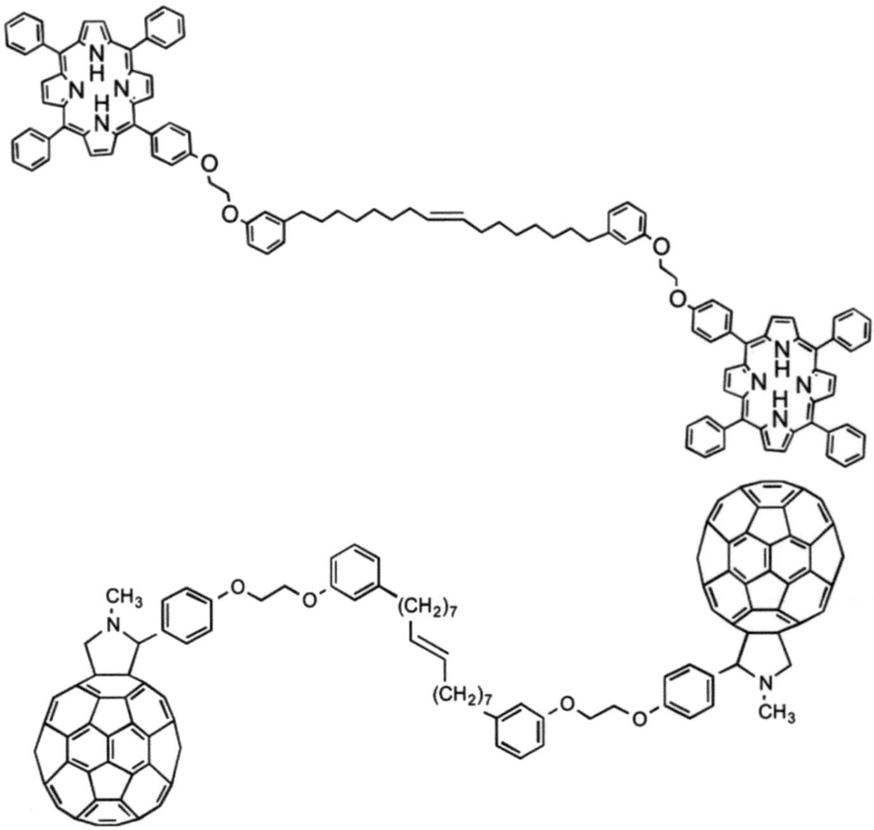

Fig. 3.8 Heterocyclic macromolecules obtained by the transformation of the unsaturated chains of cardanol by intermolecular metathesis reactions

Green nanocarriers based on the combination of natural renewable materials such as cardanol and cholesterol have been used to encapsulate lipophilic or hydrophilic compounds such as the chlorogenic acids extracted from different rowanberries such as *Sorbus americana* and *Vaccinium* sp. [12, 13].

The preparation of a superparamagnetic fluorescence nanosystem composite material by using a meso-porphyrin with side chains derived from cardanol assembled onto the Fe_3O_4 surface modified with oleic acid is shown in Fig. 3.9 [18].

Fe_3O_4 nanoparticles were also coated with anacardic acid. This nanosystem (AA-MAG) presented superparamagnetic behavior and almost the same magnetization when compared with the pure Fe_3O_4. This coating provides stability by preventing the aggregation of these nanoparticles without losing its magnetization potential. The AA-MAG demonstrates excellent and fast magneto-temperature response which can be used as high-performance hyperthermia agents [18].

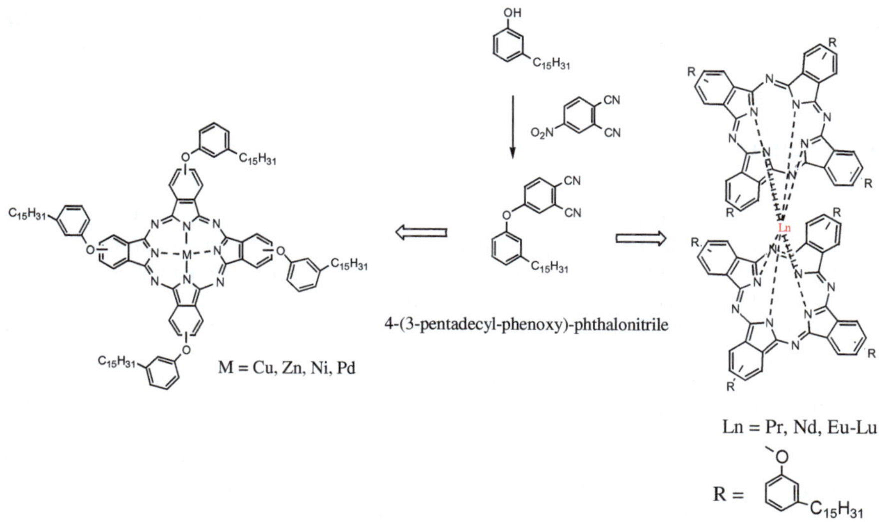

Scheme 3.7 Precursors for the preparation of cardanol-based metallophthalocyanines and bis-phthalocyanine lanthanides

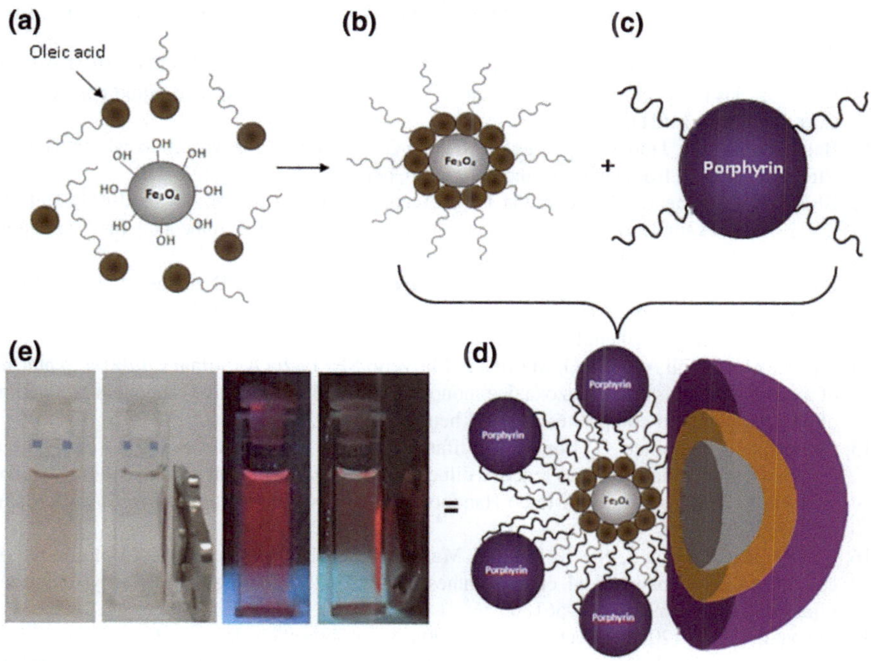

Fig. 3.9 Example of preparation of a functional superparamagnetic fluorescent nanosystem involving a cardanol-based porphyrin (published with permission from Springer-J. Nanopart Research) (Reprinted with permission from ref. [18] 2013, Springer)

Acknowledgements The authors would like to thank the Brazilian agencies CNPq (Universal 449060/2014-4, PVE 401359/2014-0 and 402566/2013-0) and Capes for the financial support.

References

1. Agag T, An SY, Ishida H (2013) 1, 3-bis (benzoxazine) from cashew nut shell oil and diaminodiphenyl methane and its composites with wood flour. J Appl Polym Sci 127(4):2710–2714
2. Agrawal JP, Satpute RS (1993) Cardanol-based epoxy flexibilizers for inhibition of composite propellants. J Macromol Sci Part A Pure Appl Chem 30(1):19–34
3. Ambrožič RR, Šebenik U, Krajnc M (2015) Synthesis, curing kinetics, thermal and mechanical behavior of novel cardanol-based benzoxazines. Polymer 76:203–212
4. Amorati R, Attanasi OA, Bassam EA, Filippone P, Mele G, Spadavecchia J, Vasapollo G (2002) Synthesis of new cardanol and cardol derivatives by allylation and regioselective cyclocarbonylation reactions. Synthesis 18:2749–2755
5. Attanasi OA, Ciccarella G, Filippone P, Mele G, Spadavecchia J, Vasapollo G (2003) Novel phthalocyanines containing cardanol derivatives. J Porphyrins Phthalocyanines 7(01):52–57
6. Attanasi OA, Del Sole R, Filippone P, Mazzetto SE, Mele G, Vasapollo G (2004) Synthesis of novel lipophilic porphyrin-cardanol derivatives. J Porphyrins Phthalocyanines 8(11):1276–1284
7. Attanasi OA, Del Sole R, Filippone P, Ianne R, Mazzetto SE, Mele G, Vasapollo G (2004) Synthesis of fullerene-cardanol derivatives. Synlett 5:799–802
8. Attanasi OA, Berretta S, Fiani C, Filippone P, Mele G, Saladino R (2006) Synthesis and reactions of nitro derivatives of hydrogenated cardanol. Tetrahedron 62(25):6113–6120
9. Attanasi OA, Mele G, Filippone P, Mazzetto SE, Vasapollo G (2009) Synthesis and characterization of novel cardanol based fulleropyrrolidines. Arkivoc 8:69–84
10. Attanasi OA, Behalo MS, Favi G, Lomonaco D, Mazzetto SE, Mele G et al (2012) Solvent free synthesis of novel mono-and bis-benzoxazines from cashew nut shell liquid components. Curr Org Chem 16(21):2613–2621
11. Balachandran VS, Jadhav SR, Vemula PK, John G (2013) Recent advances in cardanol chemistry in a nutshell: from a nut to nanomaterials. Chem Soc Rev 42(2):427–438
12. Bloise E, Carbone L, Colafemmina G, D'Accolti L, Mazzetto SE, Vasapollo G, Mele G (2012) First example of a lipophilic porphyrin-cardanol hybrid embedded in a cardanol-based micellar nanodispersion. Molecules 17(10):12252–12261
13. Bloise E, Becerra-Herrera M, Mele G, Sayago A, Carbone L, D'Accolti L et al (2014) Sustainable preparation of cardanol-based nanocarriers with embedded natural phenolic compounds. ACS Sustain Chem Eng 2(5):1299–1304
14. Calò E, Maffezzoli A, Mele G, Martina F, Mazzetto SE, Tarzia A, Stifani C (2007) Synthesis of a novel cardanol-based benzoxazine monomer and environmentally sustainable production of polymers and bio-composites. Green Chem 9(7):754–759
15. Campaner P, D'Amico D, Longo L, Stifani C, Tarzia A, Tiburzio S (2011) Study of a cardanol-based benzoxazine as reactive diluent and toughening agent of conventional benzoxazines. In: Hatsuo I, Tarek A (eds) Handbook of benzoxazine resins. Elsevier, Amsterdam, pp 365–375
16. Chen J, Liu Z, Jiang J, Nie X, Zhou Y, Murray RE (2015) A novel biobased plasticizer of epoxidized cardanol glycidyl ether: synthesis and application in soft poly(vinyl chloride) films. RSC Adv 5(69):56171–56180
17. Cheon J, Cho D (2015) Effects of the electron-beam absorption dose on the glass transition, thermal expansion, dynamic mechanical properties, and water uptake of polycardanol containing epoxy groups cured by an electron beam. J Appl Polym Sci 132(39). doi:10.1002/app.42570
18. Clemente CS, Ribeiro VGP, Sousa JEA, Maia FJN, Barreto ACH, Andrade NF et al (2013) Porphyrin synthesized from cashew nut shell liquid as part of a novel superparamagnetic fluorescence nanosystem. J Nanopart Res 15(6):1–10

19. Darroman E, Durand N, Boutevin B, Caillol S (2015) New cardanol/sucrose epoxy blends for biobased coatings. Prog Org Coat 83:47–54
20. Darroman E, Durand N, Boutevin B, Caillol S (2016) Improved cardanol derived epoxy coatings. Prog Org Coat 91:9–16
21. Dworakowska S, Cornille A, Bogdał D, Boutevin B, Caillol S (2015) Formulation of biobased epoxy foams from epoxidized cardanol and vegetable oil amine. Eur J Lipid Sci Technol 117(11):1893–1902
22. Gour RS, Kodgire VV, Badiger MV (2016) Toughening of epoxy novolac resin using cardanol based flexibilizers. J Appl Polym Sci 133(16):43318
23. Greco A, Brunetti D, Renna G, Mele G, Maffezzoli A (2010) Plasticizer for poly(vinyl chloride) from cardanol as a renewable resource material. Polym Degrad Stab 95(11):2169–2174
24. Guo YC, Mele G, Martina F, Margapoti E, Vasapollo G, Xiao WJ (2006) An efficient route to biscardanol derivatives and cardanol-based porphyrins via olefin metathesis. J Organomet Chem 691(24):5383–5390
25. Guo YC, Xiao WJ, Mele G, Martina F, Margapoti E, Mazzetto SE, Vasapollo G (2006) Synthesis of new meso-tetraarylporphyrins bearing cardanol and further transformation of the unsaturated chains. J Porphyrins Phthalocyanines 10(08):1071–1079
26. Jaillet F, Darroman E, Ratsimihety A, Auvergne R, Boutevin B, Caillol S (2014) New biobased epoxy materials from cardanol. Eur J Lipid Sci Technol 116(1):63–73
27. Kim YH, An ES, Park SY, Song BK (2007) Enzymatic epoxidation and polymerization of cardanol obtained from a renewable resource and curing of epoxide-containing polycardanol. J Mol Catal B Enzym 45(1):39–44
28. Kolanadiyil SN, Bijwe J, Varma IK (2014) Thermal properties of bisitaconimide and bisbenzoxazine blends. J Therm Anal Calorim 116(1):427–434
29. Kumar P (2011) Cathodic electrodeposition of self-curable polyepoxide resins based on cardanol. J Coat Technol Res 8(5):563–575
30. Li S, Yan S, Yu J, Yu B (2011) Synthesis and characterization of new benzoxazine-based phenolic resins from renewable resources and the properties of their polymers. J Appl Polym Sci 122(5):2843–2848
31. Li S, Zou T, Feng L, Liu X, Tao M (2013) Preparation and properties of cardanol-based polybenzoxazine/SiO_2 hybrids by sol–gel technique. J Appl Polym Sci 128(6):4164–4171
32. Li S, Yan S (2015) Synthesis and characterization of novel biobased benzoxazines from cardbisphenol and the properties of their polymers. RSC Adv 5(76):61808–61814
33. Li X, Luo X, Gu Y (2015) A novel benzoxazine/cyanate ester blend with sea-island phase structures. Phys Chem Chem Phys 17(29):19255–19260
34. Lochab B, Varma IK, Bijwe J (2010) Thermal behaviour of cardanol-based benzoxazines. J Therm Anal Calorim 102(2):769–774
35. Lochab B, Varma IK, Bijwe J (2012) Cardanol-based bisbenzoxazines. J Therm Anal Calorim 107(2):661–668
36. Mele G, Del Sole R, Vasapollo G, Garcìa-Lòpez E, Palmisano L, Mazzetto SE et al (2004) Polycrystalline TiO_2 impregnated with cardanol-based porphyrins for the photocatalytic degradation of 4-nitrophenol. Green Chem 6(12):604–608
37. Mele G, Vasapollo G (2008) Fine chemicals and new hybrid materials from cardanol. Mini-Rev Org Chem 5(3):243–253
38. Mele G, Li J, Margapoti E, Martina F, Vasapollo G (2009) Synthesis of novel porphyrins cardanol based via cross metathesis. Catal Today 140(1):37–43
39. Mgaya JE, Mubofu EB, Mgani QA, Cordes DB, Slawin AM, Cole-Hamilton DJ (2015) Isomerization of anacardic acid: a possible route to the synthesis of an unsaturated benzolactone and a kairomone. Eur J Lipid Sci Technol 117(2):190–199
40. Oliveira TI, dos Santos VN, Lomonaco D, Correia AN, Mazetto SE, de Lima-Neto P (2013) Gold electrode modified with cu-porphyrin derived from cardanol as electrochemical sensor for nitric oxide. J Electrochem Soc 160(8):B113–B118
41. Patel MB, Patel RG, Patel VS (1988) Curing kinetics of epoxy resin using epoxidized cardanol as diluent with/without fortifier. Thermochim Acta 129(2):277–284

42. Puangmalee S, Petsom A, Thamyongkit P (2009) A porphyrin derivative from cardanol as a diesel fluorescent marker. Dyes Pigm 82(1):26–30
43. Rao BS, Palanisamy A (2011) Monofunctional benzoxazine from cardanol for bio-composite applications. React Funct Polym 71(2):148–154
44. Rao BS, Palanisamy A (2012) A new thermo set system based on cardanol benzoxazine and hydroxy benzoxazoline with lower cure temperature. Prog Org Coat 74(3):427–434
45. Rao BS, Palanisamy A (2013) Synthesis of bio based low temperature curable liquid epoxy, benzoxazine monomer system from cardanol: Thermal and viscoelastic properties. Eur Polym J 49(8):2365–2376
46. Rao HSP, Kamalraj M, Swain J, Mishra AK (2014) Characterization and phase transition study of a versatile molecular gel from a glucose-triazole-hydrogenated cardanol conjugate. RSC Adv 4(24):12175–12181
47. Sahila S, Jayakumari LS (2015) Development of in situ generated graphene/benzoxazine-epoxy nanocomposite for capacitor applications. Polym Compos 36(1):1–7
48. Sandrino B, da Silva Clemente C, Oliveira TMBF, Ribeiro FWP, Pavinatto FJ, Mazzetto SE et al (2013) Amphiphilic porphyrin-cardanol derivatives in Langmuir and Langmuir–Blodgett films applied for sensing. Colloids Surf A 425:68–75
49. Selvaraj V, Jayanthi KP, Lakshmikandhan T, Alagar M (2015) Development of a polybenzoxazine/TSBA-15 composite from the renewable resource cardanol for low-k applications. RSC Adv 5(60):48898–48907
50. Sethuraman K, Alagar M (2015) Thermo-mechanical and dielectric properties of graphene reinforced caprolactam cardanol based benzoxazine–epoxy nanocomposites. RSC Adv 5(13):9607–9617
51. Sharma P, Shukla S, Lochab B, Kumar D, Roy PK (2014) Microencapsulated cardanol derived benzoxazines for self-healing applications. Mater Lett 133:266–268
52. Shukla R, Kumar P (2011) Self-curable epoxide resins based on cardanol for use in surface coatings. Pigm Resin Technol 40(5):311–333
53. Shukla S, Mahata A, Pathak B, Lochab B (2015) Cardanol benzoxazines–interplay of oxazine functionality (mono to tetra) and properties. RSC Adv 5(95):78071–78080
54. Słota R, Dyrda G, Hofer M, Mele G, Bloise E, Sole RD (2012) Novel lipophilic lanthanide bis-phthalocyanines functionalized by pentadecylphenoxy groups: synthesis, characterization and UV-photostability. Molecules 17(9):10738–10753
55. Silverman JR, Samateh M, John G (2016) Functional self-assembled lipidic systems derived from renewable resources. Eur J Lipid Sci Technol 118(1):47–55
56. Swain J, Kamalraj M, Rao HSP, Mishra AK (2014) Thermotropic gelation induced changes in micropolarity and microviscosity of hydrogel derived from glucose-triazole-hydrogenated cardanol conjugate: a study using fluorescent molecular probe. RSC Adv 4(98):55377–55382
57. Swain J, Kamalraj M, Rao HSP, Mishra AK (2015) Effect of a glucose-triazole-hydrogenated cardanol conjugate on lipid bilayer membrane organization and thermotropic phase transition. J Mol Struct 1081:124–127
58. Vasapollo G, Mele G (2006) Synthesis of heterocycles by transition metals-catalyzed cyclocarbonylation reactions. Curr Org Chem 10(12):1397–1421
59. Vasapollo G, Mele G, Del Sole R (2011) Cardanol-based materials as natural precursors for olefin metathesis. Molecules 16(8):6871–6882
60. Vasapollo G, Mele G, Del Sole R, Pio I, Li J, Mazzetto SE (2011) Use of novel cardanol-Porphyrin hybrids and their TiO_2-based composites for the photodegradation of 4-Nitrophenol in water. Molecules 16(7):5769–5784
61. Xu GM, Shi T, Liu J, Wang Q (2014) Preparation of a liquid benzoxazine based on cardanol and the thermal stability of its graphene oxide composites. J Appl Polym Sci 131(11). doi:10.1002/app.40353
62. Xu G, Shi T, Wang Q, Liu J, Yi Y (2015) A facile way to prepare two novel DOPO-containing liquid benzoxazines and their graphene oxide composites. J Appl Polym Sci 132(11). doi:10.1002/app.41634

Chapter 4
Facile Green Strategy for Preparation of Advanced Structured Materials Based on Amphiphilic Cardanol

Ayman M. Atta and Hamad A. Allohedan

Abstract Cashew nutshell liquid oil among different renewable resources is one of the agricultural by-products available in tropical countries such as India that is used to produce cardanol as naturally occurring fatty phenols. It has attracted great attention in advanced chemical industries to produce biopolymers, bio-surfactants, bio-composites and nanomaterials. It represents an excellent alternative feedstock for the elaboration of chemicals and polymers. Moreover, cardanol is regarded as a versatile and valuable raw material for wide applications in the form of brake linings, surface coatings, paints and varnishes as well as in polymer production. In this chapter, the possible modifications of cardanol to produce monomeric and polymeric surfactants have been discussed in different reaction schemes. It can be recommended that because of their renewable nature and structural characteristics, cardanol is likely candidates for preparing "green" surfactant species like anionic (sulfonates), non-ionic (ethoxylated and cardanol–formaldehyde ethoxylated polymers) and cationic (quaternary ammonium, pyridinium and imidazolium derivatives) to replace the petroleum based on nonylphenol surfactants. Moreover, it can be used to prepare biopolymer, polymer nanocomposites and semisynthetic processes to prepare derivatives with biological and pharmaceutical applications.

A.M. Atta (✉)
Petroleum Application Department, Egyptian Petroleum Research Institute, Cairo 11727, Egypt
e-mail: aatta@ksu.edu.sa

A.M. Atta · H.A. Allohedan
Surfactants Research Chair, Department of Chemistry, College of Science, King Saud University, P.O. Box 2455, Riyadh 11451, Kingdom of Saudi Arabia

P. Anilkumar (ed.), *Cashew Nut Shell Liquid*,
DOI 10.1007/978-3-319-47455-7_4

4.1 Cardanol as Renewable Feedstock for the Chemical Industry

Significant research efforts of the academic and industrial world directed to use renewable resources to replace the petrochemicals for production of commercial materials due to the increasing prices of petrochemical products and growing environmental concerns. Cashew nutshell liquid (CNSL) is represented as an important bio-based natural material due to its availability, cheaper cost and unique structural features. Cardanol is the main phenolic component produced from the cashew nutshell liquid or oil obtained as a by-product of the cashew (*Anacardium occidentale* L.) nut processing. Figure 4.1 represents the chemical constituents of commercial-grade CNSL, in which cardanol is produced in percentages of 62.8 % due to decarboxylation of anacardic acid to cardanol [138]. The CNSL (shell oil extracted from the brown mesocarp) constitutes 18–27 % of the total raw nut weight. On an average, 15–20 % by weight of unshelled nut in Africa and 25–30 % by weight in India and 25 % overall contain CNSL. Four constituents differing in the side chain unsaturation represented the components of CNSL (Fig. 4.1). Different methods have been used to extract CNSL from the nuts such as hot oil process and roasting process [81, 133] or solvent extraction of the oil using organic solvents [64, 134]. Oligomeric and polymeric materials have been reported by Tyman et al. [135, 136, 138].

The strategies used to modify the chemical structure of cardanol are based on the three reactive functional groups. First functional group is the reactive phenolic –OH group offering synthetic flexibility to prepare hydrophilic moieties that were used to prepare specialty polymers such as novolac or resole resins, polyurethane thermosets and liquid crystal epoxy polymers in different reaction media. The second reaction group is the meta alkyl chain with non-isoprenoic cis double bonds attributing amphiphilic and lipidic character. The presence of an unsaturated hydrocarbon chain with one to three double bonds at the meta position of the phenol group attracted great attention to produce green biopolymers. Lastly, the aromatic ring allows for π–π stacking and functionalization. Cardanol-based biomaterial polymers have been used as ionic and non-ionic surfactants in the various industrial applications due to

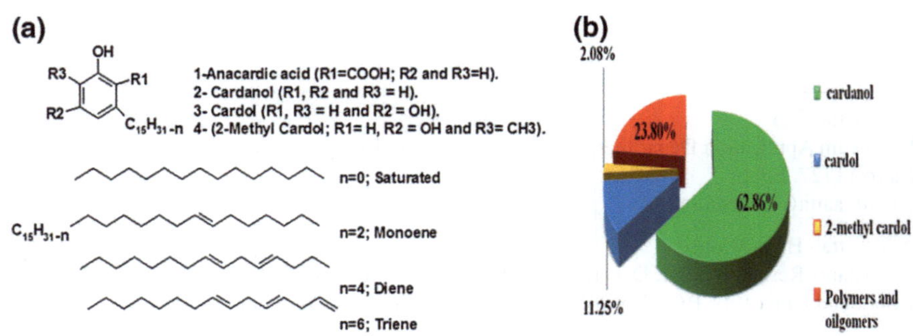

Fig. 4.1 Chemical compositions of CNSL

its antibacterial and chemically modifiable properties. Cardanol-based surfactants are just emerging as green alternative to conventional petroleum-derived nonylphenols for the preparation of surfactants. Cardanol has naturally an ideal chemical structure for the synthesis of amphiphilic derivatives and surfactants. The alkyl $C_{15}H_{27}$ alkyl side chain provides an essential hydrophobicity, while the slightly polar phenoxy group allows a wide variety of chemical modifications. This section aims to report the conversion of cardanol to polymeric and monomeric amphiphilic surfactants to produce advanced materials based on polymers, nanomaterials and composites.

The preparation of new molecules based on natural materials by environmental friendly and cost-effective means is the target for advanced researches. Cardanol as one of the renewable bio-resource feedstocks attracted great attention to produce different types of amphiphiles that can be used to produce advanced materials such as biopolymers and nanomaterials [67, 68, 111, 132].. Further, it is assumed that an increased fraction of natural building blocks in the chemical structure will be beneficial for reducing aquatic toxicity and increasing the rate of biodegradation. In this section, we represent two decades of research focused on turning cardanol into value-added materials due to inherent 'programmed,' or 'built-in' functionalities. We highlight the efficient conversion of cardanol into a wide array of surfactant monomers, polymers and self-assembled systems to produce nanomaterials and polymer composite materials. The presence of different functional groups became target for different chemical modifications to generate a library of amphiphiles and functional monomers that remained a vital building block for the synthesis of different chemicals that find applications as described in Fig. 4.2 [80].

Fig. 4.2 Cardanol-modified materials including traditional polymer, fine chemical synthesis and soft nanomaterial synthesis

4.2 Cardanol Surfactants

4.2.1 Non-ionic Surfactants

Cardanol resembles in its chemical structure alkyl phenol surfactants that ethoxylated to produce non-ionic surfactants that attracted great attentions in the petroleum and chemicals industries as emulsifier, demulsifier, dispersant and wetting agents [28, 29, 101, 102]. Different routes can be used to obtain non-ionic surfactants as illustrated in Scheme 4.1. Route 1 is based on the preparation of polyethylene glycol derivatives as described by Tocco et al. [131]. The monomethyl ether of poly-(ethylene glycols), MPEG, was selected because it can be easily reacted with cardinal and it is soluble in a wide range of solvents, non-toxic, inexpensive, commercially available, easy to functionalize and also resistant to drastic operative conditions [39]. Moreover, PEG chemistry has shown wide applications as a large part used to deliver drugs, oligonucleotides or enzymes [49].

Route 2 (Scheme 4.2) shows the preparation of polymeric non-ionic surfactants that can be prepared by forming cardanol–formaldehyde polymers by reaction of cardanol with formalin or trioxane as described by Atta et al. for preparing rosin formaldehyde resins [9, 10, 13, 17]. The produced rosin formaldehyde resins can be reacted with MPEG or ethylene oxide to produce non-ionic polymeric surfactants as reported on conversion of rosin to polymeric surfactants [9, 10, 13, 17]. The phenolic nature of resulting cardanol makes it possible to react with formaldehyde under a variety of conditions to form novolac or resole resins [38, 98, 146]. Compared with conventional alkyl phenol ethoxylated surfactants, ethoxylated cardanol–formaldehyde surfactants have improved flexibility. Moreover, the side chain imparts a hydrophobic nature to the polymer, making it water repellent and resistant to biodegradation. Non-ionic surfactants have also been synthesized by reacting cardanol with alkylene oxide (ethylene or propylene oxides), polyalkylene glycol or halohydrin [30, 120].

Scheme 4.3 is used to prepare a benzoxazine surfactant based on cardanol as a renewable resource's ability to perform its functions as non-ionic surfactants. The synthesis procedure was described as solventless method to react aniline,

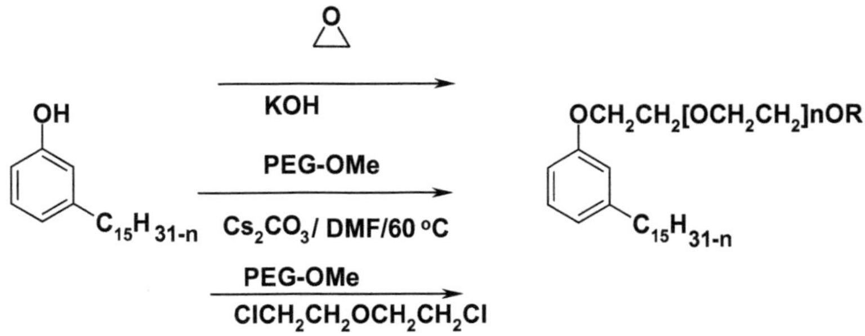

Scheme 4.1 Preparation of ethoxylated cardanol

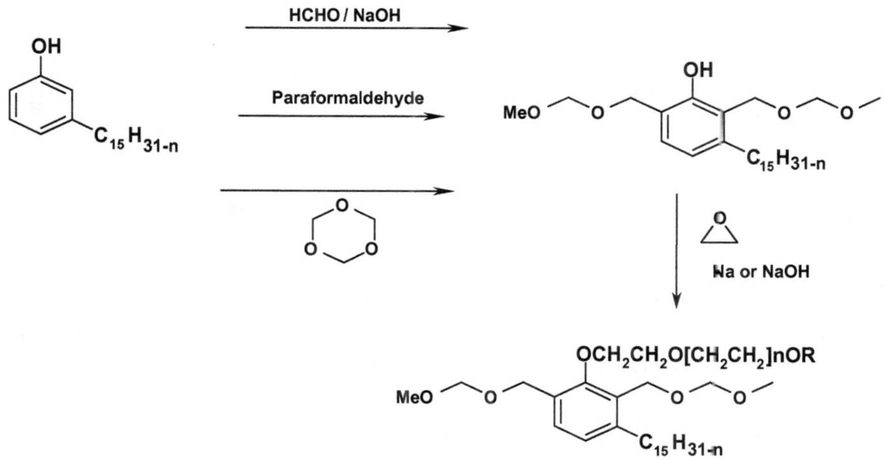

Scheme 4.2 Preparation of cardanol–formaldehyde condensates

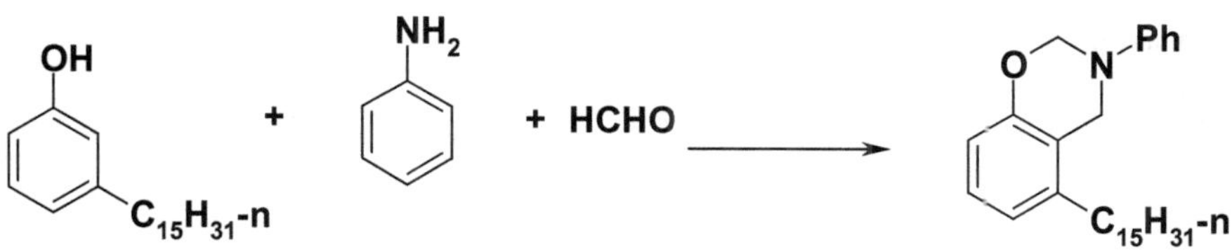

Scheme 4.3 Preparation of cardanol–benzoxazine surfactant

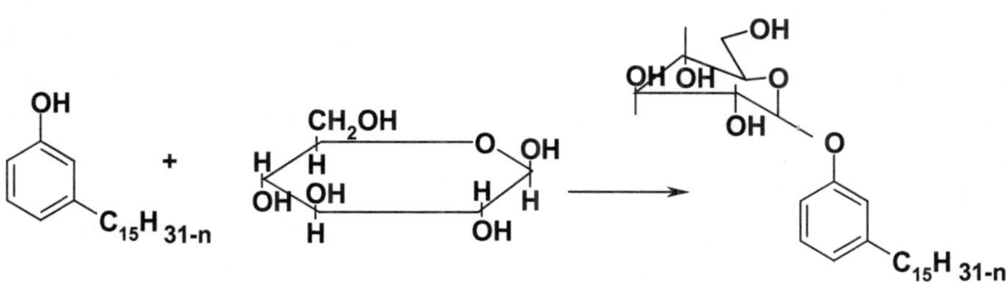

Scheme 4.4 Preparation of cardanol glycosides

formaldehyde and cardanol to prepare non-ionic surfactants that have wide application to prepare composites [31, 79, 103].

Cardanol glycosides are special type of glycolipids that can be prepared by reacting phenol group with sugar or by attaching glucopyranose to cardanol (Scheme 4.4). The presence of sugar moiety adds non-ionic polar head group to cardanol and bestows chirality to the molecule. These materials have a tendency to

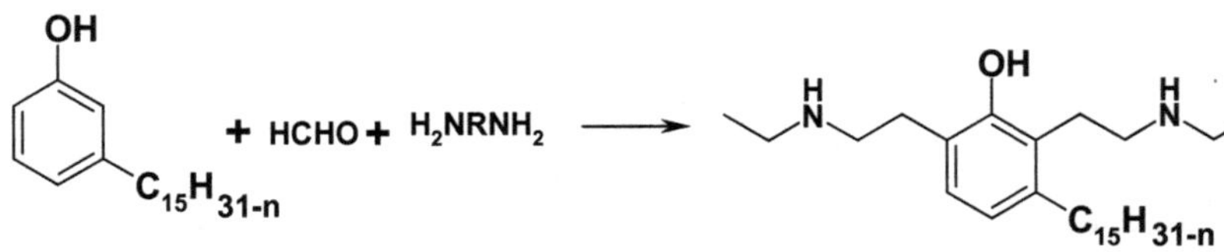

Scheme 4.5 Mannich reaction of cardanol to produce non-ionic surfactants

Table 4.1 Surface activity and application of non-ionic surfactants based on cardanol

Cardanol surfactants	Surface properties		Applications	References
	Critical micelle concentrations (CMC) (mM)	Surface tension at CMC (mN m^{-1})		
Ethoxylated cardanol acrylate copolymers	0.08–0.22	35.4–38.2	Biological and pharmaceutical applications	[34–36]
Ethoxylated cardanol phenol–formaldehyde	0.03–0.53	32.3–38.5	Petroleum demulsifiers and dispersants	[37]
Cardanyl-b-D-glucopyranoside	0.23	38.13	Nanotube architectures	[38]

form self-assembled layers to surround the colloidal particles to produce soft materials, such as twisted fibers, helical coils and high-axial-ratio nanotubes [21, 66].

The modification of CNSL with polyamines is used to produce non-ionic surfactants as represented in Scheme 4.5. This technique is based on Mannich reaction that involves the reaction of the CNSL with formaldehyde and diethylenetriamine, triethylenetetramine, tetraethylenepentamine, or pentaethylenehexamine in methanol solution. Under these conditions, the dihydric phenol, cardol, forms high molecular weight polymeric non-ionic surfactants, while the cardanol remains substantially unreacted [40].

The surface activity parameters of the non-ionic surfactants produced from modified cardanol and their industrial applications in the preparation of polymers, nanocomposites and nanomaterials are represented in Table 4.1.

4.2.2 Anionic Surfactants

It is well established that anionic surfactants based on dodecyl benzene sulfonate have greater applications as detergents in the household and chemical industries. As previously discussed, the structure of cardanol is similar to dodecyl benzene which can easily be sulfonated to produce surfactants that have similar properties to dodecyl benzene sulfonate [21, 97, 129]. The surfactant from cardanol can be

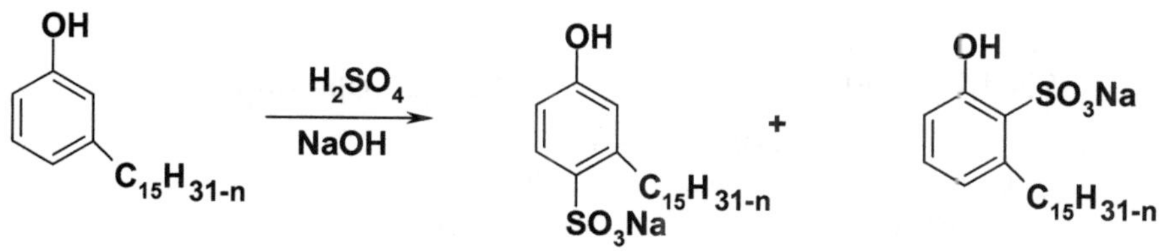

Scheme 4.6 Sulfonation of cardanol

prepared as illustrated in Scheme 4.6. The physicochemical properties such as surface activity, solubility in sea water and detergency indicated the superior properties when it was compared to dodecyl benzene sulfonate [97]. PDPSA surfactants that have sulfonate group within in O– or P– position (Scheme 4.6) have reduced the surface tension of water to be 32.25 mN m^{-1} at 20 % w/v and 28.00 mN m^{-1} for dodecylbenzene sulfonate at 15 % w/v. Critical micelle concentrations (CMC) of dodecyl benzene sulfonate and cardanol sulfonate were found to be 0.435 and 0.372 M, respectively [97]. The results confirmed that cardanol sulfonate can be applied as an alternative bio-based anionic surfactant. The bio-based surfactant produced from sulfonation of cardanol commonly known as 3-pentadecylphenol 4-sulfonic acid (PDPSA) is having amphiphilic character and can act as stabilizing and capping agents to control the shape, size and stabilization of the nanomaterials [121, 122].

Anionic surfactants can also be obtained from non-ionic surfactants of cardanol prepared by ethoxylation of cardanol with ethylene oxide as represented in Scheme 4.7a, b [137]. The hydroxyl end group of oxyethylene unit was converted to anionic groups by reacting either with chlorosodium acetate or with ammonium sulfonic acid [97].

The reactive surfactants (surfmers) behave unlike conventional non-reactive surfactants and can act as part in the polymerization with the main monomer to produce new polymeric material for emulsion polymers applied in useful protective coatings and adhesives [51]. Polymerizable reactive anionic surfactants can also be obtained by introducing reactive groups into amphiphilic surfactants [7, 8]. In this respect, acrylate and sulfonate groups can be introduced to cardanol by reacting with acryloyl chloride in the presence of triethyl amine followed by sulfonation with chlorosulfonic acid as described in Scheme 4.8 [129]. It was also reported that acrylate group can be introduced to cardanol after converting the hydroxyl phenol to alcoholic group by etherification with chloroethanol in the presence of sodium hydroxide as reported in Scheme 4.9 [129]. The surface activity measurements of the prepared surfactants indicated that the micellization of the surfmers with an ethylene spacer between the phenyl ring and the acrylate group is more active at lower concentrations like polymeric surfactants [129].

Cardanol can be converted to anionic surfactant and to trithiocarbonate used as chain transfer in radical fragmentation chain transfer (RAFT) in emulsion polymerization of methyl methacrylate (MMA) [34]. The resulting cardanol-terminated

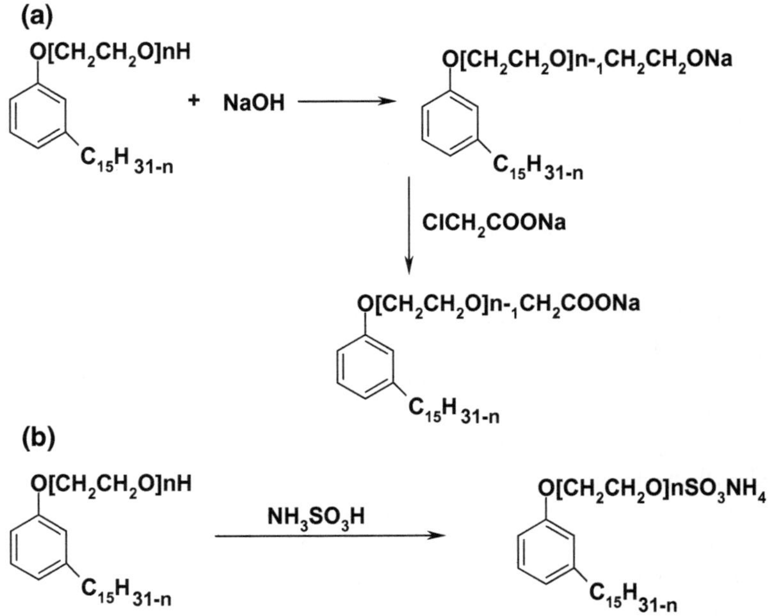

Scheme 4.7 Preparation of cardanol anionic surfactants based on **a** ethoxylated carboxylate and **b** ethoxylated ammonium sulfate salts

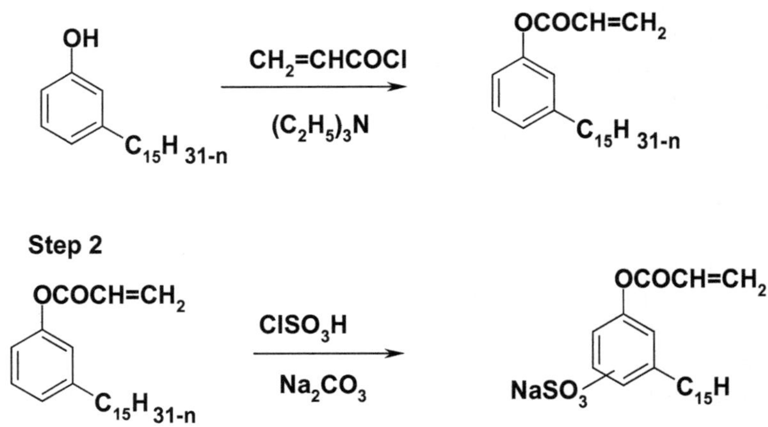

Scheme 4.8 Synthesis of polymerizable cardanol anionic surfactant

poly(methyl methacrylate) (PMMA) was photo-polymerized to produce UV-cured resin within 20 min. The great advantages of using these materials over conventional curing systems are their high efficiency, energy saving, environmental friendliness, cost economy, etc. The cardanol-based RAFT chain transfer agent

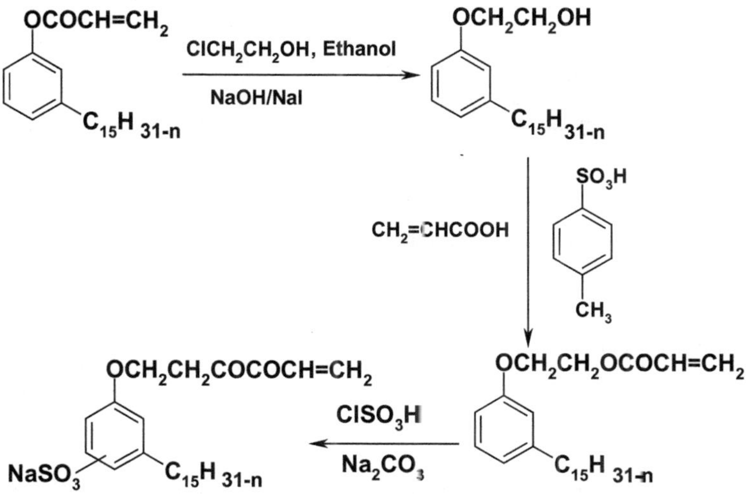

Scheme 4.9 Synthesis of ethoxylated polymerizable cardanol anionic surfactant

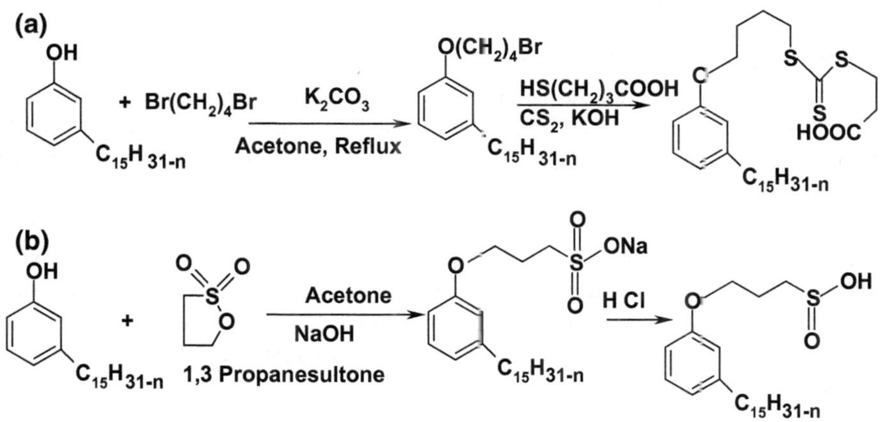

Scheme 4.10 **a** Chain transfer and **b** anionic surfactants

and a new cardanol-based anionic surfactant were prepared as represented in Scheme 4.10 [34].

Amphiphilic anionic surfactants can also prepared by reacting cardanol with 1,4-butanesultone, 1,3-propanesultone as sulfonation reagents using potassium tertiary butoxide as base catalyst. The scheme of preparation is represented in Scheme 4.11 [129].

Cardanol was also useful starting material for synthesizing anionic amphiphilic surfactants called cardanol azophenylsulfonic acid (CAPSA) (Fig. 4.3) [6].

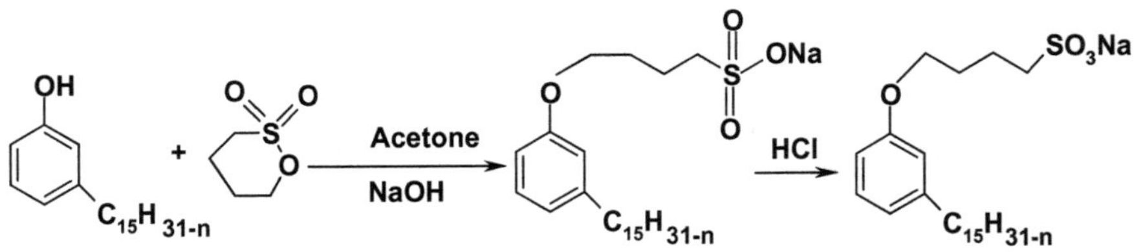

Scheme 4.11 Preparation of cardanol alkyl sulfate anionic surfactant

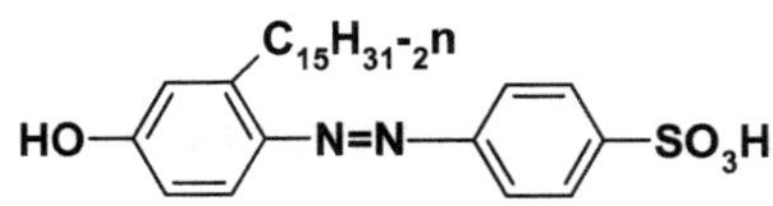

Fig. 4.3 Cardanol azophenylsulfonic acid surfactant

The anionic surfactants showed superior amphiphilic properties and are used as a template for preparation of polyaniline conducting polymers and their nanocomposites with clay as gathered in Table 4.2. It was also found that the morphology and size of the nanomaterials proved that the anionic surfactants based on cardanol have ability to self-assemble with rod-like micelles that are responsible for directing the formation of oriented nanoparticles.

The surface activity of anionic amphiphiles derived from cardanol can be modified by preparing Gemini surfactants that based on amphiphiles have more than two head and two tails linked by spacer groups. Several steps of chemical reaction and treatments were employed to prepare Gemini surfactants as illustrated in Scheme 4.12 [116]. In the first procedure, the etherification reaction was completed in the presence of 1,3-dibromopropane to generate cardanol separated with spacer groups. In the next step, the dicardanyl propyl ether was sulfonated using chlorosulfonic acid to form cardanol sulfonic acid Gemini surfactant. The last step used to neutralize the sulfonated cardanol dimmer using sodium hydroxide solution. The CMC for cardanol sulfonate Gemini surfactant was 6.20×10^{-2} mM, and surface tension at CMC was 36.92 mN m^{-1}.

Table 4.2 Surface activity and application of anionic surfactants based on cardanol

Cardanol surfactants	Surface activity		Applications	References
	CMC (mM)	Surface tension at CMC ($mN.m^{-1}$)		
PDPSA	0.372	31.2	Conducting polyaniline–clay nanocomposite	[40]
PDP	0.1	28.5	Conducting polyaniline nanomaterials	[48]
CAPSA	0.49	35.2	Material for supercapacitors	[49]

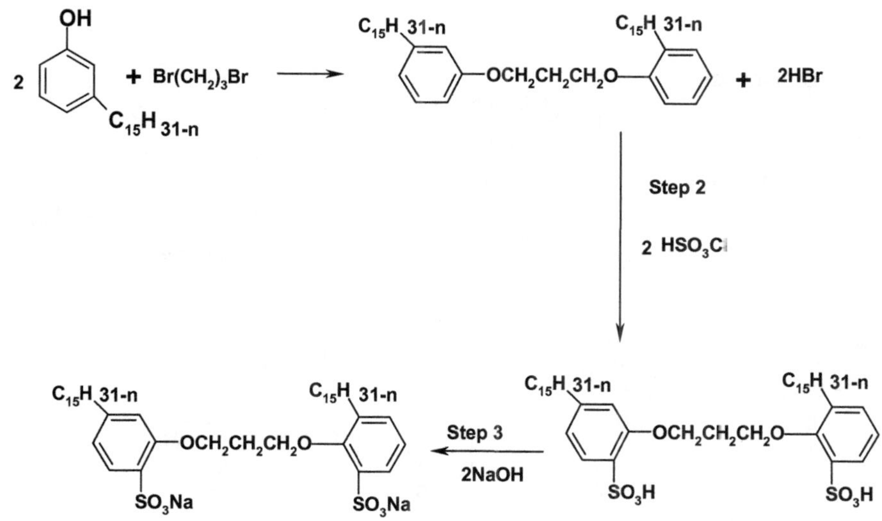

Scheme 4.12 Preparation of cardanol Gemini surfactants

4.2.3 Cationic Surfactants

Cationic amphiphiles can be produced from cardanol by attaching directly or indirectly to positively charged nitrogen atom. Moreover, synthesis of Gemini amphiphiles from cardanol as the new-generation amphiphilic molecules increases the efficiency and effectiveness as compared to their monomeric counterparts. In this respect, pyridinium and imidazolium Gemini amphiphiles are the most valuable and effective as recent class of cationic Gemini amphiphiles. It is necessary at the present time to develop of new amphiphilic molecule having better surface and interfacial properties to replace the conventional cationic surfactants. The cationic and Gemini amphiphiles have been used widely in production of biopolymers, nanomaterials, fabric softener, wetting agent, dispersant, emulsifier, foaming agent, bactericide, corrosion inhibitor, and the number of other applications continues to increase with time.

It was reported that the series of phenoxy ring containing long-chain imidazolium and pyridinium based on cardanol Gemini amphiphiles have been synthesized using different spacers (i.e., –S–(CH2)$_n$–S–, where n is 2, 3, 4 and 6) as represented in Scheme 4.13 [22]. The results of this work proved that the phenoxy group containing Gemini imidazolium and pyridinium amphiphiles has the potential to be used as gene delivery agent.

A cardanol can also be linked with quaternary ammonium salt to produce cationic surfactants. It was reported that [50] quaternary ammonium salt surfactant (3-cardanoxy-2-hydroxy)propyl-*N*,*N*,*N*-trimethylammonium chloride was prepared by linking cardanol with glycidyl trimethyl ammonium chloride (GTA)

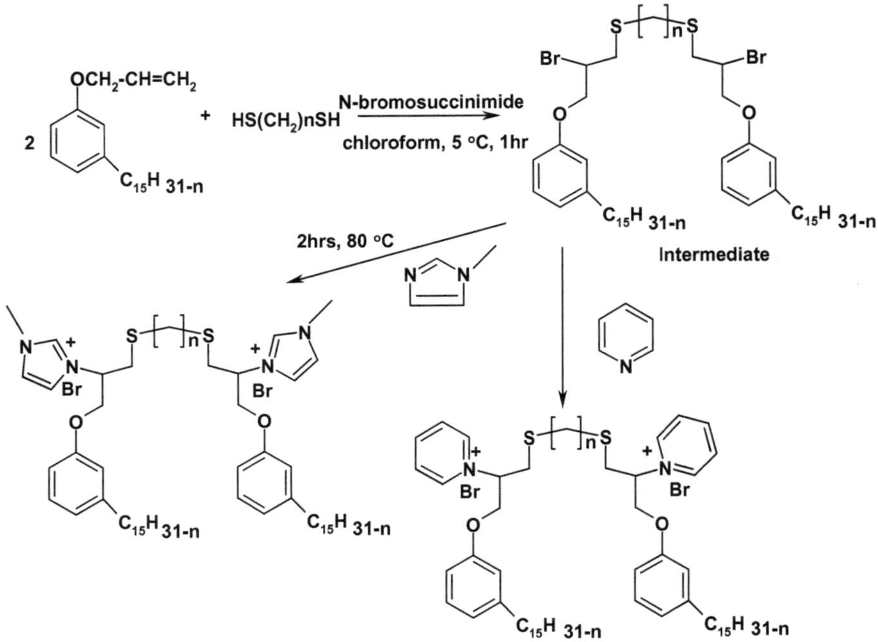

Scheme 4.13 Preparation of cardanol containing Gemini imidazolium and pyridinium amphiphiles

which can be prepared from reaction of epichlorohydrin and trimethylamine as starting materials. The produced amphiphiles showed greater surface activity characteristics through surface tension and conductivity measurements and reduced the water surface tension to 38.86 mN m^{-1}.

Cardanol was modified to produce Gemini surfactant based on phenoxy ring containing long-chain imidazolium and pyridinium amphiphiles which can be used as gene delivery agent. It is well established that the chemical structure of surfactants plays a major role to produce well-controlled size, shape and dispersed nanomaterials used in advanced polymer nanocomposites preparation. The bio-based surfactants based on cardanol have been used to exfoliate montmorillonite nanomaterials in epoxy. Sulfonated cardanol can be functionalized to form phenolic ethers, which are used as polymer additives or in nanofibers. Polyaniline–cardanol sulfonate was used to design a nanostructured electromagnetic iron–clay composite as sensor for the detection of ascorbic acid. Moreover, cardanol glucosides are used to prepare efficient nanotubes. Cardanol also attracted more attention in the production of silver, alumna, silica and magnetite nanomaterials having superior physicochemical characteristics and advanced applications.

4.3 Polymers and Resins

The chemical structure of cardanol represented in Scheme 4.1 shows great reactivity toward chemical reagents due to the presence of phenolic groups that are ortho and para directing groups for electrophilic aromatic substitution. Moreover, the presence of unsaturated alkyl groups increased the reactivity of cardanol to addition reactions to produce new valuable polymeric materials and resins having special characteristics as biopolymers. Figure 4.4 shows the chemical reactions based on phenolic nature of cardanol that is responsible for production of different chemicals, polymers and resins [44, 104, 145]. The aim of this section is hence to provide an overview of the field by highlighting and discussing the state-of-the-art, as well as selected perspectives of thermosetting materials derived not only from plant oils, but also from other renewable resources.

Cardanol can be polymerized directly either with using acids such as HCl, H_2SO_4, HNO_3 and H_3PO_4 at temperature of 150–190 °C [54] or with other catalysts such as $TiCl_4$, $SnCl_4$, $AlCl_3$ or $POCl_3$ [33] to obtain polymers that find several applications. Different metals, their oxides, hydroxides or carbonates, such as Cu, Al, Co, Ni, Pb, Mn, Sb, Zn and Fe, have been used to polymerize cardanol at temperatures ranging from 120 to 150 °C [52, 53].

Different reaction modifications can be carried out on phenolic hydroxyl, aromatic ring and unsaturation(s) in the alkenyl side chain to obtain advanced polymer as described in the forthcoming section.

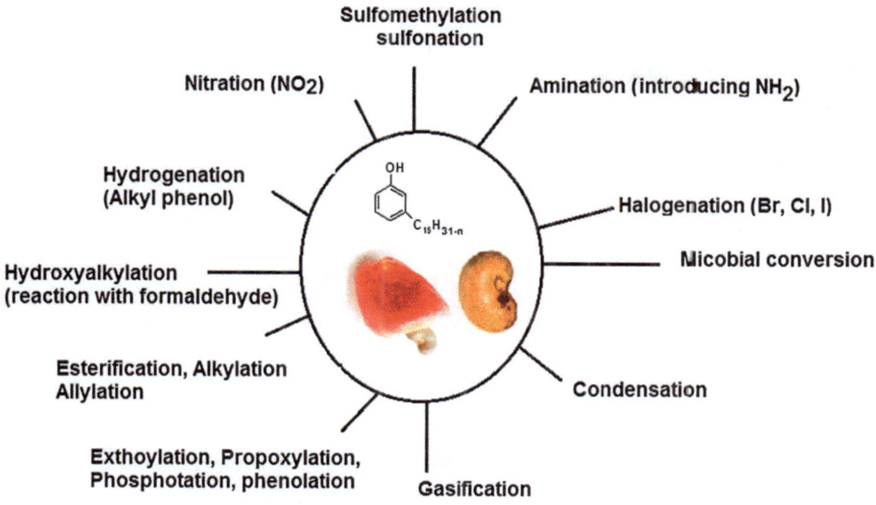

Fig. 4.4 Chemical reactions of cardanol

4.3.1 Thermosetting Cardanol Resins

There is of course increasing awareness of the need for green chemistry in chemical industries. Cardanol phenol attracted great attention to produce thermosetting materials having advanced properties and applications [104]. These materials were cured by light, heat or chemical initiators. The thermoset materials produced from cardanol such as phenolic, epoxy, polyester and polyurethane resins have merits over petroleum-based thermosetting materials as a novel and appropriate sustainable feedstock, a cost–performance basis with high eco-friendliness values [73, 119]. In previous works, rosin as natural product based on rosin was used to prepare advanced epoxy resins, curing agent, vinyl ester resins, polyester, alkyd and polyurethane [11, 14–16, 18].

Scheme 4.14 represents the use of cardanol as renewable resources in the design of thermosetting materials, such as epoxy, phenolic and vinyl ester resins. The properties of cardanol were increased by converting phenol cardanol to polyphenols using formaldehyde, paraformaldehyde and trioxane in the presence of sodium hydroxide or triethylamine as a catalyst to link phenol with methylene bridge or methyloyl groups [20, 56]. It was found that [57] when ammonia was used as the catalyst in the cardanol–formaldehyde reaction the products were solid, but they were viscous liquids when sodium hydroxide, sulfuric acid or oxalic acid was used. Moreover, the condensation reactions between cardanol and hexamine, which is used as a source of formaldehyde and ammonia, are generally carried out at higher temperatures [2].

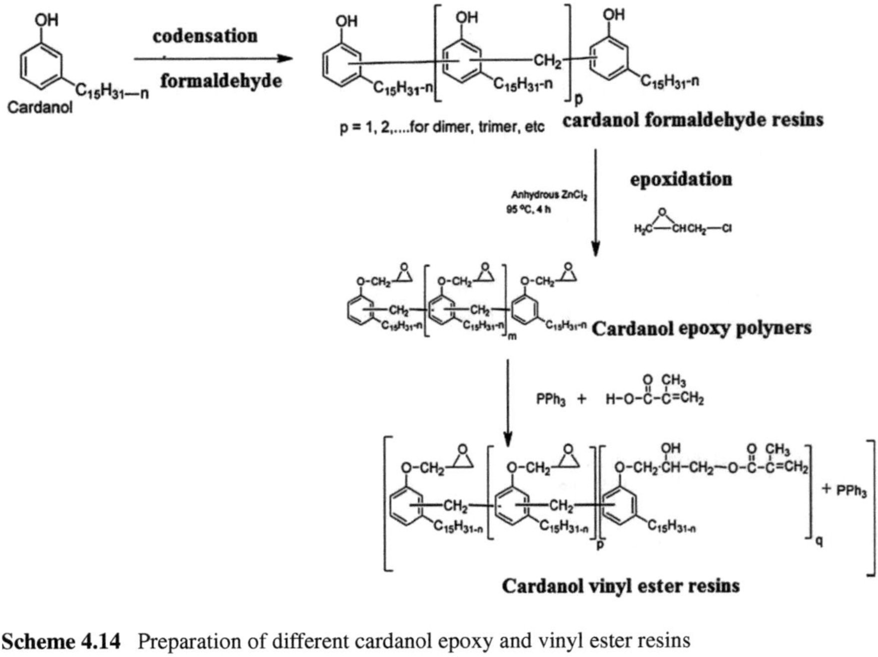

Scheme 4.14 Preparation of different cardanol epoxy and vinyl ester resins

It was reported that cardanol and its polymers have interesting structural features for chemical modification and polymerization into specialty polymers. Cardanol–formaldehyde resins have superior properties when compared to conventional phenolic resins, such as flexibility (due to the internal plasticization effect of the long chain), that improved their processability. The presence of long alkyl groups increased the weathering resistance of cardanol polymers because it acts as water-repellent group [146]. In addition, cardanol polymers have useful characteristics such as heat and electrical resistance, antimicrobial properties and insect resistance. However, in some cases natural fibers have been added, such as ramie, flax, and hemp fibers were added to cardanol–formaldehyde resins to resist the steric hindrance and reduced intermolecular interactions imparted by the C15 side chain for fabric structural applications (Maffezzoli et al. 2004). It was also reported that epoxy resin based on diglycidyl ether of bisphenol A was added with the cardanol–formaldehyde resins in order to reduce the amount of water released during the process and to reduce the porosity of cardanol–formaldehyde resins (Maffezzoli et al. 2004).

Recently, a novel benzoxazine prepolymer derived from cardanol was used to prepare phenolic resins in the presence of formaldehyde and an amine as represented in Scheme 4.15. A faster polymerization of cardanol-based benzoxazine monomer can be carried out at ambient temperature in the presence of cationic catalysts such as PCl_5. The benzoxazine-based family of phenolics showed interesting characteristics as advanced composites such as good thermal properties and flame retardancy originated from phenolics, together with the mechanical performances and molecular design flexibility [31].

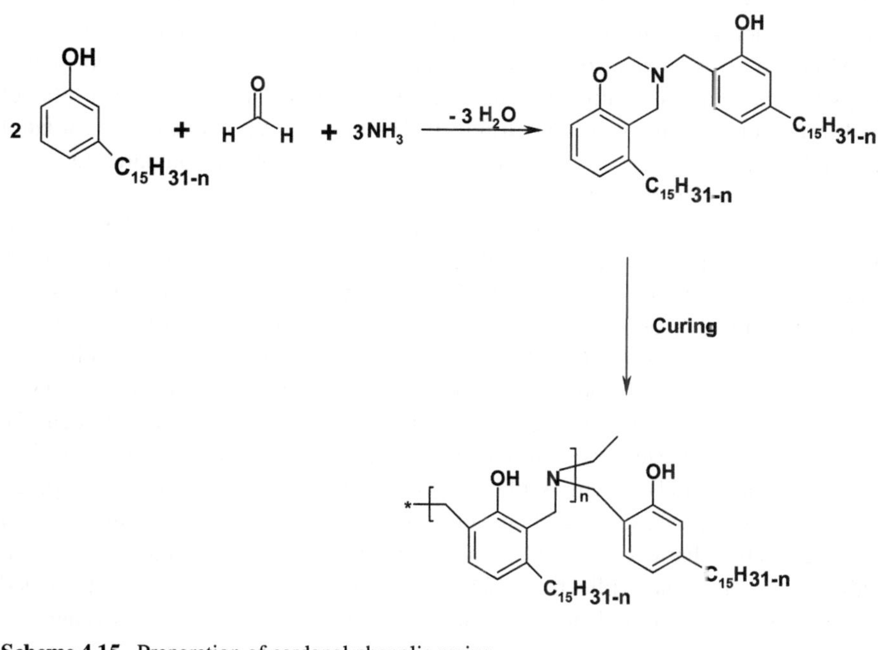

Scheme 4.15 Preparation of cardanol phenolic resins

Epoxy thermosets are the most common polymers that were produced more than 3 million tones and have different applications in coating, adhesives and composites. Epoxy resins are most commonly produced from derivatives of petroleum. More recently, bio-based feedstocks were used with great attention to produce epoxy from plastic waste and natural products as cheap sources [11–14, 16–18]. The phenolic hydroxyl group of cardanol or unsaturated double bonds on the C15 alkyl side chain can be reacted either with epichlorohydrin (ECH) or with epoxidized, respectively, to produce epoxy resins. Moreover, production of polyphenol from cardanol used to prepare epoxy resins as represented in Scheme 4.14. Cardanol was used in partial or total substitution of phenol in thermoset resins such as novolac resins [32, 59], vinyl esters [125, 126] and also in epoxy resins modification [139]. The literature reports a number of workers [35, 36, 63] who have reported the synthesis of various types of epoxy compounds based on CNSL and its derivatives. Hydroxyl group of cardanol was reacted with ECH to produce epoxy using $ZnCl_2$ as catalyst in basic condition at temperature 95 °C [144]. This product is commercially available at Cardolite Corporation. The Cardolite (commercial epoxy based on cardanol) resins have been modified with bisphenol A to obtain adhesives having greater impact strength and shear resistance [44]. The curing agents of epoxy resins were also prepared from cardanol such as polyamines (Lubi and Thachil 2000) and self-cured epoxy based on cardanol [117] to apply in surface coatings. The double bonds of alkyl group attached to cardanol phenyl group can be epoxidized by reacting with chloroperbenzoic acid, perbenzoic acid and performic acid to produce polyepoxy cardanol [48, 71, 95, 128]. The epoxidation of these double bonds could also be performed with enzymes such as Candida antarctica lipase, with acetic acid and hydrogen peroxide, in toluene (yield 95 %) [76]. The chemical structure of epoxidized cardanol is represented in Fig. 4.5.

Modified epoxy resins based on cardanol (Fig. 4.6) were synthesized in two steps by phenolation of aliphatic chain ($C_{15}H_{31}$), followed by reaction of phenol hydroxyl groups with epichlorohydrin [60]. The modified epoxy based on cardanol showed good performances after curing with polyamines when compared to diglycidyl ether based on bisphenol A [60].

Vinyl ester resins can be prepared as illustrated in Scheme 4.14. The functionalization of cardanol with acrylic/methacrylic monomer can be obtained to produce an eco-friendly vinyl ester resin [1, 45, 77, 93]. The esterification of epoxy resin of cardanol–formaldehyde resins with methacrylic or acrylic acid can be obtained at temperature ranging from 80 to 100 °C in the presence of triphenylphosphine as a catalyst [5, 123–126]. The produced cardanol vinyl ester resin can be cured with styrene as monomer and benzoyl peroxide as initiator at 120 °C at faster rate than the classical poly(vinyl ester) resin with high thermal stability up to 285 °C [125, 126].

Polyurethane (PU) polymers play an important role for their diverse applications in coatings, adhesives, sealants, elastomers and plastics because they exhibit many desirable characteristics. Cardanol derivatives have been explored in polyurethanes for different applications [24, 25, 27, 47, 74, 90]. It is well established that PUs were produced from reactions between polyols and polyisocynates. In

Fig. 4.5 Epoxidized cardanol

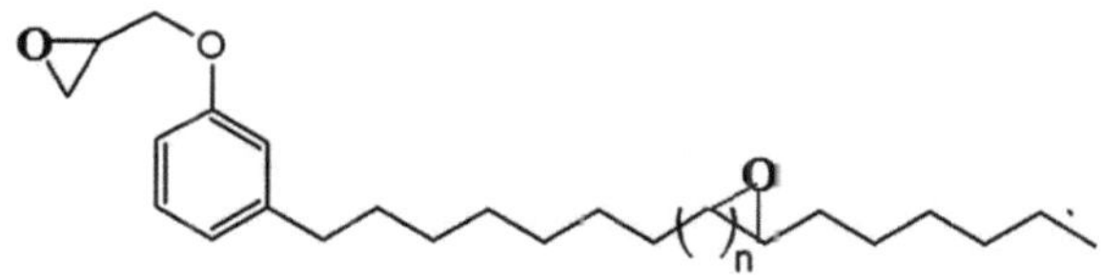

Fig. 4.6 Cardanol diglycidyl ether

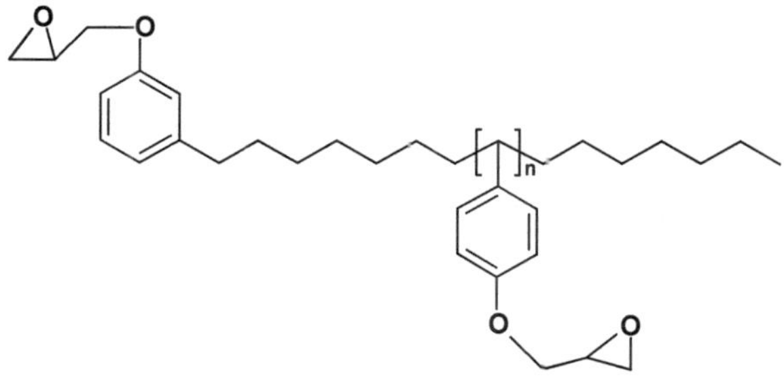

this respect, different polyols can be produced from cardanol phenols as represented in Scheme 4.16.

The PU produced from cardanol polyol derivatives has better thermal, mechanical and chemical characteristics [24, 25, 90, 91, 112, 113, 130]. Low-viscosity polyols with a range of hydroxyl values have been prepared [127]. Scheme 4.16 shows the production of trihydroxy cardanol from reaction of cardanol with ECH and diethanol amine. For synthesizing polyols, the authors have prepared first monoglycidyl ether of cardanol, followed by ring opening to prepare the diol or reaction with diethanol amine to give a triol (Scheme 4.16). The cardanol-based polyols have better hydrolytic stability compared to the triglyceride oil-based polyols. The cardanol-based PU exhibited excellent thermal stability. The second route to produce cardanol-based triol was based on the reaction of the glycerol monochlorohydrin with cardanol [127].

Mannich [61, 96] polyols cardanol can be produced from its reaction with formaldehyde and aliphatic amines (ethylene diamine, diethylene triamine and triethylenetetramine) (Scheme 4.17). These cardanol-based Mannich polyols produced rigid polyurethane foams that have excellent good physico-mechanical and fireproofing properties and applied for coating of different substrates [3].

Melamine cardanol-derived Mannich base polyol (MCMP) can also be prepared (Scheme 4.18) via a two-step process. First, cardanol was allowed to react with paraformaldehyde melamine and diethanolamine to yield mixed Mannich bases which were subsequently propoxylated to give the MCMP [147]. Polyisocyanate reacted with MCMP to produce rigid PU. It was demonstrated that the melamine incorporated into the MCMP molecular structure increased the thermal stability and the flame retardancy of the PU.

Cardanol (Fig. 4.1) contains alkyl groups that have different degrees of unsaturation, namely monoene, diene and triene. However, phenol reacted with cardanol via Friedel–Crafts alkylation (Scheme 4.19) to produce resole-type prepolymer after reaction with formaldehyde under basic conditions [55].

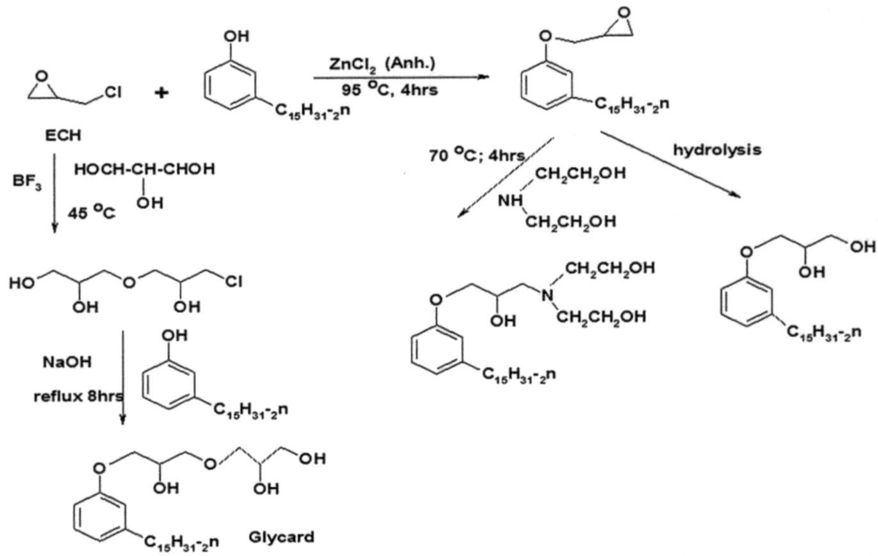

Scheme 4.16 Preparation of different cardanol polyols

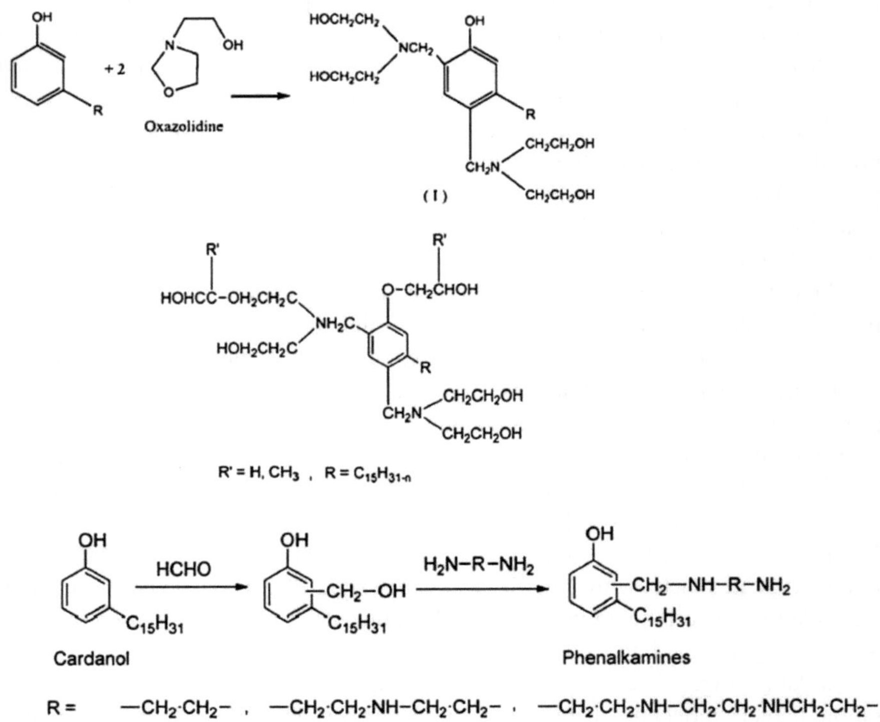

Scheme 4.17 Preparation of cardanol-based Mannich polyols

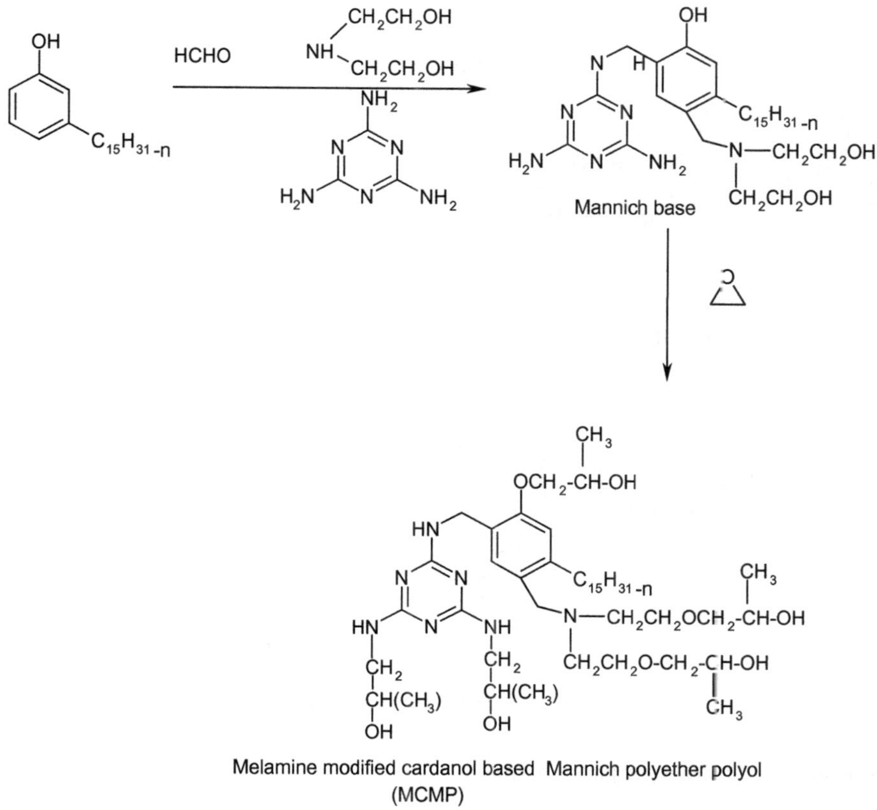

Scheme 4.18 Preparation of melamine cardanol-derived Mannich base polyol

These prepolymers were used to synthesize phenolic foams [65]. The phenolic foams based on cardanol increased the flexural strength as compared to pure phenolic foams because the incorporation of cardanol does improve the toughness of phenolic foams.

Synthesis of new polyphenols with attractive properties from natural phenolic compounds such as cardanol by oxidative polymerization using enzymes has been reviewed by Uyama et al. [140, 141]. Enzymic homo- and copolymerization of alkylphenols derived from cashew nutshell gave homopolymers that are soluble in organic solvents, but the copolymers were cross-linked, with negligible solubility [4]. The enzymic polymerization was found to be dependent on the solvent mixture used. Polymerization in a dioxane–water solvent mixture resulted in spherical particles in the case of homopolymerization, while structures without distinctive morphologies were obtained in the case of copolymerization.

A unique polyol (Scheme 4.20) has been designed and synthesized by condensation of cardanol with epoxy resin using ethyltriphenylphosphonium bromide

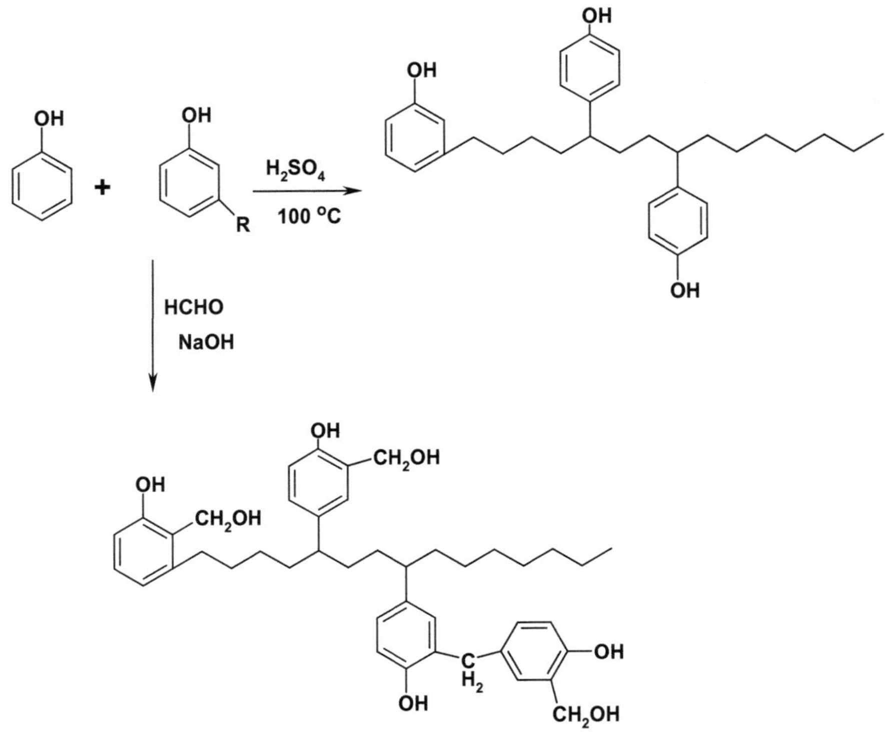

Scheme 4.19 Cardanol polyphenol polyols

(ETPPB) as catalyst and subsequently been used as polyol component (soft segment) in synthesis of polyurethane dispersion. The clear coating compositions prepared from these PUs, with and without drier (catalyst) additive, have been applied on metal substrate and cured under ambient conditions [94].

Rekha and Asha [105, 106] reported the synthesis of UV-curable telechelic urethane–methacrylate cross-linkers based on cardanol. They demonstrated that the presence of both the hydrogen bonding units and bulky hydrophobic terminal units was required for the formation of three-dimensional honeycomb patterns or vesicles from a solvent mixture of tetrahydrofuran (THF) and water or THF alone [37]. Cardanol and the saturated analogue, viz. 3-pentadecylphenol, were incorporated as pendant units to the side chain urethane–methacrylate through hydrogen bondable urethane linkage (Scheme 4.21), and the authors observed various morphologies such as tubes, spheres and honeycomb patterns in drop-cast films [37].

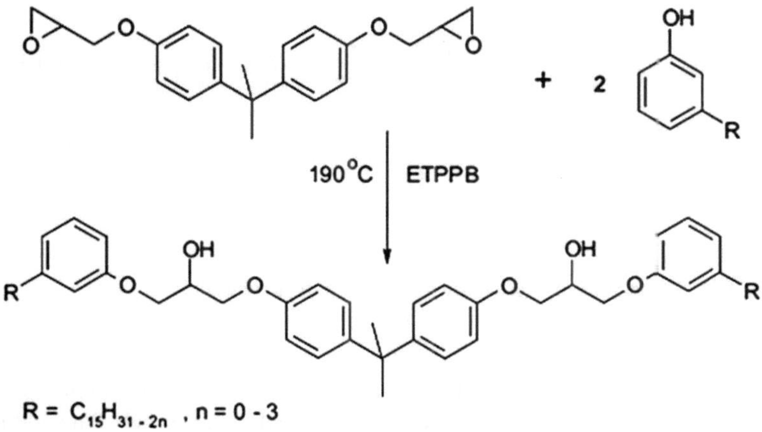

Scheme 4.20 Preparation of cardanol polyols soft segments

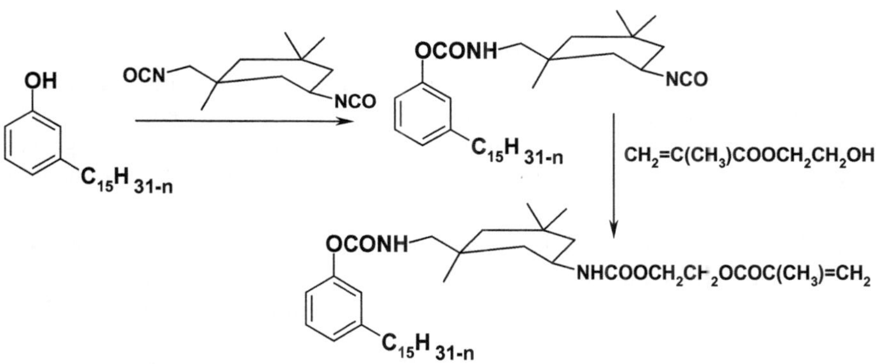

Scheme 4.21 Preparation of cardanol polyurethane methacrylate

4.3.2 Production of Biopolymers

There is of course increasing awareness of the need for green chemistry in chemical industries. For example, high-content cardanol, by-product derived from the cashew industry, has recently been used to produce bio-composites [31]. The bio-composites can be produced either by introducing of biodegradable groups to conventional polymers or by producing composites based on naturally occurred polymers based on renewable materials. Natural polymers are produced from feedstocks derived from renewable sources, while synthetic polymers are produced from non-renewable petroleum resources. To provide added value to biodegradable polymers, some advanced technologies have been applied. They include active packaging technology and natural fiber reinforcements. Nanoclay has been used with biodegradable polymers, especially with starch and aliphatic polyesters

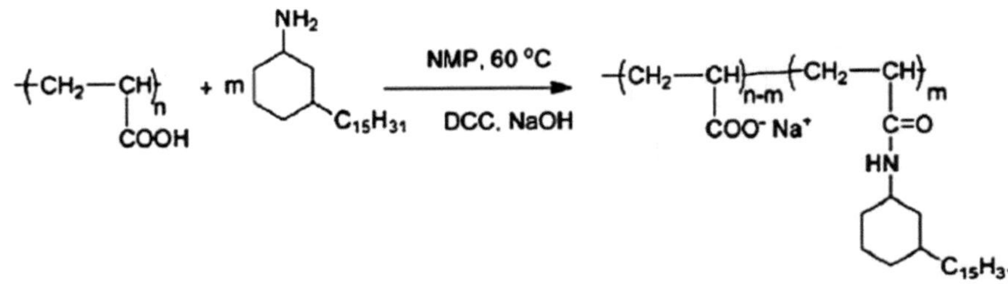

Scheme 4.22 Preparation of hydrogenated cardanol poly(acrylic acid)

to prepare nano-biocomposites. It is well reported that the conventional polymers based on petroleum resources are resistant to degradation [26, 89]. The addition of biodegradable polymeric materials has encouraged the development of such materials from readily available, renewable inexpensive natural sources [41–43, 75, 78, 114]. These renewable resources hold beneficial characteristics for being non-toxic, biodegradable and environmentally friendly.

There are different methods based on random and block copolymerization or grafting used to improve both the biodegradation rate and the mechanical properties of the final products of the biodegradable polymers. Shedge et al. [115] incorporated 3-pentadecylcyclohexylamine (3-PDCA) derived from hydrogenated cardanol onto poly(acrylic acid) to prepare a series of hydrophobically modified poly(acrylic acid) (Scheme 4.22).

Kadam et al. [70] reported the preparation method of 1-isocyanato-3-pentadecyl-cyclohexane (1-IPDC) from 3-pentadecylcyclohexylamine (3-PDCA) (Scheme 4.21). It was end-capped with poly(ethylene oxide) (PEO) of different molecular weights to improve the self-assembling properties of these associating polymers.

Hourdet et al. [58] reported synthesis of hydrophobically modified polyacrylamides, used in petroleum applications, containing random distribution of hydrophobes and studied their thermodynamic properties. These copolymers were obtained by free radical solution copolymerization of trimethylsilylacrylamide with a hydrophobic comonomer, namely 3-pentadecylcyclohexylacrylamide [3-PDCAM] (Scheme 4.23).

Flexible thermoset polymer can be produced from cardanyl acrylate or 3-pentadecenyl phenyl acrylate, obtained by the base-catalyzed reaction of cardanol and acryloyl chloride (Scheme 4.24) [69, 92, 118].

Preparation of liquid crystal based on renewable cardanol moieties was studied by modification of cardanol with methacrylate groups followed by radical polymerization as described in Scheme 4.25 [85]. The poly(methyl methacrylate) derivatives containing the cardanol moieties were prepared using 2-hydroxy-3-cardanylpropyl methacrylate (HCPM) and methacrylate (MMA) as monomers. These polymer films exhibited good optical transparency in the visible light region. The LC cells made from the polymer having 100 mol% of

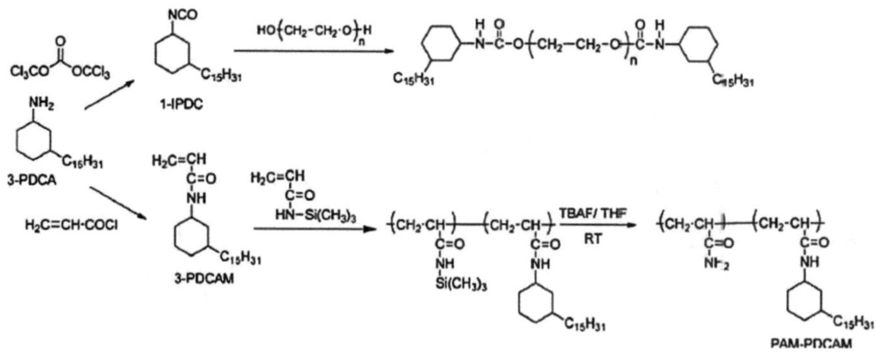

Scheme 4.23 Preparation of hydrophobically modified polycardanol–co-polyacrylamides

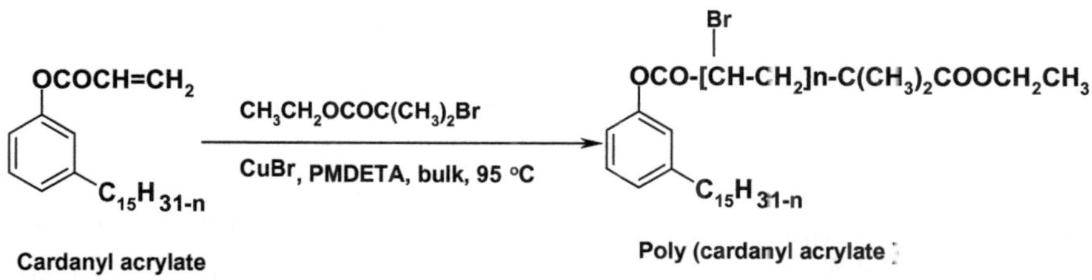

Scheme 4.24 Preparation of flexible polymer of 3-pentadecenyl phenyl acrylate

2-hydroxy-3-cardanylpropyl methacrylate (HCP100) as the alignment layer showed vertical LC alignment. The thermal stability of the LC cell made from UV-irradiated modified cardanol film was enhanced due to the formation of the cross-linked structure of the unsaturated hydrocarbon in the cardanol group.

Cardanol has been utilized as a starting material for the synthesis of bifunctional monomers that can be used to produce biopolymers using step growth method (Fig. 4.7) [19, 23, 46, 62, 72, 82–84, 86–88, 99, 107–110, 143].

These bifunctional monomers were utilized for the synthesis of a variety of high-performance polymers such as aromatic polyimides, polyamides, poly(azomethine)s, poly(amideimide)s, poly(esterimide)s, polyesters, polyhydrazides, polyoxadiazoles having higher thermal and mechanical properties. The results confirmed that the presence of the C15 alkyl chain along the polymer backbone disrupted the packing of polymer chains, as well as provided the additional "handle" for interaction with solvents, thereby enhancing the solubility of these polymers when compared to control polymers, i.e., polymers without the C15 alkyl chain. The polymers containing pendent pentadecyl chains indicated the depression in glass transition temperature (T_g) which could be attributed to the presence of pentadecyl chains, which decreases the intermolecular interactions and increases the segmental motion in the polymer backbone. A large window between glass transition temperatures and initial decomposition temperatures of

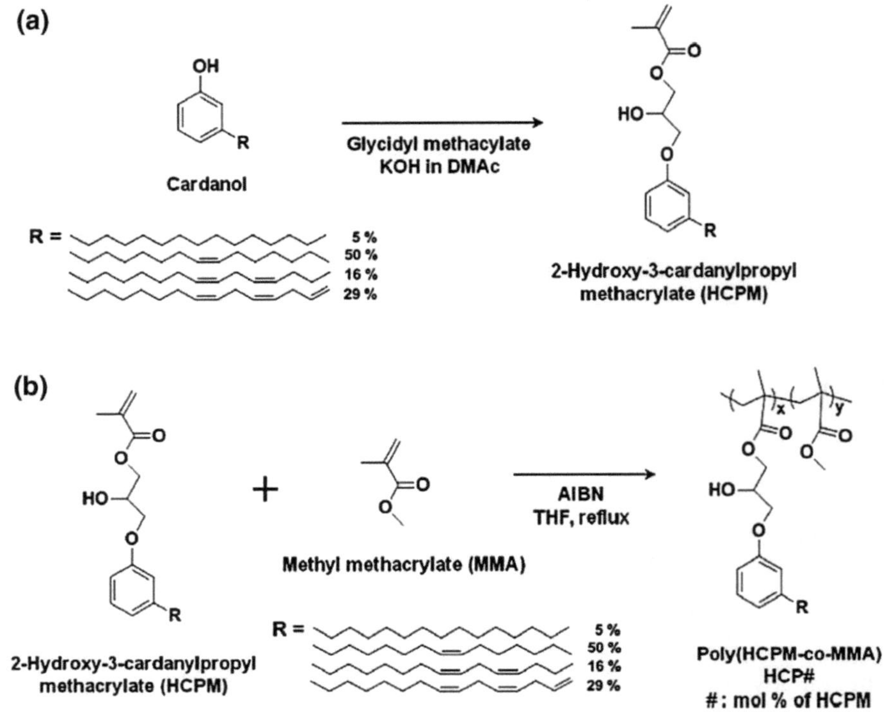

Scheme 4.25 Preparation of poly(methyl methacrylate) derivatives containing the cardanol moieties

the polymers containing pendent pentadecyl chains was observed, which gives an opportunity for these polymers to be melt-processed or compression-molded.

The application of amphiphilic materials based on cardanol used for preparation of nanomaterials and exfoliation of clay has an important role in the preparation of nanocomposites with superior characteristics. The clay minerals were completely exfoliated after organic modification with cardanol, incorporating reactive groups: $–N(CH_3)_2$, \NH_2 or two NH_2 functional groups as illustrated in Scheme 4.26 [100]. The exfoliated organic modified clay was used to prepare epoxy nanocomposites.

It was also reported that cardanol derivatives can be used as capping agent for iron oxide nanomaterials to act as oil spill collectors [142]. It was found that cardanol–furfuraldehyde/maghemite composites, represented in Scheme 4.27, used to produce magnetic green resins able to collect petroleum spilled on the water due to their lower density allowing their easy flotation and magnetic behaviors.

The cardanol and furfural resins were polymerized through acid catalysis (Scheme 4.27) to produce resins that act as capping agent for magnetic materials and have ability to easily disperse in crude oil.

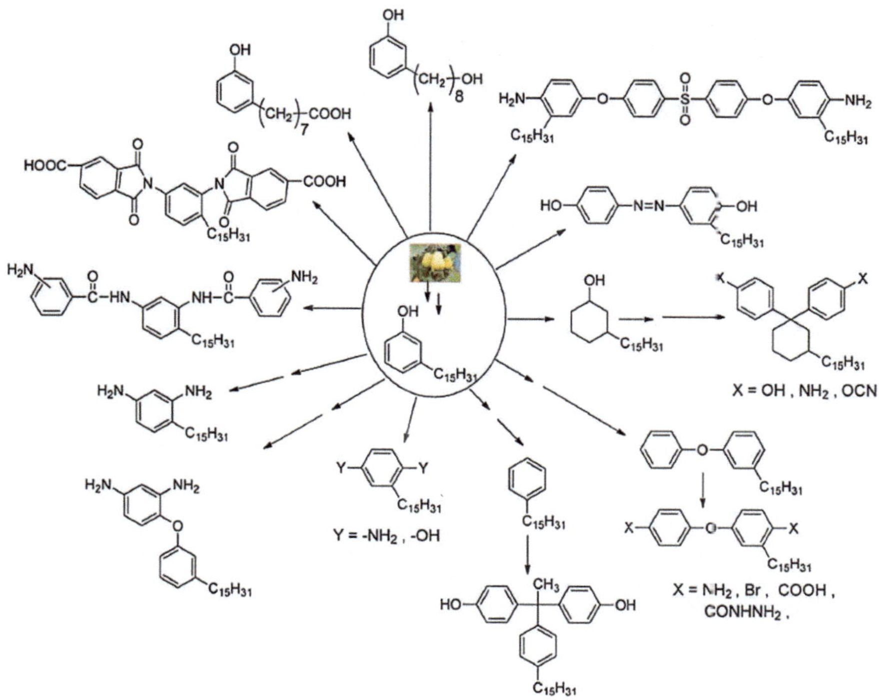

Fig. 4.7 Bifunctional bio-cardanol monomers

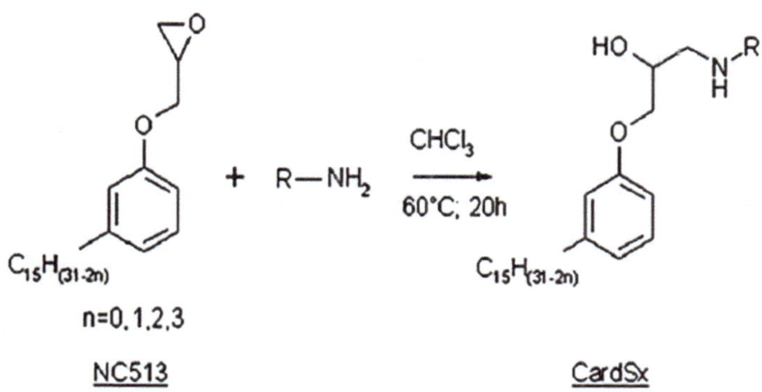

R = $-(CH_2)_2-N-[(CH_2)_2-NH_2]_2$ (TAEA = CardS1)

$-(CH_2)_2-NH_2$ (EDA = CardS2)

$-(CH_2)_2-N(CH_3)_2$ (DMEDA = CardS3)

Scheme 4.26 Preparation of cardanol alkyl amines

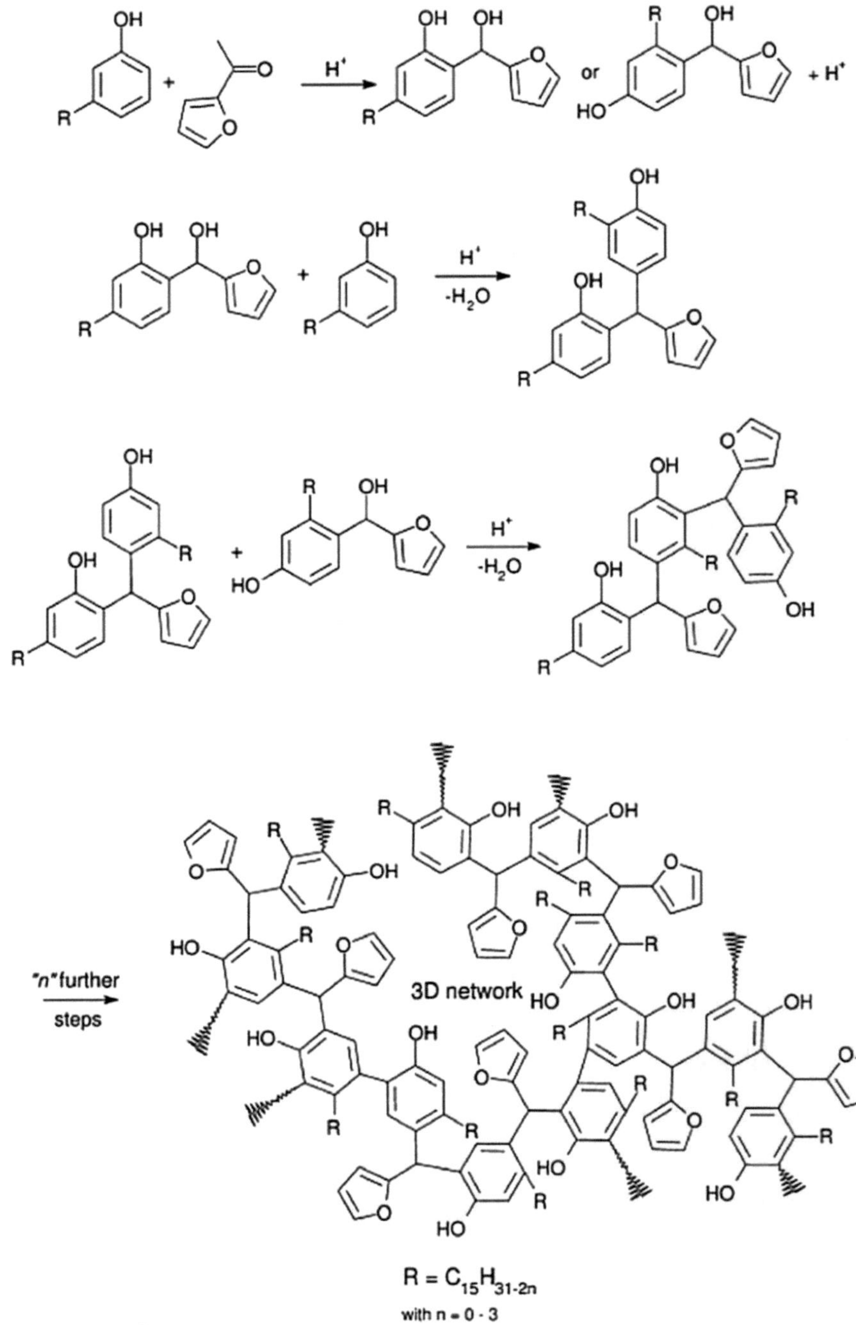

Scheme 4.27 Preparation of cardanol–furfuraldehyde composites

4.4 Conclusions and Outlook

The cardanol attracted great attention to produce advanced chemicals based on different routes of chemical modification of the alkyl side chain in the *meta* position, aromatic moieties and phenol groups, which allow the formation of products with a potential industrial value. The cheap price and abundantly available (world availability: 450,000 metric tons per year) versatile industrial raw material of great promise for the replacement of petrochemicals. Considerable attention has been devoted to transformations of the cardanol to different types of surfactants. Cardanol surfactants have outstanding antimicrobial properties which can assist mankind fight parasites and harmful microorganisms in the agricultural and medicinal fields. It can be a valuable alternative to many of the troublesome chemical agents in use today. The development of surfactants to reactive surfactants based on assist to develop new materials with superior properties and added values such as conducting nanomaterials and high-performance polymers to address the issue of processability by taking advantage of the flexible pentadecyl chain. The polymeric materials based on cardanol such as epoxy resins, phenalkamines, friction materials, brake lining agents have already found place in the market (M/s Cardolite).

In spite of the massive literature accumulated on the applications of cardanol, many areas remain which have yet to utilize this attractive raw material. The use of cardanol in the field of engineering polymers has not been widely reported. It could be a promising candidate for making bio-based composites. Moreover, cardanol is a valuable raw material for generating a variety of soft nanomaterials such as nanotubes, nanofibers, gels and surfactants. Novel bio-based surfactants, glycolipids and polymers can be derived from cardanol, and their subsequent self-assembly into functional soft materials. It is expected from these studies that new soft materials can be produced from using cardanol surfactants. Therefore, the prepared surfactants contribute to the environment encouraging nobler uses to some of the available renewable resources. A number of water-based resins can be synthesized with cardanol surfactants to apply in the coating industry due to a number of advantages they possess, including eco-friendly technology, room temperature synthesis, chemical inertness, high oxidation and abrasion resistance, excellent thermal stability, very low health hazard. The commercialization of all these technologies, however, will require further research and development for cost-effective solutions.

References

1. Agarwal S, Mishra A, Rai JSP (2003) Effect of diluents on the curing behavior of vinyl ester resin. J Appl Polym Sci 87:1948–1951
2. Agarwal SN, Murthy BGK, Sivasamban MA, Aggarwal JS (1969) Improvement in or relating to the production of pale coloured coating material from cardanol. Indian Pat 112:36

3. Alexander M, Thachil ET (2006) A comparative study of cardanol and aromatic oil plasticisers for carbon black filled natural rubber. J Appl Polym Sci 102:4835–4841. doi:10.1002/app.24811
4. Alva KS, Nayak PL, Kumar J, Tripathy SK (1997) Enzymatic polymerization of phenolic biomonomers derived from cashew nut shell liquid. J Macromol Sci, Pure Appl Chem A34:665–674. doi:10.1080/10601329708014992
5. Amendola E, Giamberini M, Carfagna C, Ambrogi V (2002) Self-toughing liquid crystalline vinyl ester adhesives. Macromol Symp 180:153–167
6. Anilkumar P, Jayakannan M (2006) New renewable resource amphiphilic molecular design for size-controlled and highly ordered polyaniline nanofibers. Langmuir 22:5952–5957. doi:10.1021/la060173n
7. Atta AM, Dyab AKF, Al-Lohedan HA (2013) Surface activity of novel polymerizable anionic polyoxyethylene 4-nonyl -2-prpylene-phenyl ether ammonium sulfate succinate surfactants. J Disp Sci Technologies 34:1211–1222. doi:10.1080/01932691.2012.739942
8. Atta AM, Dyab AKF, Al-Lohedan HA (2013) Micellization and adsorption behaviors of new reactive polymerizable surfactants based on modified nonyl phenol ethoxylates. J Surf Deterg 16:343–355. doi:10.1007/s11743-012-1413-5
9. Atta AM, Abdel-Rauf ME, Maysour NE, Gafer AK (2010) Water based oil spill dispersants based on rosin formaldehyde resins. J disp sci technol 31:583–595. doi:10.1080/01932690903212313
10. Atta AM, El-Kafrawy A, Abdel-Rauf ME, Maysour NE, Gafer AK (2010) Surface and Thermodynamic Properties of Nonionic Surfactants Based on Rosin- Maleic Anhydride and Acrylic Acid Adducts. J disp sci technol 31:567–576. doi:10.1080/01932690903192689
11. Atta AM, El-Kafrawy AF, Aly MH, Abdel-Azim AA (2007) New epoxy resins based on recycled poly(ethylene terephthalate) as organic coatings. Prog Org Coat 58:13–22. doi:10.1016/j.porgcoat.2006.11.001
12. Atta AM, Elnagdy SI, Abdel-Raouf ME, Elsaeed SM, Abdel-Azim AA (2005) Compressive properties and curing behaviour of unsaturated polyester resins in the presence of vinyl ester resins derived from recycled poly(ethylene terephthalate). J Polym Res 12:373–383. doi:10.1007/s10965-005-1638-3
13. Atta AM, Elsaeed AM (2011) Use of rosin-based nonionic surfactants as petroleum crude oil sludge dispersants. J Appl Polym Sci 122:183–192. doi:10.1002/app.34052
14. Atta AM, Elsaeed AM, Farag RK, El-Saeed SM (2007) Synthesis of unsaturated polyester resins based on rosin acrylic acid adduct for coating applications. React Funct Polym 67:549–563. doi:10.1016/j.reactfunctpolym.2007.03.009
15. Atta AM, El-Saeed SM, Farag RK (2006) New vinyl ester resins based on rosin for coating applicationsmers. React Funct Polym 66:1596–1608. doi:10.1016/j.reactfunctpolym.2006.06.002
16. Atta AM, Mansour R, Abdou MI, Sayed AM (2004) Epoxy resins from rosin acids: synthesis and characterization. Polym Adv Technol 15:514–522. doi:10.1002/pat.507
17. Atta AM, Elsaeed AM (2011) Preparation and characterization of epoxy binders based on rosin. Recent Pat Corros Sci 3:132–143
18. Atta AM, Mansour R (2005) Synthesis and characterization of tetra-functional epoxy resins from rosin. J Polym Res 12:127–138. doi:10.1007/s10965-004-2936-x
19. Avadhani CV, Wadgaonkar PP, Sivaram S (2001) 1,1-Bis(4-hydroxyphenyl)-3-alkylcyclohexanes, method for their preparation and polycarbonates prepared therefrom. US6255439
20. Bakshi SH, Krishnaswamy N (1965) Indian J Chem 3:503
21. Balachandran VS, Jadhav SR, Vemul PK, John G (2013) Recent advances in cardanol chemistry in a nutshell: from a nut to nanomaterials. Chem Soc Rev 42:427–438. doi:10.1039/C2CS35344J
22. Bhadani A, Kataria H, Singh S (2011) Synthesis, characterization and comparative evaluation of phenoxy ring containing long chain Gemini imidazolium and pyridinium amphiphiles. J Colloid Interface Sci 361:33–41. doi:10.1016/j.jcis.2011.05.023

23. Bhunia HP, Jana RN, Basak A, Lenka S, Nando GB (1998) Synthesis of polyurethane from cashew nut shell liquid (CNSL), a renewable resource. J Polym Sci, Part A: Polym Chem 36:391–400. doi:10.1002/(SICI)1099-0518(199802)36:3<391:AID-POLA3>3.0.CO;2-V
24. Bhunia HP, Nando GB, Basak A, Lenka S, Nayak P (1999) Synthesis and caracterization of polymers from cashewnut shell liquid (CNSL), a renewable resource III. Synthesis of a polyether. Eur Polym J 35:1713–1722. doi:10.1016/S0014-3057(98)00244-4
25. Bhunia HP, NandoGB Chaki TK, Basak A, Lenka S, Nayak PL (1999) Synthesis and characterization of polymers from cashewnut shell liquid (CNSL), a renewable resource II. Synthesis of polyurethanes. Eur Polym J 35:381–1391
26. Bisio AL, Xanthos M (1995) How to manage plastics wastes. Hanser Gardner Publication, USA, pp 253–280
27. Biswal S, Satapathy JR, Achary PGR, Pal NC (2012) The synthesis and FTIR, kinetics and TG/DTG/DTA study of inter penetrating polymer networks (IPNs) derived from polyurethanes of glycerol modified castor oil and cardanol based dyes. J Polym Env 20:788–793
28. Biswas AK, Roy AB (1961) Studies on sodium anacardate. J Proc Inst Chem 33(II):81–85
29. Biswas AK, Roy AB (1958) Surface-active characteristics of sodium anacardate isolated from cashew nut shell. Oil Nat 182:1299–1300. doi:10.1038/1821299b0
30. Dixit C, Sharma BL (1969) Indian Pat. I11326
31. Calo E, Maffezzoli A, Mele G, Martina F, Mazzetto SE, Tarzia A et al (2007) Synthesis of a novel cardanol-based benzoxazine monomer and environmentally sustainable production of polymers and bio-composites. Green Chem 9:754–759. doi:10.1039/B617180J
32. Campaner P, Amico DD, Longo L, Stifani C, Tarzia A (2009) Cardanol-based novolac resins as curing agents of epoxy resins. J Appl Pol Sci 114:3585–3591
33. Caplan S (1940) Treatment of cashew nut shell liquid. US Patent 2,176,059
34. Cheng C, Bai X, Liu S, Huang Q, Tu Y, Wu H, Wang X (2013) UV cured polymer based on a renewable cardanol derived RAFT agent. J Polym Res 20:197–204. doi:10.1007/s10965-013-01972
35. Darroman E, Bonnot L, Auvergne R, Boutevin B, Caillol S (2015) New aromatic amine based on cardanol giving new biobased epoxy networks with cardanol. Eur J Lipid Sci Technol 117:178–189
36. Darroman E, Durand N, Boutevin B, Caillol S (2015) New cardanol/sucrose epoxy blends for biobased coatings. Prog Org Coat 83:47–54
37. Deepak VD, Asha SK (2006) Self-organization-induced three-dimensional honeycomb pattern in structure-controlled bulky methacrylate polymers: synthesis, morphology, and mechanism of pore formation. J Phys Chem B 110:21450–21459. doi:10.1021/jp063469a
38. Devi A, Srivastava D (2007) Studies on the blends of cardanol-based epoxidized novolac type phenolic resin and carboxyl-terminated polybutadiene (CTPB). Mater Sci Eng A458:336–347. doi:10.1016/j.msea.2006.12.081
39. Dickerson TJ, Reed NN, Janda KD (2002) Soluble polymers as scaffolds for recoverable catalysts and reagents. Chem Rev 102:3325–3344. doi:10.1021/cr010335e
40. Durrani AA, Davis GL, Sood SK, Tychopoulos V, Tyman JHP (1982) Long-Chain phenols*. Part 231 practical separations of the component phenols in technical cashew nut-shell liquid (*Anacardium occidentale*): distillation procedures for obtaining cardano. J Chem Tech Biotechnol 32:681–690. doi:10.1002/jctb.5030320704
41. Dwanisa JPL, Mohanty AK, Misra M, Drazal LT (2004) Polyurethane and its composite with glass fiber. J Mater Sci 39:2081–2087
42. Fengkui LL, Larock RC (2001) New soybean oil-styrene-divinylbenzene thermosetting copolymers. III. Tensile stress–strain behavior. J Polym Sci Part B Polym Sci 39:60–77. doi:10.1002/1099-0488(20010101)39:1<60:AID-POLB60>3.0
43. Figen Balo H, Yuce L, Ucar A (2010) Physical and mechanical properties of materials prepared using Class C fly ash and soybean oil. J Porous Mater 17:553–564. doi:10.1007/s10934-009-9324-1

44. Gadem PH, Sampathkumaran PS (1986) Cashew nut shell liquid: extraction, chemistry and applications. Prog Org Coat 14:115–157. doi:10.1016/0033-0655(86)80009-7
45. Gaur B, Rai JSP (1993) The spectral and magnetic properties of some chloro and thiocyanato transition metal complexes of the aminopyridines and a study of their performance as colouring materials for poly(vinyl chloride) Eur. Polym J 29:1149–1153
46. Ghatge ND, Maldar NN (1984) Polyimides from dianhydride and diamine: structure property relations by thermogravimetric analysis (t.g.a.). Polymer 25:1353–1356. doi:10.1016/0032-3861(84)90390-2
47. Gopalakrishnan S, Fernando TL (2012) Influence of polyols on properties of bio-based polyurethanes. Bull Mater Sci 35:243–251
48. Greco A, Brunetti D, Renna G, Mele G, Maffezzoli A (2010) Plasticizer for poly(vinyl chloride) from cardanol as a renewable resource material. Pol Degrad Stab 95:2169–2174
49. Greenwald RB (2001) PEG drugs: an overview. J Control Release 74:159–171. doi:10.1016/S0168-3659(01)00331-5.20
50. Guo-peng S, Jiu-zhu L, Ming-jian W, Qian H (2013) Synthesis and surface activity of quaternary ammonium salt surfactant prepared from cardanol. China Surf Deterg Cosmet 3:188–191
51. Guyot A (2004) Advances in reactive surfactants. Adv Colloid Interface Sci 108:3–22. doi:10.1016/j.cis.2003.10.009
52. Harvey MT (1929) Resin form cashew nut shell oil. US Patent 1,725,791
53. Harvey MT (1929) Substitute for Shellac and the Lik. US Patent 1,725,793
54. Harvey MT, Damitz FM (1941) Polyiherization of cashew shell liquid and products thereof mortimer. US Patent 2,240,038
55. He JL, Zhang XL, Wang LL, Yang Y (1989) J China Plast Ind 4:14–21
56. Hemalatha IN, Yaseen M, Aggarwal JS (1972) Kinetics of reaction between meta-substituted long chain alkyl phenols and formaldehyde. Angew Macromol Chem 24:163–169. doi:10.1002/apmc.1972.050240111
57. Hemalatha IN, Dakshinamurthy H, Aggarwai JS (1968) Paint Manuf 38(6):29
58. Hourdet D, Ducouret G, Varghese S, Badiger MV, Wadgaonkar PP (2013) Thermodynamic behavior of hydrophobically modified polyacrylamide containing random distribution of hydrophobes: Experimental and theoretical investigations. Polymer 54:2676–2689. doi:10.1016/j.polymer.2013.03.039
59. Huong NL, Nieu NH, Tan TTM, Griesser U (1996) Cardanol-phenol-formaldehyde resins. Thermal analysis and characterization. Angew Makromol Chem 243:77–85
60. Ionescu M, Petrovic ZS (2011) Phenolation of vegetable oils. J Serb Chem Soc 76:591–606
61. Ionescu M, Wan X, Bilic N, Petrovic ZS (2012) Polyols and rigid polyurethane foams from cashew nut shell liquid. J Polym Environ 20:647–658. doi:10.1007/s10924-012-0467-9
62. Jadhav AS, Vernekar SP, Maldar NN (1993) Synthesis and characterization of new aromatic sulfone ether polyamides containing pendant pentadecyl groups. Polym Int 32:5–11. doi:10.1002/pi.4990320103
63. Jaillet F, Darroman E, Ratsimihety A, Auvergne R et al (2014) New biobased epoxy materials from cardanol. Eur J Lipid Sci Technol 116:63–73
64. Jain PK, Sivala KJ (1997) Development of a cashew nut sheller. J Food Eng 32:339–345. doi:10.1016/S0260-8774(97)00020-4
65. Jing S, Li T, Li X, Xu Q, Hu J, Li R (2014) Phenolic foams modified by cardanol through bisphenol modification. J Appl Polym Sci 131:39942. doi:10.1002/app.39942
66. John G, Masuda M, Okada Y, Yase K, Shimizu T (2001) Nanotube formation from renewable resources via coiled nanofibers. Adv Mater 13:715–718. doi:10.1002/1521-4095(200105)13:10<715:AID-DMA715>3.0.CO;2-Z
67. John G, Pillai CKS (1992) Self-crosslinkable monomer from cardanol: crosslinked beads of poly(cardanyl acrylate) by suspension polymerization. Macromol Rapid Commun 13:255–259. doi:10.1002/marc.1992.030130502

68. John G, Pillai CKS (1993) Synthesis and characterization of a self-crosslinkable polymer from cardanol: Autooxidation of poly(cardanyl acrylate) to crosslinked film. J Polym Sci A Polym Chem 31:1069–1073. doi:10.1002/pola.1993.080310429
69. John G, Shali KT, Pillai CKS (1994) Cardanyl acrylate/methacrylate based cross-linked copolymers as novel supports: synthesis and characterization. J Appl Polym Sci 53:1415–1423. doi:10.1002/app.1994.070531103
70. Kadam VS, Badiger MV, Wadgaonkar P, Ducouret G, Hourdet D (2008) Synthesis and self-assembling properties of α, ω-hydroxy-poly(ethylene oxide) end-capped with 1-isocyanato-3-pentadecylcyclohexane. Polymer 49:4635–4646. doi:10.1016/j.polymer.2008.07.064
71. Kanehashi S, Yokoyama K, Masuda R, Kidesaki T et al (2013) Preparation and characterization of cardanol-based epoxy resin for coating at room temperature curing. J Appl Polym Sci 130:2468–2478
72. Kang H, Choi Y-S, Hong H, Ko T, Kang D, Lee JC (2014) Vertical alignment of liquid crystals on polymer films containing renewable cardanol moieties. Eur Polym J 61:13–22. doi:10.1016/j.eurpolymj.2014.09.019
73. Kaplan DL (1998) Biopolymers from renewable resources. Springer, New York, pp 1–417
74. Kathalewar M, Sabnis A, D'Melo D (2014) Polyurethane coatings prepared from CNSL based polyols: Synthesis, characterization and properties. Prog Org Coat 77:616–626
75. Kavitha V, Radhakrishnan N (2010) Biopolymer from microbial assisted in situ hydrolysis of triglycerides and dimerization of fatty acids. Bioresour Technol 101:337–343. doi:10.1016/j.biortech.2009.08.022
76. Kim YH, An ES, Park SY, Song BK (2007) Enzymatic epoxidation and polymerization of cardanol obtained from a renewable resources and curing of epoxide-containing polycardanol. J Macromol Cat B Enzymatic 45:39–44
77. Launikitis MB (1982) Handbook of composites. In: G. Lubin (ed) Van Nostran Reinhold Company Inc., New York, pp 38–49
78. Lee N, Kwon O-J, Chun BC, Cho JW (2009) Characterization of castor oil/polycaprolactone polyurethane biocomposites reinforced with hemp fibers. Fibers Polym 10:154–160. doi:10.1007/s12221-009-0154-1
79. Lochab B, Varma IK, Bijwe J (2010) Thermal behavior of cardanol-based benzoxazines. Monom Polym 102:769–774
80. Mahesh S, Raju D, Arathi AS, Kuruvilla J (2014) Self-assembly of cardanol based supramolecular synthons to photoresponsive nanospheres: light induced size variation at the nanoscale. RSC Adv 4:42747–42750. doi:10.1039/C4RA07406H
81. Manjula S, Pillai CKS (1990) Thermal characterization of cardanol-formaldehyde resins and cardanol-formaldehyde/poly(methyl methacrylate) semi-interpenetrating polymer networks. Thermochim Acta 159:255–266. doi:10.1016/0040-6031(90)80114-E
82. Mathew JS, Vernekar SP, Mercier R, Kerboua R(2002) Polyimides, process for the preparation thereof and use thereof as alignment films for liquid crystal devices. US6500913
83. More AS, Patil AS, Wadgaonkar PP (2010) Poly(amideimide)s containing pendant pentadecyl chains: synthesis and characterization. Polym Degrad Stab 95:837–844. doi:10.1016/j.polymdegradstab.2010.01.030
84. More AS, Sane PS, Patil AS, Wadgaonkar PP (2010) Synthesis and characterization of aromatic polyazomethines bearing pendant pentadecyl chains. Polym Degrad Stab 95:1727–1735. doi:10.1016/j.polymdegradstab.2010.05.017
85. More AS, Menon SK, Wadgaonkar PP (2012) New poly(1,3,4-oxadiazole)s bearing pentadecyl side chains: synthesis and characterization. J Appl Polym Sci 124:1281–1289. doi:10.1002/app.34682
86. More AS, Naik PV, Kumbhar KP, Wadgaonkar PP (2010) Synthesis and characterization of polyesters based on 1,1,1-[bis(4-hydroxyphenyl)-4′-pentadecylphenyl]ethane. Polym Int 59:1408–1414. doi:10.1002/pi.2883

87. More AS, Pasale SK, Wadgaonkar PP (2010) Synthesis and characterization of polyamides containing pendant pentadecyl chains. Eur Polym J 46:557–567. doi:10.1016/j.eurpolymj.2009.11.014
88. More AS, Pasale SK, Honkhambe PN, Wadgaonkar PP (2011) Synthesis and characterization of organo-soluble poly(ether ether ketone)s and poly(ether ether ketone ketone)s containing pendant pentadecyl chains. J Appl Polym Sci 121:3689–3695. doi:10.1002/app.34181
89. Mustafa N (1993) Plastics waste management disposal, recycling and reuse. Marcel Dekker, New York, pp 40–68
90. Mythili CV, Retna AM, Gopalakrishnan S (2004) Synthesis, mechanical, thermal and chemical properties of polyurethanes based on cardanol Bull. Mater Sci 27:235–241
91. Mythili CV, Malar Retna A, Gopalakrishnan S (2005) Physical, mechanical, and thermal properties of polyurethanes based on hydroxyalkylated cardanol–formaldehyde resins. J Appl Polym Sci 98:284–288. doi:10.1002/app.22016
92. Nguyen LH, Koerner H, Lederer K (2003) Free radical Co- and terpolymerization of styrene, hydrogenated cardanyl acrylate, and cardanyl acetate. J Appl Polym Sci 88:1399–1409. doi:10.1002/app.11635
93. Pal N, Srivastava A, Agrawal S, Rai JSP (2005) Mater Manuf Process 20:317–327
94. Patel J, Mannaric V (2014) Air-drying bio-based polyurethane dispersion from cardanol: synthesis and characterization of coatings Chintankumar. Prog Org Coat 77:997–1006. doi:10.1016/j.porgcoat.2014.02.006
95. Patel MB, Patel RG, Patel VS (1989) Effects of reactive diluent diepoxidized cardanol and epoxy fortifier on curing kinetics of epoxy resin. J Therm Anal 35:447–457
96. Pathak SK, Rao BS (2006) Structural effect of phenalkamines on adhesive viscoelastic and thermal properties of epoxy networks. J Appl Polym Sci 102:4741–4748. doi:10.1002/app.25005
97. Peungjitton P, Sangvanich P, Pornpakakul S, Petsom A, Roengsumran S (2009) Sodium cardanol sulfonate surfactant from cashew nut shell liquid. J Surfact Deterg 12:85–89. doi:10.1007/s11743-008-1082-6
98. Pillai CKS, Prasad VS, Sudha JD, Bera SC, Menon ARR (1990) Polymeric resins from renewable resources. II. Synthesis and characterization of flame-retardant prepolymers from cardanol. J Appl Polym Sci 41:2487–2501. doi:10.1002/app.1990.070410947
99. Pillai CKS, Sherrington DC, Sneddon A (1992) Thermotropic liquid crystalline copolyester based on 8-(3-hydroxyphenyl) octanoic acid and *p*-hydroxybenzoic acid. Polymer 33:3968–3970. doi:10.1016/0032-3861(92)90390-I
100. Puchot L, Verge P, Toniazzo V, Ruch D, Fouquet T, Bomfim AS (2014) Tailoring the reactivity of bio-surfactants as a tool for montmorillonite exfoliation in epoxy thermosets. Appl Clay Sci 99:35–41. doi:10.1016/j.clay.2014.06.003
101. Ramaiingamt T, Saitur B (1987) Synthesis and biological activity of 3-pentadecylaryloxyacetic. Acids, their hydrazides and cyclic derivatives: oxadiazoles and pyrroles. Ind J Chem 26B:1204–1208
102. Ramalingam T, Sattur PB (1989) Synthesis and biological activity of 0.-(3-pentadecylaryloxy)isobutyric acids, their hydrazidesand cyclic derivatives: oxadiazoles and pyrroles. J Chem 28B:611–613
103. Rao BS, Palamisamy AW (2011) Monofunctional benzoxazine from cardanol for biocomposite applications. React Funct Polym 71:148–154. doi:10.1016/j.reactfunctpolym.2010.11.025
104. Raqueza J-M, Deléglisea M, Lacrampe M-F, Krawczak P (2010) Thermosetting (bio) materials derived from renewable resources: a critical review. Polym Sci 35:487–509. doi:10.1016/j.progpolymsci.2010.01.001
105. Rekha N, Asha SK (2008) Synthesis and FTIR spectroscopic investigation of the UV curing kinetics of telechelic urethane methacrylate crosslinkers based on the renewable resource—Cardanol. J Appl Polym Sci 109:2781–2790. doi:10.1002/app.28342

106. Rekha N, Asha SK (2009) Solvent-induced self-assembly in cardanol-based urethane methacrylate comb polymers. J Polym Sci, Part A: Polym Chem 47:2996–3009. doi:10.1002/pola.23383
107. Sadavarte NV, Halhalli MR, Avadhani CV, Wadgaonkar PP (2009) Synthesis and characterization of new polyimides containing pendent pentadecyl chains. Eur Polym J 45:582–589. doi:10.1016/j.eurpolymj.2008.11.013
108. Sadavarte NV, Avadhani CV, Naik PV, Wadgaonkar PP (2010) Regularly alternating poly(amideimide)s containing pendent pentadecyl chains: synthesis and characterization. Eur Polym J 46:1307–1315. doi:10.1016/j.eurpolymj.2010.03.007
109. Sadavarte NV, Avadhani CV, Wadgaonkar PP (2011) High performance polyamides based on s-Triazine ring: synthesis and characterization. High Perform Polym 23:494–505. doi:10.1177/0954008310378054
110. Sadavarte NV, Patil SS, Avadhani CV, Wadgaonkar PP (2013) New organosoluble aromatic poly(esterimide)s containing pendent pentadecyl chains: synthesis and characterization. High Perform Polym 25:735–743. doi:10.1177/0954008313483344
111. Saladino R, Neri V, Mincione E, Marini S, Coletta M, Fiorucci C, Filippone P (1995) A new and efficient synthesis of ortho- and para-benzoquinones of cardanol derivatives by the catalytic system $MeReO_3$–H_2O_2. J Chem Soc Perkin Trans 1:581–586. doi:10.1039/A908073B
112. Sathiyalekshmi K, Gopalakrishnan S (2000) Synthesis and characterisation of rigid polyurethanes based on hydroxyalkylated cardanolformaldehyde resin. Plast, Rubber Compos 29:63–69
113. Sathiyalekshmi K, Gopalakrishnan S (2004) Evaluation of performance of polyurethanes based on hydroxy alkylated cardanol formaldehyde resins under aging conditions. Adv Polym Technol 23:91–102. doi:10.1002/adv.20005
114. Sharma V, Kundu P (2006) Addition polymers from natural oils—a review. Prog Polym Sci 31:983–1008. doi:10.1016/j.progpolymsci.2006.09.003
115. Shedge AS, Lele AK, Wadgaonkar P, Hourdet D, Perrin P, Chassenieux C, Badiger MV (2005) Hydrophobically modified poly(acrylic acid) using 3-pentadecylcyclohexylamine: synthesis and rheology. Macromol Chem Phys 206:464–472. doi:10.1002/macp.200400392
116. Shi W, Wang P, Li C, Li J, Li H, Zhang Z, Wu S, Wang J (2014) Synthesis of cardanol sulfonate gemini surfactant and enthalpy-entropy compensation of micellization in aqueous solutions. Open J Appl Sci 4:360–365. doi:10.4236/ojapps.2014.46033
117. Shukla R, Kumar P (2011) Self-curable epoxide resins based on cardanol for use in surface coatings. Pigm Resin Technol 40:311–333
118. Sitaramam BS, Chatterjee PC (1989) Synthesis, polymerization, and end-use evaluation of 3-pentadecylphenyl acrylate and methacrylate. J Appl Polym Sci 37:33–37. doi:10.1002/app.1989.070370103
119. Stewart R (2008) Going green: eco-friendly materials and recycling on growth paths. Plastics Eng 64:16–23
120. Subba Rao R, Harigopal VP (1975) Analysis of ethylene oxide adducts of cardanol and 3-pentadecylphenol by nuclear magnetic resonance and thin-layer chromatography procedures. Fette Seifen Anstrichm 77:197–199. doi:10.1002/lipi.19750770509
121. Sudha JD, Reena VL (2007) Structure-directing effect of renewable resource based amphiphilic dopants on the formation of conducting polyaniline-clay nanocomposite. Macromol Symp 254(2007):274–284. doi:10.1002/masy.200750841
122. Sudha JD, Reena VL, Pavithran C (2007) Facile green strategy for micro/nano structured conducting polyaniline-clay nanocomposite via template polymerization using amphiphilic dopant, 3-pentadecylphenol 4-sulphonic acid. J Polym Sci, Part B: Polym Phys 45:2664–2673. doi:10.1002/polb.21273
123. Sultania M, Rai JSP, Srivastava D (2011) Process modeling, optimization and analysis of esterification reaction of cashew nut shell liquid (CNSL)-derived epoxy resin using response surface methodology. J Hazard Mater 185:1198–1204

124. Sultania M, Rai JSP, Srivastava D (2012) Modeling and simulation of curing kinetics for the cardanol-based vinyl ester resin by means of non-isothermal DSC measurements. Mater Chem Phys 132:180–186
125. Sultania M, Rai JSP, Srivastava D (2010) Kinetic modeling of esterification of cardanol-based epoxy resin in the presence of triphenylphosphine for producing vinyl ester resin: Mechanistic rate equation. J Appl Polym Sci 118:1979–1989
126. Sultania M, Rai JSP, Srivastava D (2010) Studies on the synthesis and curing of epoxidized novolac vinyl ester resins from nenewable resource material. Eur Polym J 46:2019–2032
127. Suresh KI, Kishanprasad VS (2005) Synthesis, structure, and properties of novel polyols from cardanol and developed polyurethanes. Ind Eng Chem Res 44:4504–4512. doi:10.1021/ie0488750
128. Suresh KI (2013) Rigid polyurethane foams from cardanol: synthesis, structural characterization, and evaluation of polyol and foam properties. ACS Sustain Chem Eng 1:232–242
129. Suresh KI, Gesche F, Rolf S, Eckhard B (2012) Synthesis and micellization properties of new anionic reactive surfactants based on hydrogenated cardanol. J Surf Deterg 15:207–215. doi:10.1007/s11743-011-1294-z
130. Tan TTM (1997) Cardanol–glycols and cardanol–glycol-based polyurethane films. J Appl Polym Sci 65:507–510. doi:10.1002/(SICI)1097-4628(19970718)65:3<507:AID-APP10>3.0.CO;2-U
131. Tocco G, Fais A, Meli G, Begala M, Podda G, Fadda MB, Corda M, Attanasi OA, Filippone P, Berretta S (2009) PEG-immobilization of cardol and soluble polymer-supported synthesis of some cardol–coumarin derivatives: preliminary evaluation of their inhibitory activity on mushroom tyrosinase. Bioorg Med Chem Lett 19:36–39
132. Tyman JHP (1979) Non-isoprenoid long chain phenols. Chem Soc Rev 8:499–537. doi:10.1039/CS9790800499
133. Tyman JHP (2008) Synthetic and natural phenols. Elsevier, Amsterdam, ch. 3, pp 53–98
134. Tyman JHP (1989) The extraction of natural cashew nut-shell liquid from the cashew nut (*Anacardium occidentale*), development of a cashew nut sheller. J Am Oil Chem Soc 66:553–557. doi:10.1007/BF02885447
135. Tyman JHP, Wilczynski D, Kashani MA (1978) Compositional studies on technical cashew nutshell liquid (CNSL) by chromatography and mass spectroscopy. J Am Oil Chem Soc 55:663–668
136. Tyman JHP (1976) Determination of the component phenols in natural and technical cashew nut-shell liquid by gas-liquid chromatography. Anal Chem 48:30–34
137. Tyman JHP, Bruce IE (2003) Synthesis and characterization of polyethoxylate surfactants derived from phenolic lipids. J Surf Deterg 6(4):291–297. doi:10.1007/s11743-003-0272-3
138. Tyman JHP (1978) Long-chain phenols: XII. Compositional studies: The polymeric material in the unsaturated phenols of *Anacardium occidentale*. J Chromatogr 156:255–266
139. Unnikrishnan KP, Thomas Thachil EBY (2008) Studies on the modification of commercial epoxy resins using cardanol based phenolic resins. J Elastomers Plast 40:271–286
140. Uyama H, Kobayashi S (1999) Enzymatic polymerization yields useful polyphenols. ChemTech 29:22–28
141. Uyama H, Ikeda R, Sugihara J, Kobayashi S (1998) Polym Prepr (Am Chem Soc Div Polym Chem) 39:106
142. Varela A, Oliveira G, Souza FG Jr, Rodrigues CHM, Costa MAS (2013) New petroleum absorbers based on cardanol-furfuraldehyde magnetic nanocomposites. Polym Eng Sci 53:44–51. doi:10.1002/pen.23229
143. Varma AJ, Sivaram S (2002) (Hydroxyalkyl)phenols, method for their preparation, and uses thereof US 6451957
144. Vernekar SP (1980) Epoxy compound of cashew nut shell liquid as a stabilizer for polyvinyl-chloride. Ind J Technol 18:170

145. Voirin C, Caillol S, Sadavarte NV, Tawade BV, Boutevin B, Wadgaonkar PP (2014) Functionalization of cardanol: towards biobased polymers and additives. Polym Chem 5:3142–3162. doi:10.1039/C3PY01194A
146. Yadav R, Srivastava D (2007) Kinetics of the acid-catalyzed cardanol–formaldehyde reactions. Mat Chem Phys 106:74–81. doi:10.1016/j.matchemphys.2007.05.020
147. Zhang M, Zhang J, Chen S, Zhou Y (2014) Synthesis and fire properties of rigid polyurethane foams made from a polyol derived from melamine and cardanol. Polym Degrad Stab 110:27–34. doi:10.1016/j.polymdegradstab.2014.08.009

Chapter 5
Liquid Crystalline Polymers and Molecules Derived from Cardanol, A Natural Phenol Derivative: A Summary

K.Y. Sandhya and C.K.S. Pillai

5.1 Introduction

Over a period of millions of years, nature has evolved a variety of polymers of complex structures and highly specific functions from a small set of subunits such as amino acids, nucleobases, and hexoses through the combination of covalent bonds and noncovalent assemblies [1]. The structural diversity arising out of these combinations are enormous which offers ample opportunities for better development and use. In addition, there are a number of other factors that favor the use of natural monomers or polymers as a source for the production of polymers for various applications. The resistance of synthetic polymers to chemical, physical, and biological degradation has raised serious disposal problems [2]. The concerns on environmental pollution including the increasing difficulties of waste disposal (caused by the non-biodegradability of synthetic polymers in general), the deepening threat of global warming (due to carbon dioxide released during production and incineration), tensions in the gulf countries over oil, and the fear of a possible future depletion of oil add to the move to a bio-based material policy [3–5], which has gained importance for polymers from bioresources. Apart from the availability of various natural polymers, there are also a variety of natural monomers available among agricultural and forest products lending further possibilities to have a fresh look into their utilization. Therefore, the synthesis of polymers from renewable resources has attracted considerable attention of researchers throughout the world. In this respect, the phenolic unsaturated hydrocarbon material called cashew nut shell liquid (CNSL), a reddish brown viscous fluid extracted from the cashew nut shell, the waste product obtained

K.Y. Sandhya (✉)
Indian Institute of Space Science and Technology (IIST), Valiamala, Thiruvananthapuram, Kerala, India
e-mail: sandhya@iist.ac.in

C.K.S. Pillai
National Institute for Interdisciplinary Science and Technology (NIIST), Thiruvananthapuram, India

P. Anilkumar (ed.), *Cashew Nut Shell Liquid*,
DOI 10.1007/978-3-319-47455-7_5

from *Anacardium occidentale* (cashew tree), holds considerable promise as a source of unsaturated hydrocarbon phenol which serves as an excellent monomer for polymer production. CNSL consists of four major components namely 3-pentadecenyl phenol (cardanol), 2-methyl, 5-pentadecenyl resorcinol (2-methyl cardol), 5-pentadecenyl resorcinol (cardol), and 6-pentadecenyl salicylic acid (anacardic acid), and this paper focuses on cardanol. The cardanol is widely used in the manufacture of special phenolic resins for coatings, lamination, and friction materials and recently many specialized polymers such as liquid crystalline polymers (LCPs) and nonlinear optical (NLO) polymers. This chapter reviews the state of the art of NLO polymers and LCPs developed using cardanol carried out as part of the program on specialty polymers synthesized or prepared, at the Polymer Division of NIIST by Dr. C. K. S. Pillai and his team, to study and develop high-performance and functional polymers and composites based mostly on the regionally available natural resources such as bio monomers, natural polymers, and natural fibers.

The natural monomer, cardanol, has shown to possess special structural features for chemical modifications to design high-value polymers [6, 7] and functional materials. This involves a value addition of many orders of magnitude, and the chemical transformation includes intermediates that are 100 % chemically pure such that all the problems of CNSL utilization can at a stretch be overcome. Cardanol belongs to the family of long-chain hydrocarbon phenols which are phenolic lipids present in plants. They are wide spread in tropical temperate climates in trees, shrubs, many small plants, and certain bacterial sources. They are monohydric or dihydric phenols or phenolic acids with a hydrocarbon side chain at the *meta* position. In general, many of these monomers are found to be mixtures of four components varying in the degree of unsaturation of the side chain [8]. Recently Pillai et al. formed a new polymer based on cardanol. Cardanol was oxidized with phase transfer catalyzed permanganate, and the oxidized product 8-(3-hydroxy phenyl) octanoic acid was copolymerized with *p*-hydroxy benzoic acid (*p*-HBA) to produce a thermotropic liquid crystalline copolymer [9], and Saminathan et al. reported a polymer containing 2-pentadecyl phenyl azo group with lower phase transition temperature made from cardanol [10] and suggested it to be a potential NLO material.

Thus, the structure of cardanol as described is suitable for making variety of polymers including functional polymers such as LCPs and NLO polymers and many other functional properties and assembled structures and can be achieved by modifying the monomer. Being unsaturated increases the versatility of the monomer, and these phenols can undergo oxidative coupling polymerization giving rise finally to cross-linked polymer films and hence leading to their applications in surface coatings.

5.2 What Are Liquid Crystals?

Before we go into details of LCPs, let us see a little bit about the basics of liquid crystals. The word “Liquid Crystal” is coined from liquids and crystals. Crystalline materials on melting completely lose their (a) long-range positional order and (b) long-range orientational order. LCs on melting do not completely lose their long-range orders. They retain partial order and hence the anisotropy of the crystals while maintaining the mobility of the liquid phase. Thus, materials that possess both the flow of the liquid and the anisotropy and order of the crystal or, more clearly those materials which exhibit an intermediate phase between the crystal and the liquid, are called LCs. And of course the polymers which possess this property are termed LCPs. The LCs possess the typical properties of a fluid such as mobility, flow, and inability to support shear and certain crystalline characteristic properties such as periodic arrangement of molecules in certain order and anisotropy. Certain structural features are often found in molecules forming LC phases and are termed mesogens. LC behavior originates from the presence of mesogens which are chemical entities with structures akin to rod-like, disc-like, or lath-like structures. The ordering of the mesogens in the liquid crystals can be of the type of nematic, smetic, or cholesteric. LCs are classified mainly into two: thermotropic: the LC phase is formed during heating; and lyotropic: the LC phase is obtained by changing the concentration of the molecules in a solvent.

LCPs rose into prominence because of (a) the impressive high-performance properties obtained through spontaneous orientation during melting resulting from a pronounced anisotropy in shape and (b) the applications in imaging technology, nonlinear optical, telecommunications, etc., arising from the electro-optical and magnetic properties of the mesophase. The high-performance LCPs are characterized by their exceptional mechanical properties, spontaneous orientation, ease of processing, etc. The industry faced problems such as high processing temperature and high cost of the monomer. This laboratory (NIIST) addressed strategies to reduce the melting temperature (T_m) and make the process cost-effective. Naturally occurring nontoxic and cost-effective monomers that have in-built short methylene spacers attached to electrophilic carbonyl were used with an added advantage of controlled biodegradability. LCPs containing azobenzene mesogen are emerging as an attractive area because of their potential applications in LC displays, NLO materials, information storage, devices, etc. The long aliphatic lateral side chain and hydrogen bonding stabilize the mesophase.

5.3 Liquid Crystalline Polymers (LCPs) from Cardanol

Thermotropic LC transition is brought out by temperature, and the high T_m of thermotropic LCPs was one of the major problems hampered the development of LCPs in industrial application. Processable LCPs were highly desired due to their high-performance applications. This was hindered ironically due to the structural requirement of LC formation, which requires rigid structures and consequently were prepared from aromatic monomers. Many approaches have been adopted to lower the T_m of LCPs to processable range, and the approaches include the introduction of rigid kink structures into the main chain of LCPs or flexible spacers into the structure without destroying the mesogenic order. Incidently, the structure

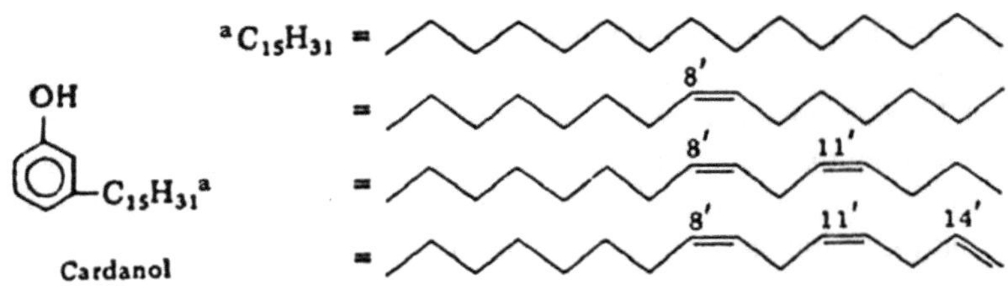

Fig. 5.1 Structure of cardanol

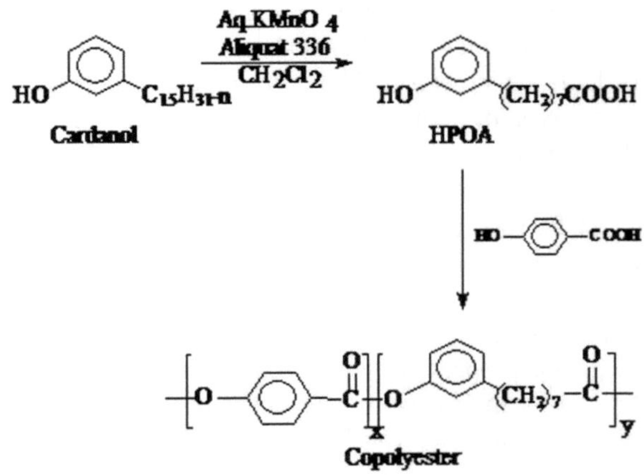

Scheme 5.1 Synthesis of LC copolyester from cardanol

of cardanol (Fig. 5.1) with its rigid phenolic group and flexible side chain at the meta position had the structural features suitable for lowering the T_m of LCPs when introduced as comonomers. In this context, Pillai et al. [9] prepared a new monomer, 8-(3-hydroxyphenyl) octanoic acid (3-HPOA), from cardanol and synthesized the homopolyester of 3-HPOA and its thermotropic LC copolyester with *p*-HBA (Scheme5.1). The DSC scan showed endothermic peaks at 256 and 342 °C, respectively. When viewed under crossed polarized light, the copolyester was anisotropic between 230 and 409 °C. However, decomposition of the polymer was observed above 395 °C. On cooling from 400 °C, no exothermic transition to a mesophase was observed and assumed as due to the partial decomposition. However, when the polyester was heated to 350 °C and cooled, a single transition was observed at 275 °C, and Fig. 5.2 shows the optical micrograph of the mesophase observed at 390 °C.

In order to tackle this problem, a strategy of using comonomers consisting of hydroxyphenylalkanoic acids such as HPAA, HPPA, and HPOA (Fig. 5.3) which are characterized by the presence of a short flexible aliphatic group directly attached to the aromatic moiety (this structure is expected to reduce the T_m without affecting the thermal stability deleteriously) and an aliphatic carbonyl (which are expected to be more sensitive to hydrolytic degradation to induce

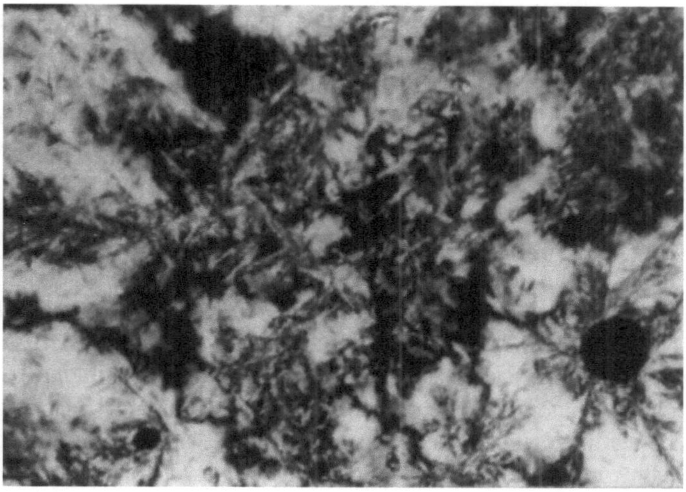

Fig. 5.2 Optical micrograph of the copolyester of m-HPOA and *p*-HBA captured under crossed polarized light at 290 °C (Pillai et al. Thermotropic LC copolyester based on 8-(3-hydroxyphenyl) octanoic acid and *p*-hydroxybenzoic acid, Polymer Communications 1992) (Reprinted with permission from ref. [9] 2003, Elsevier)

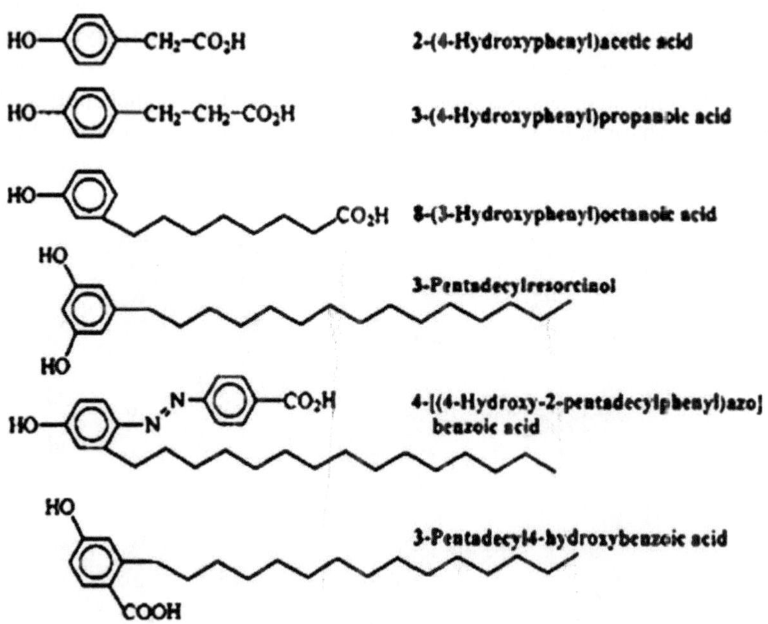

Fig. 5.3 Structure of the monomers used for LCP preparation (HPOA is hydroxyl phenyl octanoic acid prepared from cashew nut shell liquid)

biodegradability) was adopted [11]. As mentioned earlier, HPOA was prepared from cardanol by permanganate oxidation [9]. The terpolyesters synthesized by acidolysis polycondensation of the acetates of the monomers exhibited liquid-like flow with nematic Schlieren texture under polarized light microscopy (PLM). However, its thermal properties were not encouraging.

In an attempt to improve the thermal stability of the polymer, another polymer [poly(2-bromo-4- hydroxybenzoate-co-3-hydroxyphenyl octanoate)] containing Br group was synthesized. This polymer also showed clear nematic LC behavior. Composition variation study carried out on these polymers indicated that the bromo derivative has a good range of nematic phase for processability studies. Melt polycondensation techniques were employed to prepare the semiflexible random thermotropic copolymers containing HPOA with either 3-chloro-4-hydroxy benzoic acid (CHBA) or 3,5-dibromo-4-hydroxy benzoic acid (DBHBA) [12].

The copolyesters were characterized by Fourier transform infrared (FTIR) spectroscopy, dilute solution viscometry, hot stage PLM, differential scanning calorimetry (DSC), thermogravimetric analysis (TGA), and wide-angle X-ray diffraction (XRD). Studies revealed that the amount of HPOA as disruptor incorporated into the backbone of substituted 4-hydroxy benzoic acids has a detrimental effect on the LC behavior. Mesophase transition temperatures observed were between 210 and 250 °C, and the optical textures were of typical nematic phases (Fig. 5.4). The degree of crystallinity decreased with increase in HPOA content. The thermal stability of the copolymers was in the range of 310–370 °C. The WAXD studies reveal the existence of small semicrystalline region embedded in the matrix of amorphous structure.

In another attempt [13] aromatic–aliphatic thermotropic LC terpolyesters derived from 2,6-naphthalenedicarboxylic acid, hydroquinones, and HPOA, a monomer with built-in kink and flexible moieties, were prepared by a melt polycondensation reaction. These polymers were characterized for their thermotropic LC properties. The study of effect of variation in the composition of the polyester on the

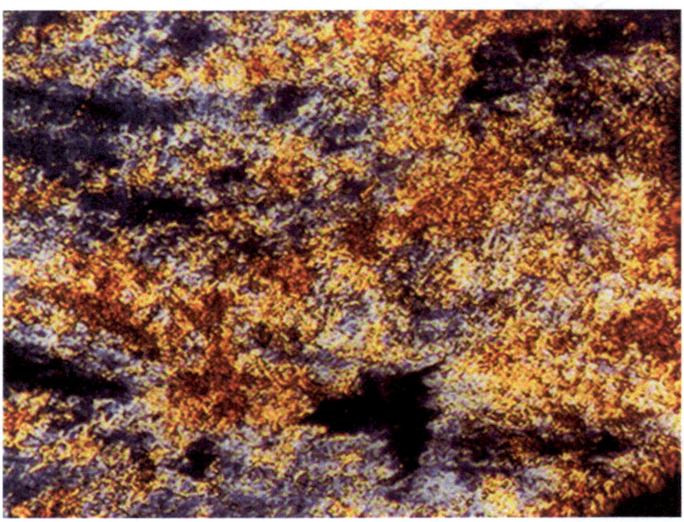

Fig. 5.4 Nematic threaded texture of polymer (Reprinted with permission from ref. [12] 2008, Wiley)

LC behavior and thermal properties suggested that increasing the content of kinked units increased the amount of structural irregularity and decreased the isotropization temperature and the T_m, and thermal stability of the polyesters were found to decrease with increase in the HPOA content. The formation of the LC phase was intact up to 40 mol% of HPOA in the copolyester. In a study of the effect of substituents on the thermal properties of the terpolyesters, it was found that introduction of a phenyl substituent on the hydroquinone ring lowered the transition temperature compared to substituents such as methyl or methoxy. The possible applications of these LCPs could be in the areas of eco-friendly plastics and biomedical implants.

5.4 LCPs Containing Azobenzene Groups Made from Modified Cardanol

A novel LC polymer containing azophenyl group was synthesized by performing a diazotization reaction between hydrogenated cardanol or 3-pentadecyl phenol and *p*-aminobenzoic acid and polymerization of the resulting monomer to form poly 4-[4-hydroxy-2-pentadecyl phenyl)azo]benzoic acid (Vb) [10], and compared with its unsubstituted counterpart poly 4-[4-hydroxyphenyl)azo]benzoic acid Va. The DSC curve of polymer Va exhibited endotherms at 262, 337, and 382 °C. At 265 °C, a threaded nematic texture was observed under PLM which remained up to the iisotropization temperature of 382 °C. The effect of the pendant flexible substituent of –$C_{15}H_{31}$ in Vb decreased its transition temperatures as compared to those of Va. When observed under PLM, the polymer melted and exhibited a Schlieren texture of the nematic phase at 135 °C (Fig. 5.5) and isotropized at 262 °C which did not appear in the DSC scan. Thus the presence of the pendant flexible hydrocarbon substituent in polymer Vb lowers its transition temperature of crystal to mesophase by 130 °C and its isotropization temperature by 120 °C. Earlier it was shown that the presence of a side chain enhances the solubility of polymer in common organic solvents, and similarly, the 2-pentadecyl-substituted *p*-phenyleneazo-group-containing LC polyester had better solubility than the unsubstituted counterpart. Azo-based monomers and polymers are well known to give NLO behavior, and for such applications the monomers/polymers are supposed to not absorb beyond 800 nm, and the polymer Vb and its monomer did not absorb beyond 500 and 715 nm, respectively, indicating possibly that these materials or related structures might show useful NLO properties.

5.5 LCPs from Cardanol with Cross-linked Networks

It was shown further that cross-linked LCPs, which are attractive due to the possibilities of freezing the LC phase, could be synthesized from cardanol, and the unsaturated C15 hydrocarbon side chain of cardanol can be utilized for

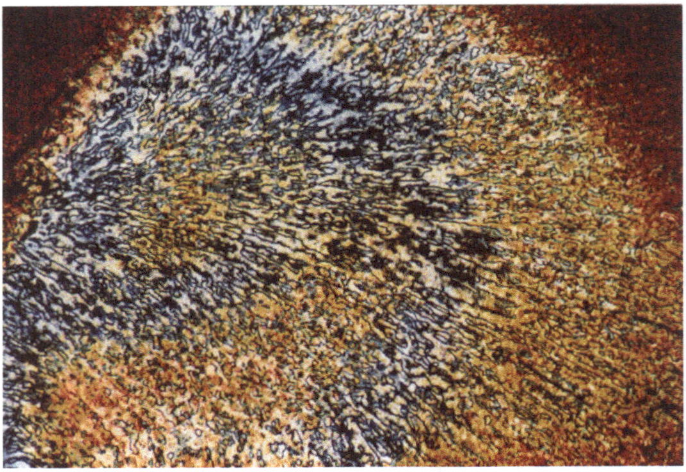

Fig. 5.5 LC nematic threaded texture of poly 4-[4-hydroxy-2-pentadecyl phenyl)azo]benzoic acid

cross-linking reactions. [14]. These type of cross-linked polymers with LC properties pass through a mesophase during curing and hence can retain the mesophase in the final state. An earlier work of the synthesis of poly(cardanyl acrylate) from cardanol yielded cross-linked polymer beads [15], directly, due to the auto cross-linking of the unsaturated side chain of cardanyl unit [16]. Therefore, this property of the cardanyl side chain was used to form cross-linked network structures of LC polymers. Here in this work, a series of LC polymers based on azobenzene mesogen (Scheme 5.2, 5.3, and 5.4) were synthesized from cardanol. In one of the polymer (IV), the -C15 unsaturated hydrocarbon side chain of cardanol was utilized for cross-linking reactions during curing. Another polymer (V) was synthesized by using the unsaturation in the side chain of cardanol which was able to cast into a film from solution and later thermally converted to an insoluble transparent film that retains the LC phase. The azobenzene group was introduced by the diazo coupling reaction between cardanol and 4-aminobenzoic acid (Scheme 5.2). The resulting monomer, 4-[(4-cardanyl)azo]benzoic acid (**I**) was polymerized by self-polycondensation using thionyl chloride and pyridine to get poly[4-[(4-cardanyl)azo]benzoic acid] (**III**) (Scheme 5.2). (**I**) was also converted to poly[4-[(4-acryloyloxycardanyl)azo]benzoic acid] (**IV**) through acryloylation followed by free radical polymerization (scheme 5.3). Cationic polymerization of (**I**) gave poly {4-[(4-cardanyl)azo]benzoic acid} (**V**) (Scheme 5.4). The monomers and polymers were characterized using elemental analysis, IR, ^{1}H and ^{13}C NMR, and UV–visible spectroscopy. Polymer (**III**) was neither soluble in any solvent nor fusible. Polymer (**IV**) was insoluble, but swelling was observed with many solvents, and polymer (**V**) was soluble in polar solvents indicating that it is not cross-linked at this stage. On slow heating, polymers **IV** and **V** also got converted to totally insoluble products. The IR spectra indicated that polymer **III** is a fully cross-linked

Scheme 5.2 Synthetic route of the monomer (I) and its homopolymer (III)

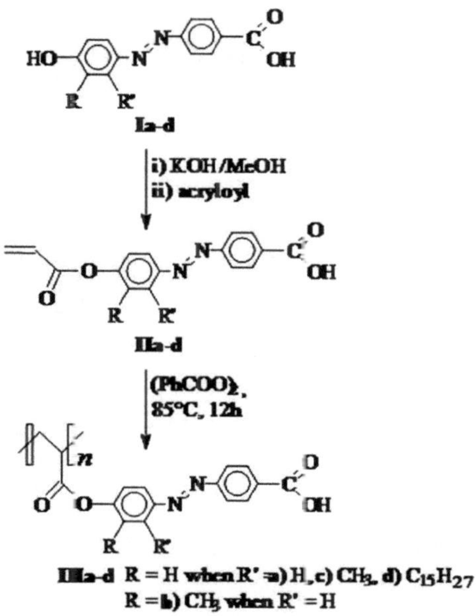

Scheme 5.3 Schematic of the synthesis of polymer IV

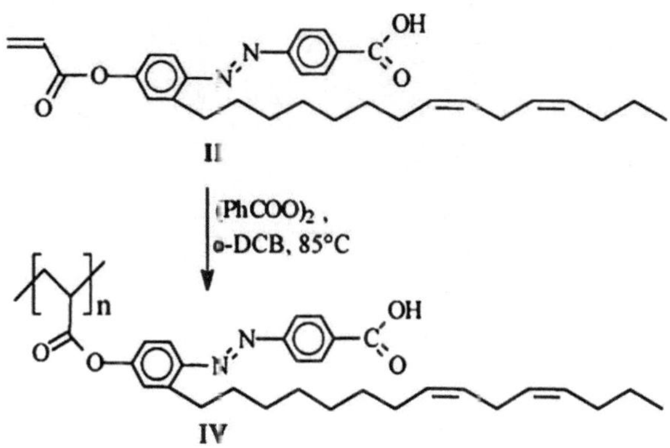

polymer and the involvement of side-chain unsaturation in bond formation leading to cross-linking. Threaded nematic textures were obtained when polymer (**III**) was heated to below 200 °C and rapidly quenched to room temperatures and observed under PLM. On slow heating, polymers **IV** and **V** got converted to fully cross-linked products. While polymer **III** gave a nematic phase with threaded texture, polymers **IV** and **V** exhibited nematic phases with schlieren textures (Fig. 5.6). The significance of polymer **V** lies in its capability to be cast into a film from solution and later thermally converted to an insoluble transparent film that retains the LC phase. The important property of these polymers is that the cross-linking in this case is developed from an unsaturated hydrocarbon side chain and was the first of the kind of network developed through a side-chain unsaturation.

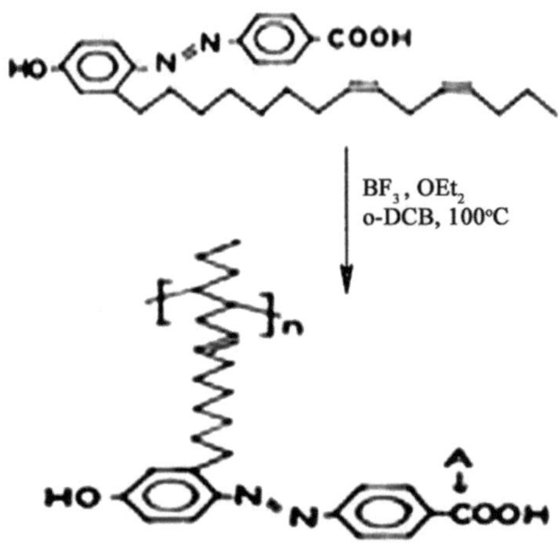

Scheme 5.4 Synthesis of polymer V

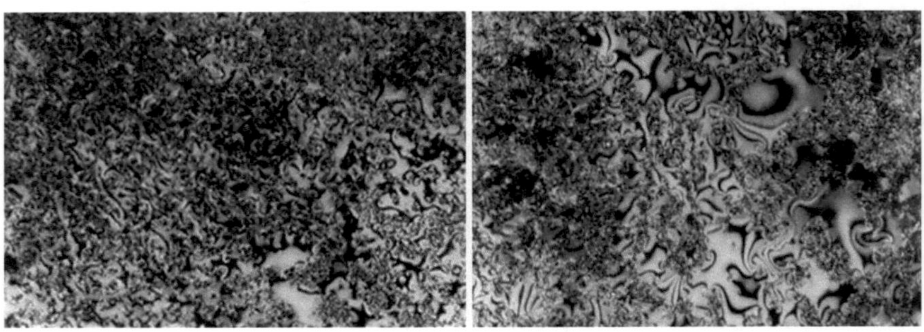

Fig. 5.6 Nematic Schlieren texture of polymer IV and V

Further work involved synthesis and characterization of main-chain and side-chain LCPs containing substituted (methyl and long alky side chain) azobenzene mesogens [17], and the representative structure of the polymers is given in Fig. 5.7. A series of side-chain LC polyacrylates containing substituted (methyl/long alkyl) azobenzene mesogens attached directly to the polymer backbone using monomers, 4-[(4-acryloyloxyphenyl)azo]benzoic acid, 4-[(4-acryloyloxy-3-methylphenyl)azo]benzoic acid, 4-[(4-acryloyloxy-2-methylphenyl)azo]benzoic acid, and 4-[(4-acryloyloxy-2-pentadecylphenyl)azo]benzoic acid, in which the latter one is made from cardanol, in order to investigate the combined effects of the laterally substituted alkyl and the hydrogen-bonded carboxyl groups on the phase behavior of azobenzene mesogen containing side-chain polymers. The pentadecyl substituent did not produce the expected level of decrease in the transition temperature, and this effect was explained as due to alignment of the side chain along the major axis of the mesophase thus giving rise to stabilization of the mesophase.

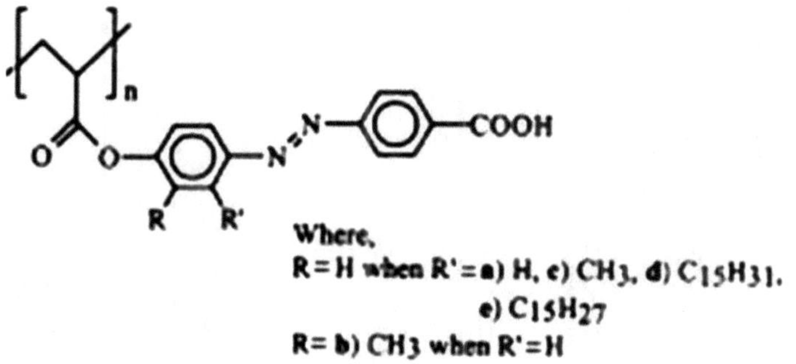

Fig. 5.7 Representative structure of the azobenzene side-chain polyacrylates. All the polymers exhibited LC phase. The azobenezene side-chain polymers with R' = $C_{15}H_{27}$, $C_{15}H_{31}$ are made from cardanol

While the polyacrylates containing methyl substituted azobenzene mesogen exhibited stir opalescence and low viscosity characteristic of nematic phase and exhibited only nematic phase, a polymorphic phase behavior with additional smectic phase was observed in the case of the side-chain polymer with the long alkyl side chain ($C_{15}H_{31}$), made from cardanol. The carboxylic acid groups of the azophenyl mesogen are found to exist as hydrogen-bonded dimers as indicated by IR and NMR and finally confirmed by WAXD. It was proposed in the study that the long alkyl substituent might be involved in the stabilization of the mesophase through its alignment parallel to the long axis of the molecule and might probably be supporting the formation of smectic phase by reducing the transition temperature low enough to maintain the layered arrangement of hydrogen-bonded dimers. The schematic of the formation of the smectic phase aided by hydrogen-bonded carboxylic acid dimers is given in Scheme 5.5.

In the case of main-chain polymers, a short alkyl substitution such as a methyl group brings down the transition temperature in an expected way, but the long alkyl side chain, $C_{15}H_{31}$, instead of destroying the mesophase stabilizes it and was thought to be possibly aligning parallel to the molecular axis of the polymer. In the case of the side-chain polymer, the long alkyl chain induces formation of a smectic phase in addition to the nematic phase. The presence of the unsaturated side chain induces formation of cross-linked network structures without destroying the mesophase.

5.6 Attempts to Prepare LCPs with NLO Properties from Cardanol

It was earlier mentioned that the Poly[4-[hydroxy-2-pentadecyl)azo]benzoic acid] derived from 3-*n*-pentadecyl phenol, the hydrogenated derivative of cardanol, is LC and is potentially a NLO material [10]. Azobenzene derivatives and polymers

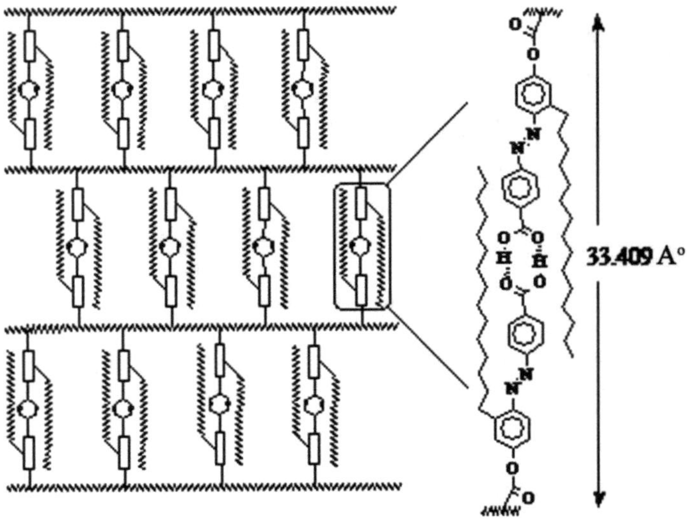

Scheme 5.5 Schematic proposal for the probable smectic layer arrangement in side-chain LCP, poly[4-[(4-acryloyloxy-2-pentadecylphenyl)azo]benzoic acid, (the rigid rod units are carboxylic acid dimers)

containing donor–acceptor groups in conjugation are known for their NLO activity [18, 19]. While the rigidity and the planarity of azobenzene group provide the shape required for LC property, it has structural features that can be modified to incorporate property profiles for photoresponsive behavior. LCPs containing azobenzene mesogen are interesting because they combine the properties of anisotropy and photoresponsive behavior in a single polymer [20]. Thus, azobenzene-based LCPs have a number of applications in areas such as LC displays, NLO materials, and information storage devices [21, 22]. A number of side-chain and main-chain polymers having second-order NLO properties have been synthesized starting from cardanol and have been characterized using spectral and thermal analysis. A novel LCP containing azophenyl group was synthesized by performing a diazotization reaction between cardanol or pentadecyl phenol and *p*-aminobenzoic acid and polymerization of the resulting monomer to get poly 4-[4-hydroxy-2-pentadecyl phenyl)azo]benzoic acid [10]. Though the polymer exhibited clear nematic threaded texture under PLM, it did not have the thermal stability. Relaxation of the alignment of dipoles is an issue with NLO polymers, to retain the NLO property in a polymer, and it is important to prevent the dipole reversal or relaxation from their dipolar alignment.

One of the methods recommended to arrest the relaxation is the introduction of cross-linking during poling which "lock in" the dipole alignment due to the restricted molecular mobility. Therefore, instead of using hydrogenated monomer, Saminathan et al. later synthesized LC polymers with unsaturated side chains, and the unsaturated C_{15} hydrocarbon side chain of cardanol was utilized for cross-linking reactions to produce cross-linked network structures [14]. Azo-based LC polymers are well known to give NLO behavior, and for such applications the polymers should not absorb in the UV range of >800 nm. The LC polymer prepared did not absorb in this region indicating possibly that this polymer or related structures might show useful NLO properties.

Polyesters having an NLO chromophore with a dipole moment aligned transversely to the backbone with LC properties, and when poled at the temperature region of LC phase, transition showed larger NLO activity than that of the similar polymer with no LC phase. The LC polymer possessed better temporal stability at a longer time region of storage and found to possess excellent stability during the whole process of NLO studies. It was thought that the orientational stability at the initial time after poling can be improved by introducing flexible groups in the main chain. NLO rigid main chain polyesters containing azobenzene mesogens with high thermal and temporal stabilities were synthesized from derivatives of hydroxyphenylazobenzoic acid [23]. The NLO properties of the homopolymer, poly[4-(4-hydroxy-3-methyl phenyl)azo]benzoic acid, and copolymers of 4-[(4-hydroxy-3-methylphenyl)azo]benzoic acid, 4-[(4-hydroxy-2-methylphenyl)azo]benzoic acid, and 4-[(4-hydroxy-2-pentadecyl phenyl)azo]benzoic acid (PSCpHBA) (Scheme 5.6) with *p*-HBA were measured by the Maker fringe technique (Fig. 5.8). The thermal and LC phase behaviors of the polymers were examined by DSC and PLM. The polymers except that of PSCpHBA exhibited nematic threaded and Schlieren textures. The LC orientations gave rise to an enhanced NLO response. This suggests that the LC phase contribute to the alignment of chromophores [20, 24]. This study suggests that the rigid aromatic main chain exhibiting an LC phase is a promising simple method to synthesize highly stable NLO polymers. The polymers had high thermal and temporal stabilities for second-harmonic generation activity because of their rigid aromatic backbone. The SHG activity of the polymer PSCpHBA poled at 280 °C did not undergo any loss even when stored at temperatures of 100 and 200 °C for 4.5 days (Fig. 5.8). The nonlinear coefficient (d_{33}) value is higher when poled at 280 °C than when

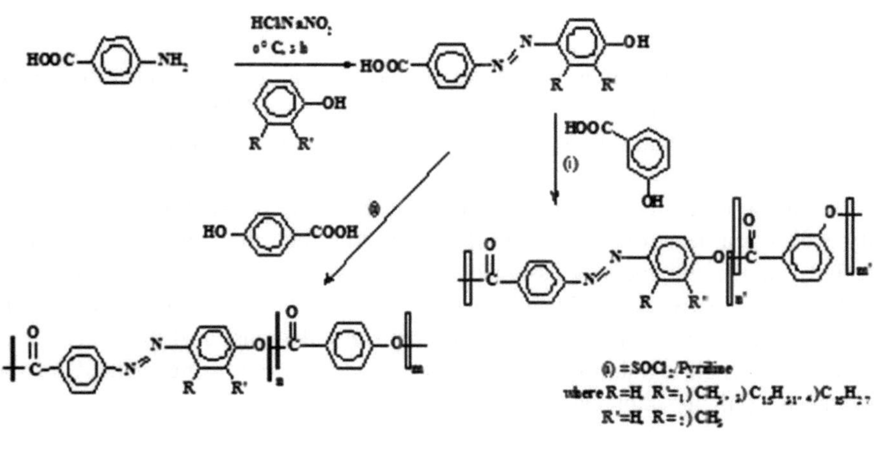

Scheme 5.6 Schematic of the synthesis of NLO polymers the azo derivative from cardanol

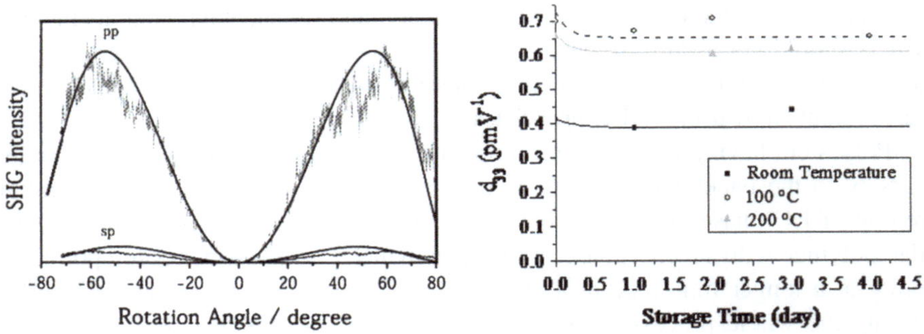

Fig. 5.8 Representative Maker fringe pattern obtained for the polymers and dependence of d_{33} value on storage time at various temperatures for polymer PSCpHBA. (poling condition: 280°C, 60 min, 7 kV for the sample stored at 100 and 200°C; 240°C, 60 min 7 kV for the sample stored at room temperature.)

Fig. 5.9 Dependence of d_{33} value on the annealing temperature of the polymers

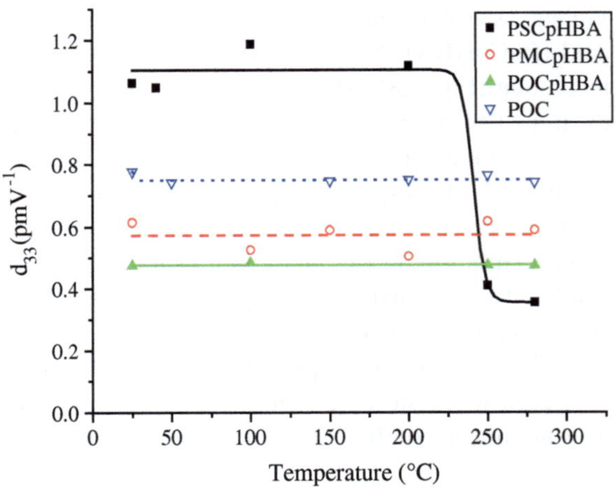

poled at 240 °C. This study demonstrates that on contrary to the previous report [10] polymer made from cardanol possessed stability for NLO measurements and excellent orientational stability which is attributed to the rigid and aromatic nature of the polymer chain.

It was observed that the polymers did not show any decay in their SHG activity even when heated up to 220 °C (Fig. 5.9). The high thermal stability of SHG activity of the polymers is assigned to their aromatic rigid structure. Polymer PSCpHBA underwent loss in SHG activity at a comparatively lower temperature possibly due to thermal decomposition. This study demonstrated that a rigid aromatic main chain exhibiting an LC phase could be a promising method for the synthesis of highly stable NLO polymers.

5.7 Conclusions

The chapter gives a summary of various monomers and polymers synthesized from cardanol and discusses their various properties giving focus to their LC properties. The study has established that main- and side-chain polyesters and azobenzene-based polyesters containing LC mesogens could be prepared from cardanol by appropriate chemical modifications. The cardanol side chain was found to stabilize the mesophase, instead of destroying it, by aligning themselves and had a tendency toward polymorphic phase behavior by exhibiting a smectic phase in addition to nematic phase. The unsaturation in the side chain of cardanol led to cross-linked cardanol-based polymer films with LC phases. The homopolymers prepared from modified cardanol did not show good thermal stability, and the main-chain NLO co-polymers containing azobenzene mesogens showed thermal stability above 250 °C, though no LC property. The LC phase poling was found to enhance the NLO response of polymer due to the better alignment of dipoles. The SHG activity of the polymers poled at LC transition temperatures possessed excellent orientational stability, indicating the assistance of LC phase in better alignment of dipoles. Also, the polymers exhibited high thermal stability to SHG activity. This study demonstrates that a variety of structures of polymers such as rigid aromatic main-chain polymers and side-chain azobenzene polymers with aliphatic main chain exhibiting LC phase(s) can be prepared from cardanol and aromatic main-chain copolymers is a promising method for the synthesis of highly stable NLO polymers.

Acknowledgements Thanks are due to the members of the polymer group for support. Thanks are also due to Prof. Tsutsumi for discussions and the SHG measurements.

References

1. Cunliffe D, Pennadam S, Alexander C (2004) Synthetic and biological polymers: merging the interface. Eur Polym J 40:5–25
2. Kobayashi S, Uyama H (2003) Biomacromolecules and bio-related macromolecules. Macromol Chem Phys 204(2):235–256
3. Pillai CKS (2014) Recent advances in biodegradable polymeric materials. Mater Sci Technol 30(5):558–566
4. Luckachan GL, Pillai CKS (2011) Biodegradable polymers—a review on recent trends and emerging perspectives. J Polym Environ 19:637–676
5. Sandhya KY, Saritha A, Joseph K (2015) Liquid crystalline polymers from renewable resources: synthesis and properties. In: Liquid crystalline polymers, Springer Int, pp 273–306
6. Pillai CKS (1996) High performance polymers from natural monomers and polymers. In: Cheremisinoff NP (ed) Handbook of advanced polymeric materials. Marcel Dekker, New York
7. Pillai CKS (2010) Challenges for natural monomers and polymers: novel design strategies and engineering to develop advanced polymers. Des Monomers Polym 13:87–121

8. Bhuniaa HP, Basakb A, Chakia TK, Nandoa GB (2000) Synthesis and characterization of polymers from cashewnut shell liquid: a renewable resource: V. Synthesis of copolyester. Eur Polymer J 36:1157–1165
9. Pillai CKS, Sherrington DC, Sneddon A (1992) A thermotropic liquid crystalline copolyester from 8-(3-hydroxy phenyl) octanoic acid and *p*-hydroxybenzoic acid. Polymer 33:3968
10. Saminathan M, Krishna C, Pillai S, Pavithran C (1993) Macromolecules 26:7103–7105
11. Pillai CKS (1998) Pure Appl Chem 70:1249
12. Mitra S, Pillai CKS (2008) Semiflexible random thermotropic copolymers from 8-(3-hydroxy phenyl) octanoic acid and 3-chloro-4-hydroxy benzoic acid/3,5-dibromo-4-hydroxy benzoic acid. J Appl Polym Sci 107(2):778–783
13. Vijayanathan V, Prasad VS, Pillai CKS (1997) J Appl Polym Sci 82(4):1021–1029
14. Saminathan M, Pillai CKS (2000) Synthesis of novel liquid crystalline polymers with cross-linked network structures. Polymer 41(8):3103–3108
15. John G, Pillai CKS (1992) Macromol Chem Rapid Commun 13:255
16. John G, Pillai CKS (1993) J Polym Sci, Part A: Polym Chem 31:1069
17. Saminathan M, Pillai CKS (2000) Macromol Chem Phys 201:2475–2483
18. Yesodha SK, Pillai CKS, Tsutsumi N (2004) Stable polymeric materials for nonlinear optics: a review based on azobenzene systems. Prog Polym Sci 29(1):45–74
19. Sandhya KY, Pillai CKS, Sato M, Tsutsumi N (2003) Non-linear optically active liquid crystalline polymers containing azobenzene mesogen. Fibre Preprints Japan 58:71–73
20. Sandhya KY, Pillai CKS, Sato M, Tsutsumi N (2000) Nonlinear optical properties and liquid-crystalline behavior of new polyesters with dipole moments aligned transverse to the backbone. Macromol Chem Phys 203(8):1126–1134
21. Sutherland RL (1996) Handbook of nonlinear optics. Marcel Dekker, New York
22. Pillai CKS, Sandhya KY, Sudha JD, Saminathan M (2003) Influence of hydrogen bonding on the generation and stabilization of liquid crystalline polyesters, poly(esteramide)s and polyacrylates. Pramana 61:417–423
23. Sandhya KY, Pillai CKS, Sato M, Tsutsumi N (2002) Highly stable rigid main-chain nonlinear optical polymers with nematic phase: effect of liquid-crystalline phase on nonlinear optical response. J Polym Sci, Part A: Polym Chem 41(10):1527–1535
24. Heldmann C, Warner M (1998) Macromolecules 31:3519–3531

Chapter 6
Cardanol-Derived-Amphiphiles-Based Soft Templates for Conducting Polymer Nanoarchitectures

Menachery Jinish Antony and Parambath Anilkumar

Abstract In recent times, there is a growing trend among researchers in utilizing plant based starting materials to synthesize functional nanomaterials. Toward this goal, cashew nut shell oil-derived cardanol caught great attention due to its wide availability, low cost, easy isolation and unique molecular architecture. In this chapter, we summarize the literature based on soft template approach of cardanyl amphiphiles for fabricating conducting polymer (polyaniline and polypyrrole) nanostructures. The amphiphiles synthesized from cardanol has typical surfactant structure with sulfonic acid polar head and hydrophobic aliphatic tail, often referred as dopants due to its post-polymerization doping effect on polymer chain. The soft- templates were generated by selectively mixing monomer and dopants for different polymerization conditions such as emulsion, dilute, interfacial, dispersion and gel phase. The templates upon treatment with polymerization initiator produce conducting polymer morphology such as nanofibers, nanorods, nanotubes, nanospheres, hollow nanospheres and nanotapes. The cardanyl amphiphilic dopants electrostatically complex with conducting polymer nanomaterials and significantly improve the solubility, solid-state ordering, conductivity and optical properties.

M.J. Antony (✉)
Department of Chemistry, St. Thomas College (Autonomous Under University of Calicut), Thrissur, Kerala 680001, India
e-mail: jinish06@yahoo.co.in

P. Anilkumar (✉)
Department of Pathology and Laboratory Medicine, Life Sciences Centre, University of British Columbia, 2350 Health Sciences Mall, Vancouver, BC V6T 1Z3, Canada
e-mail: p.anilkumar@live.in

P. Anilkumar (ed.), *Cashew Nut Shell Liquid*,
DOI 10.1007/978-3-319-47455-7_6

6.1 Conducting Polymers and Its Nanostructures

Conducting polymers are class of π-conjugated materials that can conduct electricity via the π-electron delocalization through alternative single and double bonds present in the polymer backbone [1–5]. Conducting nature was intrinsic to the π-conjugated electrons present in the conducting polymer backbone. Some important members of conducting polymer family are polyacetylene (PA), polypyrrole (PPy), polythiophene (PT), poly (p-phenylene) (PPP), poly (p-phenylenevinylene) (PPV) and polyaniline (PANI), etc. [5–9]. Among various types of conducting polymers, polyaniline and polypyrrole receive a special attention owing to its easy synthesis, low cost of production, acid-base and redox control on structures and conductivity, etc. [6–9]. Polyaniline and polypyrrole are generally synthesized by chemical or electrochemical oxidative polymerization of monomers in acidic aqueous solution [10–15]. Usually acid-base treatment (often referred as doping–dedoping process) brings about changes in charge and electronic states of the polyaniline and polypyrrole, while retaining the same the structural backbone. The three commonly known forms of polyaniline are (i) fully reduced colorless leucoemeraldine, (ii) fully oxidized brown-colored pernigraniline and (iii) half-oxidized blue-colored emeraldine base (see Fig. 6.1). All these three forms are found to be insulators of electricity in normal conditions. Green-colored conducting emeraldine salt form of polyaniline is obtained by *in situ* or post-doping using mineral/organic acid [16–20]. Similar to polyaniline, neutral polypyrrole upon doping produces charged polypyrrole (with quinoid like structures) which remarkably changes the conductivity from insulating states to metallic with the formation of polaron and bipolaron charge carriers [6, 7, 21–23]. The optical, electrical, thermal, electrochemical and solid-state properties of these interesting polymers were extensively studied and well documented

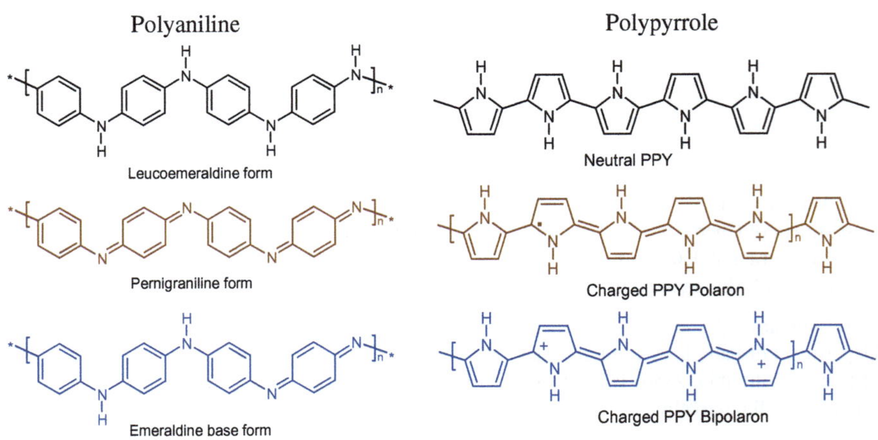

Fig. 6.1 Various forms of polyaniline and polypyrrole

by various groups around the world [23–27]. Polyaniline and polypyrrole found to have promising applications in different fields such as sensors, actuators, catalysis, memory devices, modified electrodes, DNA sensing, metallic corrosion protection, electromagnetic interference shielding, super capacitors and many other applications [27–40].

Ever since the emergence of the nanotechnology, there are tremendous efforts directed toward restricting the morphology of conducting polymers into nanometric scale for the expected supreme performance over bulk conducting polymer materials [41, 42]. The miniaturization of conducting polymeric materials yields profound changes in physical properties such as optical, electrical, mechanical and dimensional properties (such as surface area, volume). [41–52]. Different physical or chemical methods such as electrospinning, templating, emulsion and interfacial polymerization methods were utilized for the development of conducting polymer nanostructures [53–56]. Templating method utilizing hard or soft templates was extensively used for fine tuning the size, shape and other physiochemical properties of the resulting conducting polymer nanomaterials [56–60]. The use of hard templates such as porous silica or alumina often encounters difficulty in post-polymerization removal of template and purification [41, 60]. On the other hand, the use of soft templates such as micelles, gels and liquid crystals is easily applied for nanomaterial synthesis without any tedious post-polymerization steps [58–66]. This chapter summarizes the utilization of soft templates derived from cashew nut shell liquid (CNSL) as amphiphilic dopants for constraining the morphology of polyaniline and polypyrrole in nano/micro-dimensions.

6.2 Design and Synthesis of Cardanyl Amphiphiles

The synthesis of nanomaterials utilizing renewable resource feed stocks is highly attractive in both academic and industrial point of view because of its ability to replenish non-renewable petroleum-based starting materials [67, 68]. CNSL is a natural organic plant-based renewable resource and a by-product from cashew industry, often considered as an industrial waste and pollutant. Technical CNSL contains a variety of phenolic derivatives with anacardic acid as the major chemical constituent [67, 68]. Preheating and subsequent vacuum distillation (230 °C at 5 mm of Hg) of CNSL result in the decarboxylation of anacardic acid to yield cardanol. Cardanol is a unique amphiphilic phenolic molecule with *meta*-pentadecenyl chain with one or two or three unsaturated bonds [67, 68]. Recently, there is a renewed interest in utilizing the unique chemical features of cardanol to synthesize cardanyl-based amphiphiles for developing various self-assembled nanostructures. For example, G. John et al. synthesized glycolipids by tethering glucopyranose to cardanol, and the sugar moiety acts as a nonionic polar head group and bestows chirality to the molecule [69, 70]. Soft materials, such as twisted fibers, helical coils and high axial ratio nanotubes, were formed through the self-assembly of amphiphilic glycolipids. Pillai and co-workers synthesized amphiphilic sulfonic and phosphonic acids as dopants from cardanol derivatives and utilized for bulk polyaniline synthesis [71–73]. They observed that cardanol upon treatment with conc. H_2SO_4 produces a resin rather than ring sulfonation due to the presence of unsaturated bond in pentadecyl side chain. Hence, in the first step hydrogenation of cardanol was carried out to saturate the side chain and the saturated cardanol (3-pentadecylphenol) was sulfonated to give the sulfonic acids. These saturated aromatic cardanyl sulfonic acids were found to be good dopants for bulk polyaniline and their blends. But these saturated cardanol-based dopants have limited solubility in water at room temperatures, and it does not act as suitable soft templates for conducting polymer nanomaterials. To address this issue Jayakannan and co-workers came up with new chemical approaches for the direct utilization of cardanol to synthesize water-soluble dopants which can act as soft templates for the growth of polyaniline nanomaterials and related polymeric systems [74–89].

The design of cardanyl amphiphiles is very simple; the in-built long *meta*-pentadecenyl side chain attached to the cardanol acts as the flexible hydrophobic tail component, and sulfonic acid group introduced either through diazotization reaction or by direct ring opening reaction serves as a polar head. The molecular design is optimum for imparting dopant functionality and amphiphilic surfactant properties. The chemistry employed for the synthesis of the amphiphilic molecules is direct single-step reaction, which gives fairly good yield. In the first synthetic approach, cardanol under basic conditions reacted with diazotized sulphanilic acid to yield 4-[4-hydroxy-2(Z)-pentadec-8-enyl)phenylazo]-benzenesulfonic acid (dopant-1) (see Fig. 6.2). Dopant-1 is an amphiphilic anionic surfactant molecule having a flexible tail and polar azobenzenesulfonic acid head group [76, 77]. Orange red-colored dopant-1 shows molecular dye properties such molecular light absorption and emission properties which were valuable tools for tracing the molecular self-assembly. In the second approach of cardanol derivatization, direct coupling reaction of butane sultone with cardanol and saturated pentadecyl phenol using potassium tert-butoxide as base produced two amphiphilic molecules (Z)-4-(3-(pentadec-8-enyl)phenoxy) butane-1-sulfonic acid (dopant-2) and 4-(3-pentadecylphenoxy)butane-1-sulfonic acid (dopant-3), respectively [78, 79] (see Fig. 6.2). Dopant-2 is having optimum

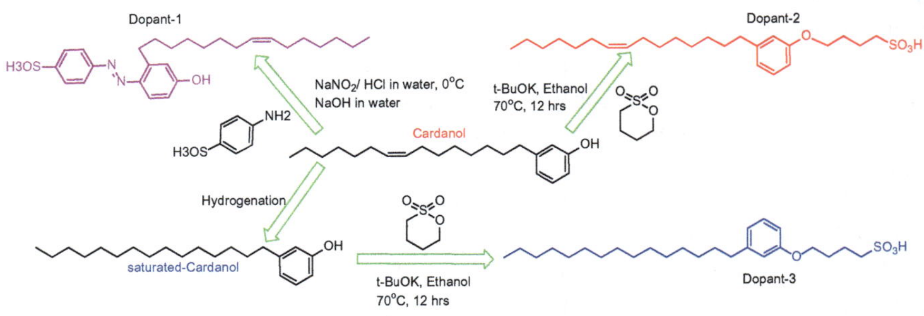

Fig. 6.2 Synthetic strategies for cardanol-based amphiphiles (dopant-1, dopant-2 and dopant-3). Reprinted with permission from ref [74] (2009, http://ir.niist.res.in:8080/xmlui/handle/123456789/878), [78] (2009, American Chemical Society), 81 (2008, American Chemical Society)

polar head to tail balance and capable of showing typical micellar characteristics and emulsification behavior with different aromatic monomers. The dopant-3 derived from 3-pentadecyl phenol is partially soluble in water, but it has the advantage of forming organogel in ethanol and other polar solvents.

6.3 Self-Assembly Properties of Amphiphilic Dopants

The two cardanol-based sulfonic acid amphiphiles dopant-1 and dopant-2 were soluble in water, whereas the saturated cardanol derivative dopant-3 was only sparingly soluble in water. The dopant-1 and dopant-2 have typical surfactant structure and forms micelles in aqueous solution. The critical micellar concentration values of the dopants determined by multiple techniques such as UV–visible absorbance spectroscopy, fluorescence studies, ionic conductance and surface tension techniques were shown good agreement, and the values were obtained in millimolar range [74, 77, 81].

The dopant-1 upon dissolution in water resulted in pale yellow-to-brown-colored clear solutions [77] (see Fig. 6.3). Dynamic light scattering (DLS) studies of dopant-1 showed that the cardanyl azobenzenesulfonic acid molecule exists in the form of the micelles (at 1×10^{-3} M) in water with an average diameter of

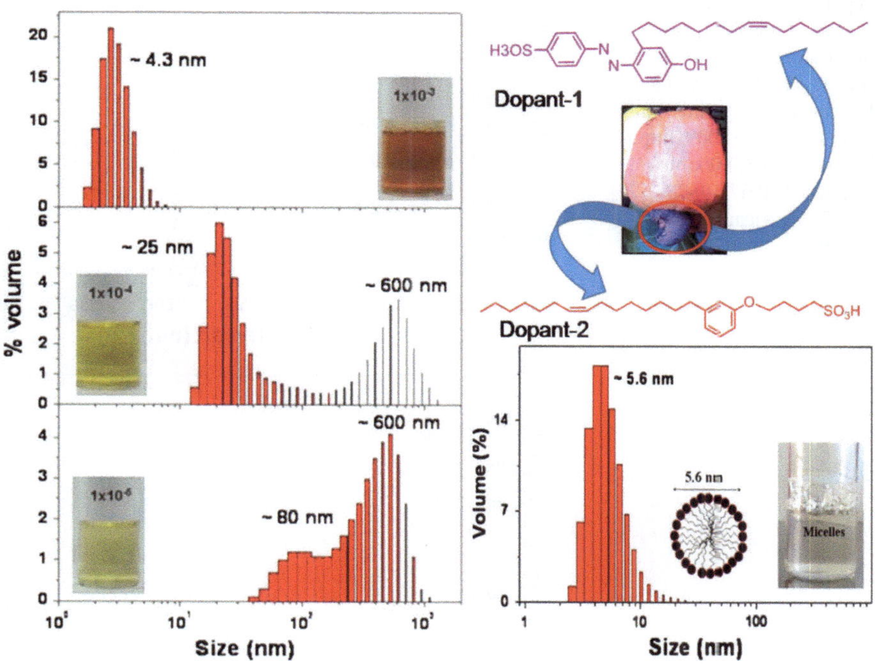

Fig. 6.3 DLS histograms of dopant-1 and dopant-2 in water. Reprinted with permission from Ref. [74] (2009, http://ir.niist.res.in:8080/xmlui/handle/123456789/878), [78] (2009, American Chemical Society), [81] (2008, American Chemical Society)

4.29 nm. DLS profile of 1×10^{-4} M dopant solution showed a bimodal distribution with maxima centered at 25 and 600 nm. A further decrease in dopant concentration (1×10^{-5} M) increases the aggregate size to 600 nm. DLS study indicated that at lower concentrations the dopant molecules were existing as open aggregates (like layers) in water and completely transformed to micelles as the concentration of dopant becomes 1×10^{-3} M [74, 81]. The dopant-2 upon dissolution in water forms a colorless clear solution at lower concentration and become clear soapy solution at higher concentration above CMC. DLS study of the dopant-2 in water for the concentration of 1.1×10^{-2} M shows that the amphiphilic cardanyl butanesulfonic acid molecule existed as 5.6-nm-size micelles in water [78] (see Fig. 6.3).

Dopant-1 possesses fluorescence property which is peculiar and rarely seen in amphiphilic dopants. Fluorescence property of the dopant-1 provides great deal of information regarding the self-assembly behavior in aqueous solution [77]. The fluorescence spectra of the dopant-1 at different concentrations in water were obtained by exciting molecule at 360 nm wavelength. The fluorescent intensity of the dopant-1 is highly dependent on its concentration in water, and the emission maximum is centered at 450 nm. The emission intensity of the dopant increases with the increase in the dopant concentration (1×10^{-6} to 1×10^{-5} M) and

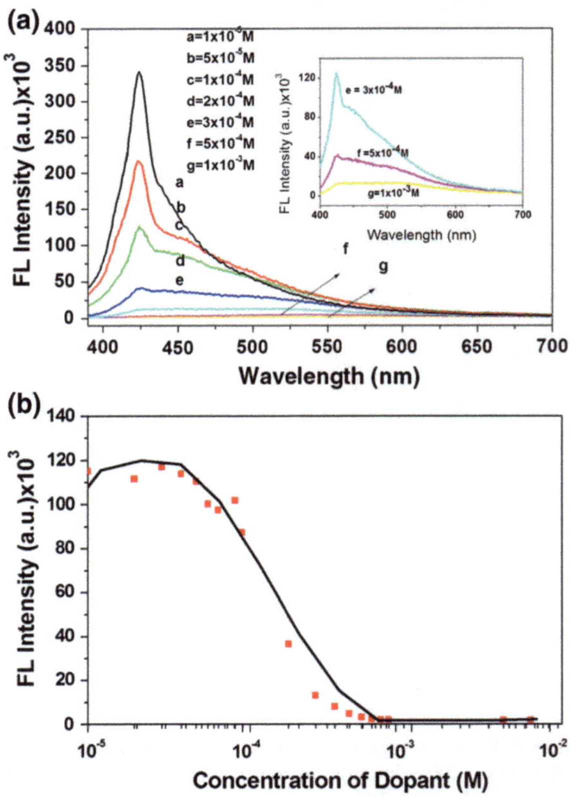

Fig. 6.4 Concentration depended fluorescence studies of dopant-1: (**a**) fluorescence emission spectra, (**b**) intensity change. Reprinted with permission from Ref. [74] (2009, http://ir.niist.res.in:8080/xmlui/handle/123456789/878), [77] (2007, American Chemical Society)

attained maxima at 1×10^{-5} to 7×10^{-5} M (see Fig. 6.4). Further increase in the dopant concentration (6×10^{-5} to 6×10^{-4} M) decreases the emission intensity, and the fluorescence nature is completely lost above 1×10^{-3} M. This nonlinear trend is due to the existence of dopant-1 aggregates in more than one form depending on its concentration in water. At low concentration, the dopant-1 molecules existed as isolated molecules, and with an increase in concentration, they form fluorescent layer structures. At higher concentration, the dopant-1 molecules forms well-defined spherical micelles and the fluorescence property is quenched [74, 77]. Concentration-dependent DLS and fluorescence studies of dopant-1 pointed out the involvement self-assembly which eventually leads to spherical micelles at higher concentration.

The dopant-3 molecule showed organogel formation in various organic solvents such as ethanol, methanol, chloroform, DMSO. [79]. Among different solvents, ethanol produced stable opaque white gel for wider concentration range. Differential scanning calorimetric (DSC) analysis of the gel in ethanol showed a broad melting peak at 65 °C with respect to gel-to-sol transitions, and upon cooling the sol-to-gel transition phase occurred at 53 °C [79]. SEM analysis of the xerogels showed the presence of long self-assembled fibrous tape-like morphology characteristic of the typical organogel texture, which is further supported by AFM analysis (see Fig. 6.5) [79]. TEM analysis revealed that the inner parts of the fibrils possess hollow space similar to nanotubes. The average inner and outer diameters of the nanotubes were 60 and 175 nm, respectively. The morphological studies of nanotubes conducted support the logical conclusion that nanotubes are formed by a multilayer packing of self-assembled dopant-3 molecules [79].

6.4 Soft-Template Formation with Monomers

Synthesis of conducting polymer nanostructures with various types (like nanofibers, nanorods, nanotubes and vesicles) has been achieved through polymerisation of soft template produced by aggregating monomers (aniline or pyrrole) into the self-assembly of dopants (dopant-1, dopant-2 and dopant-3). These dopant plus monomer soft templates produced in solution phase or gel phase upon treatment with suitable initiators induce polymerization of the monomers trapped inside the molecular templates and thereby translate the template morphology into the resultant polymer morphology. The polymerization method utilized were emulsion, interfacial, dispersion, gel-assisted routes and copolymerization employing ammonium persulfate or ferric chloride as the initiator [74–88].

The soft-template formation very much depends on factors such as concentration, types of monomer with which it combines, polymerization routes, pH. In the emulsion route, the dopant-1 plus aniline template mostly existed in more than one form of assembly such as layered membranes and cylindrical micelles depending on the concentration of the dopant-1 in solution. SEM and TEM studies of the dopant-1 plus aniline template confirm the formations of cylindrical micellar

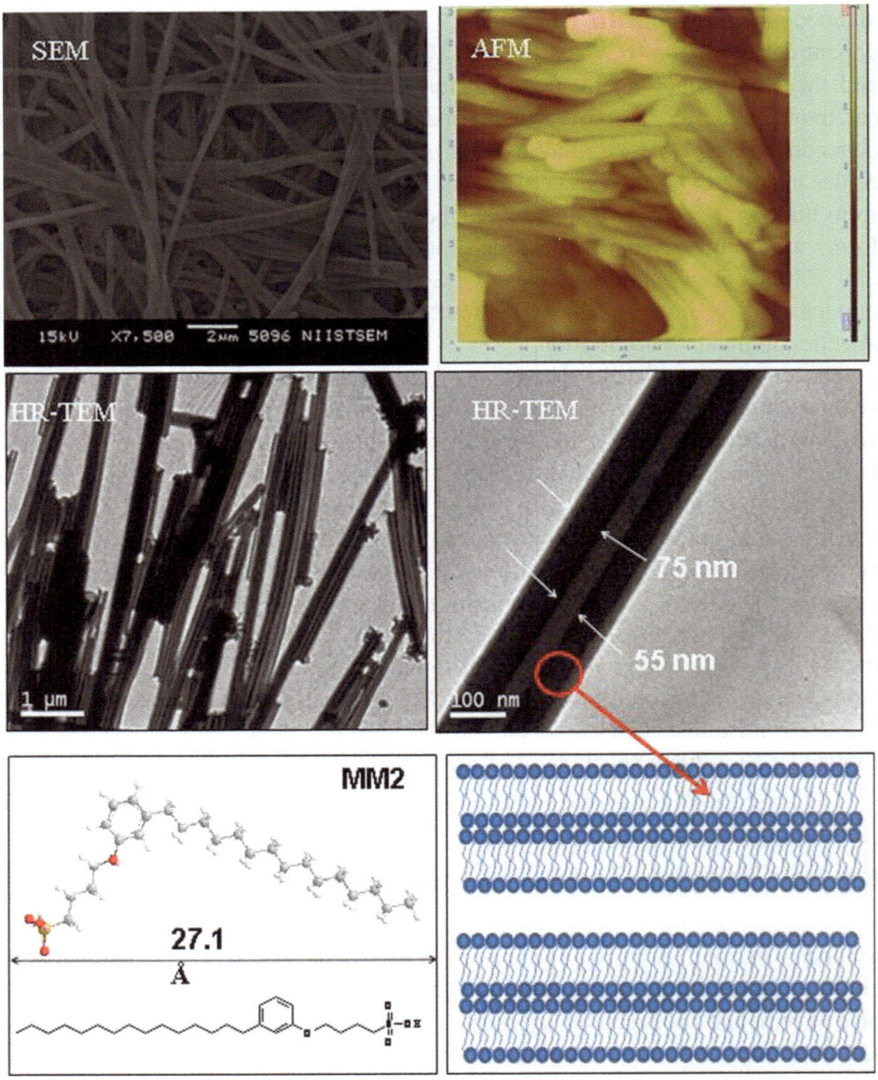

Fig. 6.5 SEM, AFM and TEM images of xerogel of dopant-1 in ethanol (*top* and *middle rows*) and theoretical structure and packing of dopant-1 in gel (*bottom row*). Reprinted with permission from Ref. [79] (2009, American Chemical Society)

templates exclusively at higher dopant concentrations [76, 77]. Interestingly, the emulsion of pyrrole showed a completely different aggregation pattern with dopant-1, and spherical aggregates of 650 nm were formed at higher concentration of dopant-1. The sizes of the dopant-1 + pyrrole micelle aggregates were reduced to 150–200 nm and even less at dilute concentration of dopant-1 [84, 85]. Soft templates of aniline and pyrrole together with dopant-1 exhibited short flaky rods at higher concentration and hollow spheres and nanotubes at lower concentration

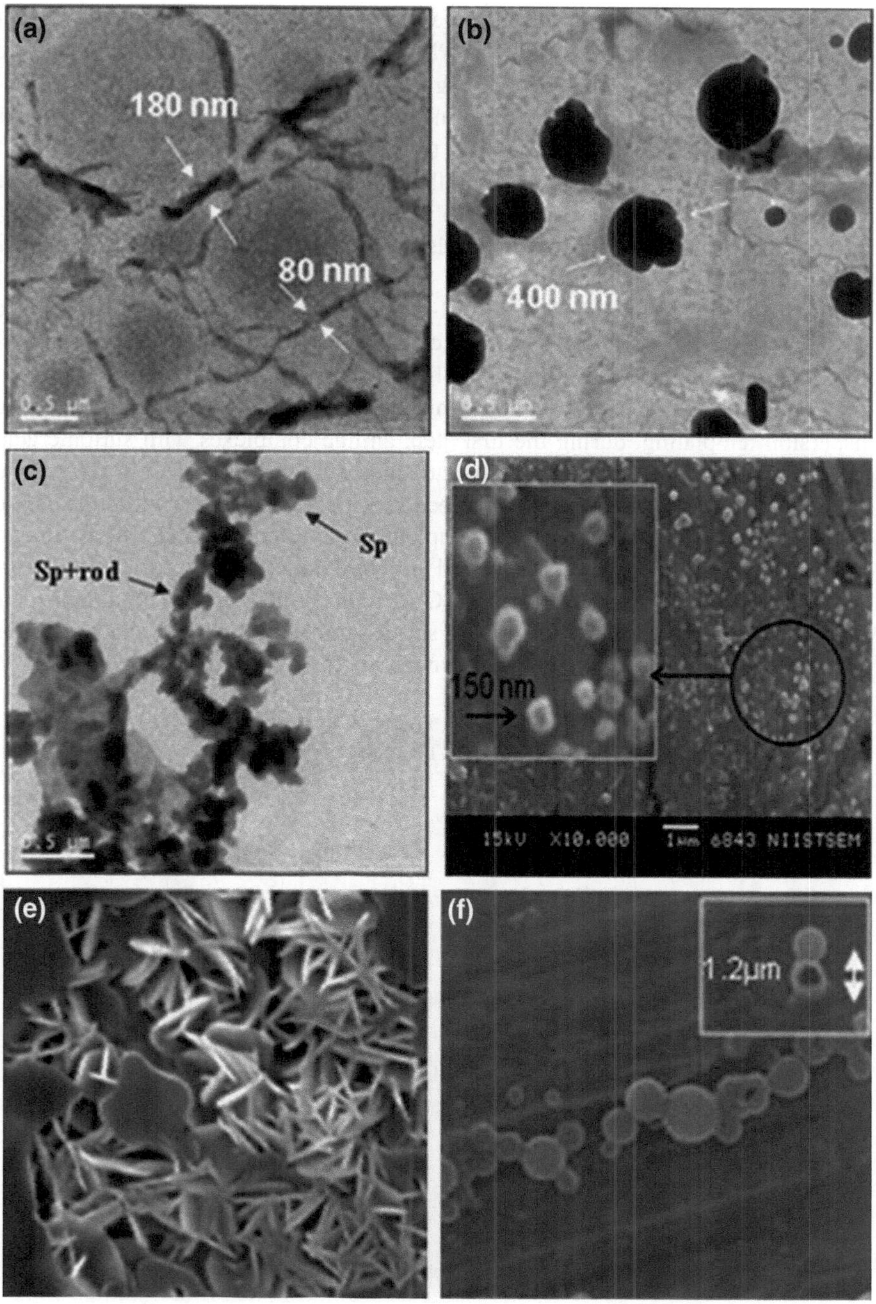

Fig. 6.6 Self-assembled molecular templates of dopant-1 with monomers: (**a**) aniline + dopant-1 in emulsion route, (**b**) aniline + dopant-1 in interfacial route, (**c**) aniline + dopant-1 in dilute route, (**d**) pyrrole + dopant-1 in emulsion route, (**e**) aniline + pyrrole + dopant-1 in emulsion route, (**f**) aniline + pyrrole + dopant-1 in dilute route. Reproduced with permission from Refs. [81] (2008, American Chemical Society), [85] (2009, John Wiley and Sons) and [86] (2010, American Chemical Society)

[86] (see Fig. 6.6a–f). In Interfacial route, the templates were assembled by introducing ammonium peroxydisulfate (APS) into dopant-1 micelles. DLS showed that the 4.3-nm-size micelles undergo aggregation in the presence of APS to produce larger micrometer size. The TEM images of the dopant-1 plus APS templates exclusively consist of spheres of 0.5 μm to 200 nm.

The dopant-2 molecule is structurally analogous to commercially available surfactant dodecylbenzenesulfonic acid (DBSA), which is widely used dopant for polyaniline and polypyrrole. In emulsion polymerization conditions, dopant-2 exhibits typical micellar behavior and forms cylindrical micellar templates of ~150 nm diameter with length up to 1–3 μm with aniline monomer [78] (see Fig. 6.7). In the case of biphasic (organic/water) dispersion polymerization conditions, the template formed is mostly spherical vesicles with size in the range of ~450 ± 200 nm in diameter and wall thickness of 20 ± 10 nm (see Fig. 6.7). In dispersion templating, during the first step, aniline complexes with sulfonic acid part of the micelles in water to produce cylindrical aggregates, and in the second stage, the organic solvent (toluene) molecules get stacked in the hydrophobic core of the micelles which transforms the templates into vesicular shape [78]. Saturated cardanol-based dopant-3 retained the gel structure upon cooling a solution of aniline and dopant-3 in ethanol from 60 °C to room temperature. The aniline plus dopant-3 gel template is in the form of self-assembled cylindrical nanostructures with average diameters 200 nm and length up to 4–5 μm (see Fig. 6.7). TEM

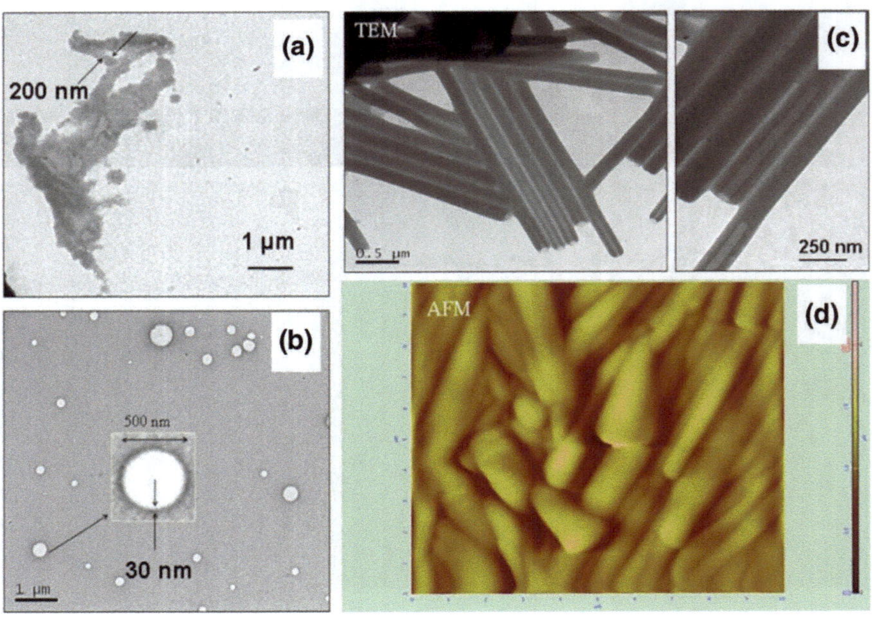

Fig. 6.7 Self-assembled molecular templates of dopant-2 and dopant-3 with aniline monomer: (**a**) dopant-2 ⥥←aniline in emulsion route, (**b**) dopant-2 ⥥← aniline dispersion route, (**c, d**) TEM and AFM of dopant-3 ⥥← aniline in gel route. Reproduced with permission from Refs. [78] (2009, American Chemical Society) and [79] (2010, American Chemical Society)

images indicated that during the gel formation the aniline monomers selectively occupied inside the cavity of the nanotubes formed by dopant molecules.

6.5 Template-Produced Conducting Polymer Nanostructures

The morphology of the dopant-1-templated polyaniline nanomaterials by emulsion polymerization route depended on the type of micellar structure (see Fig. 6.8). At very high dopant-1 concentrations (1×10^{-1} M) in the emulsion, the template formed exhibited less colloidal stability and mostly produced micron-sized polyaniline aggregates and hollow spheres. With further decrease in concentration, but still well above CMC, the cylindrical micelles templates were stable and yielded very uniform nanofibers (see Fig. 6.8). At a concentration above the CMC, a mixture of both nanotubes and nanofibers was produced [76, 77, 81]. The bilayered template formed at a dopant-1 concentration below CMC exclusively yielded nanotubes. Interestingly, thick emulsion of cylindrical micellar template (dopant-1 + aniline) formed above CMC value when diluted by adding water and oxidized using APS yielded nanorods rather than fibers. In the interfacial (biphasic) route, during the course of polymerization aniline monomers slowly diffuse

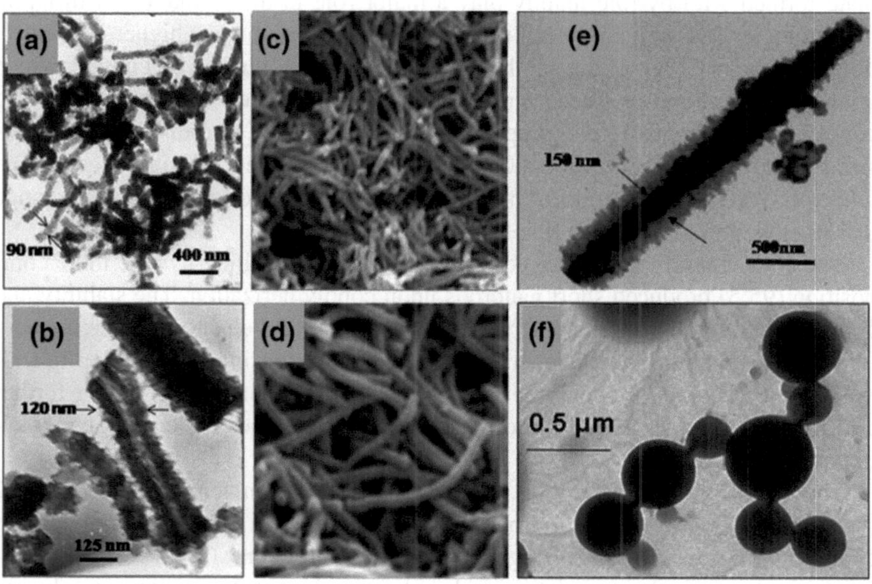

Fig. 6.8 TEM and SEM images of nanofibers by emulsion route (**a**, **c**), TEM and SEM image of nanotubes by emulsion route (**b**, **d**), TEM image of polyaniline nanorod by dilute route (**e**) and TEM image of nanospheres by interfacial route (**f**). Reprinted with permission from Refs. [77] (2007, American Chemical Society) and [80] (2007, American Chemical Society)

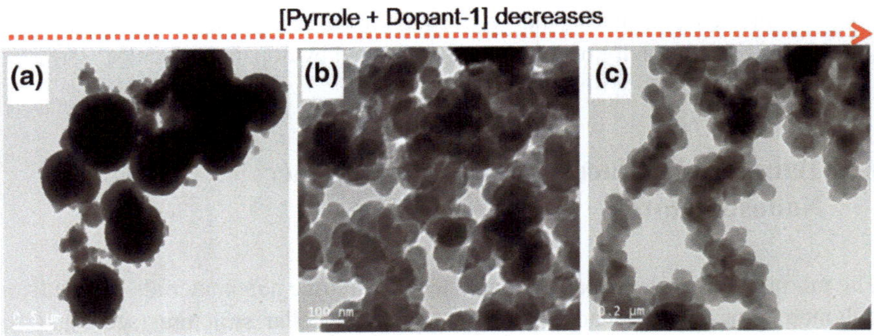

Fig. 6.9 TEM images of polypyrrole nanospheres under emulsion (**a**) and dilute route (**b, c**). Reprinted with permission from Ref. [85] (2009, John Wiley and Sons)

from organic layer and move toward the aqueous phase containing dopant-1 plus APS template through interface and get adsorbed on it. The subsequent polymerization results in the formation of nano/microspheres as evident from SEM and TEM images of the polyaniline formed [80]. The dopant-1 is unique in the sense that it self-assembles to more than one type of template by small variations in the concentration, combination of the constituents and types of polymerization employed and leads to the production of different kinds of nanostructures.

In the case of polypyrrole, the concentration of dopant-1 and stability of template formed with pyrrole mainly play a major role in the formation of different morphologies [84–86]. Emulsion template of pyrrole with very higher concentrations (far above CMC) of dopant-1 resulted in emulsion plus the precipitate, which upon polymerization produced mostly coral-like aggregated structures. Reducing concentration of pyrrole plus dopant-1 (still above CMC) produced submicron-sized polypyrrole spheres (500–600 nm), and a further decrease in concentration (close to CMC) yielded uniform polypyrrole nanospheres of diameter in the range of 150–250 nm (see Fig. 6.9) [84, 85].

The emulsion templates formed from dopant-1 with aniline-pyrrole mole composition (95:5) produced short nanorods upon copolymerization. The stability and competitive interplay of monomers (aniline and pyrrole) helps to produce molecular templates with intermediate structure, which is responsible for the fiber-to-rod-to-sphere transitions in copolymers [86]. The nanorod composition when further diluted with water produced hollow spheres and nanotubes under copolymerization route (Fig. 6.10).

The emulsion polymerization of cylindrical emulsion template of dopant-2 transforms the polyaniline morphology into a uniform mat of nanofibers of diameter ~80 nm with an average length of few microns (see Fig. 6.11) [78]. Interestingly, vesicular templates produced from dopant-2 + aniline in toluene-water dispersion yielded planar polyaniline nanotapes rather than vesicles upon adding APS and subsequent polymerization (see Fig. 6.11). During dispersion polymerization, the vesicular templates get ripped out from vesicular surface

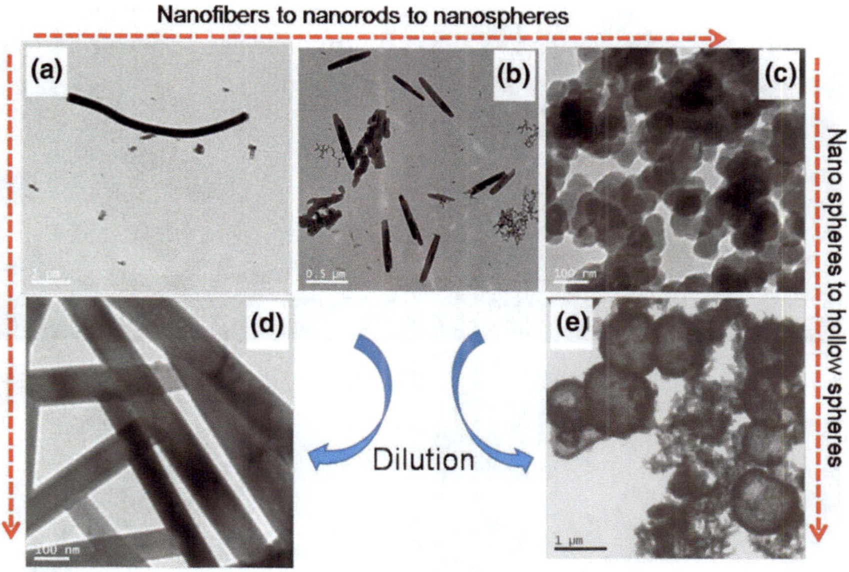

Fig. 6.10 (**a**) TEM image of polyaniline nanofibers, (**b**) polyaniline–polypyrrole copolymer nanorods, (**c**) polypyrrole nanospheres, (**d**) polyaniline–polypyrrole copolymer nanotubes and (**e**) polyaniline–polypyrrole copolymer hollow spheres. Reprinted with permission Ref. [86] (2010, American Chemical Society)

which produces nanotapes [78]. Optimum surfactant geometry of dopant-2 enables to form stable templates which upon emulsion and dispersion polymerization process produce nanofibers and nanotapes, respectively. The gel template formed by dopant-3 plus aniline in ethanol upon polymerization yielded very uniform nanofibers with an average fiber diameter of ~135 nm and length up to 5 μm long (see Fig. 6.11). On the other hand, when the polymerization was carried at sol state (at 65 °C), it has not yielded good nanostructures; instead agglomerated particles were formed. The physical state of the organogel helped to transcribe the morphology of the resulting polymer nanostructures.

6.6 Properties of Conducting Polymer Nanomaterials

Post-polymerized nanomaterials contain sulfonic acid head of amphiphilic dopants which complexes with nanomaterials (polyaniline/polypyrrole) electrostatically and makes these structures easily dispersible in solvents such as water, chloroform, methanol, DMF and DMSO [74, 75, 79]. This increased solubility/dispersibility enables the complete structural characterization by spectroscopic techniques such as NMR, UV–Vis and FT-IR [74–87]. Generally, in the literature the NMR spectra of polyaniline/polypyrrole are seldom reported due the insoluble nature of these materials. The cardanol dopant-based polymer nanostructures are

Fig. 6.11 (**a**) SEM images of polyaniline nanofibers synthesized using dopant-2 by emulsion route, (**b**) polyaniline nanotapes by dispersion route, and (**c, d**) SEM and TEM images of polyaniline nanofibers synthesized using gel dopant-3. Reproduced with permission from Refs. [78], (2009, American Chemical Society) and [79] (2010, American Chemical Society)

highly soluble in NMR solvents like d_6-DMSO. For example, ^{1}H-NMR spectra of the polyaniline nanomaterials show the peaks corresponding to the dopant molecule in addition to the peaks for the polymer structures indicating post-polymerization doping. The aromatic peaks of polyaniline backbone appeared at 7.65 and 7.52 ppm, and additionally the three equally intense peaks (triplet) at 7.1, 7.2 and 7.3 ppm corresponding to the free radical NH proton resonance due to the ^{14}N with unit spin were also observed. ^{1}H-NMR study confirms the doping associated with strong complexation of cardanyl sulfonic acid dopant to the polymer nanostructure.

Polyaniline/polypyrrole polymers are highly rigid due to their linear aromatic backbone, and lack of flexibility to chain folding prevents them acquiring crystalline nature. Therefore, generally emeraldine base forms as well as mineral acid-doped emeraldine salt forms of these polymers are highly amorphous [41]. In the presence of amphiphilic cardanyl sulfonic acid dopants, the dopant–polymer complex undergoes various interactions, which improves the solid-state packing of the polymer chains inside the nanostructures. The powder X-ray diffraction patterns of nanostructures showed lower angle peaks characteristics of the interdigitations of dopant molecules between polymer chains [74–87]. This improved solid-state packing also reflected in reasonable high electrical conductivity observed in nanostructures.

Fluorescence emission studies of polyaniline nanospheres synthesized from interfacial method and polyaniline nanofibers synthesized by emulsion route revealed that nanospheres have much more fluorescence quantum yield than fibers [74, 77]. The better fluorescence properties of nanospheres were correlated with the more expanded chain conformation and more effective layered packing of azochromophore group of dopant-1 in the polymer chain. The polyaniline nanorods obtained by dilute condition route with dopant-1 were found to be very weakly luminescent since the dilute reaction conditions will make the dopant well dispersed in polymer matrix and prevent close packing of chromophores in the resulting sample.

6.7 Conclusion

Simple chemical approaches were developed to synthesize amphiphilic dopants from the renewable resource cardanol. The cardanyl dopants were self-assembled and utilized as soft templates for the formation of polyaniline and polypyrrole nanostructures. Cardanyl dopant-1 and dopant-2 have a typical surfactant structure having a sulfonic acid polar head and pentadecenyl chain as hydrophobic tail. The concentration-dependent micellation behavior and its size in aqueous solution were characterized by various techniques like fluorescence measurements and DLS studies. Different types of stable soft templates formed by these dopants with monomers (aniline and pyrrole) under different types of polymerization route (emulsion, dilute, interfacial and dispersion) yielded various polymeric nanostructures such as nanofibers, nanotubes, nanorods and nanotapes. The gel-forming dopant-3 derived from saturated cardanol has limited solubility in water; however, it forms gel in polar solvents like ethanol. The nanotubular gel template of dopant-3 filled with aniline upon polymerization produced nanofibers. The copolymer nanorods were produced from soft emulsion templates of dopant-1 with co-monomers (aniline 95 % and pyrrole 5 %) through copolymerization. Dilute copolymerization produced hollow spheres and nanotubes from the same diluted emulsion templates. The cardanyl amphiphilic dopants acted as dopant-cum soft template and greatly improved the properties of nanomaterials such as solubility, solid-state ordering, conductivity and optical properties. In short, cardanyl amphiphilic soft-template-assisted polymerization is easy, facile and efficient for the creating the architectural library of conducting polymer nanomaterials.

Acknowledgements MJA and PA are gratefully acknowledging UGC, New Delhi, for providing financial supports as research fellowships. Both the authors are indebted to their doctoral mentor Prof. Manickam Jayakannan, Associate professor, IISER, Pune, India, for his continuous guidance and encouragements.

References

1. Kiess HG (ed) (1992) Conjugated conducting polymers. Springer, Berlin
2. Brédas JL, Silbey R (eds) (1991) Conjugated polymers. Springer, Dordrecht
3. MacDiarmid AG (2001) "Synthetic metals": a novel role for organic polymers (nobel lecture) copyright((c)) The Nobel Foundation 2001. We thank the Nobel Foundation, Stockholm, for permission to print this lecture. Angew Chem Int Ed Engl 40:2581–2590
4. Shirakawa H (2001) The discovery of polyacetylene film: the dawning of an era of conducting polymers (nobel lecture) copyright((c)) The Nobel Foundation 2001. We thank the Nobel Foundation, Stockholm, for permission to print this lecture. Angew Chem Int Ed Engl 40:2574–2580
5. MacDiarmid AG, Mammone RJ, Kaner RB et al (1985) The concept of `doping' of conducting polymers: the role of reduction potentials [and discussion]. Philos Trans R Soc A Math Phys Eng Sci 314:3–15
6. Prasanna Chandrashekar (1999) Conducting polymers, fundamentals and applications: a practical approach. Springer, Berlin
7. Skotheim TA, Reynolds J (2006) Conjugated polymers: theory, synthesis, properties, and characterization. Handbook of conducting polymers, 3rd edn. CRC Press, Boca Raton, FL
8. Müllen K, Reynolds JR, Masuda T (eds) (2013) Conjugated polymers. Royal Society of Chemistry, Cambridge
9. Heeger AJ, Sariciftci NS, Namdas EB (2011) Semiconducting and metallic polymers. Oxford Univ Press 21:391–393
10. Stejskal J, Gilbert RG (2002) Polyaniline. Preparation of a conducting polymer (IUPAC Technical Report). Pure Appl Chem 74:857–867
11. Goto H, Yoneyama H, Togashi F et al (2008) Preparation of conducting polymers by electrochemical methods and demonstration of a polymer battery. J Chem Educ 85:1067
12. Inzelt G (2012) Conducting polymers. Springer, Berlin
13. Zhang X, Lee J-S, Lee GS et al (2006) Chemical synthesis of PEDOT nanotubes. Macromolecules 39:470–472
14. Huang L, Wang Z, Wang H et al (2002) Polyaniline nanowires by electropolymerization from liquid crystalline phases. J Mater Chem 12:388–391
15. Li C, Bai H, Shi G (2009) Conducting polymer nanomaterials: electrosynthesis and applications. Chem Soc Rev 38:2397
16. Kang E (1998) Polyaniline: a polymer with many interesting intrinsic redox states. Prog Polym Sci 23:277–324
17. MacDiarmid AG, Epstein AJ (1989) Polyanilines: a novel class of conducting polymers. Faraday Discuss Chem Soc 88:317
18. Kar P (2013) Doping in conjugated polymers. In: Polymer science and plastics engineering. Wiley, New York
19. Macdiarmid AG, Chiang JC, Richter AF, Epstein AJ (1987) Polyaniline: a new concept in conducting polymers. Synth Met 18:285–290
20. MacDiarmid AG, Epstein AJ (1995) Secondary doping in polyaniline. Synth Met 69:85–92
21. Brédas JL, Chance RR, Silbey R (1982) Comparative theoretical study of the doping of conjugated polymers: polarons in polyacetylene and polyparaphenylene. Phys Rev B 26:5843–5854
22. Bredas JL, Street GB (1985) Polarons, bipolarons, and solitons in conducting polymers. Acc Chem Res 18:309–315
23. Cho YS, Yoon KH (2001) Handbook of advanced electronic and photonic materials and devices. Elsevier, Amsterdam
24. Lee K, Cho S, Heum Park S et al (2006) Metallic transport in polyaniline. Nature 441:65–68
25. Huang K, Qiu H, Wan M (2002) Synthesis of highly conducting polyaniline with photochromic azobenzene side groups. Macromolecules 35:8653–8655

26. Ng SW, Neoh KG, Sampanthar JT et al (2001) Conversion of polyaniline from insulating to conducting state in aqueous viologen solutions. J Phys Chem B 105:5618–5625
27. Saxena V, Malhotra BD (2003) Prospects of conducting polymers in molecular electronics. Curr Appl Phys 3:293–305
28. Li D, Huang J, Kaner RB (2009) Polyaniline nanofibers: a unique polymer nanostructure for versatile applications. Acc Chem Res 42:135–145
29. McQuade DT, Pullen AE, Swager TM (2000) Conjugated polymer-based chemical sensors. Chem Rev 100:2537–2574
30. Tran HD, Li D, Kaner RB (2009) One-dimensional conducting polymer nanostructures: bulk synthesis and applications. Adv Mater 21:1487–1499
31. Janata J, Josowicz M (2003) Conducting polymers in electronic chemical sensors. Nat Mater 2:19–24
32. Huang J, Virji S, Weiller BH, Kaner RB (2004) Nanostructured polyaniline sensors. Chem - A Eur J 10:1314–1319
33. Virji S, Fowler JD, Baker CO et al (2005) Polyaniline nanofiber composites with metal salts: chemical sensors for hydrogen sulfide. Small 1:624–627
34. Virji S, Huang J, Kaner RB, Weiller BH (2004) Polyaniline nanofiber gas sensors: examination of response mechanisms. Nano Lett 4:491–496
35. Huang J (2006) Syntheses and applications of conducting polymer polyaniline nanofibers. Pure Appl Chem
36. Huang J, Kaner RB (2004) Flash welding of conducting polymer nanofibres. Nat Mater 3:783–786
37. Jang J (2006) conducting polymer nanomaterials and their applications. pp 189–260
38. Prabhakar N, Arora K, Singh H, Malhotra BD (2008) Polyaniline based nucleic acid sensor. J Phys Chem B 112:4808–4816
39. Yakuphanoglu F, Basaran E, Şenkal BF, Sezer E (2006) Electrical and optical properties of an organic semiconductor based on polyaniline prepared by emulsion polymerization and fabrication of Ag/polyaniline/n-Si Schottky diode. J Phys Chem B 110:16908–16913
40. Xia L, Wei Z, Wan M (2010) Conducting polymer nanostructures and their application in biosensors. J Colloid Interface Sci 341:1–11
41. Wan M (2008) Conducting polymers with micro or nanometer structure. Springer, Berlin Heidelberg
42. Tran HD, Li D, Kaner RB (2009) One-dimensional conducting polymer nanostructures: bulk synthesis and applications. Adv Mater 21:1487–1499
43. Sergeev GB (2006) Nanochemistry, 1st edn. Elsevier, Amsterdam
44. Bandyopadhyay AK (2008) Nanomaterials, 1st edn. New Age, New Delhi
45. Brechignac C, Houdy P, Lahmani M (2007) Nanomaterials and nanochemistry. Springer, Berlin
46. Wang ZM (2008) One-dimensional nanostructures, 1st edn. Springer, New York
47. Nabok A (2005) Organic and inorganic nanostructures, 1st edn. Artec. Inc., Norwood, MA
48. Dai H (2001) Nanotube growth and characterization, 2nd edn. Springer, Berlin
49. Niemeyer CM, Mirkin CA (2004) Nanobiotechnology, 1st edn. Wiley-VCH, Weinheim
50. Rao CNR, Muller A, Cheetham AK (2004) The chemistry of nanomaterials, 1st edn. Wiley-VCH, Weinheim
51. Xia Y, Rogers JA, Paul KE, Whitesides GM (1999) Unconventional methods for fabricating and patterning nanostructures. Chem Rev 99:1823–1848
52. Ramakrishnan S (2005) Nanostructured polymers. The chemistry of nanomaterials. Wiley-VCH Verlag GmbH & Co. KGaA, Weinheim, FRG, pp 476–517
53. Armes SP, Aldissi M, Hawley M et al (1991) Morphology and structure of conducting polymers. Langmuir 7:1447–1452
54. Greiner A, Wendorff JH (2007) Electrospinning: a fascinating method for the preparation of ultrathin fibers. Angew Chemie Int Ed 46:5670–5703

55. Martin CR (1995) Template synthesis of electronically conductive polymer nanostructures. Acc Chem Res 28:61–68
56. Palaniappan S, John A (2008) Polyaniline materials by emulsion polymerization pathway. Prog Polym Sci 33:732–758
57. Jackowska K, Biegański AT, Tagowska M (2008) Hard template synthesis of conducting polymers: a route to achieve nanostructures. J Solid State Electrochem 12:437–443
58. Qiu H, Zhai J, Li S et al (2003) Oriented growth of self-assembled polyaniline nanowire arrays using a novel method. Adv Funct Mater 13:925–928
59. Wan M (2008) A template-free method towards conducting polymer nanostructures. Adv Mater 20:2926–2932
60. Martin CR (1995) Template synthesis of electronically conductive polymer nanostructures. Acc Chem Res 28:61–68
61. Zhang X, Chan-Yu-King R, Jose A, Manohar SK (2004) Nanofibers of polyaniline synthesized by interfacial polymerization. Synth Met 145:23–29
62. Wan M (2008) A template-free method towards conducting polymer nanostructures. Adv Mater 20:2926–2932
63. Zhang X, Goux WJ, Manohar SK (2004) Synthesis of polyaniline nanofibers by "nanofiber seeding". J Am Chem Soc 126:4502–4503
64. Wei Z, Wan M (2002) Hollow microspheres of polyaniline synthesized with an aniline emulsion template. Adv Mater 14:1314–1317
65. Carswell ADW, O'Rea EA, Grady BP (2003) Adsorbed surfactants as templates for the synthesis of morphologically controlled polyaniline and polypyrrole nanostructures on flat surfaces: from spheres to wires to flat films. J Am Chem Soc 125:14793–14800
66. Li C, Hatano T, Takeuchi M, Shinkai S (2004) Polyaniline superstructures created by a templating effect of organogels. Chem Commun 20:2350
67. Voirin C, Caillol S, Sadavarte NV et al (2014) Functionalization of cardanol: towards biobased polymers and additives. Polym Chem 5:3142–3162
68. Phani Kumar P, Paramashivappa R, Vithayathil PJ et al (2002) Process for isolation of cardanol from technical cashew (*Anacardium occidentale* L.) nut shell liquid. J Agric Food Chem 50:4705–4708
69. Balachandran VS, Jadhav SR, Vemula PK, John G (2013) Recent advances in cardanol chemistry in a nutshell: from a nut to nanomaterials. Chem Soc Rev 42:427–438
70. John G, Masuda M, Okada Y et al (2001) Nanotube formation from renewable resources via coiled nanofibers. Adv Mater 13:715–718
71. Paul RK, Pillai C (2001) Thermal properties of processable polyaniline with novel sulfonic acid dopants. Polym Int 50:381–386
72. Paul RK, Pillai CK (2000) Melt/solution processable conducting polyaniline with novel sulfonic acid dopants and its thermoplastic blends. Synth Met 114:27–35
73. Paul RK, Pillai CKS (2001) Melt/solution processable polyaniline with functionalized phosphate ester dopants and its thermoplastic blends. J Appl Polym Sci 80:1354–1367
74. Anilkumar P (2009) Self-assembled molecular templates for conducting polyaniline nanomaterials. Doctoral dissertation. Retrieved from http://ir.niist.res.in:8080/xmlui/handle/123456789/878
75. AntonyMJ (2010) Self-organization approach for conducting polypyrrole and their copolymer nanomaterials. Doctoral dissertation. Retrieved from http://ir.niist.res.in:8080/xmlui/handle/123456789/159
76. Anilkumar P, Jayakannan M (2006) New renewable resource amphiphilic molecular design for size-controlled and highly ordered polyaniline nanofibers. Langmuir 22:5952–5957
77. Anilkumar P, Jayakannan M (2007) Fluorescent tagged probing agent and structure-directing amphiphilic molecular design for polyaniline nanomaterials via self-assembly process. J Phys Chem C 111:3591–3600
78. Anilkumar P, Jayakannan M (2009) Self-assembled cylindrical and vesicular molecular templates for polyaniline nanofibers and nanotapes. J Phys Chem B 113:11614–11624

79. Anilkumar P, Jayakannan M (2010) A novel supramolecular organogel nanotubular template approach for conducting nanomaterials. J Phys Chem B 114:728–736
80. Anilkumar P, Jayakannan M (2007) Single-molecular-system-based selective micellar templates for polyaniline nanomaterials: control of shape, size, solid state ordering, and expanded chain to coillike conformation. Macromolecules 40:7311–7319
81. Anilkumar P, Jayakannan M (2008) Divergent nanostructures from identical ingredients: unique amphiphilic micelle template for polyaniline nanofibers, tubes, rods, and spheres. Macromolecules 41:7706–7715
82. Anilkumar P, Jayakannan M (2008) Hydroxyl-functionalized polyaniline nanospheres: tracing molecular interactions at the nanosurface via vitamin C sensing. Langmuir 24:9754–9762
83. Anilkumar P, Jayakannan M (2009) Large-scale synthesis of polyaniline nanofibers based on renewable resource molecular template. J Appl Polym Sci 114:3531–3541
84. Antony MJ, Jayakannan M (2007) Amphiphilic azobenzenesulfonic acid anionic surfactant for water-soluble, ordered, and luminescent polypyrrole nanospheres. J Phys Chem B 111:12772–12780
85. Antony MJ, Jayakannan M (2009) Self-assembled anionic micellar template for polypyrrole, polyaniline, and their random copolymer nanomaterials. J Polym Sci, Part B: Polym Phys 47:830–846
86. Antony MJ, Jayakannan M (2010) Molecular template approach for evolution of conducting polymer nanostructures: tracing the role of morphology on conductivity and solid state ordering. J Phys Chem B 114:1314–1324
87. Antony MJ, Jayakannan M (2011) Polyaniline nanoscaffolds for colorimetric sensing of biomolecules via electron transfer process. Langmuir 27:6268–6278
88. Antony MJ, Jayakannan M (2011) Role of anionic micellar template on the morphology, solid-state ordering, and unusual conductivity trend in poly(aniline- co -pyrrole) nanomaterials. J Phys Chem B 115:6427–6436
89. Jayakannan M, Anilkumar P, Sanju A (2006) Synthesis and characterization of new azobenzenesulfonic acids doped conducting polyaniline. Eur Polym J 42:2623–2631

Chapter 7
Cardanol-Based Supramolecular Gels

Thomas F. Garrison, Audrey Scholz, Heidi Grimm and Rafael L. Quirino

Abstract With the increasing interest in the development of new materials from bio-renewable resources seen in the last couple of decades, great emphasis has been put in the use of bio-renewable materials as building blocks for the preparation of bio-based polymers and composites. Lately, cardanol, a phenol derived from extracts of cashew nut shells, has gained attention due to its abundance, lack of competing applications, and great potential for the preparation of supramolecular structures through various routes. In this chapter, an initial overview of supramolecular gels is presented, covering background information on the terminology, chemical structure, properties, and the starting materials used in their synthesis. A section is dedicated to the latest advancements of cardanol-based gels, followed by a section describing applications involving cardanol-based gels and related self-assembled materials.

7.1 Background

Supramolecular gels represent a small, but important group of gels that are increasingly being used in advanced material applications. Supramolecular gels are versatile materials due to their responsiveness to different stimuli (e.g., temperature, light, pH) [1].

The number of publications related to research on supramolecular gels has grown very rapidly in recent years with major developments in areas such as the synthesis of novel gelators, improved analytical techniques, controlled responses

A. Scholz · H. Grimm · R.L. Quirino (✉)
Georgia Southern University, Statesboro, GA, USA
e-mail: rquirino@georgiasouthern.edu

T.F. Garrison
King Fahd University of Petroleum and Minerals, Dhahran, Saudi Arabia

P. Anilkumar (ed.), *Cashew Nut Shell Liquid*,
DOI 10.1007/978-3-319-47455-7_7

to stimuli, and advanced applications [2]. The advances in the field of supramolecular gels have been extensively treated in feature articles [3], perspectives [4], tutorial reviews [5–8], highlights [2], and critical reviews [9].

This chapter focuses on the development of cardanol-based supramolecular gels. The synthesis of supramolecular gels from bio-based materials, particularly for biomedical applications, is one of the major areas of research in this field [2]. To the best of our knowledge, cardanol-based supramolecular gels have not been the subject of a dedicated review.

Supramolecular gels, like all other gels, are primarily composed of a liquid phase, yet are solid, jelly-like materials due to the presence of three-dimensional networks. The materials that compose the solid continuous phase are called gelators, while solvents are the liquid dispersed phase. The solvents used in the liquid phase can either be water or organic liquids, which are used to form hydrogels or organogels, respectively [10].

While conventional gels are formed when covalently bonded, cross-linked polymeric networks are cured trapping liquid solvents in the structure, supramolecular gels are formed by the self-assembly of low molecular weight molecules [11]. Typically, the molecular mass of the gelators is less than 3000 amu [6]. Supramolecular gels can be held together by a variety of weak non-covalent forces including hydrogen bonding, pi–pi interactions, van der Waals forces, and hydrophilic-lipophilic balances (HLB) [12]. Supramolecular gels are typically made up of amphiphilic molecules, which have both hydrophilic and hydrophobic properties.

The structure of supramolecular gels is determined by the non-covalent forces. The rheological behavior and other macroscopic properties of supramolecular gels are determined by the three-dimensional networks of the nano- and microstructures [4].

An important feature of supramolecular gels is the reversibility on the weak interactions between molecules. As a result, the formation of supramolecular structures can be controlled or tuned by external stimuli. Examples of stimuli that have been used to control the formation of supramolecular gels include thermal, mechanical, electrochemical, electromagnetic, and chemical stimuli [4].

7.2 Synthesis of Supramolecular Gels

Supramolecular gels are prepared by bringing a solution of the precursor, also known as gelator, to an elevated temperature, and subsequently allowing it to cool down to room temperature [1]. During that process, the gelator forms reversible physical cross-links in organic solvents through non-covalent interactions, resulting in a consistent network of self-assembled molecules [1]. The solvent used varies depending on the desired application and properties of the gel, but usually consists of an organic liquid. In most cases, the solvent makes up approximately

99 wt% of the gel, giving it a liquid-like aspect [12]. Most gelators commonly used are polymeric materials designed to meet specific desired properties and to confer the gel with a solid-like aspect. There are three classes of gelators, namely metallo-gelators, hydrogelators, and organogelators.

In supramolecular gels, the gelators are referred to as low molecular weight gelators (LMWGs) because of their relatively small molar masses, which favor the non-covalent interactions necessary for the gel formation [13]. An interchangeable term sometimes used is low molecular mass organic gelators (LMOGs) [6]. LMWGs can be used individually or in combination with other species in order to obtain specific properties [9].

Cardanol-based LMWGs can have a variety of intermolecular forces that contribute to the formation of supramolecular gels as illustrated in Fig. 7.1 [14]. In this particular example, cardanol modified with diaminopyridine (DAP) has three types of intermolecular forces: hydrogen bonding, pi–pi stacking, and van der Waals interactions. The DAP head group forms three hydrogen bonds, the aromatic rings interact with each other via pi–pi stacking, and the hydrophobic tails have van der Waals interactions. Interestingly, unsaturation in the hydrophobic

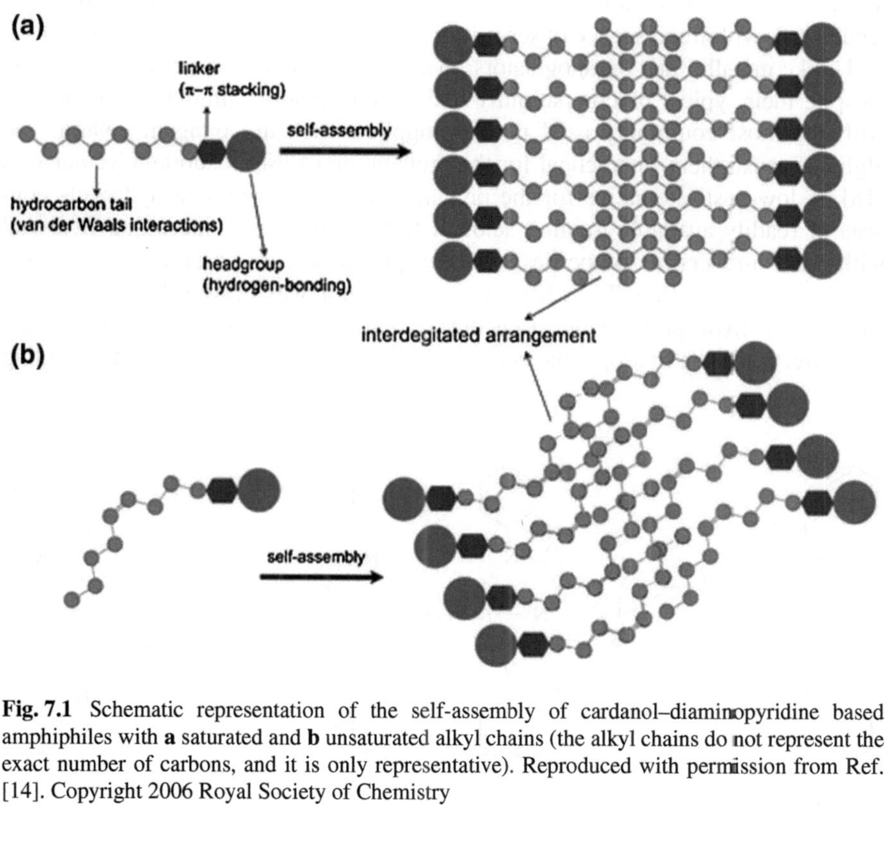

Fig. 7.1 Schematic representation of the self-assembly of cardanol–diaminopyridine based amphiphiles with **a** saturated and **b** unsaturated alkyl chains (the alkyl chains do not represent the exact number of carbons, and it is only representative). Reproduced with permission from Ref. [14]. Copyright 2006 Royal Society of Chemistry

tails influences the shapes of nanostructures formed by self-assembly. [14]. A variety of hydrophilic head groups can be used to modify cardanol for the formation of supramolecular gels.

7.2.1 Gelators

In metallo-gelators, the LMWG is coordinated to a metal center, which typically results in systems that exhibit sensitivity to certain stimuli, such as pH, temperature, solvent polarity, ionic strength [15]. For example, the combination of transition metals and lanthanides was successfully used with a pentaethylene glycol derivative gelator in chloroform [16]. Photoluminescence was achieved through the use of Zn/Eu in the gelator, and the effect was reversed by addition of formic acid [16]. Later on, Zn(II) and La(III) were successfully used with a 2,6-bis-(1-methylbenzimidazolyl)-4-oxypyridine attached to both ends of a penta(ethylene glycol) core to synthesize thermally, mechanically, and chemically responsive gels (Fig. 7.2) [15]. The addition of Fe(II), Ni(II), Cu(II), Zn(II), or Ru(III) to tripodal terpyridine gels in methanol resulted in further cross-linking in the structure, and a transition from a one-dimensional helix to a three-dimensional fibrous supramolecular gel, with fluorescence observed for the structures containing Zn(II) [17].

Unlike metallo- and organogelators, hydrogelators require water as the solvent. Despite their typical biocompatibility and easy formation, limitations associated with the most commonly used natural supplies, such as collagen, gelatin, and alginate, make them impractical for the preparation of useful supramolecular gels [18]. A low-cost alternative for the preparation of biocompatible hydrogels consists of readily available *l*-amino acids [19]. For example, a successful hydrogel with thermo-reversible properties has been prepared from an amino acid-substituted cyclohexane (Fig. 7.3) [20]. The hydrogelator was designed in a spherical shape, with hydrophilic ends to encourage H-bonding at the extremities of the structure, while the hydrophobic core results in more aggregation forces, shielding

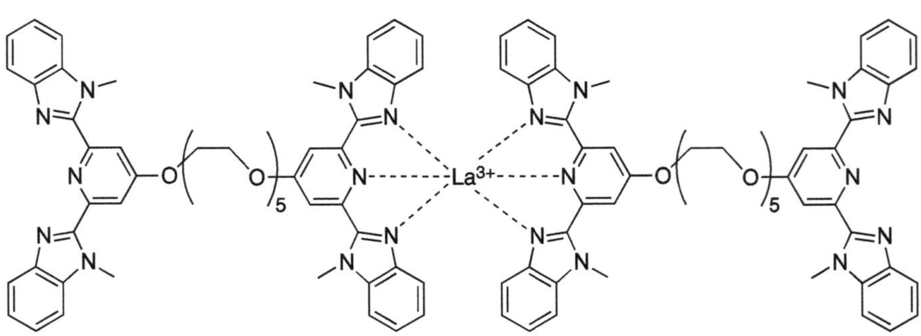

Fig. 7.2 Metallo-gelator with La(III) ion

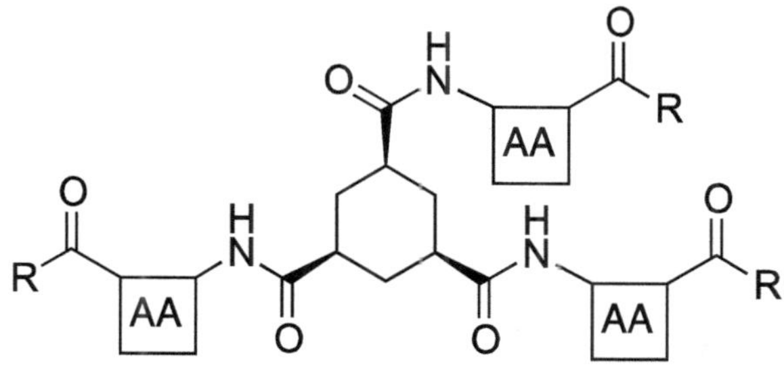

Fig. 7.3 Amino acid-substituted cyclohexane gelator. *AA* amino acid, *R* hydrophilic substituent

the inner amides from water [20]. This arrangement allows hydrogen bonding among the amide groups, thus creating the interlinking network for the gel [20].

It has been shown that hydrogelators can be designed using nearly any combination of nucleobases, amino acids, and glycosides [18]. For example, hydrogelators with potential for use in biomaterials can be prepared when glycoside derivatives are covalently bonded with a nucleotide and/or amino acid, as depicted in Fig. 7.4. The creation of novel hydrogelators is still relatively new, but it has already revealed itself as a faster method for forming gels with stronger mechanical properties [21].

Organogelators only differ from hydrogelators in that they use organic solvents instead of water. Common solvents include alkanes, alcohols, polar organic solvents in general, and vegetable oils [19]. Certain organogelators can be used to create organogel electrolytes, which allow ionic conductivity [22, 23]. The anthracene derivatives 2,3-bis(*n*-decyloxy)anthracene (Fig. 7.5a) [22] and *N*-carbobenzyloxy-L-isoleucylaminooctadecane (Fig. 7.5b) [23] were both used with a variety of polar organic solvents to produce organogel electrolytes. L-amino acids, such as L-lysine [19, 24] and L-isoleucine,[1] can also be used to make organogelators (Fig. 7.6). Like hydrogelators, organogelators can be used for biomedical purposes due to their biocompatibility and readily availability of the corresponding precursors.

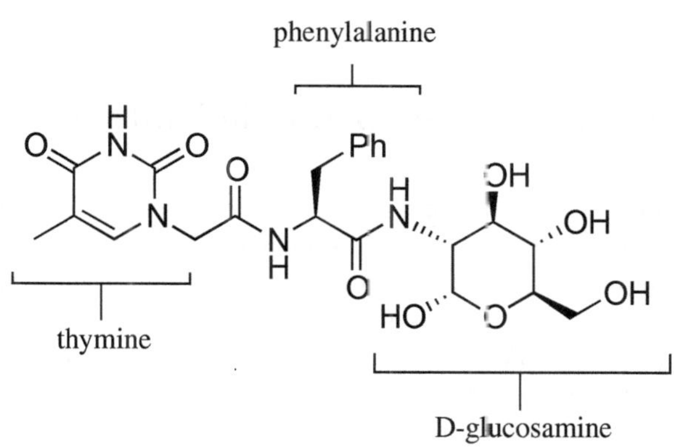

Fig. 7.4 Hydrogelator made from thymine, phenylalanine, and D-glucosamine

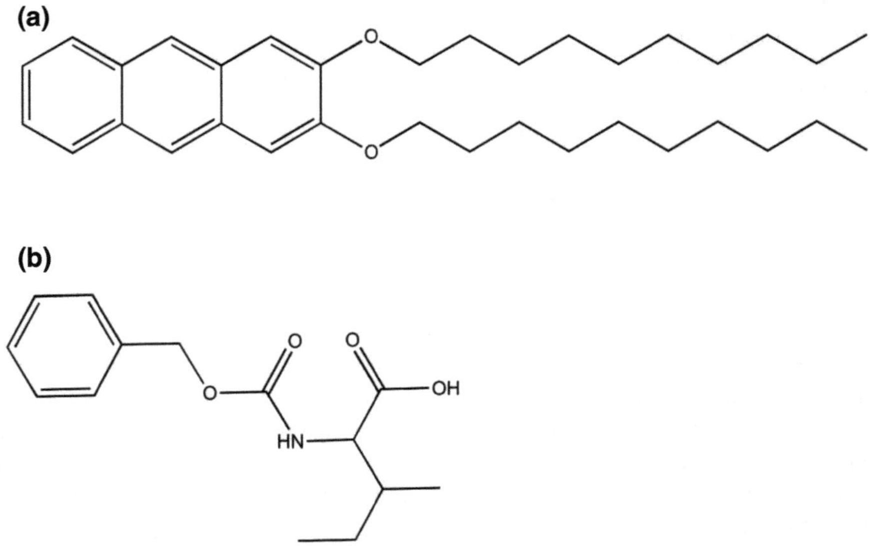

Fig. 7.5 Organogel electrolytes. **a** 2,3-Bis(*n*-decyloxy)anthracene, and **b** *N*-carbobenzyloxy-L-isoleucylaminooctadecane

Fig. 7.6 **a** L-lysine- and **b** L-isoleucine-based organogelators

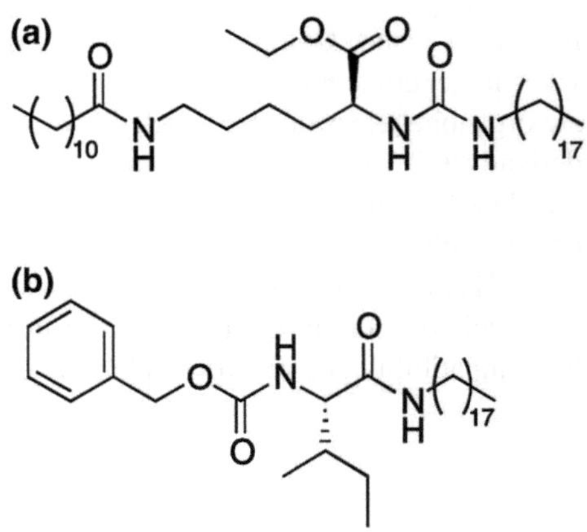

7.3 Recent Developments in Cardanol-Based Materials

In recent years, there has been a shift into the search for bio-renewable supramolecular gel precursors. Sustainable, bio-based sources of LMWGs, such as sugar derivatives, are an area of particular research interest [25]. These bio-based compounds are selected for properties that promote their ability to act as gelators (Fig. 7.7). For example, amygdalin, an agricultural by-product found primarily in kernals (seeds) of certain fruits, has been used to make amphiphilic molecules that will self-assemble into nanoaggregates [14]. Amygdalin has multiple hydroxyl

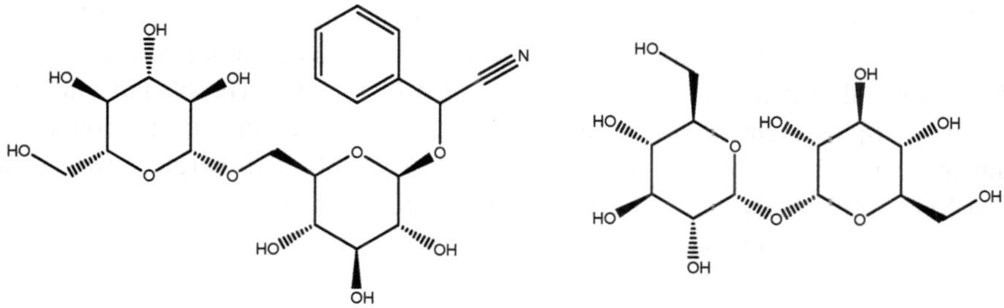

Amygdalin Trehalose

Fig. 7.7 Structure of natural precursors of LWMGs

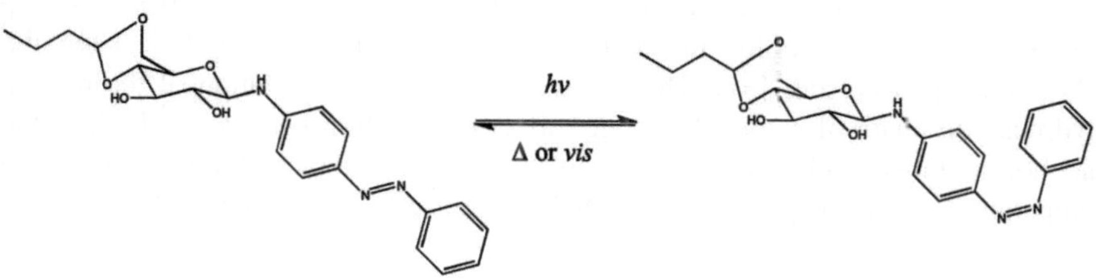

Fig. 7.8 Reversible light-induced isomerization of bio-based azobenzene derivatives [25]

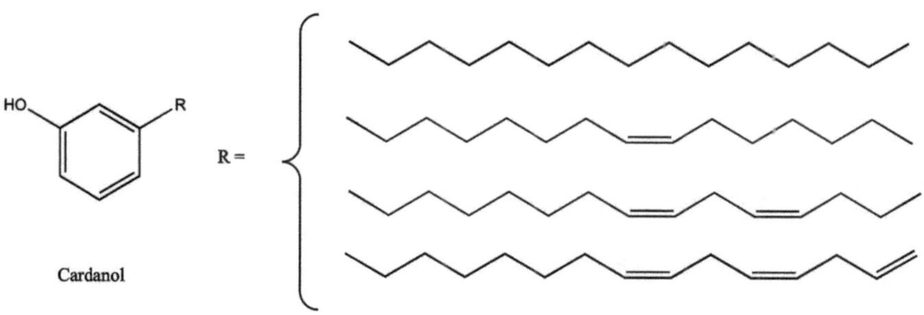

Fig. 7.9 Chemical structure of cardanol

groups, as well as an aromatic ring, that promote hydrogen bonding and pi–pi interactions, respectively. Trehalose, a naturally occurring disaccharide found in certain fungi, plants and invertebrates, has also been investigated as a feedstock for making LMWGs due to its strong affinity for water retention [26].

Reversible, light-controlled self-assembly of gel structures can be achieved by incorporating photoresponsive azobenzene functional groups onto sugar molecules. Azo groups are known to isomerize from trans to cis under UV exposure, but will reverse under visible-light exposure (Fig. 7.8) [25].

In that context, cardanol, a by-product from cashew nut processing [27, 28], is an excellent candidate as a starting material for making supramolecular gels. The chemical structure of cardanol is provided in Fig. 7.9. The phenol unit and the long aliphatic chain, which can have different degrees of unsaturation varying from 0 to 3, provide opportunities for self-assembly and therefore use as an LMWG. Additionally, cardanol also contains an aromatic ring, like amygdalin, that promotes pi–pi interactions.

7.3.1 Sugar Derivatives

Amphiphilic glycolipids made from cardanol can be prepared by linking a sugar molecule to the phenolic hydroxyl group of cardanol (Fig. 7.10) [29]. These glycolipids self-assemble to form low molecular weight gels [30], nanofibers [14], nanotubes [29], and liquid crystals [31]. The relative amounts of unsaturation of the pendant hydrocarbon chain of cardanol strongly affect the gelation behavior in both water–alcohol mixtures and in pure organic solvents [14, 30]. Notably, the phase transition temperature significantly decreased with the presence of double bonds [30].

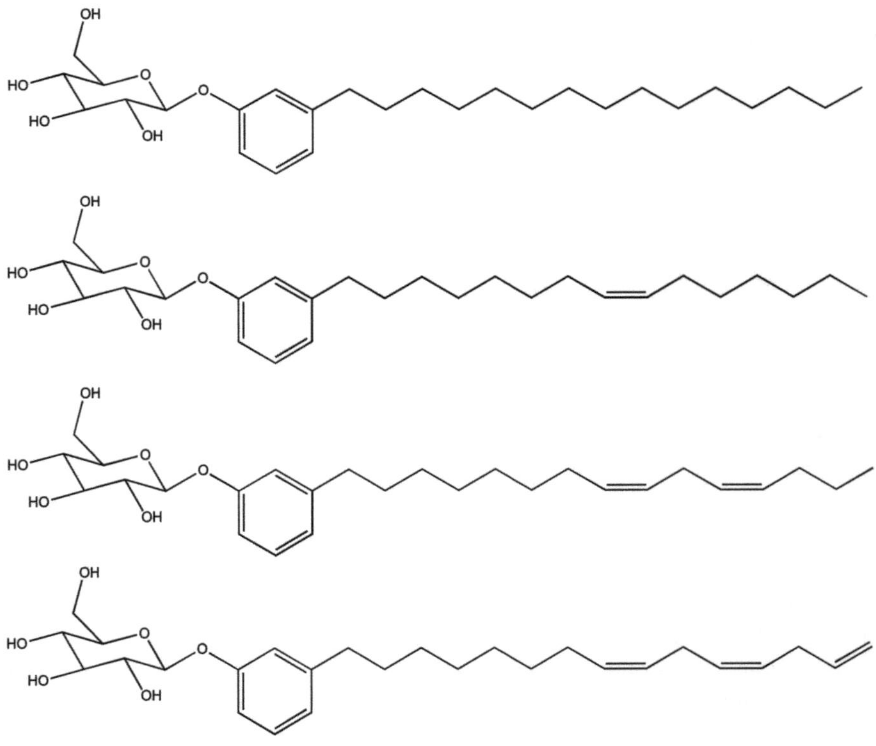

Fig. 7.10 Chemical structure of cardanol-based amphiphiles [31]

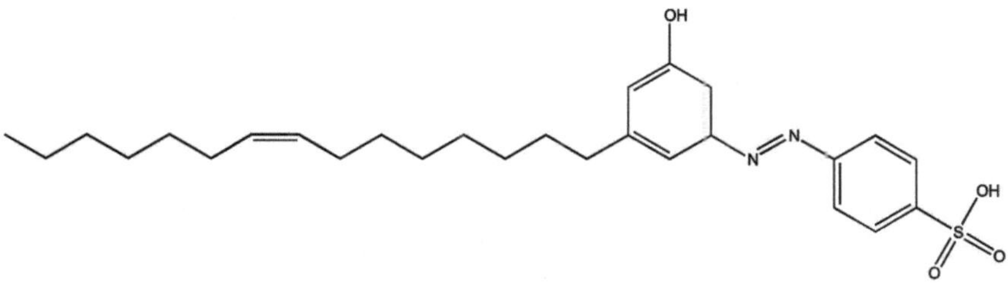

4-(3-pentadecylphenoxy) butane-1-sulfonic acid

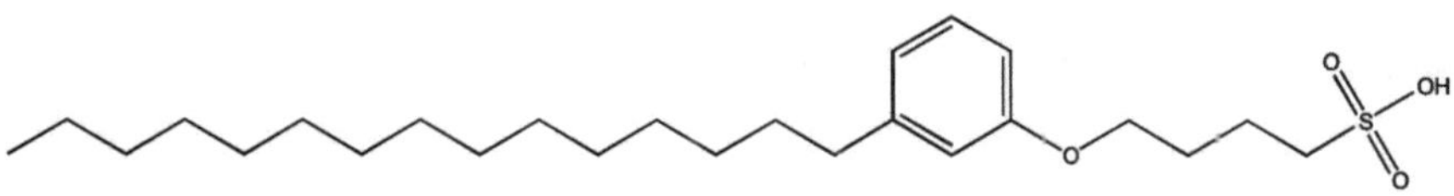

4-(3-pentadecylphenoxy) butane-1-sulfonic acid

Fig. 7.11 Azobenzene sulfonic acid derivatives used as a dopant for the preparation of polyaniline nanostructures [32, 35]

7.3.2 Sulfonic Acid Derivatives

Recently, cardanol has been used to develop water-soluble amphiphilic sulfonic acid derivatives that have been used as cylindrical, spherical, and vesicular templates for the self-assembly of various conducting polyaniline nanostructures (Fig. 7.11) [32]. Polyaniline nanomaterials have garnered considerable interest for their potential applications in electrical devices and sensors [33]. The amphiphilic sulfonic acid derivatives act as surfactants, forming stable emulsions with aniline, which allows for the formation of polyaniline nanostructures during polymerization. The sulfonic acid groups also act as a counter-anion dopant for the conductive polyaniline chains [34]. It has been observed that the carbon–carbon double bonds in the pendant chains on cardanol prevent the direct sulfonation of the phenyl ring [33]. However, the use of a diazotization coupling reaction to synthesize cardanol-based azobenzenesulfonic acid derivatives has been reported as an alternative path for sulfonation [33–35]. More recently, the use of cardanol-based butane sulfonic acid derivatives as a supramolecular nanotubular template for the synthesis of polyaniline nanomaterials has been reported [32].

7.3.3 Pyrene Derivatives

Pyrene pi-conjugated organogel systems that self-assemble due to pi–pi stacking have been synthesized by coupling cardanol-based coumarin derivatives with a

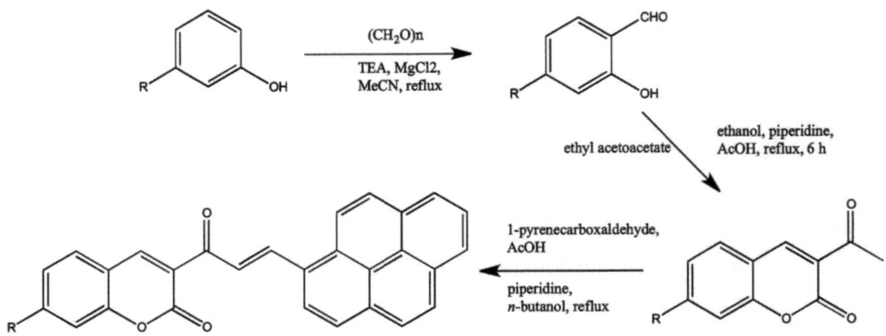

Fig. 7.12 Synthesis of multifunctional coumarin-coupled pyrene derivatives [36]

pyrene derivative. The cardanol-coumarin derivatives were prepared in a two-step process by first making a cardanol-based aldehyde via an electrophilic aromatic substitution of the phenol, followed by a Knoevenagel reaction with ethyl acetoacetate (Fig. 7.12). Coupling of that substrate to 1-pyrenecarboxaldehyde was then performed using aldol condensation [36].

7.4 Applications of Cardanol-Based Supramolecular Gels and Related Self-Assembled Structures

As mentioned previously in this chapter, cardanol-based self-assembled structures, including supramolecular gels, are gaining considerable attention in advanced applications due to their responsiveness to different stimuli, including temperature, light, and pH [1]. Triggered, fundamentally, by its intermolecular interactions, the self-assembly propensity of cardanol and any of its direct derivatives lead to a variety of nanosized structures. The design of nanostructured materials from cardanol and their corresponding proposed uses will be discussed in the final section of this chapter.

7.4.1 Biomedical Materials

The use of cardanol for the preparation of biomedical materials has been proposed for more than a decade now. Various specific applications have been proposed, with materials ranging from surface coatings to drugs. The variety of chemistries proposed with cardanol are directly related to its chemical structure, which exhibits a phenol group and a long unsaturated aliphatic chain (Fig. 7.9), allowing for various possible chemical transformations.

The antibacterial properties of cardanol against *S. aureus* have been previously studied and attributed to its phenolic and catechol units.[29] It has been shown that bio-based films made from cardanol inhibit the growth of *E. coli* and *S. aureus*, making it a suitable precursor for the preparation of materials for use in medical and healthcare applications.[29]

In an innovative approach, antifungal hydrogels have been prepared from the self-assembly of amphiphilic azo-modified/functionalized cardanol and proposed as potential antifungal drugs against *Candida albicans* [37]. After noticing a markedly inhibition of various types of cancerous human cell proliferation in the presence of propolis extracts containing cardanol, it has been suggested that cardanol and cardol may be potentially used to aid in cancer treatments [38]. A similar inhibition in prostate cancer cell (PC3) proliferation has been observed in the presence of self-assembled nanostructures derived from cardanol [36]. Interestingly, these pyrene-based supramolecular structures are non-toxic to fibroblasts, resulting in potential applications in cell imaging and/or cancer treatments. [36].

Finally, cardanol has been investigated as a potential substitute for cholesterol in the preparation of liposomes for drug delivery [39]. The presence of cardanol imparts stability to the liposome's membrane, allowing for the preparation of liposomes with a greater volume than previously obtained upon use of a non-toxic additive [39]. The use of cardanol in drug delivery will be covered in more detail in a separate chapter.

7.4.2 Nanostructured Materials

Due to its multifunctional characteristic, cardanol has been used for the preparation of a great diversity of self-assembled chemical architectures [27]. A combination of modified cardanol and cardanol-based polymers has been proposed for the preparation of nanotubes, nanofibers, supramolecular gels, and surfactants based on non-covalent interactions [27]. As an example, the incorporation of sugar units and other highly polar moieties onto cardanol via its phenol group led to amphiphilic molecules capable of self-assembling into nanotubes, twisted/helical nanofibers, low molecular weight hydrogels/organogels, and liquid crystals with potential for use in drug delivery applications [14]. It has been shown that the presence of unsaturations along the carbon chain of the cardanol unit is the key factor determining the final architecture of the soft materials and that the nanotubes obtained correspond to a compact helical structure [40, 41]. Indeed, it was shown that the higher the monoene content in the supramolecular gel, the more compact the helices obtained,[33] as explained in the scheme of Fig. 7.13. With the idea of preparing easily degradable, non-covalent, self-assembled systems, solutions of cardanyl tauramide were prepared and its behavior was investigated under different stimuli [41].

In a more application-oriented study, cardanol-based liquid crystals (LCs) have been prepared in aqueous solutions by the coupling of cardanol with perylene bisimide [42]. The self-assembled system maintained LC behavior at room

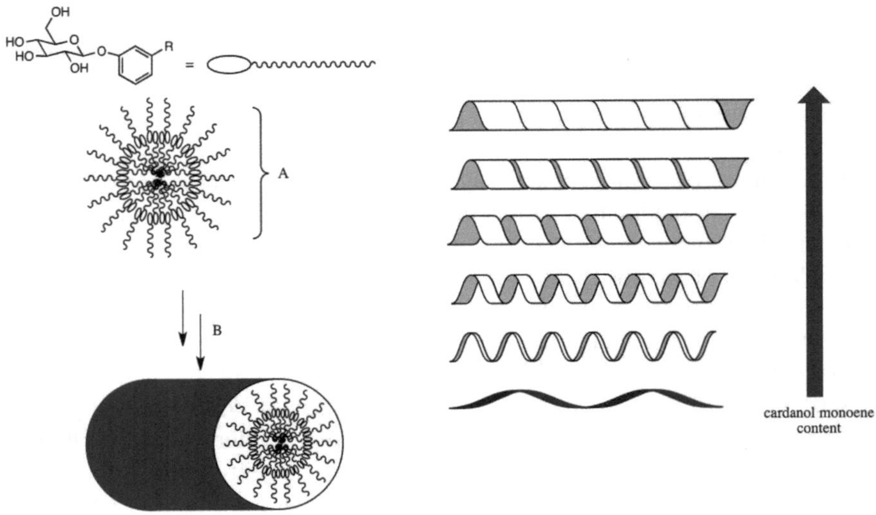

Fig. 7.13 Morphology of cardanol-based supramolecular gel as a function of cardanol monoene content. **a** Inverse micelle-like self-assembly in lipophilic solvent. **b** Stacking of structures leading to tubular morphology. *R* aliphatic chains

temperature [42]. Another application of cardanol-based self-assembled systems consists in the preparation of an amphiphilic diazo dye through the reaction of cardanol with an azobenzene sulfonic acid [43]. This system was used as a fluorescent tag and was employed in morphological and structural studies of polyanilines.

Cardanol self-assembled structures have also been used as a template for the preparation of well-defined, nanostructured polyaniline nanofibers.[19] Over the years, polyaniline nanofibers have collected wide interest as sensors, storage devices, and microelectronics.[19] Polyaniline nanotubes have been obtained by the preparation of polyaniline in the presence of a fluorescent cardanol-based surfactant and ammonium persulfate.[19] The architectures obtained with such system include bilayers or micelles that can, upon subsequent oxidation, lead to highly conductive polyaniline hollow spheres (1–2 mm), dendritic nanofibers, and linear nanofibers of 8–10 mm length with a diameter of 130–180 nm.[19] It has been proposed that the hydrophobic domains in the system's bilayer walls were able to accommodate aniline, which upon polymerization formed the nanostructures with the observed morphologies.[19]

7.5 Conclusions

The development of bio-based materials and technologies based on the use of bio-renewable resources represents a strategic approach to offset the economical impact related to the frequent oscillation in petroleum prices. Many technologies have been proposed to date based on the use of vegetable oils as building blocks for the preparation of bio-based polymers and composites. More recently, cardanol has shown potential for the preparation of supramolecular structures through various routes. In many instances, amphiphiles are prepared from the chemical modification of cardanol. The incorporation of carbohydrates, amines, sulfonate groups, and other hydrophilic units into cardanol confers it the ability to self-assemble in supramolecular architectures that depend on the chemical environment. Such structures are susceptible to changes upon specific stimuli, like pH, temperature, mechanical stress. Many applications for cardanol-based supramolecular gels have been proposed to date, including biomedical materials and templates for the preparation of nanostructured systems.

References

1. Sangeetha NM, Maitra U (2005) Supramolecular gels: functions and uses. Chem Soc Rev 34:821–836. doi:10.1039/B417081B
2. Steed JW (2011) Supramolecular gel chemistry: developments over the last decade. Chem Commun 47:1379–1383. doi:10.1039/C0CC03293J
3. Yang X, Zhang G, Zhang D (2012) Stimuli responsive gels based on low molecular weight gelators. J Mater Chem 22:38–50. doi:10.1039/C1JM13205A
4. Lloyd GO, Steed JW (2009) Anion-tuning of supramolecular gel properties. Nat Chem 1:437–442. doi:10.1038/nchem.283
5. Kumar DK, Steed JW (2014) Supramolecular gel phase crystallization: orthogonal self-assembly under non-equilibrium conditions. Chem Soc Rev 43:2080–2088. doi:10.1039/C3CS60224A
6. Dastidar P (2008) Supramolecular gelling agents: Can they be designed? Chem Soc Rev 37:2699. doi:10.1039/b807346e
7. Steed JW (2010) Anion-tuned supramolecular gels: a natural evolution from urea supramolecular chemistry. Chem Soc Rev 39:3686. doi:10.1039/b926219a
8. Segarra-Maset MD, Nebot VJ, Miravet JF, Escuder B (2013) Control of molecular gelation by chemical stimuli. Chem Soc Rev 42:7086–7098. doi:10.1039/C2CS35436E
9. Buerkle LE, Rowan SJ (2012) Supramolecular gels formed from multi-component low molecular weight species. Chem Soc Rev 41:6089. doi:10.1039/c2cs35106c
10. Piepenbrock M-OM, Lloyd GO, Clarke N, Steed JW (2010) Metal- and anion-binding supramolecular gels. Chem Rev 110:1960–2004. doi:10.1021/cr9003067
11. Smith DK (2010) Supramolecular gels: Building bridges. Nat Chem 2:162–163. doi:10.1038/nchem.566
12. Yu G, Yan X, Han C, Huang F (2013) Characterization of supramolecular gels. Chem Soc Rev 42:6697–6722. doi:10.1039/C3CS60080G
13. Abdallah DJ, Weiss RG (2000) Organogels and low molecular mass organic gelators. Adv Mater 12:1237–1247. doi:10.1002/1521-4095(200009)12:17<1237:AID-ADMA1237>3.0.CO;2-B
14. John G, Vemula PK (2006) Design and development of soft nanomaterials from biobased amphiphiles. Soft Matter 2:909–914. doi:10.1039/B609422H
15. Weng W, Beck JB, Jamieson AM, Rowan SJ (2006) Understanding the mechanism of gelation and stimuli-responsive nature of a class of metallo-supramolecular gels. J Am Chem Soc 128:11663–11672. doi:10.1021/ja063408q
16. Beck JB, Rowan SJ (2003) Multistimuli, multiresponsive metallo-supramolecular polymers. J Am Chem Soc 125:13922–13923. doi:10.1021/ja038521k

17. Kotova O, Daly R, dos Santos CMG et al (2015) Cross-linking the fibers of supramolecular gels formed from a tripodal terpyridine derived ligand with d-block metal ions. Inorg Chem 54:7735–7741. doi:10.1021/acs.inorgchem.5b00626
18. Li X, Kuang Yi, Shi J et al (2011) Multifunctional, biocompatible supramolecular hydrogelators consist only of nucleobase, amino acid, and glycoside. J Am Chem Soc 133:17513–17518. doi:10.1021/ja208456k
19. Suzuki M, Hanabusa K (2009) L-Lysine-based low-molecular-weight gelators. Chem Soc Rev 38:967–975. doi:10.1039/B816192E
20. van Bommel KJC, van der Pol C, Muizebelt I et al (2004) Responsive cyclohexane-based low-molecular-weight hydrogelators with modular architecture. Angew Chem Int Ed 43:1663–1667. doi:10.1002/anie.200352396
21. Boekhoven J, Poolman JM, Maity C et al (2013) Catalytic control over supramolecular gel formation. Nat Chem 5:433–437. doi:10.1038/nchem.1617
22. Placin F, Desvergne J-P, Lassègues J-C (2001) Organogel electrolytes based on a low molecular weight gelator: 2,3-bis(n-decyloxy)anthracene. Chem Mater 13:117–121. doi:10.1021/cm001118h
23. Hanabusa K, Hiratsuka K, Kimura M, Shirai H (1999) Easy preparation and useful character of organogel electrolytes based on low molecular weight gelator. Chem Mater 11:649–655. doi:10.1021/cm980528r
24. Hirst AR, Coates IA, Boucheteau TR et al (2008) Low-molecular-weight gelators: elucidating the principles of gelation based on gelator solubility and a cooperative self-assembly model. J Am Chem Soc 130:9113–9121. doi:10.1021/ja801804c
25. Rajaganesh R, Gopal A, Mohan Das T, Ajayaghosh A (2012) Synthesis and properties of amphiphilic photoresponsive gelators for aromatic solvents. Org Lett 14:748–751. doi:10.1021/ol203294v
26. Vemula PK, John G (2008) Crops: a green approach toward self-assembled soft materials. Acc Chem Res 41:769–782. doi:10.1021/ar7002682
27. Balachandran VS, Jadhav SR, Vemula PK, John G (2012) Recent advances in cardanol chemistry in a nutshell: from a nut to nanomaterials. Chem Soc Rev 42:427–438. doi:10.1039/C2CS35344J
28. Voirin C, Caillol S, Sadavarte NV et al (2014) Functionalization of cardanol: towards biobased polymers and additives. Polym Chem 5:3142–3162. doi:10.1039/C3PY01194A
29. John G, Masuda M, Okada Y et al (2001) Nanotube formation from renewable resources via coiled nanofibers. Adv Mater 13:715–718
30. John G, Jung JH, Masuda M, Shimizu T (2004) Unsaturation effect on gelation behavior of aryl glycolipids. Langmuir 20:2060–2065. doi:10.1021/la030177h
31. John G, Minamikawa H, Masuda M, Shimizu T (2003) Liquid crystalline cardanyl β-D-glucopyranosides. Liq Cryst 30:747–749. doi:10.1080/0267829021000047516
32. Anilkumar P, Jayakannan M (2010) A novel supramolecular organogel nanotubular template approach for conducting nanomaterials. J Phys Chem B 114:728–736. doi:10.1021/jp909016r
33. Anilkumar P, Jayakannan M (2006) New renewable resource amphiphilic molecular design for size-controlled and highly ordered polyaniline nanofibers. Langmuir 22:5952–5957. doi:10.1021/la060173n
34. Antony MJ, Jayakannan M (2009) Self-assembled anionic micellar template for polypyrrole, polyaniline, and their random copolymer nanomaterials. J Polym Sci Part B Polym Phys 47:830–846. doi:10.1002/polb.21689
35. Anilkumar P, Jayakannan M (2007) Single-molecular-system-based selective micellar templates for polyaniline nanomaterials: control of shape, size, solid state ordering, and expanded chain to coillike conformation. Macromolecules 40:7311–7319. doi:10.1021/ma071292s
36. Lalitha K, Jenifer P, Prasad YS et al (2014) A self-assembled π-conjugated system as an antiproliferative agent in prostate cancer cells and a probe for intra-cellular imaging. RSC Adv 4:48433–48437. doi:10.1039/C4RA07710E

37. Mahata D, Mandal SM, Bharti R et al (2014) Self-assembled cardanol azo derivatives as antifungal agent with chitin-binding ability. Int J Biol Macromol 69:5–11. doi:10.1016/j.ijbiomac.2014.05.017
38. Teerasripreecha D, Phuwapraisirisan P, Puthong S et al (2012) In vitro antiproliferative/cytotoxic activity on cancer cell lines of a cardanol and a cardol enriched from Thai Apis mellifera propolis. BMC Complement Altern Med 12:27. doi:10.1186/1472-6882-12-27
39. De Maria P, Filippone P, Fontana A et al (2005) Cardanol as a replacement for cholesterol into the lipid bilayer of POPC liposomes. Colloids Surf B Biointerfaces 40:11–18. doi:10.1016/j.colsurfb.2004.09.007
40. John G, Jung JH, Minamikawa H et al (2002) Morphological control of helical solid bilayers in high-axial-ratio nanostructures through binary self-assembly. Chem Eur J 8:5494–5500. doi:10.1002/1521-3765(20021202)8:23<5494:AID-CHEM5494>3.0.CO;2-P
41. Silverman JR, Samateh M, John G (2015) Functional self-assembled lipidic systems derived from renewable resources. Eur J Lipid Sci Technol 118:47–55. doi:10.1002/ejlt.201500198
42. Bhavsar GA, Asha SK (2011) Pentadecyl phenol- and cardanol-functionalized fluorescent, room-temperature liquid-crystalline perylene bisimides: effect of pendant chain unsaturation on self-assembly. Chem Eur J 17:12646–12658. doi:10.1002/chem.201101011
43. Anilkumar P, Jayakannan M (2007) Fluorescent tagged probing agent and structure-directing amphiphilic molecular design for polyaniline nanomaterials via self-assembly process. J Phys Chem C 111:3591–3600. doi:10.1021/jp066428n

Chapter 8
Anacardic Acid and Cardanol: Prospective Applications for Cancer Therapy, Drug Delivery, and Imaging

Resmi Anand and Bindu P. Nair

Abstract Versatile derivatization feasible for cardanol to generate various amphiphilic structure-directing molecules bestows this naturally occurring phenol a great promise in the field of biomedical research. This chapter discusses the use of cardanol and cardanol derivatives for cancer therapy, drug delivery, and imaging applications. Cardanol has proven anti-proliferative effects on cancer cells, which could help its derivatives to evolve as therapeutic alternatives for anticancer treatment. At the same time, the structure-directing capacity of cardanol derivatives has been successfully used to obtain various gels and nanoparticles for drug delivery and imaging applications. These findings are summarized in three sections namely cardanol as an anticancer drug, cardanol gels, and cardanol-based nanomorphologies for drug delivery applications. The chapter also discusses prospective biomedical applications of cardanol.

Keywords Anticancer drug · Drug delivery · Cytotoxicity · Green drug carrier · Cardanol · Anacardic acid

8.1 Introduction

Plant-derived traditional medicines have been practiced for thousands of years and are generally considered safe for its low toxicological risk. Among these, plant-derived chemotherapeutics are of special interests due to the side effects

---

R. Anand
Biomedical Technology Wing, Sree Chitra Tirunal Institute for Medical Sciences and Technology, Thiruvananthapuram, Kerala 695012, India

B.P. Nair (✉)
Department of Chemistry, Mahatma Gandhi College, Thiruvananthapuram, Kerala 695004, India
e-mail: bindumelekkuttu@gmail.com

P. Anilkumar (ed.), *Cashew Nut Shell Liquid*,
DOI 10.1007/978-3-319-47455-7_8

that a synthetic chemotherapeutic exert on human body. Several naturally occurring components like curcumin, allicin, lycopene, capsaicin, limonene, and beta-carotene have been effectively used and researched for anticancer therapy [1]. More recently, anticancer properties of cashew nut shell liquid (CNSL) have been explored for therapeutic purposes. The phenolic components of CNSL, viz., anacardic acid and cardanol, are known to exhibit the anticancer properties mainly. However, plant-derived remedies may also contain constituents which exert undesired side effects, and therefore a detailed research is ongoing to establish the positive benefits of CNSL components as drugs for anticancer therapy.

Another application of the CNSL components like anacardic acid and cardanol is as valuable raw materials for synthesizing soft nanoarchitectures including lipid nanotubes, twisted/helical nanostructures, and low molecular weight gels. These nanoarchitectures were of high importance for drug delivery applications, which aimed at improving the administration of active pharmaceutical agent to achieve a therapeutic effect in humans or animals. The use of an inexpensive and renewable resource like CNSL for developing new materials to gain increased bioavailability of drugs could reduce treatment cost as well as environmental concerns. CNSL derivatives were also tailored to synthesize smart or stimuli-responsive polymers which have the ability to change its physiological functions in response to its surroundings, for instance, changes in pH, temperature, chemical environment, electric field, and magnetic field [2, 3]. The reason for the stimuli-responsive behavior of such natural smart systems could be attributed to the changes at the molecular level and also the surface functionalities. Among these, the stimuli-responsive molecular gels based on CNSL have attained considerable attention as a promising material for nanofabrication using bottom–up approaches and are expected to find applications in the areas such as nanomedicine, molecular sensors, and therapeutics [3].

The present chapter consists of three sections: (1) CNSL as an anticancer drug, (2) CNSL-based nanomorphologies for drug delivery applications, and (3) cardanol- and anacardic acid-based gels.

8.2 CNSL as an Anticancer Drug

The components of CNSL like anacardic acid, cardanol, and cardol have proven cytotoxic and anti-proliferative effects on cancer cells. A few but noteworthy reports are available on the topic. These studies were intended either to elucidate the mechanism of biological action of CNSL components in exerting the anti-proliferative effect on cancer cells or to find out the methods for its localized delivery for therapeutic purposes.

A pertinent study on the topic was that by Neto et al. [4], which explored the cytotoxicity of CNSL components and its derivatives on oral squamous cell carcinoma. The cells were treated with the parental and derived compounds at 25 ng/μL for 24, 48, and 72 h. Staurosporine (300 nM) and ethanol (diluents of compounds) were used as positive and negative controls, respectively. The results of this study undoubtedly supported the cytotoxicity of CNSL and its possible use as a chemotherapeutic.

Complimentary to the findings of the above report, Teerasripreecha et al. [5] reported the antiproliferative activity of cardanol and cardol from *Thai Apis mellifera propolis,* a complex resinous honeybee product, on cancer cell lines. Even though the source of cardanol and cardol used in the study was not CNSL, it is worthwhile to mention that the observations obtained can be applicable to cardanol from CNSL as well. Cardanol and cardol were isolated from the crude resin extract, and both exhibited broadly similar in vitro antiproliferation/cytotoxicity half-maximal inhibitory concentration (IC50) values across five cancer cell lines used in the study, derived from duet carcinoma, undifferentiated lung, liver hepatoblastoma, gastric carcinoma, and colon adenocarcinoma. The IC50 values varied from 10.8 to 29.3 μg/ml for the cardanol and <3.13–5.97 μg/ml for the cardol. Figure 8.1 shows that the human colon adenocarcinoma cells after 72 h of cell culture appeared normal, while DNA condensation within the nucleus was visible in both the cells treated with the cardanol and cardol. Morphological changes and cell debris were highlighted by a red arrow, and when compared to the control, a reduced cell density was also evident. When compared to its antiproliferation response on the control non-transformed human foreskin fibroblast cell line, both cardanol and cardol induced cytotoxicity and cell death without DNA fragmentation in the cancer cells. Interestingly, cardol was projected as a promising agent for anticancer therapy when compared to cardanol, in terms of its lower IC50 values. The study also pointed out the need for further investigation to allow safe systemic administration of cardanol and cardol, and even the need for targeted delivery was emphasized.

Kubo et al. [6] studied in detail the cytotoxicity of anacardic acid and their derivatives on murine B16-F10 melanoma cells. The exploration of anacardic acid and its derivatives for the cure of melanoma was due to the poor response rate of melanoma toward chemotherapy among all the neoplasms. Anacardic acids from *A. occidentale* apple and their derivatives, cardols, cardanol, and the *E*-isomer of anacardic acid were used for cytotoxicity evaluation. The study evaluated the tyrosinase-inhibiting activity of anacardic acid and its derivatives because of the key role played by tyrosinase in the melanin biosynthetic pathway.

Figure 8.2 shows the structure of the anacardic acids with one, two, and three double bonds, designated respectively as $C_{15:1}$, $C_{15:2}$, and $C_{15:3}$, used for cytotoxicity evaluation. The IC50 values of these anacardic acids varied in the following order: $(C_{15:1}) > (C_{15:2}) > (C_{15:3})$. It was observed that unsaturation has no role in inducing the cytotoxicity. Anacardic acids were found to enhance the melanin production in cells, despite exhibiting tyrosinase-inhibiting activity. A dose-dependent suppression of melanin production was also observed, and the total melanin content was always less than that of the control, as the number of viable cells was low. The most important finding of the study was that an appropriate balance between hydrophilic and hydrophobic moieties in the derivatives of anacardic acid and cardanol was essential to exhibit cytotoxicity by these

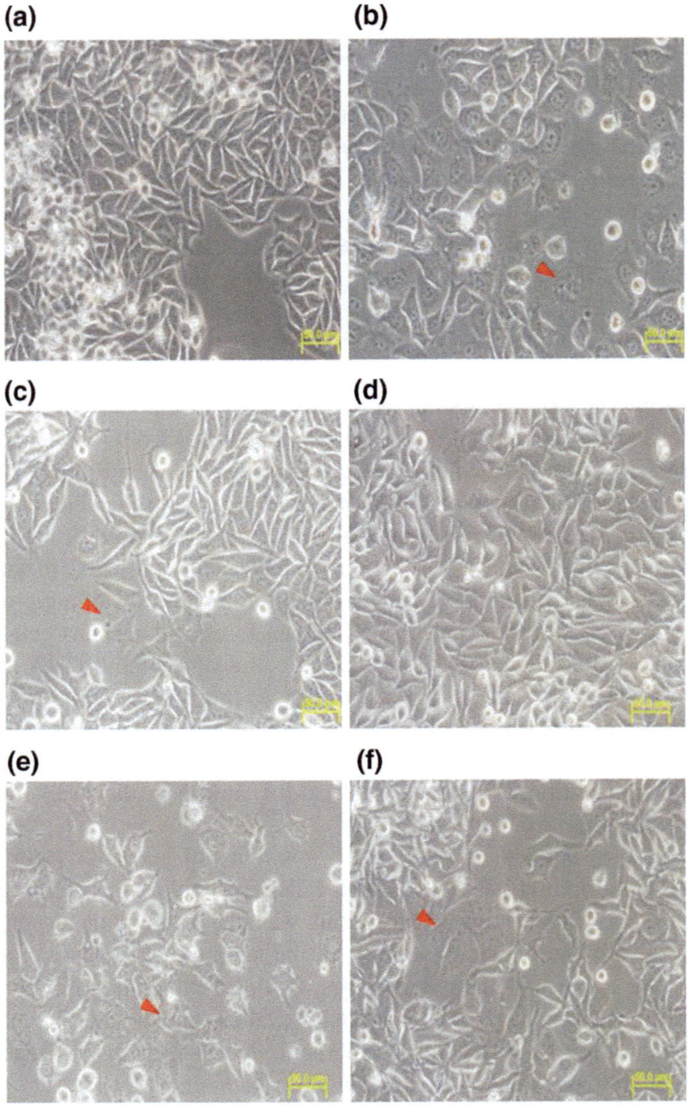

Fig. 8.1 Human colon adenocarcinoma cells after **a–c** 72 h and **d–f** 96 h of culture in **a**, **d** the DMSO alone (control) or **b**, **e** with cardanol at its IC50 value (10.76 μg/ml), or **c**, **f** cardol at its IC50 value (3.0 μg/ml; 6.54 μM). *Scale bar* represents 50 μm (Reproduced from ref. [5])

compounds. It was proposed that the hydrophilic head portion of anacardic acid binds with the cell membrane surface, and the hydrophobic tail portion disrupts the native membrane-associated function of the integral proteins non-specifically. It was also suggested that optimization of activity of anacardic acids and cardanol is attainable through the synthetic approach, where the alkyl side chain

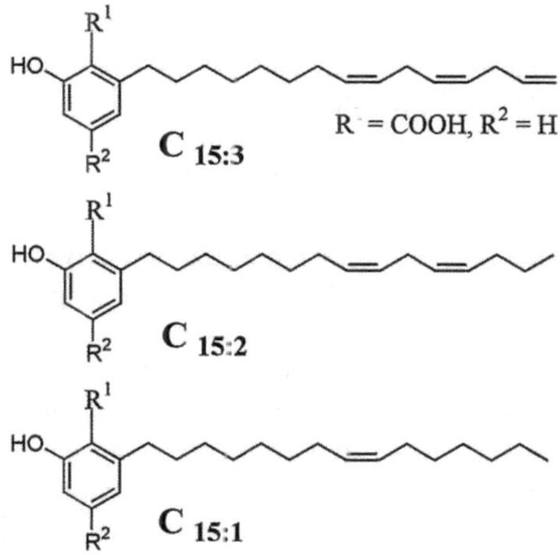

Fig. 8.2 Derivatives of anacardic acid $C_{15:1}$, $C_{15:2}$, and $C_{15:3}$

plays a direct role. As concluded by the study, additional work is necessary to clarify the exact role of alkyl side chain in inducing cytotoxicity by anacardic acid and related compounds [6].

A more detailed investigation on the pathway by which CNSL components were exerting the cytotoxic effects on cancer cells was highly essential before extending the study toward in vivo and clinical evaluation. Sung et al. proved that anacardic acid could suppress the expression of nuclear factor-κB-regulated gene products involved in cell survival, proliferation, inflammation, and invasion through inhibition of the inhibitory subunit of nuclear factor-κB kinase, which in turn lead to apoptosis. The study further demonstrated the potential of anacardic acid in inhibiting the expression of vascular endothelial growth factor (VEGF), which could make it equivalent to bevacizumab (Avastin), the approved antibody against VEGF for the treatment of cancer and macular degeneration. The inhibition of NF-κB activation by anacardic acid was explained as the reason for its antiangiogenic, antiproliferative, antimetastatic, proapoptotic, and anti-inflammatory effects. The study has given a real insight into one of the mechanisms by which anacardic acid exerts its antiproliferative effect on cancer cells. It is important to mention that further studies are needed to explore therapeutic potential of anacardic acid against cancer and other diseases [7].

Another important mechanism by which CNSL components can inhibit tumor growth is by the inhibition of histone acetyltransferases (HAT) as reported by Eliseeva et al. HAT and histone deacetylases (HDAC) are involved in chromatin structure modifications through acetylation and deacetylation of proteins, respectively. They are playing a role in important signaling pathways, and their loss of function or gain of function can lead to severe consequences for the cell and the entire organism, including malignant transformation. The study has explored anacardic acid and its derivatives as potent HAT inhibitors for anticancer therapy [8].

To increase biological activity, 28 structural analogues were derived from anacardic acid by introducing a second aromatic ring, changing the position of the alkyl chain and its length, and making various substitutions at the second aromatic ring. These compounds were investigated for their HAT inhibitory properties using human cervical cancer cell line (HeLa) nuclear extract as a source of multiple HAT activities. HAT inhibitory potency was found to vary for each of the compounds, and none of the derivatives was more efficient than anacardic acid in inhibiting total HAT activity. The single most efficient modification was the conversion of the ethyl esters to the corresponding carboxylic acids, and increased HAT inhibitory potency up to 10-fold was obtained.

The cytotoxicity of these compounds was determined using human breast cancer cell line (MCF7), prostate cancer cell line (PC3), HeLa and human T lymphocyte cell lines (Jurkat). The cytotoxicity of the derivatives was similar to or slightly greater than that of anacardic acid, and none of them were toxic to normal human cells. A clear correlation between HAT inhibitory potency of the compounds and their effects on cell proliferation was also observed. A drawback of the newly synthesized compounds in comparison with anacardic acid was suggested as the increased hydrophobicity, which could affect its cell penetration capacity. Significance of the study was that the observed wide variability of HAT inhibitory efficacy in structurally closely related compounds could facilitate the design of more potent HAT inhibitors for future studies [8].

Balasubramanyam et al. [9] evaluated anacardic acid as a potent inhibitor of HAT p300. As mentioned earlier, HAT enzymes are potential targets for antineoplastic therapy recently. The IC50 values of anacardic acid for p300 HAT and p300/CBP-associated cofactor (PCAF) HATs were 8.5 and 5 μm, respectively. An anacardic concentration of 15 μm was found to inhibit HAT activity by 90 %. The amide derivatives of anacardic acid were synthesized using substituted anilides, which showed an enhancement of p300 HAT activity. It was concluded that anacardic acid was efficient for the broad inhibition of HAT activity, without specificity. However, anacardic acid could be a potential precursor for the synthesis of novel HAT activity modulators.

Chandregowda et al. tested the cytotoxicity of several benzamide derivatives of anacardic acid using HeLa cell line with relatively high expression of HAT p300 and colorectal adenocarcinoma HCT-15, which is p300 negative. Among the compounds studied, 2-isopropoxy-6-pentadecyl-*N*-pyridin-4-ylbenzamide, 2-ethoxy-N-(3-nitrophenyl)-6-pentadecylbenzamide, and 2-ethoxy-6-pentadecyl-*N*-pyridin-4-ylbenzamide were found to be the effective cytotoxic compounds with the IC50 values of 11.02, 13.55, 15.29 mM, respectively. Interestingly, the cytotoxicity activities exhibited by these compounds were comparable with the naturally occurring and cell-permeable HAT inhibitor garcinol [10].

Sun et al. explored the sensitization of tumor cells to ionizing radiation by inhibition of HAT activity using anacardic acid. The study aimed at the inhibition of HAT Tip60, which could prevent effective repair of DNA damage and sensitize the cells to genotoxic agents such as ionizing radiation. It was verified that anacardic acid inhibits the Tip60 HAT in vitro with an IC50 of 9 μM and blocks the

Tip60-dependent activation of the ATM and DNA–PKcs protein kinases by DNA damage in vivo. Cell-specific sensitization by anacardic acid toward radiation was also observed in the study. For example, HeLa cells were found to be sensitized threefold in the presence of anacardic acid, whereas human squamous carcinoma cells showed only a twofold increase. The study demonstrated that anacardic acid could inhibit HAT, which in turn sensitize a tumor toward ionizing radiation. This observation can lead to new therapeutic inventions for anticancer treatment [11].

Sukumari-Ramesh et al. [12] evaluated the effect of anacardic acid as a radiosensitizer in reducing the growth and proliferation of pituitary adenoma cells. Although pituitary adenomas are manageable with the existing modalities, the study is highly relevant in that there are many limitations and exceptions, where the current therapeutics are not effective. It was hypothesized that the use of a novel plant-derived radiosensitizer could allow more destructive targeting using radiation and avoid collateral toxicity. This radiosensitization of the cells was also achieved through HAT-inhibiting capacity of anacardic acid.

The pituitary adenoma cells rat GH3 and mouse MMQ were used for the concentration- and time-dependent cytotoxicity evaluations. It was reported that anacardic acid significantly decreased the viability of GH3 and MMQ cells, which was dependent on both concentration and time. A non-classical apoptotic mechanism was doubted for the observed cytotoxicity. It was also noticed that anacardic acid also reduced the expression of survivin and X-linked inhibitor of apoptosis protein. These antiapoptotic proteins are associated with cellular survival and radioresistance, which in turn radiosensitized pituitary adenoma cells. As concluded by the authors, these findings definitely warrant further evaluation of anacardic acid as a single agent or as an adjunct to radiation therapy for the treatment of pituitary adenomas [12].

As an advanced study in the exploration of cardanol as an anticancer drug, Massaro et al. evaluated a method to release cardanol, at the required site by using halloysite nanotubes. The design strategy was aimed to overcome the low solubility of cardanol in physiological media, which could limit its application as an anticancer drug. Halloysite is a natural aluminosilicate clay with a hollow tubular structure, useful for loading and releasing biomacromolecules and drugs. Cardanol was encapsulated into halloysite nanotubes by mixing with both pure and triazolium salt-modified halloysite. Figure 8.3 shows the scanning electron micrographs of pure halloysite–cardanol and triazolium salt-modified halloysite–cardanol complexes. The study confirmed supramolecular interaction and complex formation between pristine or modified clay and cardanol. Extended release profile of cardanol from triazolium functionalized cardanol, in three different pH solutions, was elucidated as pH is a key factor influencing the oral drug delivery. Of the three studied pHs, i.e., 1, 7.4, and 6, maximum release was observed at a pH of 1. Because cardanol is only adsorbed on HNT outer surface, the overall release process was explained as due to only a single desorption step. Cytotoxicity assay has proven that both pure cardanol and halloysite nanotubes were noncytotoxic, whereas triazole possibly had some biological activity, which affected cell viability. The low toxicity of free cardanol was ascertained as due to the low

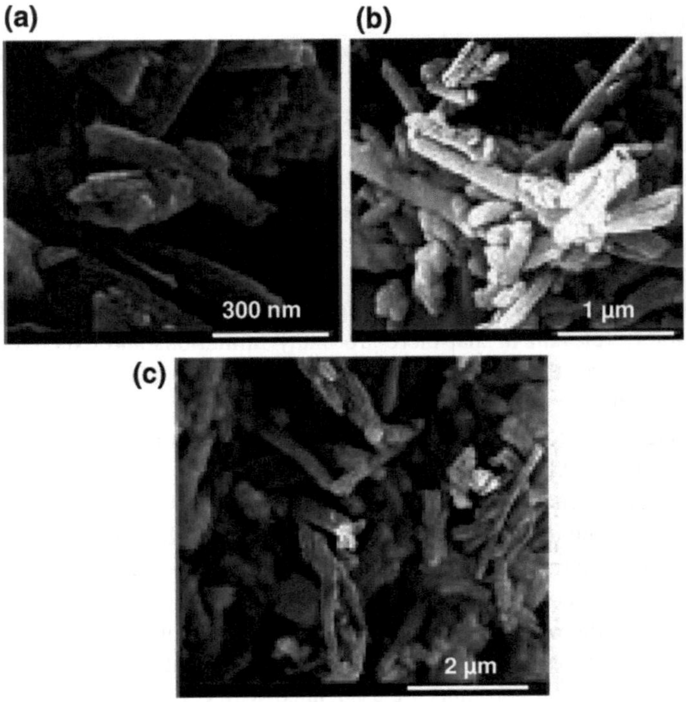

Fig. 8.3 SEM images of **a**, **b** functionalized halloysite–cardanol complex and **c** pure halloysite–cardanol complex (Reproduced from ref. [13] 2015, Elsevier)

solubility in physiological medium, which limits its clinical application. However, this observation is contradictory to many other reports, where it has proven that cardanol exhibits significant anti-proliferative activity [13].

Another interesting study was that by Al-Hazzani et al., which explored a nanoemulsion based on cashew nut shell liquid, as a potent inhibitor of human breast cancer cell proliferation. The dual aims of the study were to evaluate the potential of nanoemulsions for increasing the solubility of drugs and the in vitro cytotoxicity against human breast cancer cells. CNSL nanoemulsions were prepared using Polysorbate 80, nonionic hydrophilic surfactant as an emulsifier. Nanoemulsion of fine CSNL with cell culture medium dispersions, having droplets covering the size range of 400–1000 nm, was prepared by spontaneous emulsification mechanism. Human breast cancer cells treated with the CSNL nanoemulsion showed loss of viability and death through apoptosis and necrosis. The IC50 value for 48-h treatment groups was found in the range of 88 ± 14.2 µl/ml. Even though it was mentioned that the exact mechanism of action of CNSL on human breast cancer cells was not known, an apoptotic pathway was suggested for the observed cytotoxicity. However, the study has failed in providing adequate characterization of the nanoemulsion [14].

Contradictory to the observations that the CNSL possesses anticancer properties, reports are also there, on it as a tumor-promoting agent. Banergee et al. studied CNSL for its potency in promoting the 7,12-Dimethylbenz[a]anthracene (DMBA)-initiated cells into papillomas, in a murine two-stage skin tumorigenesis model system. It was observed that DMBA or CNSL, when applied alone, failed to induce any tumor growth, whereas a dose of DMBA with 2 % CNSL definitely exhibits a weak tumor-promoting effect [15].

The above reports clearly proved the anti-proliferation capacity of CNSL components on cancer cells. The results indicate it as a promising chemotherapeutic for anticancer therapy. Various mechanisms are suggested for the observed cytotoxicity, the HAT inhibition mechanism being the most prominent. Efforts were also initiated to evaluate the methods for the localized delivery of cardanol.

8.3 CNSL-Based Nanomorphologies for Drug Delivery Applications

In the last few decades, the study of low-cost, highly efficient, and innovative drug delivery devices has attracted many researchers around the world. The therapeutic activity of certain existing drugs is limited due to several reasons such as poor water solubility, lack of controlled drug release, poor targeting ability, low drug stability, and inability of the drug to cross the lipophilic cell membranes. These demands led to a continuous research on different drug carrying vehicles and mechanisms to carry the therapeutic agent into a specific body part or cell compartments to repair or detect the damages, called drug delivery systems.

The carrier-based drug delivery technique is the most widely studied approach due to the following reasons:

1. This method improves the bioavailability of drugs
2. The drug carrier provides necessary protection to the drug molecule
3. The encapsulation/absorption of the drug inside the carrier could reduce undesired side effects
4. The freedom of selection of appropriate carrier material according to the required drug release profile and individual medical needs.

Some of the most studied carrier systems are polymers, micelles, liposomes, hydrogels, dendrimers, cyclodextrins, and nanoparticles.

CNSL-based nanomorphologies are receiving great attention as potential drug delivery vehicles due to its unique structural advantages. The reactive phenol group and unsaturated hydrophobic alkyl side chain at the meta-position direct cardanol/anacardic acid to form amphiphilic building block with supramolecular architecture, for instance formation of cardanol/anacardic acid micellar structures. The hydrophobic drugs can be encapsulated in the hydrophobic core of micelles and transported to the required zone. Often the encapsulated drug concentrations are above the intrinsic water solubility of the drug. Moreover, the hydrophilic head of the micelle can form hydrogen bonds with the surrounding aqueous environment and form a protective shell around the core of the micelle. As a result, the encapsulated drugs at the hydrophobic core are effectively protected from degradation and harmful side effects. In addition, functionalization of cardanol/anacardic acid provides the freedom to serve as templates for the synthesis of novel nanomaterials having controlled molecular weight, size, and morphology. Further, structural modification of CNSL with specific targeting groups or pharmacologically active ligands is a very promising strategy to a broader range of sites of activity with a much higher selectivity.

A completely "green", nanomicellar system containing high percentage of cardanol was prepared by combining cardanol and cholesterol. Under alkaline condition, a mixture of cardanol and cholesterol formed a vesicle-shaped nanodispersion (Fig. 8.4). A lipophilic porphyrin functionalized with cardanol (porphyrin–cardanol nanohybrid) was encapsulated into this vesicular nanodispersion [16]. The hydrophobic layer of this vesicle acted as a host for the lipophilic porphyrin–cardanol nanohybrid. Here, cardanol acted as a solvent for the nanohybrid and cholesterol. The stability and permeability of the phospholipid bilayer of 1-palmitoyl-2-oleylphosphatydilcholine liposomes was demonstrated using self-assembled cardanol as an easily available non-toxic replacement for cholesterol in drug targeting [17].

In another study by the same research group, a natural phenolic compound derived from rowanberries called chlorogenic acid was encapsulated into the cardanol–cholesterol vesicle-shaped nanocarrier in aqueous environment and studied the encapsulation efficiency of the prepared cardanol–cholesterol green carrier [18].

Mahata et al. prepared a soft "green" drug carrier by coupling diazonium at the ortho- and para-positions of cardanol and studied the delivery of an antifungal agent, amphotericin B, into aqueous medium. In this study, an effort was made to overcome the emergence of biofilm-related multi-drug-resistant (MDR) pathogens by using the principles of combination therapy and sustained drug release. The self-assembled diazonium–cardanol enhanced antifungal activity with chitin binding ability. Among the two drug-loaded carriers, only the para-diazonium-coupled

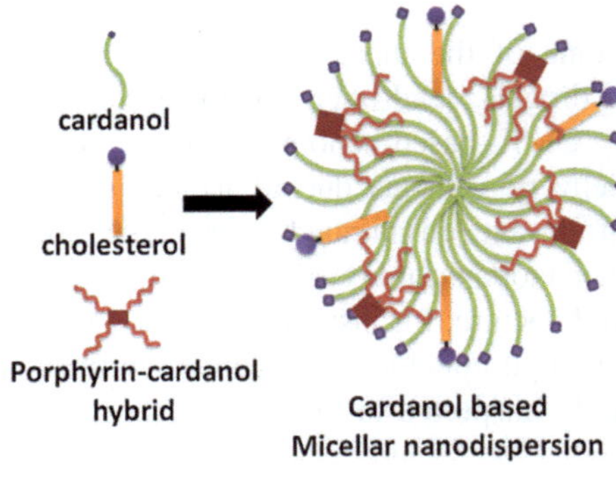

Fig. 8.4 Schematic representation of cardanol-based micellar nanodispersion (Reproduced from ref. [16])

cardanol showed a spherical morphology and the host–guest complexes (para coupled carrier: amphotericin B) were aggregated by π–π stacking interaction between polyene of amphotericin B and the unsaturated chain of para-diazonium-coupled cardanol. The system played a dual role as amphotericin B strongly bound to ergosterol as well as para-diazonium-coupled cardanol bound to chitin and enhanced the antifungal activity [19].

One of the most conventional methods of drug administration is the oral route. The oral delivery of drugs such as insulin will be a boon to diabetic patients. Sharma et al. [20] synthesized self-aggregated nanoparticles by modifying chitosan with anacardic acid, and studied its potential for oral insulin delivery. The amino group of chitosan was reacted with acetoxy derivative of anacardic acid using acid chloride chemistry to get region-specific N-amidation. Chitosan modified with anacardic acid spontaneously formed nanoparticles in aqueous insulin solution at neutral pH. The hydrophobic nature of the nanoparticles resulted in a sustained release of insulin in the intestinal environment, and the released insulin was stable with no conformational change.

In a complimentary study, Sharma et al. prepared different molecular weight anacardoylated chitosan, which can spontaneously form cytocompatible nanoparticles in aqueous protein solutions. These self-aggregated nanoparticles were further encapsulated in calcium alginate for preparing pH-responsive nanoparticles. The authors observed a sustained release profile for the protein-based drug without any conformational change. The anacardoylated chitosan nanoparticles encapsulated in alginate capsules could be an excellent carrier system for delivering protein or peptide-based drugs to the intestine [21].

8.4 Cardanol and Anacardic Acid-Based Gels

Molecular gels are a class of physical gels, formed by the self-assembly of low molecular weight compounds and are held together by non-covalent interactions such as π interaction, dipolar interaction, hydrogen bonding, coulomb interactions, van der Waals interactions, solvophobic interactions, and other supramolecular weak forces. The molecular gels formed by the self-assembly of natural systems have great interest due to their excellent biocompatibility. In addition, the hydrogel derived from small organic molecules possesses dynamic reversible property, which leads to the sol-to-gel or gel-to-sol phase transition upon a trigger by external stimuli such as light, temperature, pH, electric field, magnetic field, mechanical force, and chemical stimuli. John and Vemula developed a series of lipid amphiphiles from cardanol, having four meta-alkyl phenols, with alkyl chains differing in the degree of unsaturation. To generate such glycolipid structures, they have attached a hydrophilic group (sugar unit) on the phenolic head of cardanol (Fig. 8.5). Subsequently by utilizing the principles of self-assembly and green chemistry, a variety of soft nanomaterials such as lipid nanotubes, helical or twisted nanofibres, low molecular weight hydrogels/organogels, and liquid

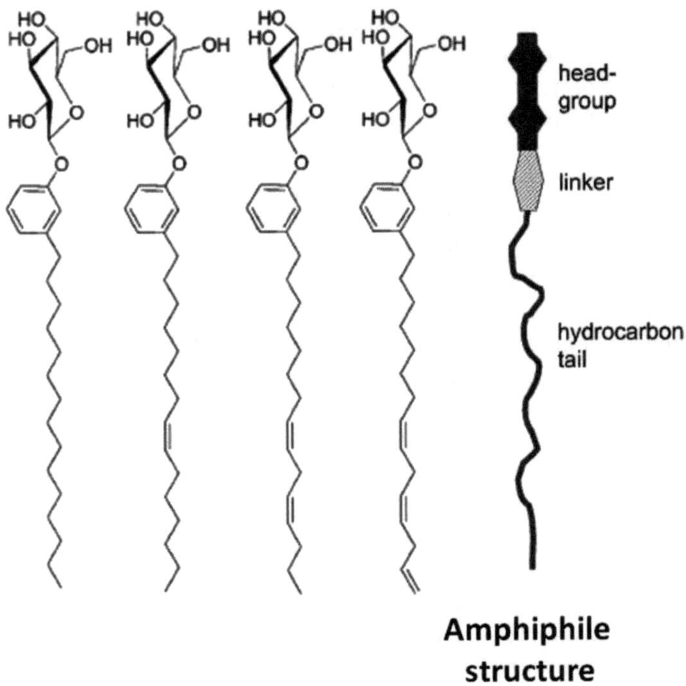

Fig. 8.5 Chemical structure of cardanol-based amphiphiles and general cartoon representation of the structure of an amphiphile (Reproduced from ref. [2] 2006, Royal Society of Chemistry)

crystals were prepared. In vitro and in vivo experiments were lacking in this study. However, the authors envisaged that these types of soft materials would find applications in biomaterials and biosensors [2].

Lalitha et al. [22] synthesized a novel multifunctional cardanol containing π-conjugated fluorescent-labeled pyrene derivative that self-assembled into supramolecular soft materials through hydrogen bonding interactions and π–π stacking interactions of pyrene units. For the synthesis, cardanol aldehyde was first prepared by electrophilic aromatic substitution on phenol. The Knoevenagel reaction of the asynthesized aldehyde with ethyl acetoacetate under ambient reaction condition resulted in the formation of 3-acetylcoumarins. Then, the organo π-gelators were synthesized by aldol condensation reaction between 3-acetyl coumarins with 1-pyrenecarboxaldehyde (Fig. 8.6).

At higher concentration, the asynthesized multifunctional coumarin-coupled pyrene derivative formed a gel, whereas at lower concentration it formed self-assembled nanostructures (Fig. 8.7). Due to the fluorescent tagging, the nanomaterial obtained at lower concentration could find applications in fibroblast and PC3 prostate cancer cell imaging (Fig. 8.8). Interestingly, this nanomaterial exhibited pharmacological activity by suppressing the proliferation of PC3 cells, which is hypothesized as the inhibition of Wnt/beta-catenin signaling [22, 23].

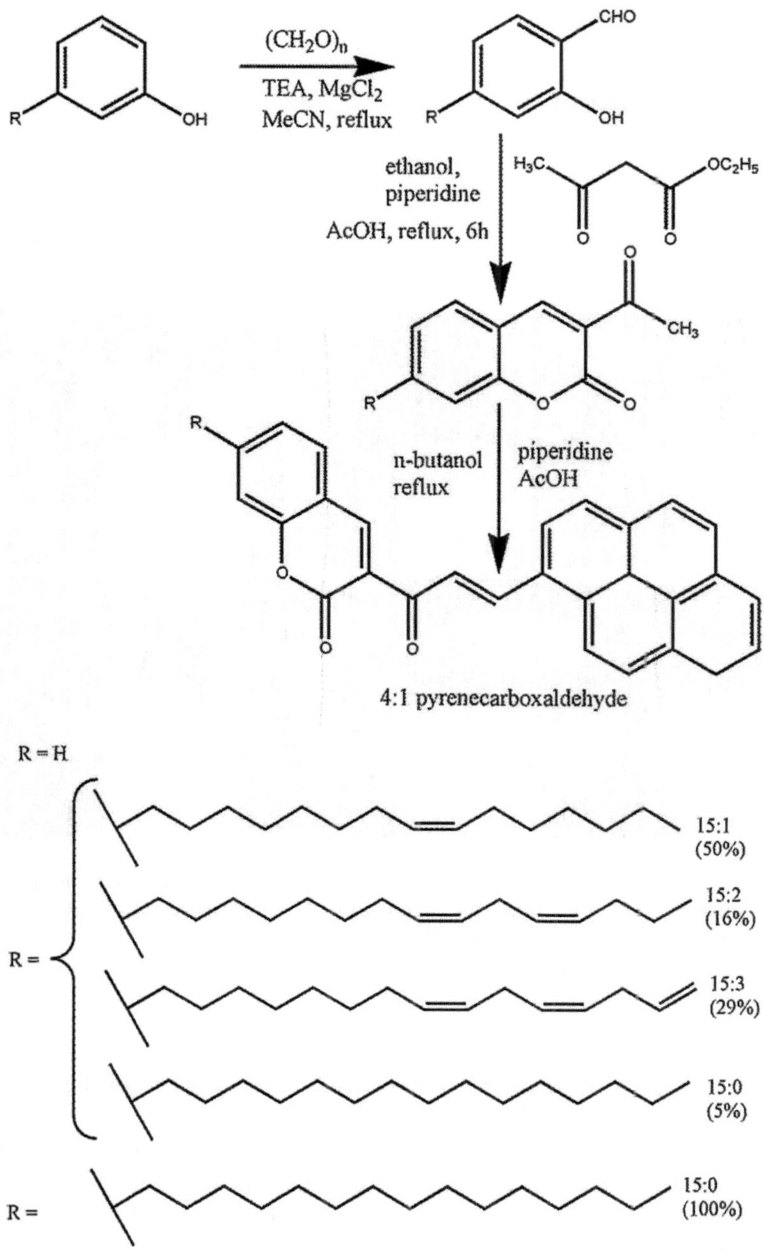

Fig. 8.6 Synthesis of multifunctional pyrene-coupled coumarin-based derivative (Reproduced from [22])

These types of fluorescent supramolecular self-assembled soft nanostructures provide a promising platform for in vitro fluorescence imaging and disease therapeutics. In an alternative study, Lalitha et al. [3] synthesized a stimuli-responsive

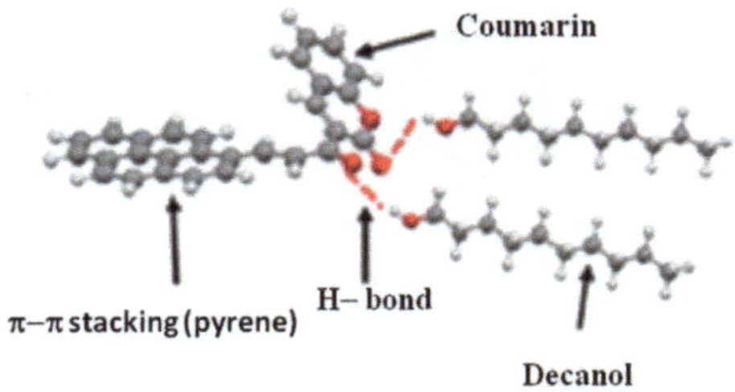

Fig. 8.7 Schematic representation of different types of interactions resulted in gel formation (Reproduced from [22])

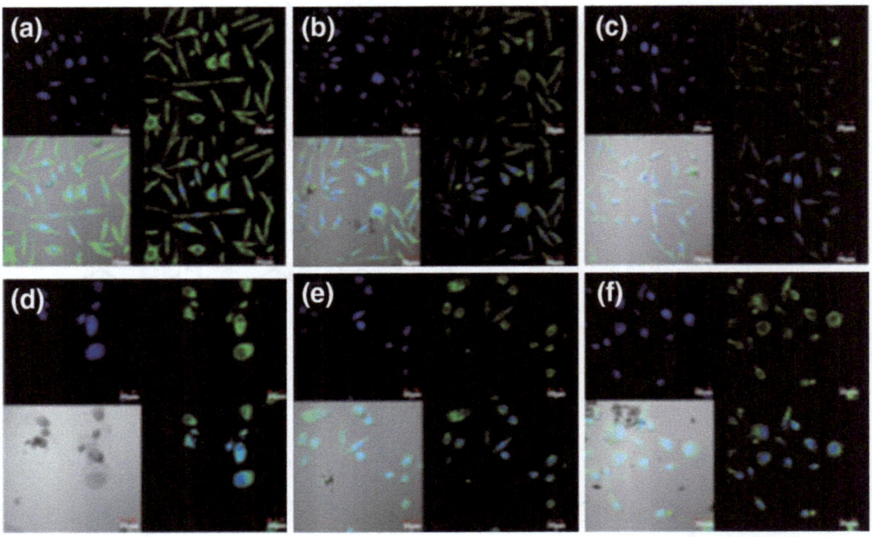

Fig. 8.8 LCSM images of **a–c** fibroblast cells and **d–f** prostate cancer cells incubated with pyrene-coupled coumarin derivative for 24 h. *Green color* is from the auto-fluorescence of the π-conjugated systems and *blue nucleus* differentiation is from Hoechst stain (Reproduced from ref. [22] 2016, Royal Society of Chemistry)

polymer coumarin-tris derivative from cardanol. The synthesis of the polymer is schematically shown in Fig. 8.9.

The presence of distinct hydrophilic part, i.e., tris-unit, directs this molecule into amphiphilic self-assembled structures. These amphiphilic molecules self-assembled into fibers in the form of gels at neutral and basic pH condition, and formed vesicles and nanotubes at acidic pH condition (Fig. 8.10). The stability of the self-assembled molecular structures was ascribed to hydrogen bonding and π–π interaction. The authors have demonstrated a multi-stimuli-responsive nature for the coumarin-tris-based gel, by exploring the reversible sol–gel phase transition of the gel in the presence of pH and Fe^{3+} ion stimuli.

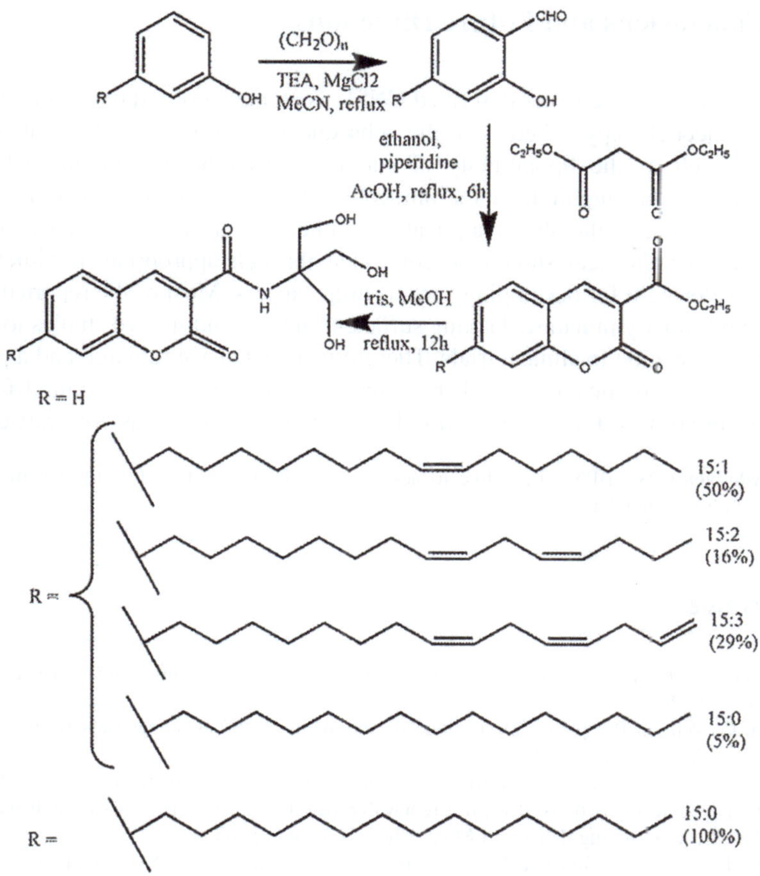

Fig. 8.9 Synthesis of coumarin-coupled tris-amphiphiles (Reproduced from [3])

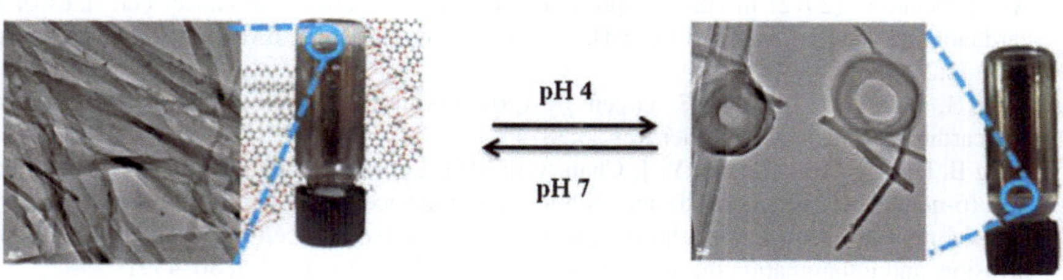

Fig. 8.10 Reversible sol–gel switching of coumarin-coupled tris-amphiphiles (Reproduced from ref. [3] 2015, Royal Society of Chemistry)

A chemopreventive drug curcumin was successfully encapsulated into this hydrogel, and the stimuli-responsive release (pH and Fe^{3+} triggered) of the encapsulated drug was studied. The reported hydrogel could contribute to the advance of new-generation stimuli-responsive drug delivery systems.

8.5 Conclusions and Future Directions

The first part of this chapter described CNSL as an anti-proliferative material useful for cancer therapy, whereas in the subsequent sections, the structural advantages as well as the lipophilicity of cardanol–anacardic acid-based molecular architectures were highlighted for intracellular/oral delivery of drugs. The contradictory nature of the above applications indicates that the cytotoxicity of pure cardanol/anacardic acid should be controlled through appropriate modifications before exploring it for developing various drug carriers. Most of the reported studies are preliminary in nature, lacking sufficient in vitro and in vivo studies to facilitate their transition to clinical trials. Therefore, an extensive research and a proper optimization (modifications of charge, lipophilicity or size) are required for the effective use of CNSL components as pharmaceutical agents or as drug carriers.

Acknowledgements BPN would like to acknowledge DST-INSPIRE Faculty scheme (IFA-11-CH-19) for financial assistance.

References

1. Bhanot A, Sharma R, Noolvi MN (2011) Natural sources as potential anti-cancer agents: a review 3:9–26
2. John G, Vemula PK (2006) Design and development of soft nanomaterials from biobased amphiphiles. Soft Matter 2:909
3. Lalitha K, Prasad YS, Maheswari CU, Sridharan V, John G, Nagarajan S (2015) Stimuli responsive hydrogels derived from a renewable resource: synthesis, self-assembly in water and application in drug delivery. J Mater Chem B 3:5560–5568
4. Neto L, Matos N, Gonzaga W, Romeiro L, Santos M, Santos D, Motoyama A (2014) Characterization of cytotoxic activity of compounds derived from anacardic acid, cardanol and cardol in oral squamous cell carcinoma. BMC Proc 8:P30
5. Teerasripreecha D, Phuwapraisirisan P, Puthong S, Kimura K, Okuyama M, Mori H, Kimura A, Chanchao C (2012) In vitro antiproliferative/cytotoxic activity on cancer cell lines of a cardanol and a cardol enriched from Thai Apis mellifera propolis. BMC Complement Altern Med 12:27
6. Kubo I, Nitoda T, Tocoli FE, Green IR (2011) Multifunctional cytotoxic agents from Anacardium occidentale. Phyther Res 25:38–45
7. Sung B, Pandey MK, Ann KS, Yi T, Chaturvedi MM, Liu M, Aggarwal BB (2008) Anacardic acid (6-nonadecyl salicylic acid), an inhibitor of histone acetyltransferase, suppresses expression of nuclear factor-κB-regulated gene products involved in cell survival, proliferation, invasion, and inflammation through inhibition of the inhib. Blood 111:4880–4891
8. Eliseeva ED, Valkov V, Jung M, Jung MO (2007) Characterization of novel inhibitors of histone acetyltransferases. Mol Cancer Ther 6:2391–2398
9. Balasubramanyam K, Swaminathan V, Ranganathan A, Kundu TK (2003) Small molecule modulators of histone acetyltransferase p300. J Biol Chem 278:19134–19140
10. Chandregowda V, Kush A, Reddy GC (2009) Synthesis of benzamide derivatives of anacardic acid and their cytotoxic activity. Eur J Med Chem 44:2711–2719
11. Sun Y, Jiang X, Chen S, Price BD (2006) Inhibition of histone acetyltransferase activity by anacardic acid sensitizes tumor cells to ionizing radiation. FEBS Lett 580:4353–4356

12. Sukumari-Ramesh S, Singh N, Jensen MA, Dhandapani KM, Vender JR (2011) Anacardic acid induces caspase-independent apoptosis and radiosensitizes pituitary adenoma cells. J Neurosurg 114:1681–1690
13. Massaro M, Colletti CG, Noto R, Riela S, Poma P, Guernelli S, Parisi F, Milioto S, Lazzara G (2015) Pharmaceutical properties of supramolecular assembly of co-loaded cardanol/triazole-halloysite systems. Int J Pharm 478:476–485
14. Al-Hazzani A, Periyasamy V, Subash-Babu P, Alshatwi AA (2012) Formulation of cashew nut shell liquid (CSNL) nanoemulsion, a potent inhibitor of human MCF-7 breast cancer cell proliferation. Med Chem Res 21:1384–1388
15. Banerjee S, Rao AR (1992) Promoting action of cashew nut shell oil in DMBA-initiated mouse skin tumour model system. Cancer Lett 62:149–152
16. Bloise E, Carbone L, Colafemmina G, D'Accolti L, Mazzetto SE, Vasapollo G, Mele G (2012) First example of a lipophilic porphyrin-cardanol hybrid embedded in a cardanol-based micellar nanodispersion. Molecules 17:12252–12261
17. De Maria P, Filippone P, Fontana A, Gasbarri C, Siani G, Velluto D (2005) Cardanol as a replacement for cholesterol into the lipid bilayer of POPC liposomes. Colloids Surf B Biointerf 40:11–18
18. Bloise E, Becerra-Herrera M, Mele G, Sayago A, Carbone L, D'Accolti L, Mazzetto SE, Vasapollo G (2014) Sustainable preparation of cardanol-based nanocarriers with embedded natural phenolic compounds. ACS Sustain Chem Eng 2:1299–1304
19. Mahata D, Mandal SM, Bharti R, Gupta VK, Mandal M, Nag A, Nando GB (2014) Self-assembled cardanol azo derivatives as antifungal agent with chitin-binding ability. Int J Biol Macromol 69:5–11
20. Shelma R, Paul W, Sharma CP (2010) Development and characterization of self-aggregated nanoparticles from anacardoylated chitosan as a carrier for insulin. Carbohydr Polym 80:285–290
21. Paul W, Shelma R, Sharma CP (2013) Alginate encapsulated anacardic acid-chitosan self aggregated nanoparticles for intestinal delivery of protein drugs. J Nanopharmaceutics Drug Deliv 1:82–91
22. Lalitha K, Jenifer P, Prasad YS, Muthusamy K, John G, Nagarajan S (2014) A self-assembled π-conjugated system as an anti-proliferative agent in prostate cancer cells and a probe for intra-cellular imaging. RSC Adv 4:48433–48437
23. Lalitha K, Nagarajan S (2015) Strongly fluorescent organogels and self-assembled nanostructures from pyrene coupled coumarin derivatives: application in cell imaging. J Mater Chem B 3:5690–5701

Chapter 9
Step-Growth Polymers from Cashew Nut Shell Liquid (CNSL)-Based Aromatic Difunctional Monomers

Deepshikha Chatterjee, Nilakshi V. Sadavarte, Rahul D. Shingte, Arvind S. More, Bhausaheb V. Tawade, Arun D. Kulkarni, Amol B. Ichake, C.V. Avadhani and Prakash P. Wadgaonkar

Abstract Cashew nut shell liquid (CNSL) is an attractive renewable resource material which is available in abundance (44,50,000 tonnes worldwide and 7,53,000 tonnes in India in 2013) at low cost (around 0.27 \$/kg in 2015) and is mainly composed of anacardic acid, cardanol, cardol and 2-methyl cardol. Cardanol is obtained as a major product during hot oil extraction or roasting process of CNSL. Cardanol possesses interesting structural features. The aromatic

D. Chatterjee · N.V. Sadavarte · R.D. Shingte · A.S. More · B.V. Tawade · A.D. Kulkarni · A.B. Ichake · C.V. Avadhani · P.P. Wadgaonkar (✉)
Polymers and Advanced Materials Laboratory, Polymer Science and Engineering Division, CSIR-National Chemical Laboratory, Dr. Homi Bhabha Road, Pune 411008, India
e-mail: pp.wadgaonkar@ncl.res.in

D. Chatterjee
e-mail: d.chatterjee@ncl.res.in

N.V. Sadavarte
e-mail: nv.sadavarte@ncl.res.in

R.D. Shingte
e-mail: rahulshingte@gmail.com

A.S. More
e-mail: arvindsmore@gmail.com

B.V. Tawade
e-mail: bvtawade@gmail.com

A.D. Kulkarni
e-mail: arundkulkarni@gmail.com

A.B. Ichake
e-mail: ab.ichake@ncl.res.in

C.V. Avadhani
e-mail: cv.avadhani@ncl.res.in

P. Anilkumar (ed.), *Cashew Nut Shell Liquid*,
DOI 10.1007/978-3-319-47455-7_9

ring of cardanol can undergo electrophilic substitution reactions; the unsaturated side chain can undergo epoxidation, hydrogenation, metathesis, etc., while the phenolic hydroxyl group can undergo various reactions such as esterification and alkylation. Such opportunities of chemical modifications offered by cardanol have been extensively explored to synthesise a range of interesting aromatic difunctional monomers that have subsequently been utilised to prepare a host of step-growth polymers. Summarised herein are research efforts that have contributed towards the synthesis of step-growth polymers based on aromatic difunctional monomers derived from cardanol. The properties of high-performance polymers, with a particular focus on processability and thermal characteristics, are highlighted.

9.1 Introduction

9.1.1 Background

Unquestionably, today, we live in the 'Plastics Age'. An estimated average worldwide production of plastics is around 250 million tonnes per year, with a yearly average growth of 10 % [1, 2]. Plastics are mostly derived from fossil resources. Almost all organic chemicals can be derived from seven basic building blocks, namely syngas, ethylene, propylene, butanes, butylenes, butadiene and BTX (mixture of benzene, toluene and xylene) which are mostly obtained from fossil resources, viz. natural gas, petroleum and coal [3, 4].

Over the past several years, spectacular progress has been made in petrochemical industry, so much so, that one might even question the need to investigate the area of renewable resources. However, the complete dependence on fossil resources has its own limitations. Fossil resources are generated through chemical and biological processes occurring over millions of years, while the rate of consumption of these resources is several times the rate of their formation [5–8]. Fossil resources being non-renewable in nature, it is logically advisable to investigate the use and consumption of renewable resources for the production of chemicals. One of the promising renewable resources is biomass. Biomass refers to any organic matter having recent biological origin, including plant materials, agricultural crops and animal residues [9]. Biomass remains an attractive resource for obtaining chemicals, especially platform chemicals [10]. Platform chemicals are defined as chemicals from which many value-added chemicals may be derived, which in turn can be used to manufacture end-use commodities such as plastics and resins [11]. A biomass-based economy will require a well-developed biorefinery system [12, 13]. The products (biochemicals) obtained from biomass can be intermediates, final products or other materials [14–16]. International Energy Agency (IEA) Bioenergy (task 42) has defined biorefinery as the sustainable processing of biomass into a spectrum of marketable products and energy. This simply implies the concept of separating the components of biomass obtained from

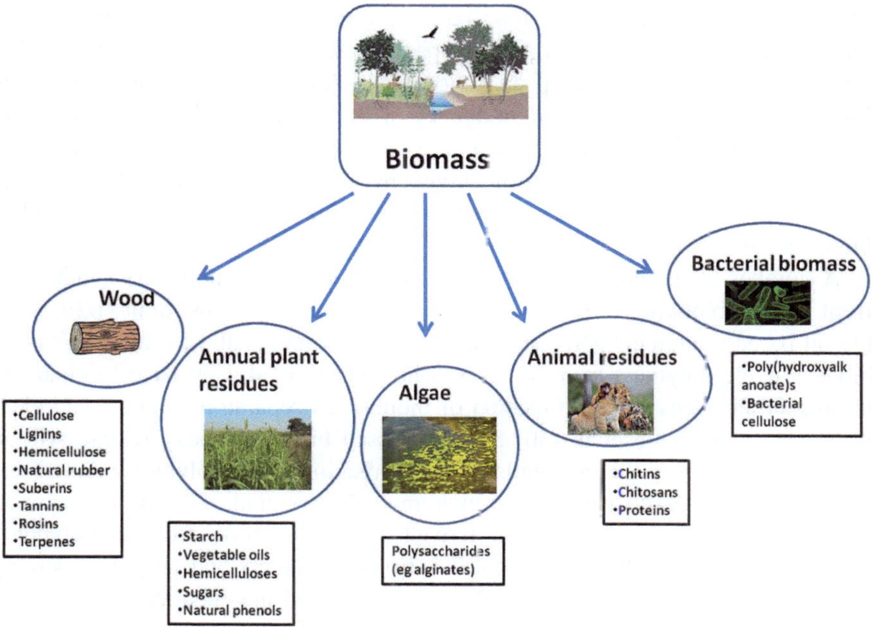

Fig. 9.1 Sources of biochemicals

a source and their valorisation. There is presently a huge potential for bioderived chemicals to share markets with fossil-based counterparts.

Major renewable resources include wood, plant residues, algae, animal residues and bacterial biomass [9, 17] (Fig. 9.1).

Edible biochemicals include starch, vegetable oils and sugars. Natural phenols, hemicelluloses and other biochemicals derived from wood, animal residues, algae and bacterial biomass are non-edible. Such non-edible bioresources are preferred as starting materials for chemicals as they do not put undue pressure on the food industry. Biochemicals are attractive sources of both aliphatic and aromatic monomers. Rich sources of aromatic monomers include lignins, tannins and natural phenols such as cashew nut shell liquid (CNSL). Lignin is an important biopolymer present in plant cell walls, which is industrially obtained as a coproduct of paper manufacture. Lignin is an important source for obtaining vanillin, which has a huge potential as a source for biobased monomers [18] Naturally occurring vegetable tannins are mainly phenolic in nature and are classified into two broad classes, namely condensed/ polyflavonoid tannins and hydrolysable tannins. 2,5-Furan dicarboxylic acid (FDCA), 5-hydroxymethyl furfural and a variety of FDCA-based materials, which can be readily and economically prepared from a vast variety of agricultural and forestry wastes [19], are important precursors for the synthesis of aromatic monomers.

Several biomass-based programmes and projects have been funded by governments and industries around the world. 'CatchBIO' is a Dutch research programme in the field of catalytic biomass conversion with 21 partners which include the research institutes and companies such as BASF [20]. 'BIO-TIC' is a project funded by the European Union [21]. 'ARENA' (Australian Renewable Energy Agency) is an independent agency of the Australian Federal Government to manage renewable energy programmes in the country [22]. Delft Advanced Biorenewables (DAB) is a private company in the Netherlands dealing with biorenewables [23]. These are just a few representative programmes involved in biorenewables.

A number of researchers have reviewed monomers and polymers synthesised from various biomass resources [16, 24, 25], including various furan and vegetable oil-based monomers and polymers [26, 27]. Tschan et al. [28] have discussed biodegradable polymers from biomass, while Zhang et al. [29] have summarised the recent advances in the utilisation of biomass as separators, binders and electrode materials. Datta et al. [30] have discussed biobased epoxy resins. Wilbon et al. [31] have reviewed the details of polymerisation and copolymerisation of terpenes, terpenoids and rosin-derived monomers. Llevot et al. [32] and Upton et al. [33] have reviewed various lignin-derived aromatic compounds and polymers, while Galbis et al. [34] have reviewed various monomers and polymers derived from carbohydrates. Arbenz et al. [35] have discussed the chemical modification and synthesis of polymeric materials from tannins.

CNSL, as an interesting bioderived chemical, with enormous potential in the field of polymer chemistry, has been discussed in the subsequent sections.

9.1.2 Cashew Nut Shell Liquid (CNSL)

CNSL belongs to the class of naturally occurring phenols, a non-edible by-product of cashew (*Anacardium occidentale* L) industry and a promising renewable resource. Cashew tree originated in Brazil, but is presently cultivated in many countries. The top five cashew producers (as of 2013) are Vietnam, Nigeria, India, Cote d'Ivoire and Brazil [36]. The actual fruit of the cashew tree is a 'drupe' out of which grows the oval- or pear-shaped cashew apple, which is an accessory (or false) fruit. It is edible and is used for making jams, jellies and alcoholic drinks. The distal end of the cashew apple is attached to the cashew nut, the edible portion of which is encased in a double-layered shell. The shell has a honeycomb structure and contains CNSL—a greenish yellow viscous liquid (Fig. 9.2). The cashew nut and cashew apple are the cash crops yielded by the cashew tree, while CNSL is an agricultural by-product [37–40].

Traditionally, CNSL is obtained as a by-product during the process of removing the cashew kernel (edible part) from the nut, by the hot oil process or roasting process [41]. More recent and improved methods of extraction include solvent extraction technique [40, 42] and supercritical extraction using CO_2 and isopropyl alcohol and vacuum pyrolysis [43, 44]. CNSL is purified by chemical treatment

Fig. 9.2 The cashew fruit

with sulphuric acid and other acid treatments before further use. Subsequently, it may be distilled under reduced pressure or hydrogenated [45].

A number of review articles have dealt with the extraction, isolation and applications of CNSL [19, 46, 47]. Bladzell [48] has discussed the usefulness of cashew nut, dealing with its history, cultivation, extraction and various products. Mele et al. [49] have reviewed CNSL-based fine chemicals as well as cardanol hybrids. Balachandran et al. [50] have discussed the generation of soft nanomaterials from cardanol. Voirin et al. [51] have reviewed purification, separation methods, reactivity and applications of CNSL.

9.1.3 Components of CNSL

Naturally occurring CNSL consists of four major components, viz. anacardic acid, cardanol, cardol and 2-methyl cardol (Fig. 9.3). Commercial-grade CNSL mainly consists of cardanol, 2-methyl cardol and cardol as anacardic acid gets almost completely converted to cardanol due to decarboxylation occurring during the extraction and refining processes [52]. Composition of natural and technical CNSL is depicted in Table 9.1. Each component of CNSL is a mixture of four constituents differing in the side chain unsaturation, namely saturated (~5–8 %), monoene (~48–49 %), diene (16–17 %) and triene (29–30 %) [53].

Cardolite Corporation, USA, is the world leader in developing and manufacturing products based on CNSL [54].

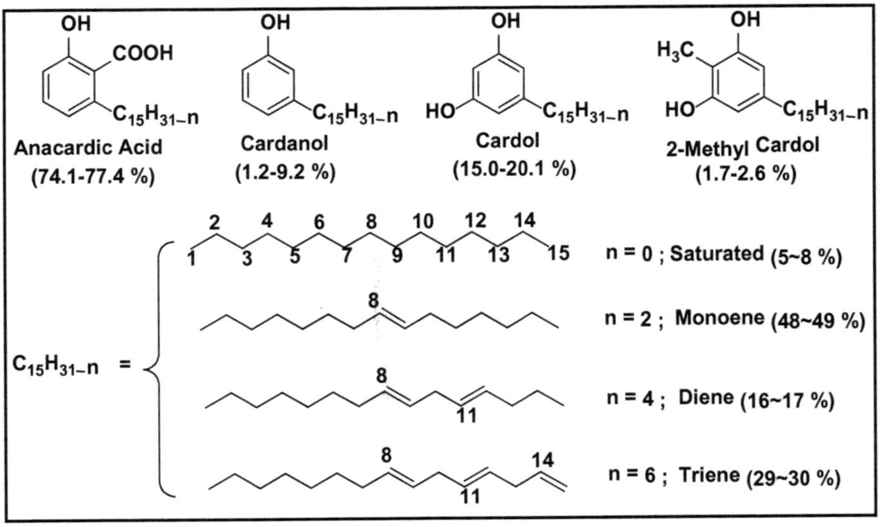

Fig. 9.3 Components of CNSL ([51] reproduced with the permission from the Royal Society of Chemistry)

Table 9.1 Components of natural and technical CNSL (in wt%) ([51]—reproduced with the permission from the Royal Society of Chemistry)

Component/other material	Natural CNSL	Technical CNSL
Cardanol (%)	1.20	62.86
Cardol (%)	11.31	11.25
2-Methyl cardol (%)	2.04	2.08
Anacardic acid (%)	64.93	N.D.*
Polymer (and minor materials)	20.3	23.8

**N.D.* not detected

9.1.4 Reactivity of Cardanol

Cardanol possesses a C-15 hydrocarbon chain *meta* to the phenolic –OH, a unique orientation which would be difficult to achieve synthetically. Moreover, cardanol provides a number of sites for functionalisation and modification which include the phenolic –OH, the aromatic ring and the unsaturation(s) in the side chain (Fig. 9.4).

The aromatic ring can undergo various electrophilic substitution reactions such as nitration, bromination, alkylation, and acylation. The unsaturated C-15 side chain provides sites for epoxidation, hydrogenation, phenolation, metathesis, etc., while the phenolic OH- group can undergo modification through esterification, alkylation, etc. Such possibilities for modifications can be utilised to synthesise designed monomers and polymers, as has been discussed in the subsequent sections.

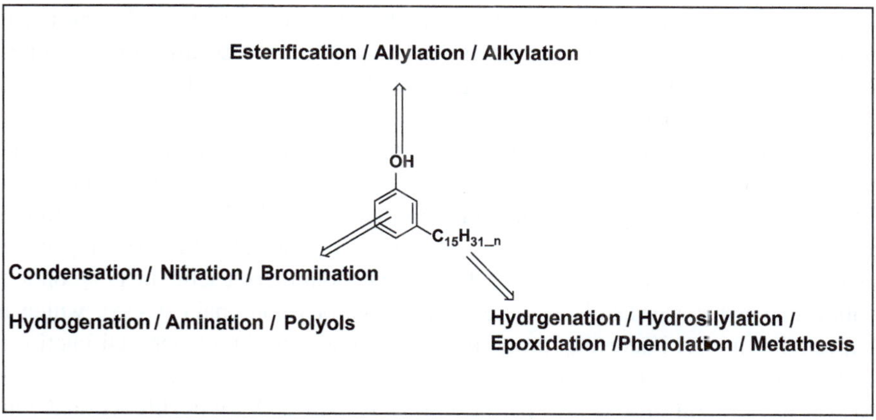

Fig. 9.4 Reactive sites in cardanol ([51] reproduced with the permission from the Royal Society of Chemistry)

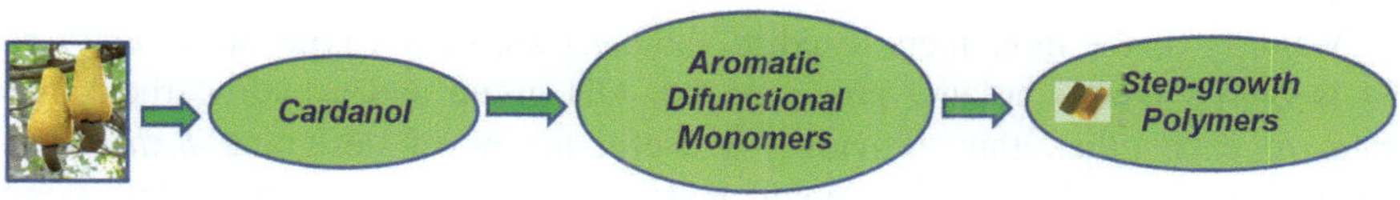

Fig. 9.5 The theme of the chapter

In this chapter, the objective is to summarise research carried out pertaining to the synthesis of aromatic difunctional monomers based on cardanol/CNSL and their utilisation for the preparation of step-growth polymers. Cardanol-based addition monomers, thermoset resins, polymer additives and other value-added chemicals are outside the scope of the present chapter and have been dealt with elsewhere [19, 46–51]. The following graphic (Fig. 9.5) suitably illustrates the theme of this chapter.

9.2 Difunctional Monomers Based on CNSL and Step-Growth Polymers Therefrom

9.2.1 CNSL-Based Aromatic Difunctional Condensation Monomers

Aromatic diamines, diacids, diphenols, diisocyanates, dialdehydes, etc., are the main classes of condensation monomers useful for the preparation of high-performance polymers. A large number of difunctional monomers possessing special structural features have been synthesised till date. Researchers still continue to synthesise new difunctional monomers which simply serve as 'drop-in substitutes' for commercial monomers already in practice for the purpose of imparting special properties to the existing polymers.

Cardanol, due to its interesting reactive sites (phenolic –OH, aromatic ring and unsaturation in the pentadecenyl chain), has been utilised as a starting material for the synthesis of difunctional condensation monomers [51]. Furthermore, cardanol is derived from biomass, which makes it an even more attractive starting material for the synthesis of aromatic difunctional monomers useful in polycondensation chemistry. Cardanol possesses a long hydrocarbon chain (C-15), which is expected to impart interesting properties to the synthesised polymers. Difunctional condensation monomers possessing long pendent hydrocarbon chains are difficult to synthesise traditionally from petrochemical resources. Various difunctional condensation monomers such as diamines, diacids, diphenols, dialdehydes, diacylhydrazides, and diisocyanates were derived from cardanol, as listed in Tables 9.2 and 9.3. The listed monomers have been synthesised either from cardanol or from 3-pentadecyl phenol, which in turn is obtained from cardanol through side chain hydrogenation.

Majority of the monomers listed in Table 9.2 are of AA type, while entry no. 34 is of AB type. Cardanol possesses an –OH group and a hydrocarbon chain *meta* to each other, thus providing a useful means of designing *meta*-substituted monomers. Besides this inbuilt *meta* relationship, *meta*-substituted monomers (Table 9.2, entries 11, 13, 14, 15, 22, 23, 31, 33, 35 and 36) have also been designed through typical electrophilic substitution reactions guided by the electron-rich –OH group. The cardanol molecule has been suitably modified to synthesise unsymmetrical monomers (Table 9.2, entries 1, 2, 3, 4, etc.). Subsequently, monomers containing flexible ether linkages (Table 9.2, entries 3, 16, 17, 20, 21, 24, 26, 28, 29, 32, 37 and 38) were synthesised. Monomers possessing aromatic ring(s) derived solely from cardanol contribute more towards their 'greenness' (Table 9.2, entries 1, 2, 5, 10–12, 31, 33–36). Cardbisphenol, i.e. 1,8-bis(hydroxyphenyl)pentadecane (Table 9.2, entry 10), was prepared from cardanol and phenol. This has been shown to consist of two isomers and is used in the preparation of resins useful in surface-coating industries [69]. Cardanol was converted to 3-(8-hydroxy-octyl) phenol (Table 9.2, entry 1) through ozonolysis, followed by reduction using sodium borohydride.

The difunctional monomers listed in Table 9.2, except the monomers represented by entries 1 and 34, have been synthesised through functional group transformations and coupling reactions, wherein the C-15 hydrocarbon chain remains unaltered with respect to carbon number. Another interesting strategy exists, wherein the unsaturation in the C-15 hydrocarbon chain of cardanol is exploited, through olefin metathesis reactions.

Olefin metathesis involves redistribution of fragments of alkenes by scission and recombination by carbon–carbon double bonds. CNSL is comprised of components which have fifteen carbon unsaturated chains. Hence, application of olefin metathesis on cardanol can give rise to interesting monomers which may be

Table 9.2 CNSL-based difunctional condensation monomers

Sr. No.	Monomers	Polymers	References
Diphenols/diols			
1.	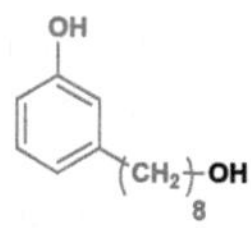	Polycarbonates	[55]
2.	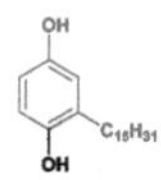	Polyesters	[56]
3.	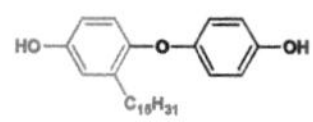	(Co)polyesters	[57]
4.	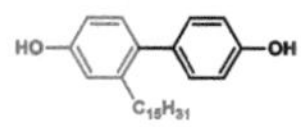	Poly(arylene ether)s	[58]
5.	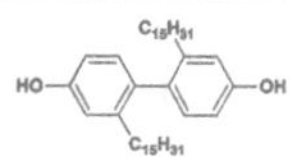	_[a]	[59]
6.	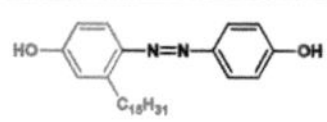	Polyurethanes	[60, 61]
7.	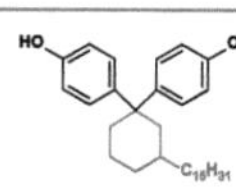	(Co)polyesters and poly(arylene ether)s	[62–64]
8.	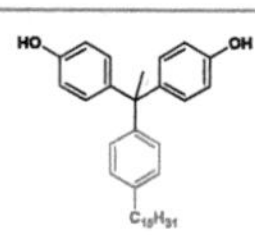	(Co)polyesters and poly(arylene ether)s	[64–67]
9.	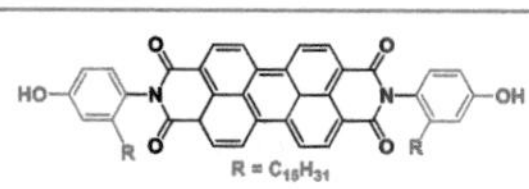	_[a]	[68]
10.	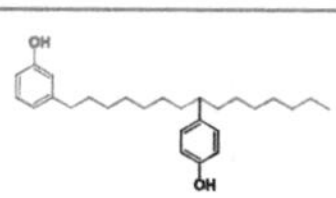 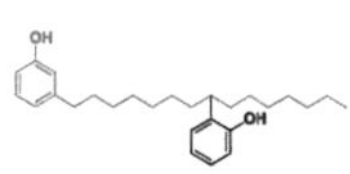	_[b]	[69]

(continued)

Table 9.2 (continued)

Sr. No.	Monomers	Polymers	References
Diamines			
11.	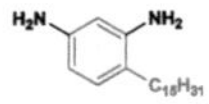	Polyimides, polyamides and poly(azomethine)s	[70, 71]
12.	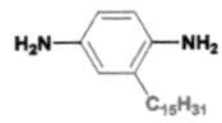	Polyimides	[72]
13.	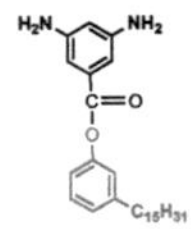	Polyimides and polyamides	[73, 74]
14.	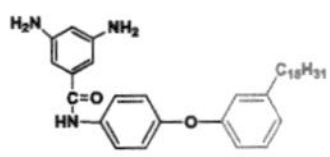	Polyimides and polyamides	[73, 74]
15.	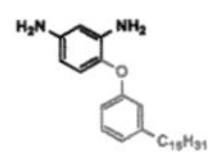	Polyimides and polyamides	[73, 74]
16.	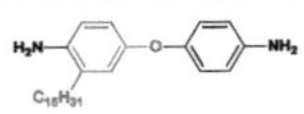	Poly(azomethine)s	[75]
17.	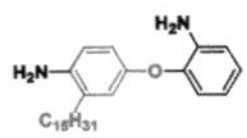	_[a]	[67]
18.	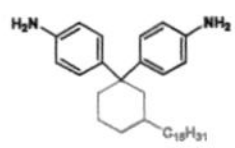	(Co)polyimides	[62, 76]
19.	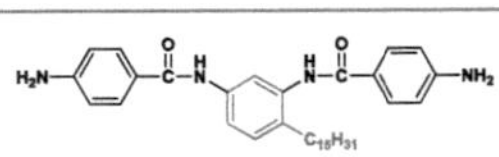	Poly(amideimide)s	[77]
20.	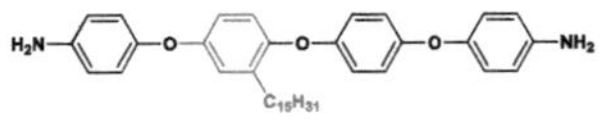	Polyimides	[78]
21.	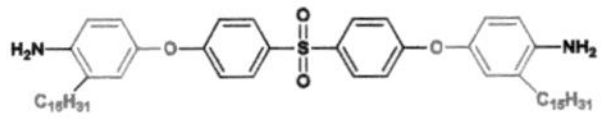	Polyamides and polyimides	[79]
22.	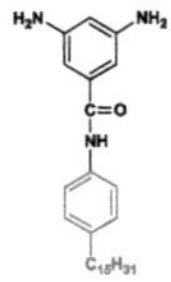	_[a]	[56]

(continued)

Table 9.2 (continued)

Sr. No.	Monomers	Polymers	References
23.	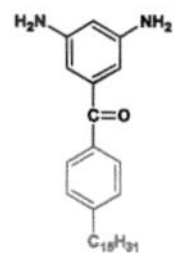	–[a]	[56]
Diacids			
24.	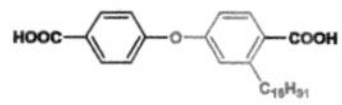	Polyamides	[80]
25.	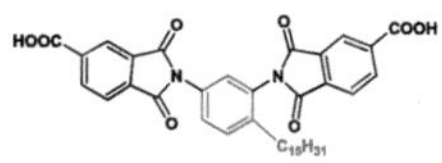	Poly(esterimide)s	[81]
26.	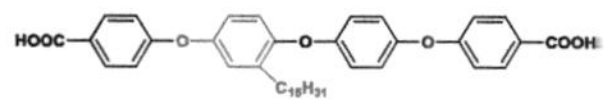	–[a]	[59]
27.	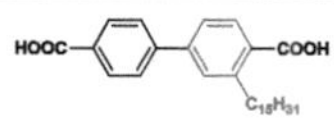	–[a]	[59]
Diacylhydrazides			
28.	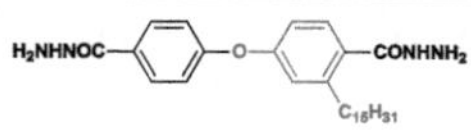	Poly(amideimide)s, polyhydrazides and poly(1,3,4-oxadiazole)s	[82–84]
29.	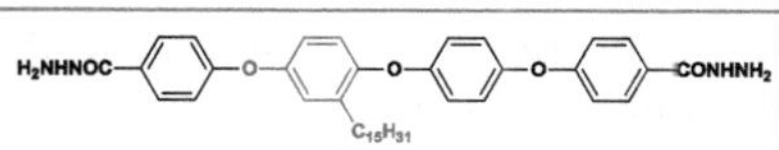	Polyhydrazides and poly(1,3,4-oxadiazole)s	[59]
30.	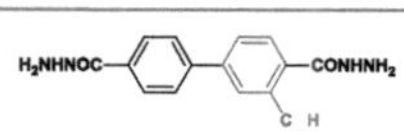	–[a]	[59]
Diisocyanates			
31.	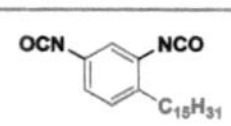	–[a]	[56]
32.	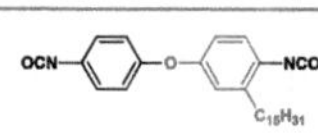	–[a]	[67]

Table 9.2 (continued)

Sr. No.	Monomers	Polymers	References
Miscellaneous monomers			
33.	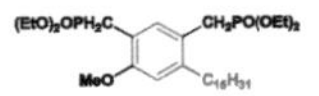	Poly(*m*-phenylenevinylene)s	[85]
34.	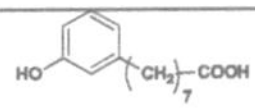	Polyesters	[86–88]
35.	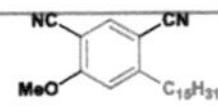	–[a]	[56]
36.	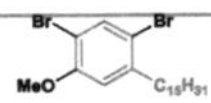	–[a]	[56]
37.	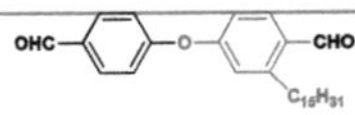	–[a]	[59]
38.	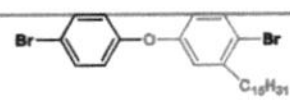	–[a]	[67]

Green colour indicates the structural components derived from cardanol
[a]Polymers not reported
[b]Not used for the synthesis of thermoplastic polymers

difficult to access by conventional routes. Monomers prepared by metathesis reaction involving cardanol are listed in Table 9.3 [89–96].

Table 9.3, entries 1–4, shows difunctional monomers obtained through self-metathesis reactions of cardanol, while monomers, represented by entries 5–12 (Table 9.3), are synthesised by cross-metathesis reaction involving cardanol [92–96]. In 2016, Mgaya et al. [96] synthesised a series of difunctional monomers of various chain lengths from cardanol by applying metathesis chemistry (Table 9.3, entries 7–12).

Metathesis thus provides a tool by which the carbon atom number of the hydrocarbon chain can be tuned (Table 9.3, entries 1–12). It also presents a means to functionalise the hydrocarbon chain (Table 9.3, entries 5–12) and provides mixed aliphatic–aromatic difunctional monomers (Table 9.3, entries 5–12).

Several monomers listed in Table 9.2 have been utilised to synthesise step-growth polymers, while some monomers have not yet been polymerised. The properties of the synthesised polymers are discussed in the subsequent sections.

Table 9.3 Condensation monomers prepared from cardanol through metathesis

Sr. No.	Metathesis products based on cardanol	References
1.	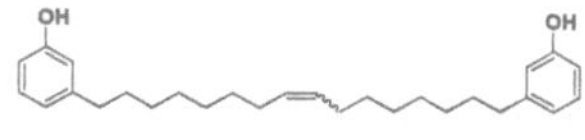	[89–91]
2.		[89–91]
3.	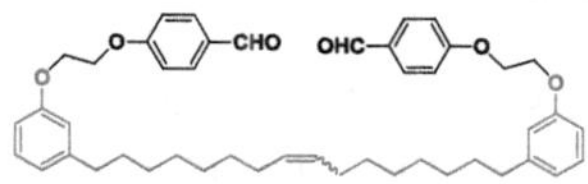	[89–91]
4.		[89–91]
5.		[92–95]
6.		[92–95]
7.	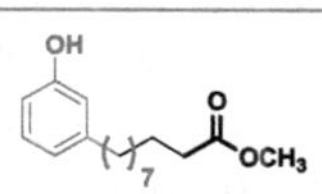	[96]
8.	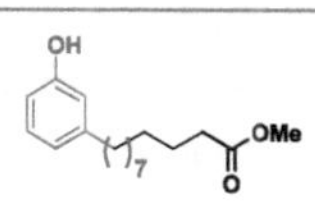	[96]
9.	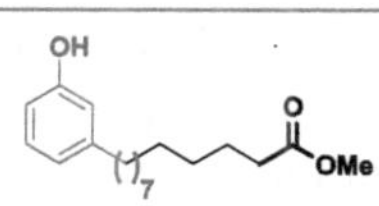	[96]
10.	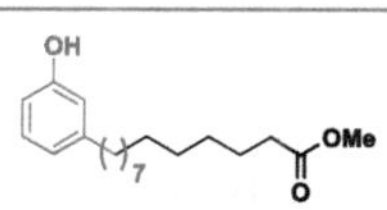	[96]
11.	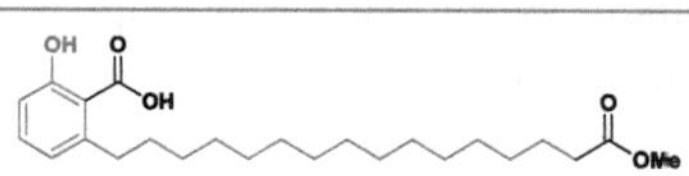	[96]
12.	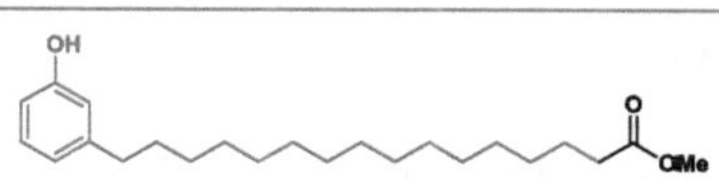	[96]

Green colour indicates the structural components derived from cardanol

9.2.2 Step-Growth Polymers from CNSL-Based Aromatic Difunctional Monomers

Aromatic polyimides, polyamides, etc., are important classes of high-performance polymers which possess excellent thermo-oxidative stability, mechanical strength, electrical properties, and high radiation and solvent resistance [97, 98]. These find applications in fields such as aerospace, automobiles, electronics and other industries as films, coatings, adhesives and membranes for separation of gases or liquids [99–103].

High-performance polymers generally possess high glass transition temperatures, which are sometimes above their decomposition temperatures, and exhibit poor solubility in common organic solvents, which results into processing difficulties. Several approaches have been used to modify/ improve the processability of high-performance polymers, which include: (1) the insertion of flexible spacers between the rigid units; (2) the insertion of bent or 'crankshaft' units or 'kinks' along the aromatic backbone; and (3) the introduction of bulky side groups or flexible side chains to the aromatic backbone [104, 105]. These structural modifications disrupt molecular symmetry, inhibit close packing of the chains and increase the free volume, consequently leading to improved solubility and processability.

Some of the difunctional monomers derived from CNSL as listed in Table 9.2 have been utilised for the synthesis of a variety of step-growth polymers such as aromatic polyimides, polyamides, poly(azomethine)s, poly(amideimide)s, poly(esterimide)s, polyesters, polyhydrazides, poly(1,3,4-oxadiazole)s and polyurethanes. The special structural features of aromatic difunctional monomers such as a long C-15 alkyl chain, *meta*-substitution and asymmetry have been exploited to synthesise processable high-performance polymers. Cardanol-based monomers possessing pendent C-15 alkyl chain reduce coplanarity of the aromatic rings and disrupt the interchain packing of the macromolecular chains, which in turn leads to processable polymers. *Meta*-substituted monomers introduce 'kinks' in the macromolecular chain resulting in processable polymers. The unsymmetrical monomers cause disruption of symmetry in the polymer chain leading to decrease in the interchain interactions between the polymer chains and also give rise to constitutional isomerism. Monomers containing flexible ether linkages provide polymers with lower internal energy of rotation and disrupt the linear progression in the polymer backbone.

Properties of high-performance polymers derived from the listed cardanol-derived difunctional monomers with a particular emphasis on processability characteristics are discussed in the following sections.

9.2.2.1 Aromatic Polyimides

Aromatic polyimides are generally synthesised by polycondensation of aromatic diamines and tetracarboxylic dianhydrides, mainly involving a two-step method via poly(amic acid) formation or by one-step high-temperature solution polymerisation [106] besides other synthetic routes [107–116]. Table 9.4 represents the repeating units of aromatic polyimides derived from CNSL-based aromatic difunctional monomers and data concerning their solubility and thermal properties.

Aromatic polyimides containing pendent pentadecyl chains (Table 9.4) were mostly synthesised through one-step high-temperature solution polycondensation, except polyimides represented by entries 10, 11 in Table 9.4, which were synthesised by two-step method involving thermal imidisation (Scheme 9.1).

Mathew [74] synthesised a series of copolyimides by polycondensation of a mixture of pentadecyl chain containing aromatic diamine corresponding to entry 15 (Table 9.2) and 4,4'-oxydianiline (ODA) at 5, 20 and 50 mol%, with a number of commercially available aromatic dianhydrides. Another series of copolyimides was synthesised by polycondensation of the same diamine (Table 9.2, entry 15) with 50 mol% of bis(3,4-dicarboxyphenyl) dimethylsilane dianhydride (SiDA) and other commercially available aromatic dianhydrides. The reported inherent viscosities (η_{inh}) of polyimides (0.23–0.70 dL/g) indicated the formation of medium to reasonably high molecular weight polymers which formed transparent and flexible films from their solutions.

^{1}H-NMR studies of polyimides (Table 9.4, entries 30–32) indicated the presence of constitutional isomerism based on the appearance of four signals corresponding to aromatic protons *ortho* to imide nitrogen, instead of the expected doublet [78]. Constitutional isomerism arises when an unsymmetrical monomer reacts with a symmetrical monomer. The presence of pentadecyl chain on the phenyl ring imparts asymmetry to the polymer backbone leading to constitutional isomerism (Fig. 9.6). Theoretically, four probable structural orientations occur, viz. (a) head-to-head, (b) tail-to-tail, (c) head-to-tail and (d) tail-to-head, out of which head-to-tail and tail-to-head would be indistinguishable by ^{1}H-NMR spectroscopy [56].

The results of solubility tests demonstrated that solubility of polyimides was considerably enhanced compared with reference polymers, i.e. polymers without C-15 alkyl chain. The C-15 alkyl chain disrupts interchain packing, thus allowing solvent molecules to penetrate inside the polymer chains, consequently improving polymer solubility. Additionally, the pentadecyl chain provides a handle for interaction with solvent molecules. For example, the polymers represented in Table 9.4 (entries 1, 2, 6, 32, etc.) were soluble in chloroform and tetrahydrofuran at room temperature in spite of possessing rigid aromatic rings in the backbone, while the reference polyimides were insoluble in the respective solvents.

Thermogravimetric (TG) analysis data indicated that polyimides possessed good thermal stability (T_{10} values above 430 °C). Differential thermogravimetric analysis (DTG) of the synthesised polyimides (Table 9.4, entries 6–9) revealed a two-stage weight loss: the first-stage weight loss was due to decomposition of pendent alkyl groups, while the second-stage weight loss was due to decomposition of polyimide backbone. Polyimides containing pendent pentadecyl chains showed a significant drop in glass transition temperature (T_g) values (113–241 °C) compared with reference polyimides (Table 9.4, entries 6–9). The lowering in T_g values

Table 9.4 Aromatic polyimides containing pendent pentadecyl chains derived from difunctional monomers based on cardanol

Sr. No.	Polyimide	η_{inh} [a] (dL/g)	Solubility[b]		T_g (°C)[c]	T_{10} (°C)[d]	References
			$CHCl_3$	THF			
1.	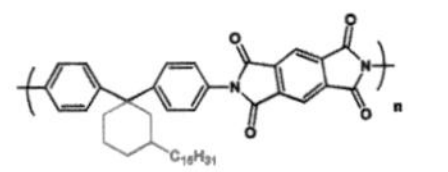	0.33	++	++	191	493	[62]
2.	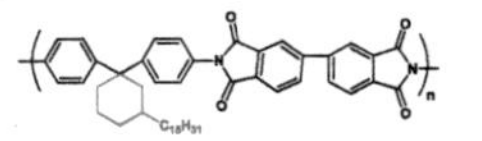	0.40	++	++	209	513	[62]
3.	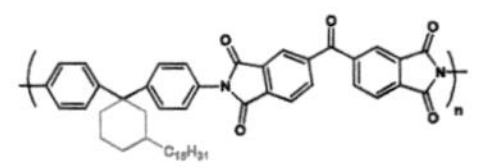	0.40	++	++	178	512	[62]
4.	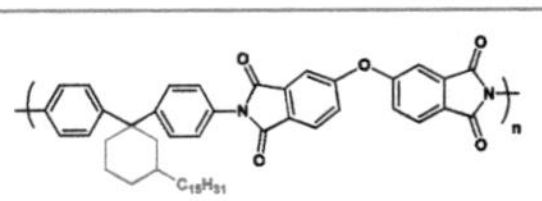	0.33	++	++	161	503	[62]
5.	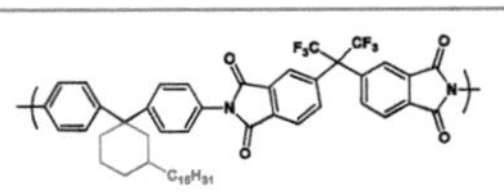	0.30	++	++	165	509	[62]
6.	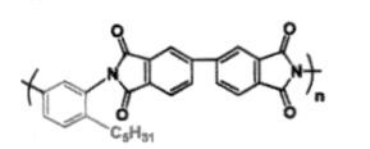	0.67	++	++	206 (330)	475	[70]
7.	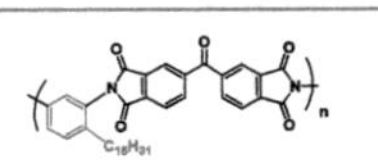	0.49	++	–	176(320)	470	[70]
8.	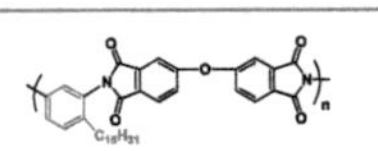	0.56	++	++	159 (305)	480	[70]
9.	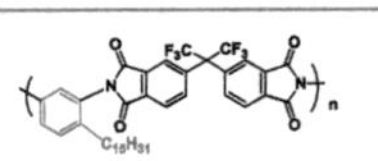	0.33	++	++	158 (297)	480	[70]
10.	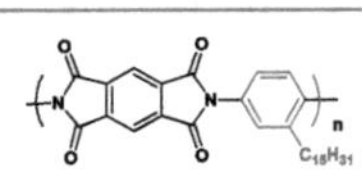	0.56*	N.R.	N.R.	N.R.	430	[72]
11.	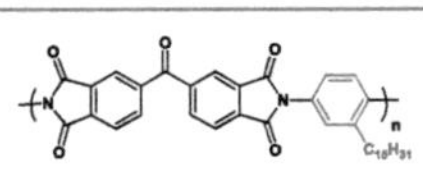	0.49*	N.R.	N.R.	N.R.	430	[72]
12.	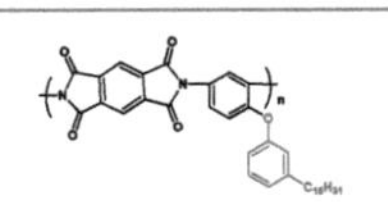	N.R.	+–	+–	212	451	[74]

(continued)

Table 9.4 (continued)

Sr. No.	Polyimide	η_{inh} [a] (dL/g)	Solubility[b]		T_g (°C)[c]	T_{10} (°C)[d]	References
			$CHCl_3$	THF			
13.	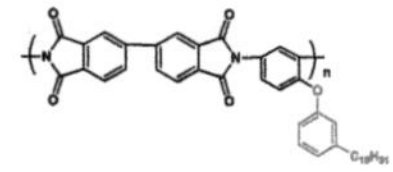	N.R.	+−	+−	226	460	[74]
14.	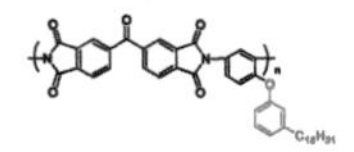	N.R.	++	++	171	468	[74]
15.	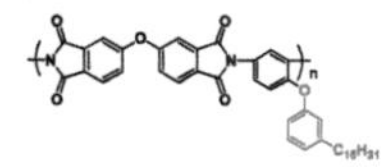	0.45	++	++	146	457	[74]
16.	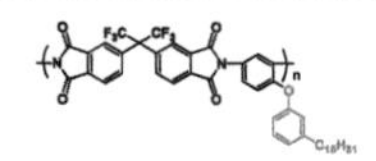	0.62	++	++	156	456	[74]
17.	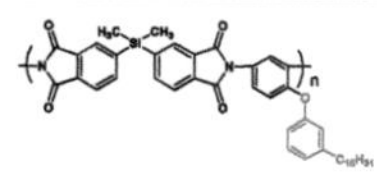	0.42	++	++	123	450	[74]
18.	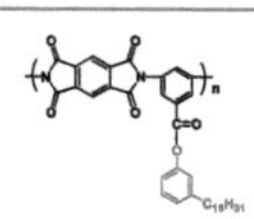	N.R.	−	−	228	514	[74]
19.	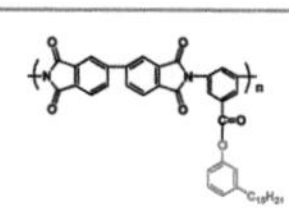	N.R.	−	+−	241	449	[74]
20.	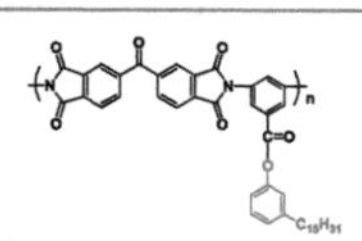	0.25*	−	++	237#	487	[74]
21.	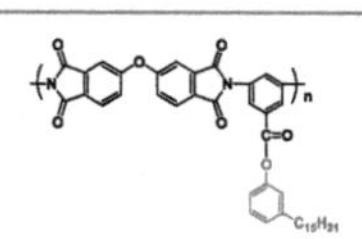	0.55*	+−	++	216	443	[74]
22.	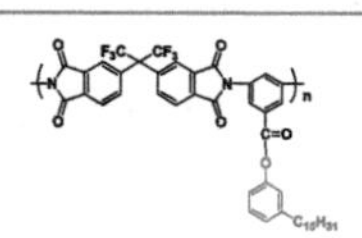	0.45*	++	++	174	488	[74]
23.	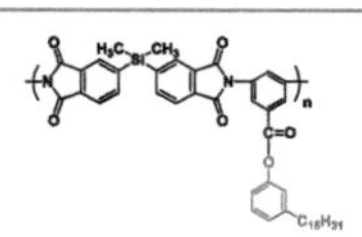	0.35*	++	++	185	436	[74]

(continued)

Table 9.4 (continued)

Sr. No.	Polyimide	η_{inh} [a] (dL/g)	Solubility[b]		T_g (°C)[c]	T_{10} (°C)[d]	References
			$CHCl_3$	THF			
24.	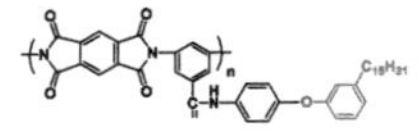	N.R.	–	–	219	445	[74]
25.	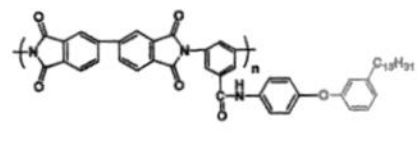	N.R.	–	–	224	443	[74]
26.	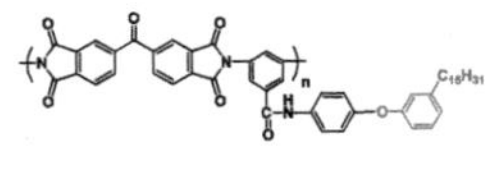	0.23*	–	–	237#	446	[74]
27.	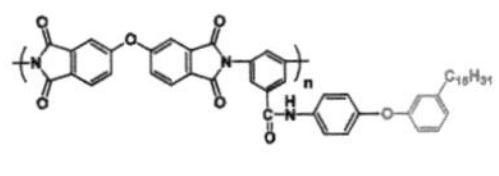	0.55*	++	+–	216#	449	[74]
28.	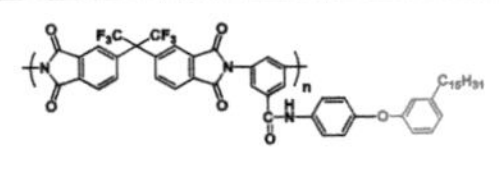	0.32	++	++	186	444	[74]
29.	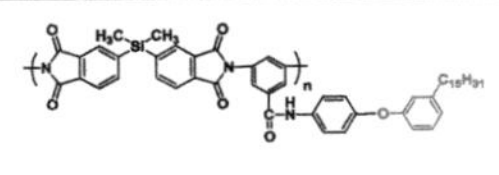	0.30	++	++	193	447	[74]
30.	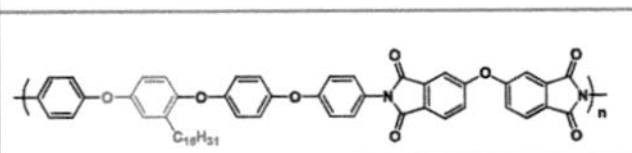	0.70	++	++	113	460	[78]
31.	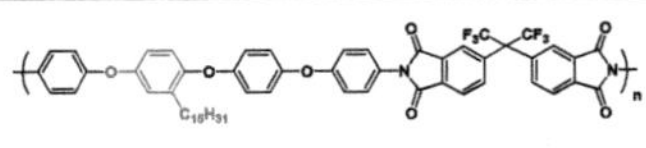	0.67	++	++	122	470	[78]
32.	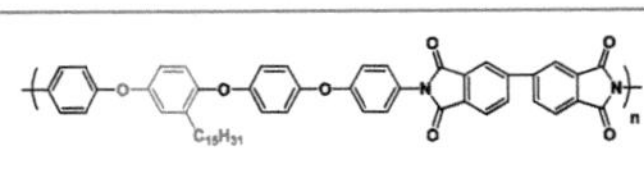	0.66	++	++	131	460	[78]

Green colour indicates the structural components derived from cardanol

*η_{inh} was measured with 0.5 % (w/v) solution of polyimides in *N,N*-dimethylacetamide at 30 ± 0.1 °C, # sample was heated to the softening point and then quenched in liquid nitrogen before determining T_g

[a]η_{inh} was measured with 0.5 % (w/v) solution of polyimides in chloroform at 30 ± 0.1 °C

[b]Solubility measured at 3 % (w/v) concentration –, insoluble; +–, partially soluble or swollen; ++, soluble at room temperature

[c]T_g values obtained from DSC curves from the second heating scans of polyimide samples, values in the bracket T_g values of reference polyimides

[d]Temperature at which 10 % weight loss was observed from TGA under nitrogen atmosphere

N.R. not reported

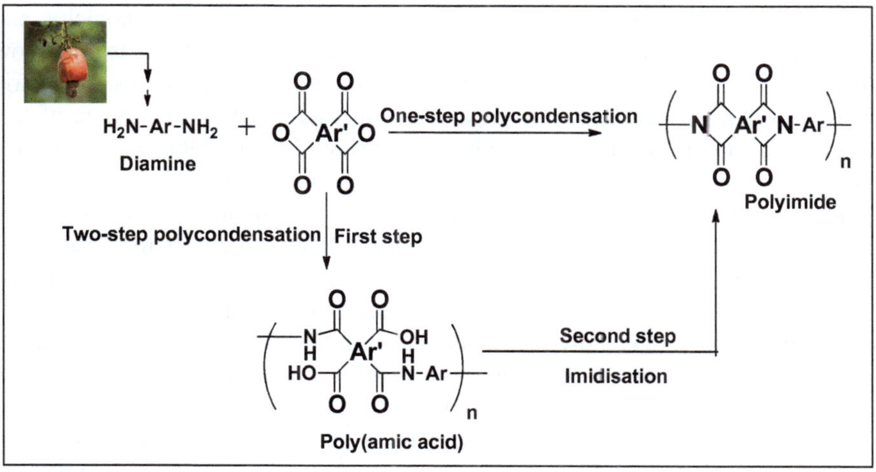

Scheme 9.1 Synthesis of polyimides based on CNSL-derived aromatic difunctional monomers

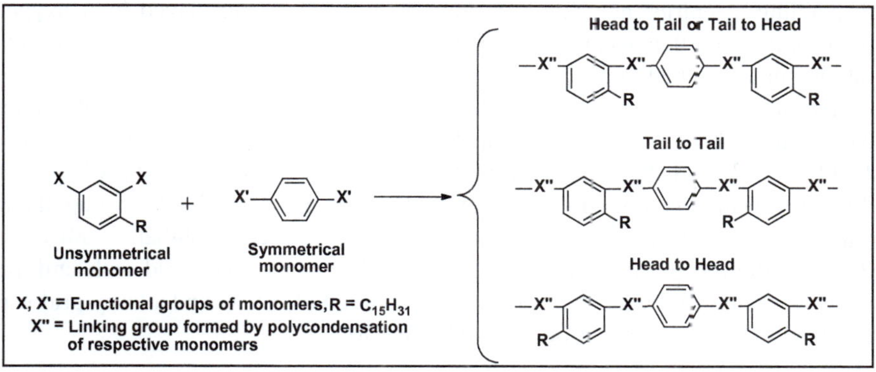

Fig. 9.6 Possible sequences arising due to constitutional isomerism

was attributed to interchain packing disruption and increased segmental mobility imparted by the pentadecyl chain.

Alkyl groups are known to improve the solubility, liquid crystal alignment properties, and dielectric and gas permeability properties of polymers. Polyimides have applications as organic alignment films due to good chemical resistance, abrasion resistance and thermal stability. Long alkyl chains and fluorinated alkyl side chains generate high pretilt angles of liquid crystals due to hydrophobic interaction with the polar liquid crystal molecules [74]. Polyimide corresponding to entry 15 (Table 9.4) provided pretilt angle in the range of 0.46°–6.6° when used in combination with various commercial liquid crystals. It showed potential as alignment layers for super-twisted nematic liquid crystal devices where a tilt angle of 5°–14° is required [74].

Dielectric properties of polymers are dependent on their hydrophobicity, and therefore, polymers containing pendent alkyl groups are expected to have low dielectric constant, which makes them useful as interlayer dielectrics. Dielectric constant (ε) was calculated from capacitance measurements over a range of frequencies for the polyimides represented by entries 15–17 of Table 9.4 and copolyimides composed of 20 mol% and 50 mol% diamine derived from cardanol (Table 9.2, entry 15) over a range of frequencies. Polyimides (Table 9.4, entries 15, 16 and copolyimide composed of 20 mol% diamine derived from cardanol) showed dielectric constant values in the range 0.4–1.7 which were lower than that of polyimide (Kapton) 'H' film ($\varepsilon = 3.5$), a preferred dielectric material in microelectronics applications. Polyimides derived from CNSL-based monomers offer a good choice for microelectronic devices as interlayer dielectrics owing to their low dielectric constant values ($\varepsilon < 3.5$) coupled with good processability [74].

The presence of pendent alkyl groups along the polymer backbone is important in gas separation applications since they cause an increase in the free volume of the resulting polymer and thereby increase the permeability. Gas permeability studies were performed on polyimides containing pendent pentadecyl chains corresponding to entries 15–16, 21–23 and 27–29 of Table 9.4. Polyimides corresponding to entries numbered 16 and 22, in Table 9.4, displayed high oxygen permeability coefficients and high O_2/N_2 selectivity, showing promise as gas separation membrane materials [74].

Transmission electron micrographs (TEMs) of grids coated with copolyimides based on diamine possessing pendent pentadecyl chain corresponding to entry 15 (Table 9.2). All the films showed holelike structures. Copolyimides based on 20 mol% of the diamine (Table 9.2, entry 15) with SiDA, 3,3',4,4'-oxydiphthalic dianhydride (ODPA) and 4,4'-(hexafluoroisopropylidene)diphthalic anhydride (6-FDA) showed resemblance with commercial grids. A holey grid with uniform holes was obtained by using a copolyimide solution of 0.1 %, and a lacey grid was obtained when the concentration was 0.07 % [74].

9.2.2.2 Aromatic Polyamides

Wholly aromatic polyamides are high-performance organic materials possessing excellent thermal and mechanical resistance arising from their stiff rodlike macromolecular chains that interact with each other via strong and highly directional hydrogen bonds [97, 117–119]. Aromatic polyamides can be prepared mainly by low-temperature interfacial polycondensation of a diamine and a diacid chloride or high-temperature solution polycondensation of an activated dicarboxylic acid and a diamine, besides other methods. Aromatic polyamides containing pendent pentadecyl chains are collected in Table 9.5.

Aromatic polyamides presented in Table 9.5, containing pendent pentadecyl chains, were synthesised by phosphorylation polycondensation of diamines with diacids (Table 9.5, entries 1–4, 10–14) or by low-temperature interfacial

Table 9.5 Aromatic polyamides containing pendent pentadecyl chains derived from difunctional monomers based on cardanol

Sr. No.	Polyamide	η_{inh} (dL/g)[a]	Solubility[b]		T_g (°C)[c]	T_{10} (°C)[d]	References
			DMAc	THF			
1	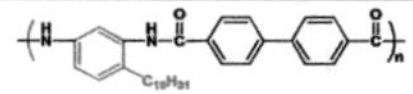	0.56	+	–	215	450 (531)	[71]
2	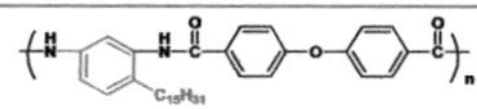	0.52	++	–	169 (232)	435 (510)	[71]
3	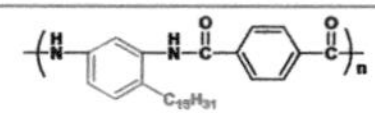	0.43	++	–	206 (295)	460 (524)	[71]
4	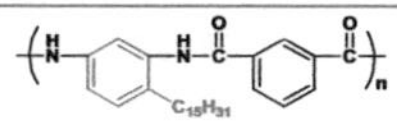	0.35	++	+–	189 (276)	430	[71]
5	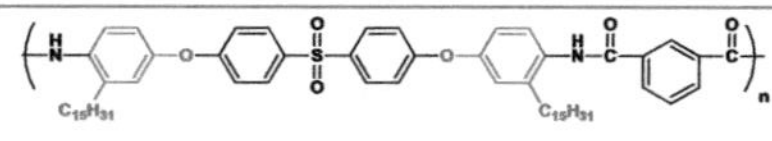	0.17*	++	++	N.R.	479##	[79]
6	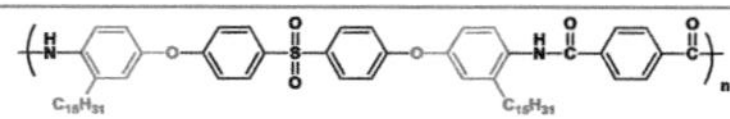	0.15	+	+–	N.R.	490##	[79]
7	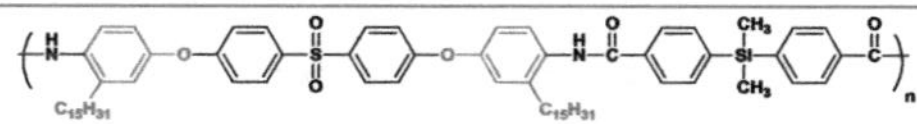	0.23*	++	++	N.R.	493##	[79]
8	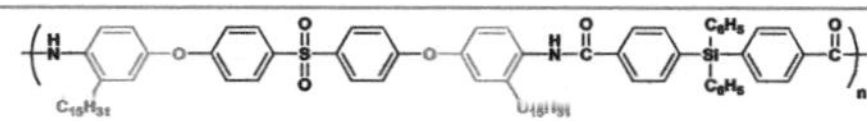	0.17*	++	++	N.R.	498##	[79]
9	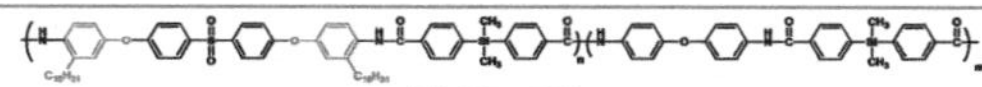 (50:50 mol%)	0.56	++	+	N.R.	482##	[79]
10	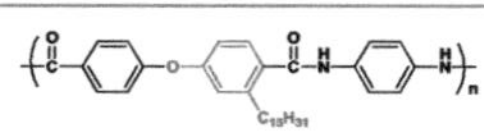	0.45	+–	N.R.	189	430	[80]

(continued)

Table 9.5 (continued)

Sr. No.	Polyamide	η_{inh} (dL/g)[a]	Solubility[b]		T_g (°C)[c]	T_{10} (°C)[d]	References
			DMAc	THF			
11	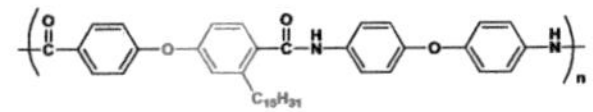	0.66	++	N.R.	160 (231)	425 (515)	[80]
12	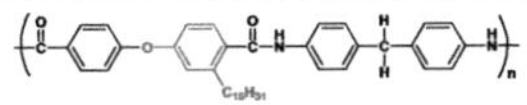	0.62	++	N.R.	139 (255)	440 (483)	[80]
13	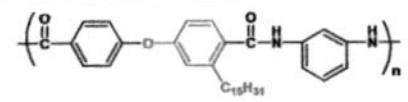	0.58	++	++	148 (250)	439 (467)	[80]
14	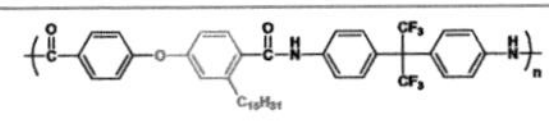	0.66	++	++	154	453	[80]
15	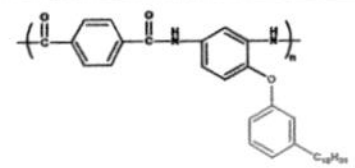	N.R.	+−	+−	164	422	[74]
16	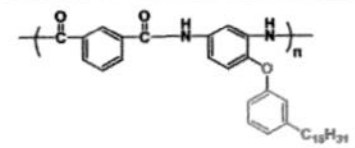	N.R.	+−	N.R.	150	422	[74]
17	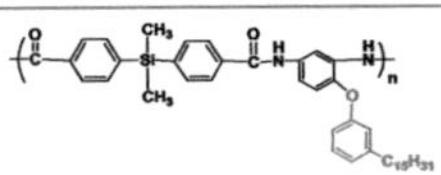	0.45	++	++	111	439	[74]
18	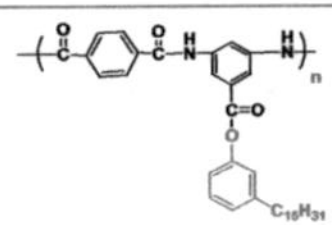	0.6**	++	N.R.	232	340	[74]

(continued)

Table 9.5 (continued)

Sr. No.	Polyamide	η_{inh} (dL/g)[a]	Solubility[b]		T_g (°C)[c]	T_{10} (°C)[d]	References
			DMAc	THF			
19	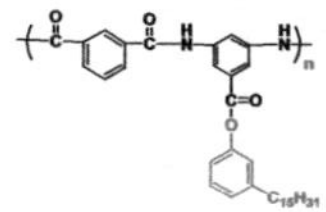	0.55**	++	++	221	365	[74]
20	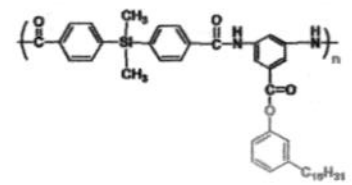	0.5**	++	++	186	354	[74]
21	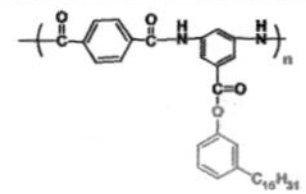	N.R.	–	–	279	417	[74]
22	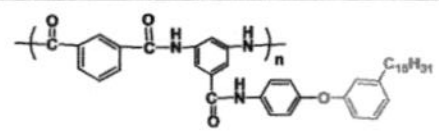	0.56**	++	++	262	401	[74]
23	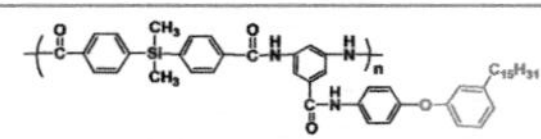	0.65**	++	++	225	437	[74]

Green colour indicates the structural components derived from cardanol

*η_{inh} was measured with 0.5 % (w/v) solution of polyamides in $CHCl_3$ at 30 ± 0.1 °C, **η_{inh} was measured with 0.5 % (w/v) solution of polyamides in NMP at 30 ± 0.1 °C, ##T_{max} maximum degradation temperature obtained from DTG curve

[a]η_{inh} was measured with 0.5 % (w/v) solution of polyamides in *N,N*-dimethylacetamide at 30 ± 0.1 °C

[b]Solubility measured at 3 % (w/v) concentration –, insoluble; +–, partially soluble or swollen; +, soluble on heating; ++, soluble at room temperature

[c]T_g values obtained from DSC curves from the second heating scans of polyamide samples, values in the bracket are T_g values of reference polyamides

[d]Temperature at which 10 % weight loss was observed from TGA under nitrogen atmosphere, values in the bracket are T_{10} values of reference polyimides

N.R. not reported

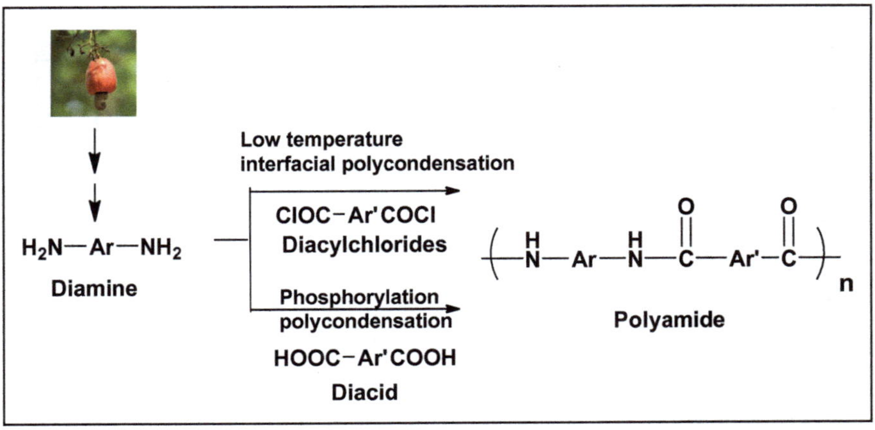

Scheme 9.2 Synthesis of polyamides based on CNSL-derived aromatic difunctional monomers

polycondensation of diamines with diacylchlorides (Table 9.5, entries 5–9, 15–23) (Scheme 9.2).

The reported inherent viscosities (0.15–0.66 dL/g) implied that the polyamides synthesised were of medium to high molecular weights. Polyamides represented in Table 9.5 formed self-standing, transparent and flexible films. ^{1}H-NMR studies of the polyamides containing pendent pentadecyl chains (Table 9.5, entries 1–4, 10–14) exhibited the appearance of four different peaks corresponding to the amide protons, indicating the existence of constitutional isomerism resulting from the use of an unsymmetrical monomer [71, 80]. Polyamides corresponding to entries 5, 7, 8, 13, 14, 17, 19, 20, 22 and 23 of Table 9.5 showed room temperature solubility in *N,N*-dimethylacetamide (DMAc) and tetrahydrofuran (THF), due to chain-disruptive effect of the pentadecyl chain. Jadhav et al. [79] synthesised a series of copolyamides by polycondensation of a mixture of pentadecyl chain containing diamine and ODA (Table 9.5, entry 9) at 25, 50 and 75 mol% with commercial diacids that demonstrated improvement in solubility with increasing incorporation of pentadecyl chain containing monomer.

The reported T_{10} values (above 340 °C) suggested that the synthesised polyamides possessed good thermal stability. The pentadecyl chain containing polyamides exhibited lower T_g values (111–279 °C) compared to reference polyamides (e.g. Table 9.5, entries 2–4, 11–13) due to increased segmental mobility imparted by the long alkyl chain. The silicon-containing polyamides (entries 17, 20, 23) showed good permeability towards gases such as nitrogen, oxygen, hydrogen and carbon dioxide [74].

9.2.2.3 Aromatic Poly(amideimide)s

Poly(amideimide)s are a class of polymers containing amide and heterocyclic imide units in the polymer backbone. Poly(amideimide)s combine the thermal stability of polyimides and ease of processability of polyamides. Aromatic poly(amideimide)s possess desirable characteristics for a variety of applications as they retain good mechanical properties at high temperatures and show easier processability [120]. Different methods have been reported for the synthesis of poly(amideimide)s which include (1) imide-forming reaction using amide-containing monomers [121, 122], (2) amide-forming reaction from imide-containing monomers such as dicarboxylic acids or diamines [123–126], and (3) by the reaction of diacylhydrazide and a dianhydride via a two-stage process [127, 128] and many other methods [129–133]. Table 9.6 lists poly(amideimide)s possessing pendent pentadecyl chains.

Poly(amideimide)s corresponding to entries 1 through 5 of Table 9.6 were synthesised by polycondensation of aromatic diacylhydrazide derived from cardanol with commercially available aromatic dianhydrides through a two-stage process (Scheme 9.3), while the poly(amideimide)s represented by entries 6–13 were prepared by polymerisation of commercially available aromatic dianhydrides and aromatic diamine monomers containing preformed amide linkage through a two-step process via a poly(amide amic acid) intermediate, followed by thermal imidisation (Scheme 9.4).

Inherent viscosities of poly(amideimide)s (0.37–1.23 dL/g), as presented in Table 9.6, indicated the formation of reasonably high molecular weight polymers. Poly(amideimide)s formed tough, transparent and flexible films from their solutions and were mostly soluble in DMAc and 1-methyl-2-pyrrolidinone (NMP) at room temperature due to the effective chain separation effect imparted by the long alkyl chain which provides a handle for interaction with solvent molecules. Poly(amideimide)s exhibited good thermal stability, as indicated by the TGA data (T_{10} values greater than 388 °C). The synthesised poly(amideimide)s showed lower T_g values (162–286 °C) compared to reference polymers (Table 9.6, entries 7–13) due to increased segmental mobility imparted by the pendent alkyl chain.

9.2.2.4 Aromatic Poly(azomethine)s

Aromatic poly(azomethine)s find applications in optoelectronics [134–140] and are high-performance fibre and film-forming polymers with remarkable thermal stability, high strength, high modulus [141] and an ability to form chelates [142–144]. However, their high melting temperatures and poor solubilities in common organic solvents make both their characterisation and processing difficult. Therefore, it is desirable to improve the processability of poly(azomethine) s. Table 9.7 represents aromatic poly(azomethine)s containing pendent pentadecyl chains synthesised from monomers derived from cardanol.

Table 9.6 Aromatic poly(amideimide)s containing pendent pentadecyl chains derived from difunctional monomers based on cardanol

Sr. No.	Poly(amideimide)	η_{inh}(dL/g)[a]	Solubility[b]		T_g (°C)[c]	T_{10} (°C)[d]	References
			DMAc	NMP			
1.	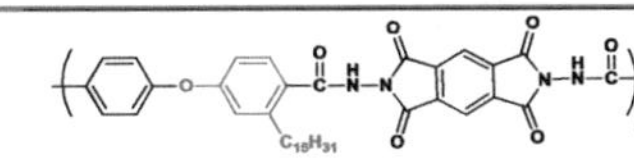	0.62*	++	++	168	388	[82]
2.	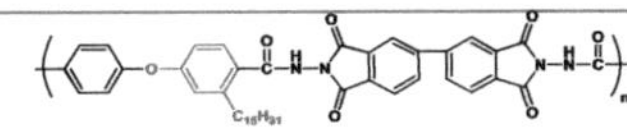	0.62*	++	++	198	410	[82]
3.	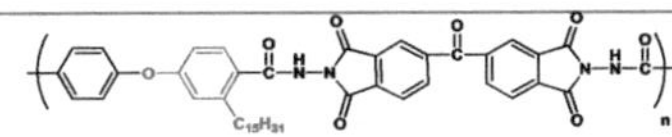	0.64*	++	+−	162	389	[82]
4.	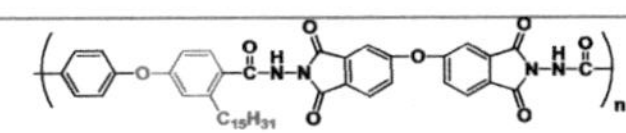	0.60*	++	++	164	398	[82]
5.	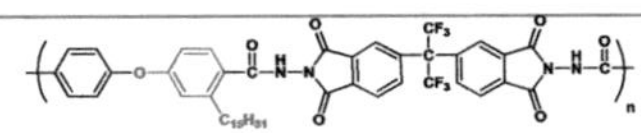	0.60*	++	++	178	394	[82]
6.	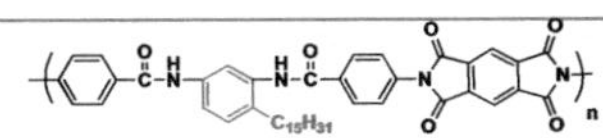	0.72	++	++	286	470 (490)	[77]
7.	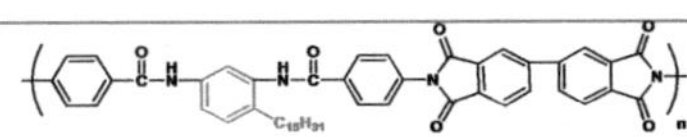	1.04	++	++	244 (316)	480 (515)	[77]
8.	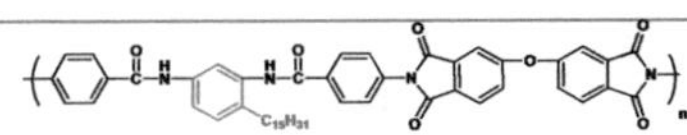	0.73	++	++	206 (306)	470 (500)	[77]

Table 9.6 (continued)

Sr. No.	Poly(amideimide)	η_{inh}(dL/g)[a]	Solubility[b]		T_g (°C)[c]	T_{10} (°C)[d]	References
			DMAc	NMP			
9.	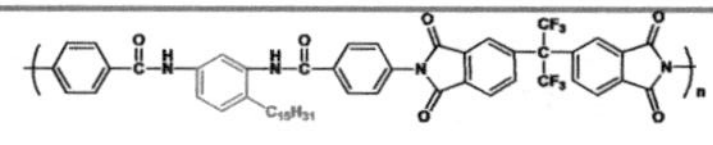	1.23	++	++	223 (330)	475 (510)	[77]
10.	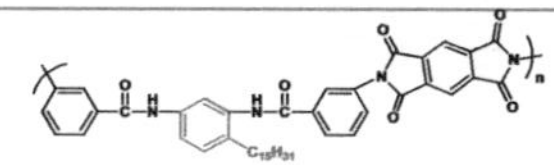	0.37	++	++	233 (331)	460 (490)	[77]
11.	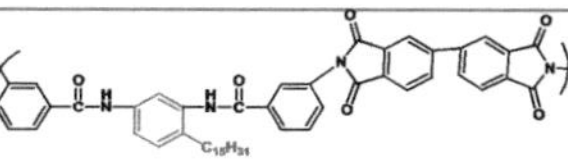	0.51	++	++	208 (301)	480 (490)	[77]
12.	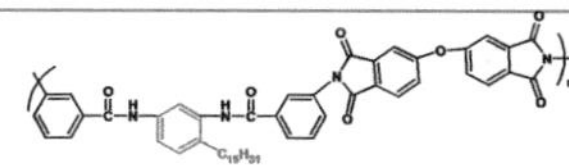	0.44	++	++	185 (275)	470 (470)	[77]
13.	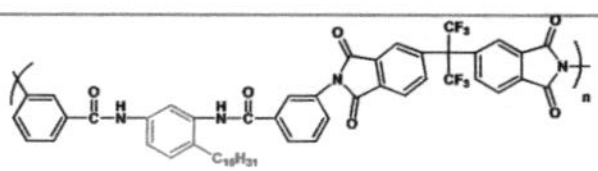	0.42	++	++	193 (287)	475 (485)	[77]

Green colour indicates the structural components derived from cardanol
*η_{inh} was measured with 0.5 % (w/v) solution of poly(amideimide)s in DMAc at 30 ± 0.1 °C
[a]η_{inh} was measured with 0.5 % (w/v) solution of poly(amideimide) s in $CHCl_3$ at 30 ± 0.1 °C
[b]Solubility measured at 3 % (w/v) concentration; +−, partially soluble or swollen; ++, soluble at room temperature
[c]T_g values obtained from DSC curves from the second heating scans of poly(amideimide) samples, values in the bracket T_g values of reference polymers
[d]Temperature at which 10 % weight loss was observed from TGA under nitrogen atmosphere, values in the bracket are T_{10} values of reference polymers

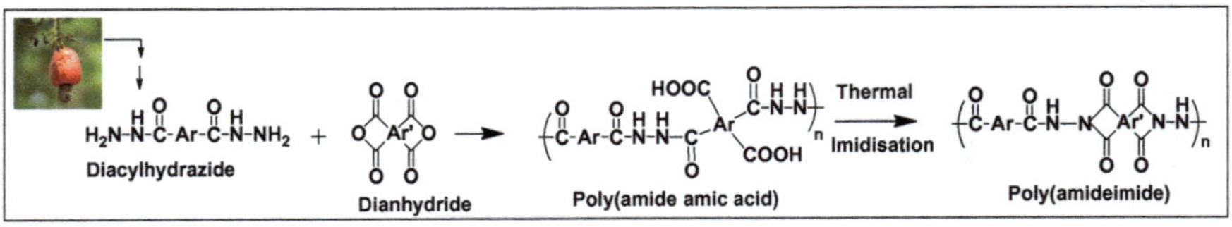

Scheme 9.3 Synthesis of poly(amideimide)s based on CNSL-derived aromatic diacylhydrazide monomers and dianhydride

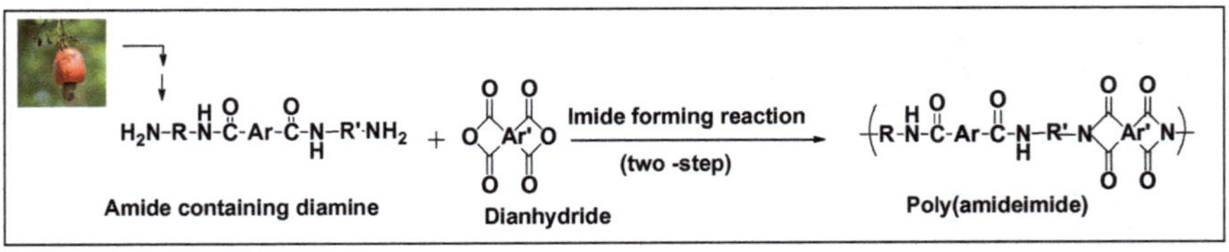

Scheme 9.4 Synthesis of poly(amideimide)s based on CNSL-derived aromatic diamine monomers containing preformed amide linkages

Poly(azomethine)s reported in Table 9.7 were synthesised by solution polycondensation of diamines containing pentadecyl chains with commercially available aromatic dialdehydes (Scheme 9.5).

The reported inherent viscosity values (0.33–0.70 dL/g) showed that the poly(azomethine)s were of medium to reasonably high molecular weights. Transparent, flexible and stretchable films of poly(azomethine)s could be cast from chloroform solution. Poly(azomethine)s corresponding to entries 4–5 of Table 9.7 exhibited constitutional isomerism due to the unsymmetrical diamine as indicated by ^{1}H-NMR studies [75]. The synthesised poly(azomethine)s exhibited good solubility in chloroform and THF at room temperature.

The T_{10} values (above 425 °C), as exhibited in Table 9.7, indicated that synthesised poly(azomethine)s possessed good thermal stability. The low T_g (16–55 °C) values are due to increase in segmental mobility and the chain-disruptive effect of the long alkyl chain.

9.2.2.5 Polyesters

Polyesters are an important class of high-performance polymers. They find applications in a wide variety of areas such as automobile, aviation and electronic industries by virtue of their attractive electrical and mechanical properties [145–147]. However, polyarylates are generally difficult to process because of their limited solubility in common organic solvents and high glass transition and melting temperatures. Polyesters are prepared by two routes, acid chloride route (three different processes: interfacial polycondensation, low-temperature solution polycondensation and high-temperature solution polycondensation) and transesterification

Table 9.7 Aromatic poly(azomethine)s containing pendent pentadecyl chains derived from diamine monomers based on cardanol

Sr. No.	Poly(azomethine)	η_{inh}(dL/g)[a]	Solubility[b]		T_g (°C)[c]	T_{10} (°C)[d]	References
			$CHCl_3$	THF			
1.	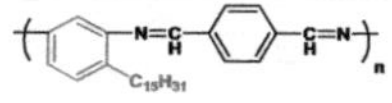	0.38	++	+−	55	440	[71]
2.	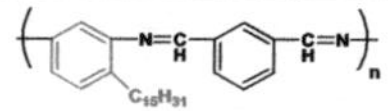	0.36	++	++	16	440	[71]
3.	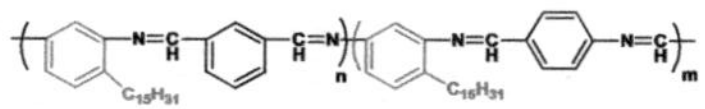 (50:50 mol%)	0.33	++	++	45	425	[71]
4.	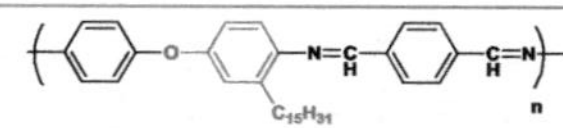	0.70	++	+−	48	435	[75]
5.	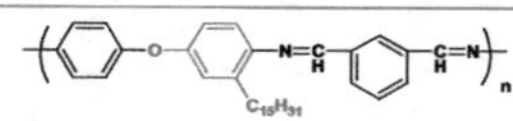	0.56	++	++	21	438	[75]
6.	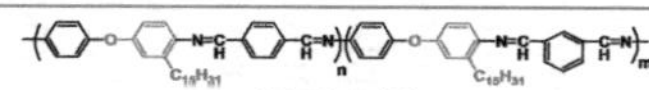 (50:50 mol%)	0.52	++	++	27	441	[75]

Green colour indicates the structural components derived from cardanol
[a]η_{inh} was measured with 0.5 % (w/v) solution of poly(azomethine)s in $CHCl_3$ at 30 ± 0.1 °C
[b]Solubility measured at 3 % (w/v) concentration; +−, partially soluble or swollen; ++, soluble at room temperature
[c]T_g values obtained from DSC curves from the second heating scans of poly(azomethine) samples
[d]Temperature at which 10 % weight loss was observed from TGA under nitrogen atmosphere

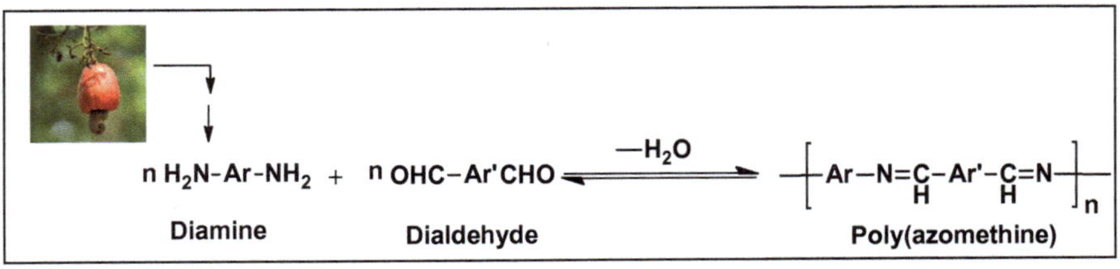

Scheme 9.5 Synthesis of aromatic poly(azomethine)s based on CNSL-derived aromatic diamine monomers

route. Polyesters synthesised from monomers containing pentadecyl chain are represented in Table 9.8.

Polyesters corresponding to entries 1, 2, 6–11 (Table 9.8) were synthesised by low-temperature interfacial polycondensation of diacylchlorides with bisphenols, while those corresponding to entries 3–5 were synthesised through direct phosphorylation of diacid and diphenol in the presence of an activating agent such as diphenyl phosphate (Scheme 9.6) and the polyesters represented by entries 12 and 13 were synthesised by melt polymerisation using acetic anhydride.

The inherent viscosities, as reported in Table 9.8 (0.47–1.74 dL/g), indicated the formation of reasonably high molecular weight polymers. The synthesised polyesters formed self-standing flexible films. Aromatic polyesters possessing pendent pentadecyl chains exhibited good solubility in chloroform and tetrahydrofuran due to the pendent alkyl chain which provides a handle for effective interaction with solvent molecules. The reported T_{10} values (above 340 °C) indicated that polyesters (Table 9.8) possessed good thermal stability. Polyesters displayed low T_g (29–183 °C) values (Table 9.8, entries 9, 10). Aromatic polyesters containing pendent pentadecyl chains demonstrated a large gap in the values of T_g and T_{10}, thus providing a wide processing window.

The attainment of solubility in common organic solvents such as chloroform and tetrahydrofuran of conventional aromatic polyester based on bisphenol A and terephthalic acid chloride was demonstrated [57, 65] by incorporation of small amount (5–15 mol%) of cardanol-derived bisphenols (Table 9.8, entries 5 and 11).

Abraham et al. [87] synthesised a series of copolyesters (Table 9.8, entry 12) by polycondensation of a mixture of AB-type monomer (Table 9.2, entry 34) and 4-hydroxybenzoic acid in concentrations of 50, 55, 60 and 65 mol%. Another series of copolyesters (Table 9.8, entry 13) were synthesised by polycondensation of mixture of the AB-type monomer (Table 9.2, entry 34) and 3-bromo-4-hydroxybenzoic acid at combinations of 50, 55, 60 and 65 mol%. These copolyesters exhibited liquid crystalline phase transitions at lower temperatures compared to homopolyesters and showed potential for melt processability.

Table 9.8 Polyesters derived from difunctional monomers based on cardanol

Sr. No.	Polyester	η_{inh}(dL/g)[a]	Solubility[b]		T_g (°C)[c]	T_{10} (°C)[d]	References
			$CHCl_3$	THF			
1.	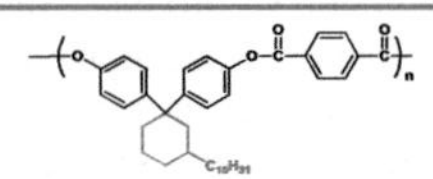	1.45	++	++	94	451 (480)	[62]
2.	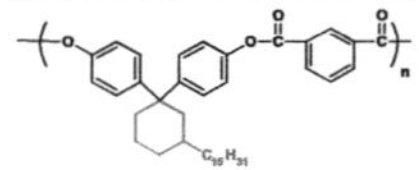	0.66	++	+	74	439	[62]
3.	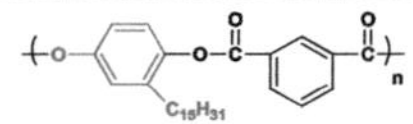	0.53	++	++	88	440	[56]
4.	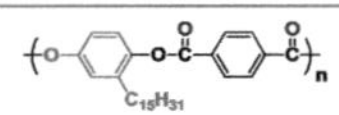	0.47	–	+	91	450	[56]
5.	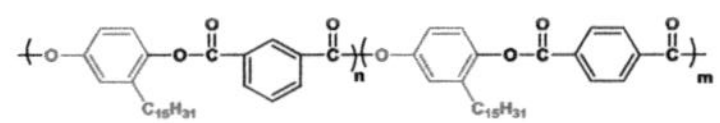 (50:50 mol%)	0.51	++	++	95	430	[56]
6.	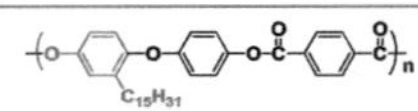	1.19	++	++	47	430	[57]
7.	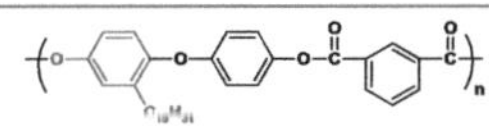	0.74	++	++	29	425	[57]
8.	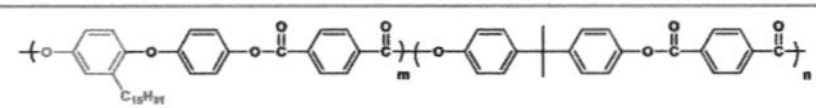 (15:85 mol%)	0.72	++	+–	160	455	[57]

Table 9.8 (continued)

Sr. No.	Polyester	η_{inh}(dL/g)[a]	Solubility[b]		T_g (°C)[c]	T_{10} (°C)[d]	References
			$CHCl_3$	THF			
9.	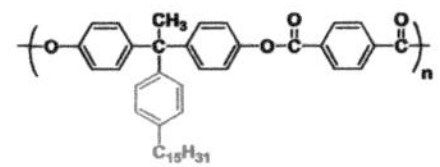	0.91	++	++	82(244)	406 (533)	[65]
10.	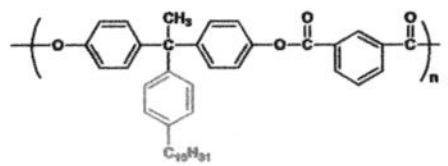	0.72	++	++	63(213)	400 (525)	[65]
11.	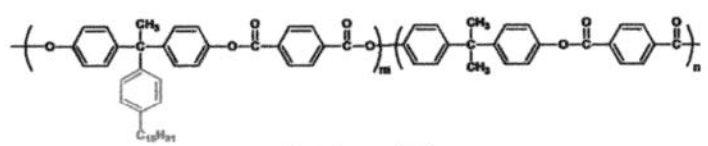 (5:95 mol%)	1.74	++	+−	183	458	[65]
12.	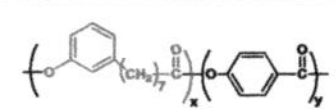 (50:50 mol%)	0.47	N.R.	N.R.	73	360*	[87]
13.	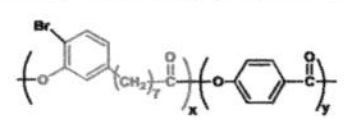 (50:50 mol%)	0.53	N.R.	N.R.	72	340*	[87]

Green colour indicates the structural components derived from cardanol
*T_5: temperature at which 5 % weight loss was observed from TGA under nitrogen atmosphere
[a]η_{inh} was measured with 0.5 % (w/v) solution of polyesters in $CHCl_3$ at 30 ± 0.1 °C
[b]Solubility measured at 3 % (w/v) concentration –, insoluble; +−, partially soluble or swollen; +, soluble on heating; ++, soluble at room temperature
[c]T_g values obtained from DSC curves from the second heating scans of polyesters, values in the bracket are T_g values of reference polymers
[d]Temperature at which 10 % weight loss was observed from TGA under nitrogen atmosphere, values in the bracket are T_{10} values of reference polymers
N.R. not reported

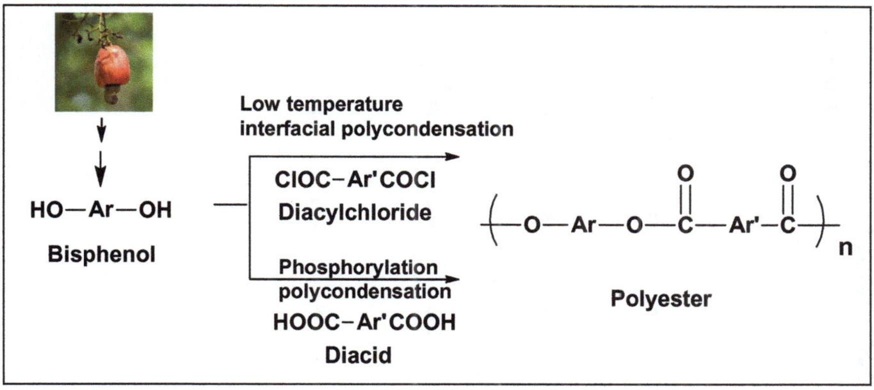

Scheme 9.6 Synthesis of polyesters based on CNSL-derived difunctional monomers

9.2.2.6 Aromatic Poly(esterimide)s

Poly(esterimide)s possess ester and imide functionality at regular intervals and combine the thermal stability of polyimides with the processability of polyesters [148–151]. The five general methods used in the synthesis of poly(esterimide)s are (1) polycondensation of an ester group-containing dianhydride with a diamine [152], (2) polycondensation of an ester group-containing diamine with a dianhydride [153], (3) reaction between a dicarboxylic acid containing an imide ring with a bisphenol [154–159], (4) reaction between a bisphenol containing an imide ring with a dicarboxylic acid and (5) self-reaction of an imide-containing phenol acid monomer [160–162]. Table 9.9 lists poly(esterimide)s containing pendent pentadecyl chains.

Poly(esterimide)s presented in Table 9.9 were synthesised by polycondensation of diacid containing preformed imide linkages and commercially available bisphenols in the presence of diphenyl phosphate as an activating agent (Scheme 9.7).

The reported inherent viscosities of poly(esterimide)s containing pendent pentadecyl chains (0.54–0.83 dL/g in $CHCl_3$) indicated the formation of medium to high molecular weight polymers. Poly(esterimide)s displayed excellent solubility in chloroform and tetrahydrofuran and formed tough, transparent and flexible films from their solutions. Poly(esterimide)s exhibited good thermal stability (T_{10} values above 450 °C), and the T_g values were in the range of 145–198 °C, which indicated a large depression due to increased mobility imparted by the pendent pentadecyl chain. A large difference in T_g and T_{10} values of poly(esterimide)s offers a broad processing window. This provides an opportunity for these poly(esterimide)s to be melt-processed or compression-moulded.

Table 9.9 Aromatic poly(esterimide)s containing pendent pentadecyl chains derived from difunctional monomers based on cardanol

Sr. No.	Poly(esterimide)	η_{inh} (dL/g)[a]	Solubility[b]		T_g (°C)[c]	T_{10} (°C)[d]	References
			$CHCl_3$	THF			
1.	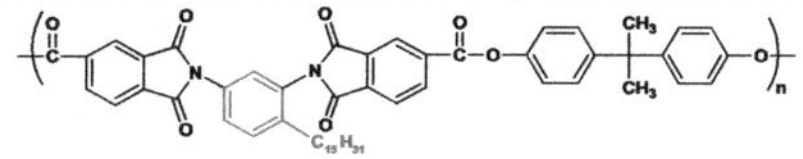	0.66	++	++	151	455	[81]
2.	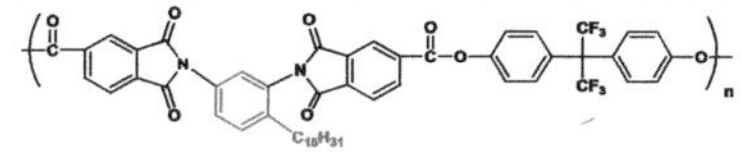	0.58	++	++	153	455	[81]
3.	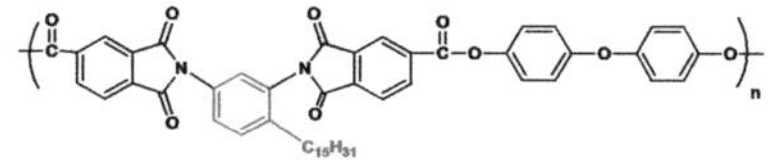	0.54	++	++	145	450	[81]
4.	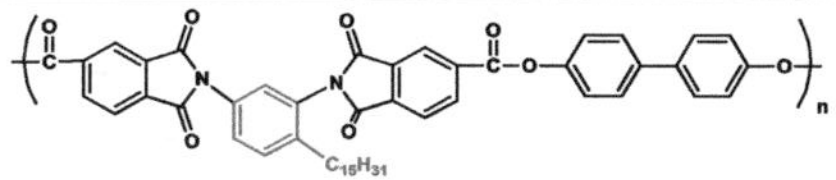	0.72	++	++	183	460	[81]
5.	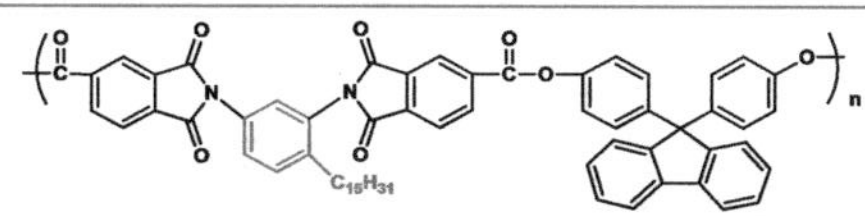	0.83	++	++	198	470	[81]

Green colour indicates the structural components derived from cardanol
[a]η_{inh} was measured with 0.5 % (w/v) solution of poly(esterimide)s in $CHCl_3$ at 30 ± 0.1 °C
[b]Solubility measured at 3 % (w/v) concentration; ++, soluble at room temperature
[c]T_g values obtained from DSC curves from the second heating scans of poly(esterimide) samples
[d]Temperature at which 10 % weight loss was observed from TGA under nitrogen atmosphere

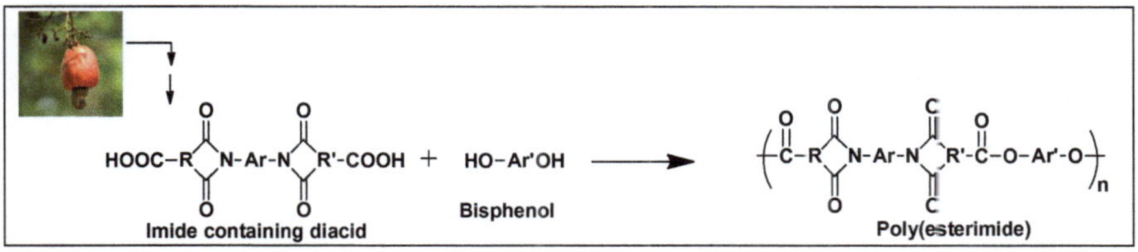

Scheme 9.7 Synthesis of poly(esterimide)s based on CNSL-derived aromatic difunctional monomers

9.2.2.7 Poly(arylene ether)s

Poly(arylene ether)s such as poly(ether ketone)s and poly(ether sulphone)s constitute an important class of high performance thermoplastics because of their attractive properties such as high thermo-oxidative and chemical stability and excellent mechanical strength. Poly(arylene ether)s are therefore extensively used in aviation, spacecraft, automobile, food processing sectors, medical devices, gas separation membranes, ultrafiltration, fuel cells, electronic/electrical industries, etc. Poly(arylene ether)s can be synthesised by a number of methods such as (1) electrophilic substitution, (2) nucleophilic aromatic substitution [163, 164] and (3) other methods [165–169]. Table 9.10 summarises the solubility and thermal properties of poly(arylene ether)s containing pendent pentadecyl chains.

Poly(arylene ether)s presented in Table 9.10 were synthesised through nucleophilic aromatic substitution polycondensation of aromatic biphenols and activated aromatic difluorides (Scheme 9.8).

The reported inherent viscosity values of poly(arylene ether)s (0.50–1.27 dL/g) suggested the formation of reasonably high molecular weight polymers. Poly(arylene ether)s exhibited good solubility in chloroform and THF and formed transparent, tough and self-standing films from their solutions. The existence of constitutional isomerism in poly(arylene ether)s (Table 9.10 entries 7–9) was evidenced from detailed NMR spectroscopy studies [58]. The synthesised poly(arylene ether)s possessed bulky pendent aryl groups (Table 9.10 entries 1–3) as well as cardo moieties such as cyclohexylidene groups (Table 9.10 entries 4–6) along with pendent pentadecyl chains. Consequently, they exhibited excellent solubility in common organic solvents such as chloroform and tetrahydrofuran due to packing disruptive effects of the pendent groups. The pendent pentadecyl chains along the polymer backbone act as a handle for interaction with solvent molecules. T_{10} values (above 410 °C) indicated that these polymers possessed good thermal stability. Poly(arylene ether)s exhibited low glass transition temperatures (35–95 °C) due to increased polymer chain mobility imparted by the pendent pentadecyl chains. The data on the properties of polymers suggest that bisphenols containing pentadecyl chains are useful comonomers to tune the thermal and other properties of commercial poly(arylene ether)s.

Table 9.10 Aromatic poly(arylene ether)s containing pendent pentadecyl chains derived from difunctional monomers based on cardanol

Sr. No.	Poly(arylene ether)	η_{inh} (dL/g)[a]	Solubility[b]		T_g (°C)[c]	T_{10} (°C)[d]	References
			$CHCl_3$	THF			
1.	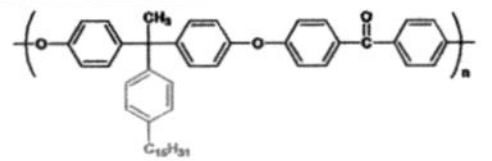	1.07	++	++	68	440	[64]
2.	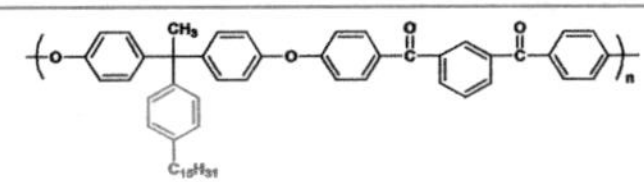	0.70	++	++	75	429	[64]
3.	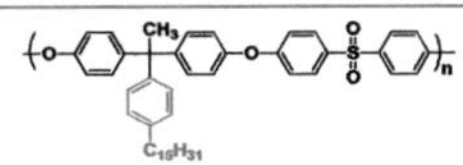	0.64	++	++	93	416	[64]
4.	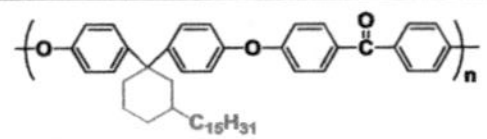	1.27	++	++	69	459	[64]
5.	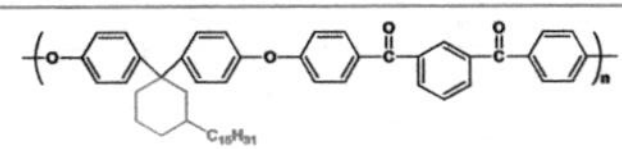	1.0	++	++	78	416	[64]

Table 9.10 (continued)

Sr. No.	Poly(arylene ether)	η_{inh} (dL/g)[a]	Solubility[b]		T_g (°C)[c]	T_{10} (°C)[d]	References
			$CHCl_3$	THF			
6.	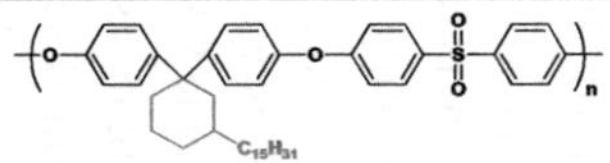	1.12	++	++	95	443	[64]
7.	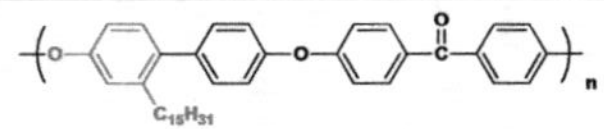	0.81	++	++	35	455	[58]
8.	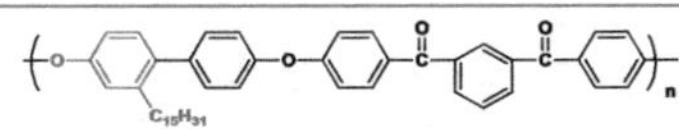	0.50	++	++	51	455	[58]
9.	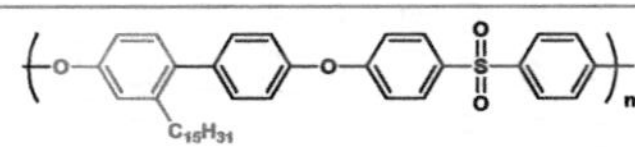	0.67	++	++	60	410	[58]

Green colour indicates the structural components derived from cardanol
[a] η_{inh} was measured with 0.5 % (w/v) solution of poly(arylene ether)s in $CHCl_3$ at 30 ± 0.1 °C
[b] Solubility measured at 3 % (w/v) concentration; ++, soluble at room temperature
[c] T_g values obtained from DSC curves from the second heating scans of poly(arylene ether) samples
[d] Temperature at which 10 % weight loss was observed from TGA under nitrogen atmosphere

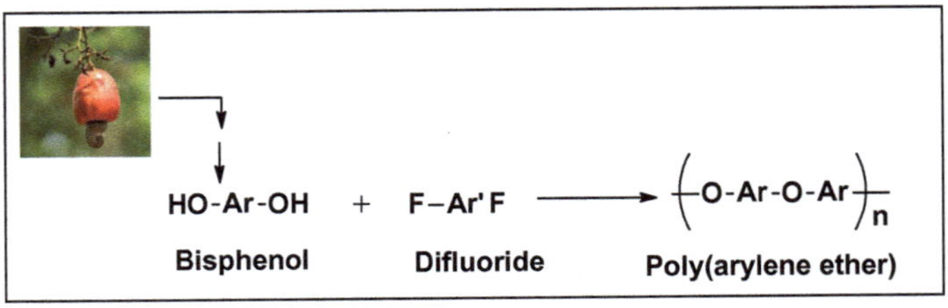

Scheme 9.8 Synthesis of poly(arylene ether)s based on CNSL-derived aromatic biphenol monomers

9.2.2.8 Polyhydrazides and Poly(1,3,4-oxadiazole)s

Polyhydrazides are interesting polymeric materials that have wide applications as fibres and membranes due to their excellent dyeability, improved elasticity and fair absorption characteristics. Furthermore, polyhydrazides serve as precursors to poly(1,3,4-oxadiazole)s and polytriazoles. Poly(1,3,4-oxadiazole)s are a useful class of high-performance polymers which find wide applications as polymer light-emitting diodes (PLEDs) and other polymer electronic devices by virtue of their optoelectronic properties [170, 171]. The most common method for the preparation of poly(1,3,4-oxadiazole)s is thermal (heating up to 300 °C) or chemical (heating in the presence of dehydrating agents such as phosphorous oxychloride) cyclodehydration of polyhydrazides, which in turn are synthesised via polycondensation of diacylhydrazides with dicarboxylic acids or their derivatives such as diester and diacid chloride [172, 173]. Another method which has been extensively used to prepare poly(1,3,4-oxadiazole)s is a one-step process: polycondensation of aromatic dicarboxylic acids or their derivatives such as ester, nitrile, and amide, with hydrazine (H_2NNH_2) or hydrazine sulphate ($H_2NNH_2 \cdot H_2SO_4$) in fuming sulphuric acid (oleum) [174] or in a mixture of strong acids [175, 176]. Table 9.11 depicts the solubility and thermal properties of various aromatic polyhydrazides and poly(1,3,4-oxadiazole)s synthesised from monomers derived from cardanol.

Aromatic polyhydrazides and poly(1,3,4-oxadiazole)s containing pendent pentadecyl chains, as depicted in Table 9.11, were synthesised by polycondensation of diacylhydrazides with diacyl chlorides to form polyhydrazides which were chemically cyclodehydrated to afford poly(1,3,4-oxadiazole)s (Scheme 9.9).

Poly(1,3,4-oxadiazole)s were of reasonably high molecular weights as implied by the reported inherent viscosities (0.49–0.62 dL/g) and they formed self-standing flexible films. Poly(1,3,4-oxadiazole)s showed good solubility in chloroform and tetrahydrofuran due to disruption of interchain packing caused by pendent pentadecyl chains. The reported T_{10} values (above 425 °C) indicated the formation of polymers with good thermal stability. Poly(1,3,4-oxadiazole)s displayed low glass transition temperatures (90–192 °C) due to increased chain mobility imparted by the pentadecyl chain. Electrochemical fluorescence switching studies were carried out using poly(1,3,4-oxadiazole)s corresponding to entries 4 and 5 of

Table 9.11 Aromatic polyhydrazides and poly(1,3,4-oxadiazole)s containing pendent pentadecyl chains derived from difunctional monomers based on cardanol

Sr. No.	Polyhydrazide and poly(1,3,4-oxadiazole)	η_{inh} (dL/g)[a]	Solubility[b]		T_g (°C)[c]	T_{10} (°C)[d]	References
			$CHCl_3$	THF			
1.	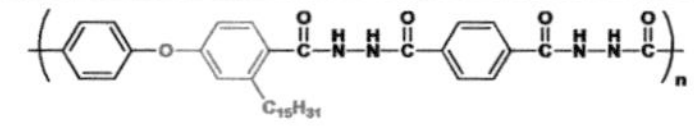	0.64*	–	–	166	N.R.	[83]
2.	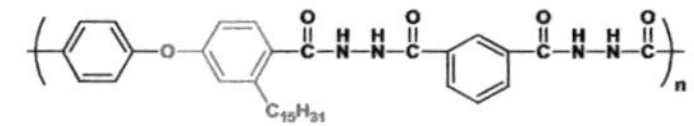	0.53*	–	–	143	N.R.	[83]
3.	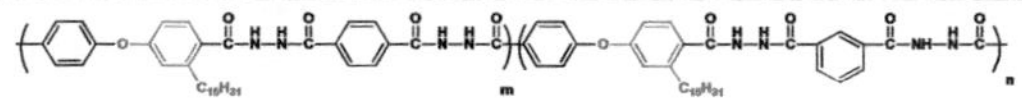 (50:50)	0.66*	–	–	148	N.R.	[83]
4.	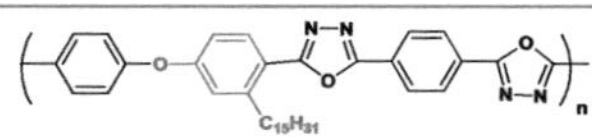	0.49	+–	+–	102	435	[83]
5.	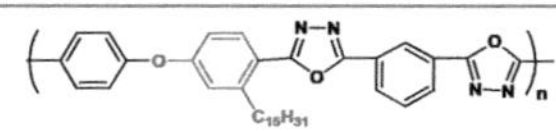	0.50	++	++	90	433	[83]
6.	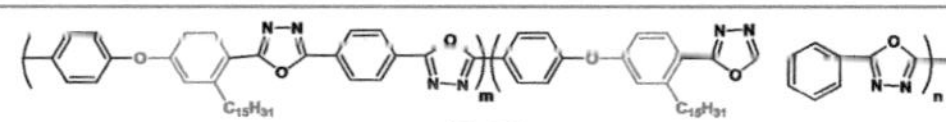 (50:50)	0.53	++	++	100	449	[83]
7.	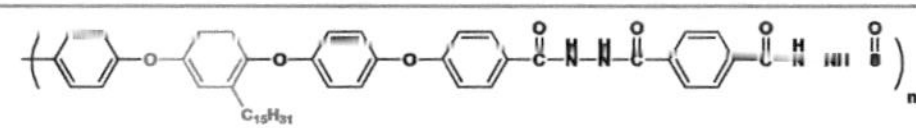	0.72*	–	–	192	N.R.	[59]

Table 9.11 (continued)

Sr. No.	Polyhydrazide and poly(1,3,4-oxadiazole)	η_{inh} (dL/g)[a]	Solubility[b]		T_g (°C)[c]	T_{10} (°C)[d]	References
			$CHCl_3$	THF			
8.	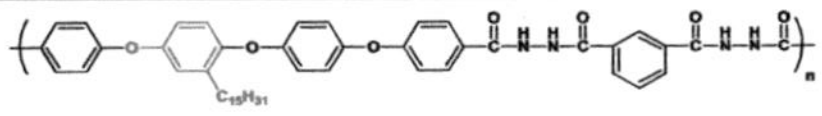	0.69*	–	–	175	N.R.	[59]
9.	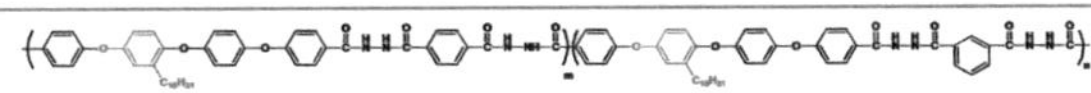 (50:50)	0.65*	–	–	183	N.R.	[59]
10.	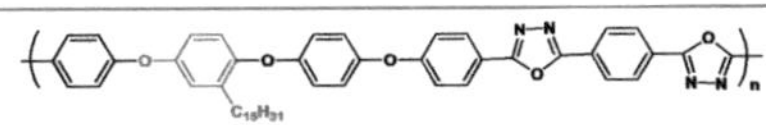	0.56	++	++	103	440	[59]
11.	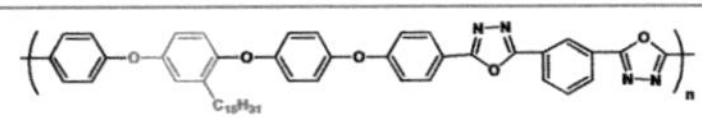	0.54	++	++	92	430	[59]
12.	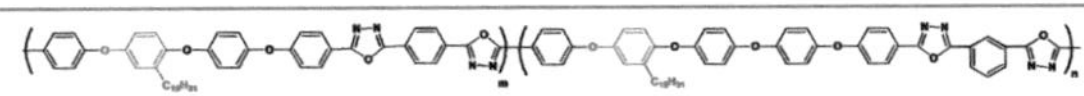 (50:50)	0.62	++	++	97	425	[59]

Green colour indicates the structural components derived from cardanol

*η_{inh} was measured with 0.5 % (w/v) solution of polyhydrazides in DMAc at 30 ± 0.1 °C

[a]η_{inh} was measured with 0.5 % (w/v) solution of poly(1,3,4-oxadiazole)s in $CHCl_3$ at 30 ± 0.1 °C

[b]Solubility measured at 3 % (w/v) concentration –, insoluble; +–, partially soluble or swollen; ++, soluble at room temperature

[c]T_g values obtained from DSC curves from the second heating scans of polyhydrazides/poly(1,3,4-oxadiazole)s samples

[d]Temperature at which 10 % weight loss was observed from TGA under nitrogen atmosphere

N.R. not reported

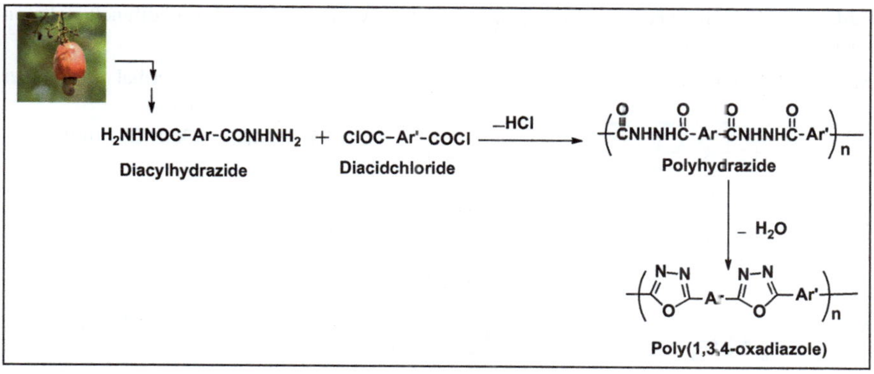

Scheme 9.9 Synthesis of polyhydrazides and poly(1,3,4-oxadiazole)s based on CNSL-derived aromatic difunctional monomers

Table 9.11 which showed reversible fluorescence switching between dark (n-doping) and bright (neutral) states with maximum on/off ratio of 2.5 and cyclability longer than 1000 cycles [84].

9.2.2.9 Polyurethanes

Polyurethanes are one of the most versatile plastic materials. Polyurethanes can be either rigid or flexible and are the materials of choice for a broad range of end-use applications such as manufacture of non-flexible, high-resistance foam seating, rigid foam insulation panels, microcellular foam seals and gaskets, synthetic fibres (e.g. Spandex), carpet underlay, and hard-plastic parts. Polyurethanes are traditionally and most commonly formed by reaction of di- or polyisocyanates with diols/polyols possessing two or more functional groups per molecule.

A high ortho novolac resin, synthesised by reacting cardanol with formaldehyde, was condensed with diphenylmethane diisocyanates to produce rigid polyurethanes [177]. Mannich polyols, excellent replacements for petrochemical-derived polyols, were synthesised from cardanol which had good physico-mechanical and fire-retardant properties [178]. Subsequently, rigid polyurethane foams were made from polyols synthesised through epoxidation of side chain unsaturation of cardanol [179]. Polyols bearing urethane groups were synthesised from cardanol [180]. Similarly, several other polyols were also synthesised from cardanol [181, 182].

Polyurethanes depicted in Table 9.12 were synthesised by solution polycondensation of bisphenols derived from cardanol and diisocyanates (Scheme 9.10).

Bhunia et al. [60], in 1998, synthesised polyurethanes based on cardanol (Table 9.12, entry 1), which exhibited semi-crystalline behaviour as evidenced by differential scanning calorimetry (DSC) and dynamic mechanical thermal analysis

Table 9.12 Polyurethanes containing pendent pentadecyl chains derived from difunctional monomers based on cardanol

Sr. No.	Polyurethane	η_{inh} (dL/g)	T_g(°C)[a]	Thermal data[b]	References
1.	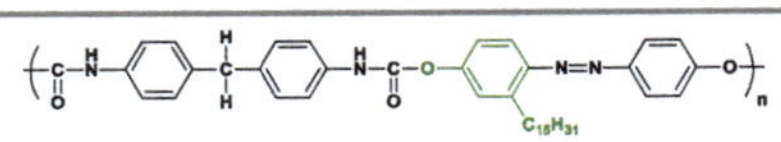	1.85*	N.R.	360 °C#	[60]
2.	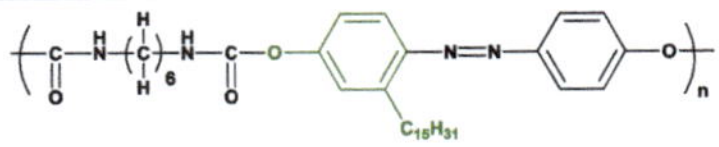	1.68	−70	340 °C#	[61]
3.	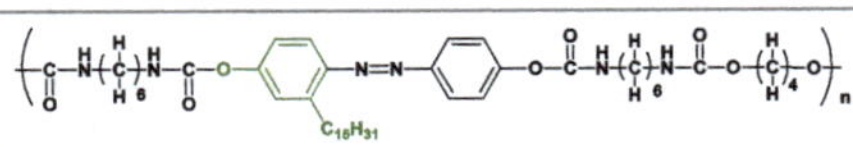	1.75	−57, −20	326 °C#	[61]

Green colour indicates the structural components derived from cardanol
*η_{inh} was determined at 35 °C in DMF using Ostwald's viscometer
#T_{max} maximum degradation temperature obtained from DTG curve
[a]T_g values obtained from DSC curves from the second heating scans of polyurethane samples
N.R. not reported
[b]Thermal measurements were performed by employing a 20 ml/min flow of dry nitrogen as purge gas [61]

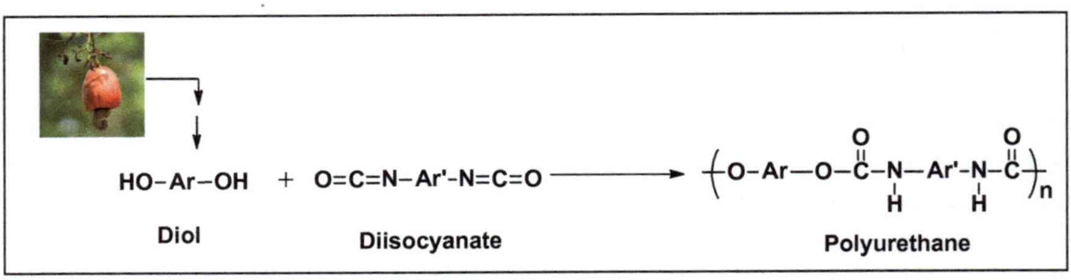

Scheme 9.10 Synthesis of polyurethanes based on CNSL-derived bisphenols and diisocyanates

(DMTA). Subsequently, polyurethanes (Table 9.12, entries 2, 3) were synthesised which exhibited semi-crystallinity as supported by DSC and wide-angle X-ray diffraction.

Aromatic diisocyanates derived from cardanol as depicted in Table 9.2 (entries 31, 32) are potential monomers for polyurethanes and polyureas. However, they have not yet been utilised for polymer synthesis.

9.2.3 Poly(m-phenylenevinylene)s

Poly(*m*-phenylenevinylene)s are a class of π-conjugated polymers that have important applications in polymeric light-emitting diodes. *Meta*-substitution

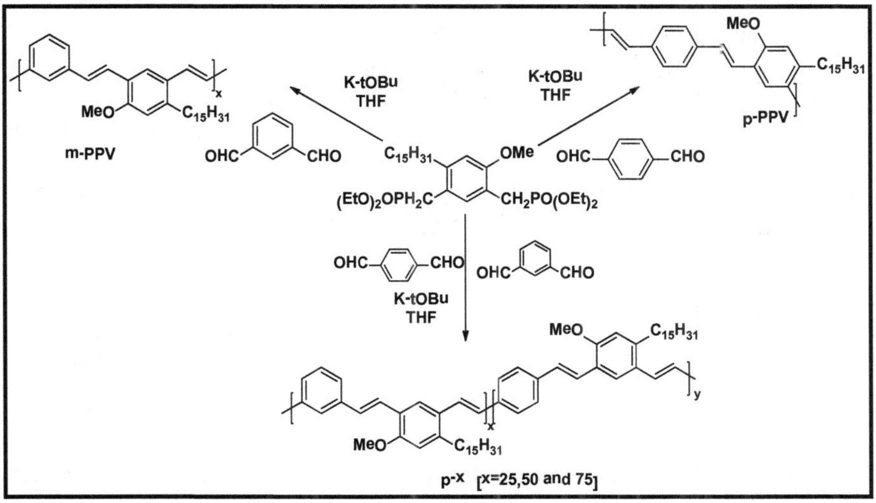

Scheme 9.11 Synthesis of poly(*m*-phenylenevinylene)s based on cardanol-derived diphosphonate monomers

allows efficient colour tuning, improvement in solubility and better control of molecular aggregation in the polymer chain.

Cyriac et al. [85] reported synthesis of bisylide monomer based on 3-pentadecyl phenol which in turn is obtainable from cardanol. The bisylide monomer was polymerised with bisaldehyde under Wittig–Horner reaction conditions to prepare poly[(4-methoxy-6-pentadecyl-1,3-phenylenevinylene)-alt-(1, 3-phenylenevinylene)] (*m*-PPV) and its *para*-counterpart, viz. poly [(4-methoxy-6-pentadecyl-1, 3-phenylenevinylene)-alt-(1, 4-phenylenevinylene)] (*p*-PPV) (Scheme 9.11).

The cardanol-based *m*-PPV and *p*-PPV polymers were used to trace energy transfer process and molecular aggregation phenomena in the *p*-conjugated system. The detailed photophysical studies by absorption and emission spectroscopes indicated efficient Forster energy transfer from the *m*-conjugated segments to the *p*-counterpart in solution as well as in the solid state [85].

9.3 General Trends in Properties of High-Performance Polymers from CNSL-Based Aromatic Difunctional Monomers

As has been discussed in the previous sections, the C-15 alkyl chain present in the aromatic difunctional monomers derived from cardanol influences the properties of polymers synthesised therefrom in several ways. Overall, the polymer processability/ solubility was improved by the incorporation of pendent pentadecyl chains.

Table 9.13 Influence of the pentadecyl chain on polymer properties

Property	Remark
Crystallinity	Decrease in crystallinity leading to amorphous polymers and enhancement in solubility characteristics
Solubility in (common) organic solvents	Improvement in solubility characteristics and solution processability
Temperature at 10 wt% decomposition in TGA (T_{10})	Decrease in T_{10} (TGA) and T_{max} (DTG)
Glass transition temperature (T_g)	Depression in T_g compared to reference polymers, a large gap between T_g and T_{10}, wider processing window and possibilities of compression moulding

The presence of C-15 alkyl chain disrupts interchain packing of the polymers, enhances chain mobility and facilitates interaction of solvent molecules with the polymer chains. All these effects are manifested in interesting changes in polymer properties as summarised in Table 9.13.

A large gap between T_g and T_{10} values was observed with the studied high-performance polymers. This offers polymers a wide processing window. Thus, improved solubility coupled with reasonably good thermal stability of these polymers makes them attractive candidates for high-performance polymer applications.

9.4 Conclusions and Future Outlook

Cardanol, a major constituent of CNSL, is a by-product of cashew industry and is a non-edible oil. Cardanol possesses a number of reactive sites for functionalisation and chemical modification, which makes it an attractive biobased starting material for the synthesis of various types of monomers and polymers. This chapter has summarised the library of difunctional monomers synthesised from cardanol/CNSL. Olefin metathesis reactions have opened up interesting opportunities to synthesise new cardanol-derived monomers and polymers possessing side chains of variable lengths and functionalised alkyl chains. These difunctional monomers are a welcome addition to the existing portfolio of difunctional condensation monomers and represent useful (co)monomers for the synthesis of step-growth polymers with attractive processability characteristics. Polymers derived from cardanol bear pendent C-15 chains which have been amply exploited by researchers to modify polymer properties and thereby improve solubility and processability of various polymer classes, viz. polyimides and polyamides. The present work provides a comprehensive account of several difunctional monomers synthesised from cardanol/ CNSL and the preparation, solubility and thermal properties of various step-growth polymers. The incorporation of pendent flexible pentadecyl chains resulted in significant improvement in solubility characteristics of polymers and induced the depression of T_g. High-performance polymers derived from CNSL-based aromatic difunctional monomers exhibited reasonably good thermal stability. However, T_{10} values were lower than the corresponding reference polymers without pentadecyl chains. Furthermore, a large gap between T_g and T_{10} values of polymers based on CNSL-derived monomers offers a broad processing window. The data on solubility and thermal properties of (co)polymers revealed that even at low concentration of cardanol-derived monomer, there is a significant enhancement in solubility and reduction in T_g.

With growing demand for renewables, it is only a matter of time for CNSL-based difunctional condensation monomers to find widespread commercial applications.

The studies on mechanical properties such as tensile strength and elongation of polymers containing pendent pentadecyl chains have not been reported in the literature. There is still scope for such investigations. Isothermal TGA studies of polymers containing pendent pentadecyl chains need to be undertaken to assess the long-term thermal stability of the polymers. The extent of depression of T_g values of polymers containing pendent pentadecyl chains, compared to reference polymers, is different in each polymer class and is governed by the nature of polymer backbone. This aspect needs to be further looked into.

References

1. Munoz-Guerra S, Lavilla C, Japu C, Martinez de Ilarduya A (2014) Renewable terephthalate polyesters from carbohydrate-based bicyclic monomers. Green Chem 15:1716
2. Meier M (2014) Sustainable polymers: reduced environmental impact, renewable raw materials and catalysis. Green Chem 16:1672
3. Xu Y, Hanna MA, Isom L (2008) "Green" Chemicals from Renewable Agricultural Biomass–A Mini Review. Open Agric J 2:54
4. Big Chemical Encyclopedia (2015) http://chempedia.info/info/151307/. Accessed 13 July 2016
5. Berner RA (2003) The long-term carbon cycle, fossil fuels and atmospheric composition. Nature (London) 426:323
6. Brooks JD, Smith JW (1969) The diagenesis of plant lipids during the formation of coal, petroleum and natural gas—II. Coalification and the formation of oil and gas in the Gippsland Basin. Geochim Cosmochim Acta 33:1183
7. Tissot BP, Welte DH (1984) Petroleum formation and occurrence. Springer, Berlin
8. Mango FD (1997) The light hydrocarbons in petroleum: a critical review. Org Geochem 26:417
9. Crocker M, Crofcheck C (2006) Biomass conversion to liquid fuels and chemicals. Energeia 17:1
10. Okkerse C, van Bekkum H (1999) From fossil to green. Green Chem 1:107
11. U.S. Biobased Products: Market Potential and Projections Through 2025 (2008). In: United States Department of Agriculture, Office of the Chief Economist. Available via http://www.usda.gov/oce/reports/energy/BiobasedReport2008.pdf. Accessed 13 July 2016
12. de Jong Ed, Jungmeier G (2015) Industrial biorefineries and white biotechnology, Chap 1, p 3
13. Cherubini F (2010) The biorefinery concept: using biomass instead of oil for producing energy and chemicals. Energy Convers Manage 51:1412

14. FitzPatrick M, Champagne P, Cunningham MF, Whitney RA (2010) A biorefinery processing perspective: treatment of lignocellulosic materials for the production of value-added products. Bioresour Technol 101:8915
15. Naik SN, Goud VV, Rout PK, Dalai AK (2010) Production of first and second generation biofuels: a comprehensive review. Renewable Sustainable Energy Rev 14:578
16. Stocker M (2008) Biofuels and biomass-to-liquid fuels in the biorefinery: catalytic conversion of lignocellulosic biomass using porous materials. Angew Chem Int Ed 47:9200
17. Belgacem MN, Gandini A (eds) (2008) Monomers, polymers and composites from renewable resources. Elsevier, Amsterdam
18. Fache M, Darroman E, Besse V, Auvergne R, Caillol S, Boutevin B (2014) Vanillin, a promising biobased building-block for monomer synthesis. Green Chem 16:1987
19. Lochab B, Shukla S, Varma IK (2014) Naturally occurring phenolic sources: monomers and polymers. RSC Adv 4:21712
20. CatchBio. http://www.catchbio.com/. Accessed 13 July 2016
21. BIO-TIC. http://www.industrialbiotech-europe.eu/. Accessed 13 July 2016
22. ARENA. http://arena.gov.au/. Accessed 13 July 2016
23. Delft Advanced Biorenewables. http://delftab.com/. Accessed 13 July 2016
24. Gandini A, Lacerda TM (2015) From monomers to polymers from renewable resources: recent advances. Prog Polym Sci 48:1
25. Vilela C, Sousa AF, Fonseca AC, Serra AC, Coelho JFJ, Freire CSR, Silvestre AJD (2014) The quest for sustainable polyesters–insights into the future. Polym Chem 5:3119
26. Gandini A, Lacerda TM, Carvalho AJF, Trovatti E (2016) Progress of polymers from renewable resources: furans, vegetable oils, and polysaccharides. Chem Rev 116:1637
27. Sousa AF, Vitela C, Fonseca AC, Matos M, Freire CSR, Gruter G-JM, Coelho JFJ, Silvestre AJD (2015) Biobased polyesters and other polymers from 2,5-furandicarboxylic acid: a tribute to furan excellency. Polym Chem 6:5961
28. Tschan MJ-L, Brule E, Haquette P, Thomas CM (2012) Synthesis of biodegradable polymers from renewable resources. Polym Chem 3:836
29. Zhang L, Liu Z, Cui G, Chen L (2015) Biomass-derived materials for electrochemical energy storages. Prog Polym Sci 43:136
30. Datta J, Wloch M (2014) Selected biotrends in development of epoxy resins and their composites. Polym Bull 71:3035
31. Wilbon PA, Chu F, Tang C (2013) Progress in renewable polymers from natural terpenes, terpenoids, and rosin. Macromol Rapid Commun 34:8
32. Llevot A, Grau E, Carlotti S, Grelier S, Cramail H (2016) From lignin-derived aromatic compounds to novel biobased polymers. Macromol Rapid Commun 37:9
33. Upton BM, Kasko AM (2016) Strategies for the conversion of lignin to high-value polymeric materials: review and perspective. Chem Rev 116:2275
34. Galbis JA, Garcia-Martin MG, de Paz MV, Galbis E (2016) Synthetic polymers from sugar-based monomers. Chem Rev 116:1600
35. Arbenz A, Averous L (2015) Chemical modification of tannins to elaborate aromatic biobased macromolecular architectures. Green Chem 17:2626
36. Food and Agriculture Organisation of the United Nations Statistics Division (FAOSTAT) (2013). http://faostat3.fao.org/home/E. Accessed 13 July 2016
37. Tyman JHP (1996) Synthetic and natural phenols. Elsevier, Amsterdam
38. Attanasi OA (1983) Chemistry and industrial utilization of polyketide natural long chain phenols. Chim Oggi 8:11
39. Harvey MT, Caplan S (1940) Cashew nut shell liquid. Ind Eng Chem 32:1306
40. Paramashivappa R, Phani Kumar P, Vithayathil PJ, Srinivasa Rao A (2001) Novel method for isolation of major phenolic constituents from cashew (*Anacardium occidentale* L.) nut shell liquid. J Agric Food Chem 49:2548
41. Manjula S, Pillai CKS, Kumar VG (1990) Thermal characterization of cardanol-formaldehyde resins and cardanol-formaldehyde/poly(methyl methacrylate) semi-interpenetrating polymer networks. Thermochim Acta 159:255

42. Tyman JHP, Johnson RA, Muir RA, Rokhgar R (1989) The Extraction of natural cashew nut-shell liquid From the cashew nut (*Anacardium occidentale*). J Am Oil Chem Soc 66:553
43. Jain RK, Kumar S (1997) Development of a cashew nut sheller. J Food Eng 32:339
44. Patel RN, Bandyopadhyay S, Ganesh A (2011) Extraction of cardanol and phenol from bio-oils obtained through vacuum pyrolysis of biomass using supercritical fluid extraction. Energy 36:1535
45. Gandhi T, Patel M, Dholakiya BK (2012) Studies on effect of various solvents on extraction of cashew nut shell liquid (CNSL) and isolation of major phenolic constituents from extracted CNSL. J Nat Prod Plant Resour 2:135
46. Lubi MC, Thachil ET (2000) Cashew nut shell liquid (CNSL)–a versatile monomer for polymer synthesis. Des Monomers Polym 3:123
47. Gedam PH, Sampathkumaran PS (1986) Cashew nut shell liquid: extraction, chemistry and applications. Prog Org Coat 14:115
48. Bladzell P (2000) The mighty cashew. Interdisciplinary Sci Rev 25:220
49. Mele G, Vasapollo G (2008) Fine chemicals and new hybrid materials from cardanol. Mini-Rev Org Chem 5:243
50. Balachandran VS, Jadhav SR, Vemula PK, John G (2013) Recent advances in cardanol chemistry in a nutshell: from a nut to nanomaterials. Chem Soc Rev 42:427
51. Voirin C, Caillol S, Sadavarte NV, Tawade BV, Boutevin B, Wadgaonkar PP (2014) Functionalization of cardanol: towards biobased polymers and additives. Polym Chem 5:3142
52. Tyman JHP, Wilczynski D, Kashani MA (1978) Compositional studies on technical cashew nut shell liquid (CNSL) by Chromatography and Mass Spectroscopy. J Am Oil Chem Soc 55:663
53. Trevisan MTS, Pfundstein B, Haubner R, Wurtele G, Spiegelhalder B, Bartsch H, Owen RW (2006) Characterization of alkyl phenols in cashew (Anacardium occidentale) products and assay of their antioxidant capacity. Food Chem Toxicol 44:188
54. Cardolite. https://www.cardolite.com/. Accessed 13 July 2016
55. Varma AJ, Sivaram S (2002) U.S. Patent US6451957B1
56. Sadavarte NV (2012) Ph.D. thesis, University of Pune, India
57. Tawade BV, Salunke JK, Sane PS, Wadgaonkar PP (2014) Processable aromatic polyesters based on bisphenol derived from cashew nut shell liquid: synthesis and characterization. J Polym Res 21:617
58. Tawade BV, Shaligram SV, Valsange NG, Kharul UK, Wadgaonkar PP (2016) Synthesis and properties of poly(arylene ether)s based on 3-pentadecyl 4,4'-biphenol. Polym Int 65:567
59. Tawade BV (2015) Ph.D. thesis, Savitribhai Phule Pune University, India
60. Bhunia HP, Jana RN, Basak A, Lenka S, Nando GB (1998) Synthesis of polyurethane from cashew nut shell liquid (CNSL), a renewable resource. J Polym Sci A Polym Chem 36:391
61. Bhunia HP, Nando GB, Chaki TK, Basak A, Lenka S, Nayak PL (1999) Synthesis and characterization of polymers from cashew nut shell liquid (CNSL), a renewable resource II. Synthesis of polyurethanes. Eur Polym J 35:1381
62. Shingte RD (2006) Ph.D. thesis, University of Pune, India
63. Avadhani CV, Wadgaonkar PP, Sivaram S (2001) U. S. US6255439B1
64. More AS, Pasale SK, Honkhambe PN, Wadgaonkar PP (2011) Synthesis and characterization of organo-soluble poly(ether ether ketone)s and poly(ether ether ketone ketone)s containing pendant pentadecyl chains. J Appl Polym Sci 121:3689
65. More AS, Naik PV, Kumbhar KP, Wadgaonkar PP (2010) Synthesis and characterization of polyesters based on 1,1,1-[bis(4-hydroxyphenyl)-4′-pentadecylphenyl]ethane. Polym Int 59:1408
66. More AS, Wadgaonkar PP (2008) U.S. US7446234B2
67. More AS (2009) Ph.D. thesis, University of Pune, India
68. Bhavsar GA, Asha SK (2011) Pentadecyl Phenol- and Cardanol-Functionalized Fluorescent, Room-Temperature Liquid-Crystalline Perylene Bisimides: effect of pendant chain unsaturation on self-assembly. Chem Eur J 17:12646

69. Ramasri M, Srinivasa Rao GS, Sampatkumaran PS, Shirsalkar MM (1987) Synthesis & Identification of 1,8-Bis(hydroxyphenyl)pentadecane from 3-Δ^8-Pentadecenylphenol. Ind J Chem 26B:683
70. Sadavarte NV, Halhali MR, Avadhani CV, Wadgaonkar PP (2009) Synthesis and characterization of new polyimides containing pendent pentadecyl chains. Eur Polym J 45:582
71. Sadavarte NV, Avadhani CV, Wadgaonkar PP (2011) Synthesis and characterization of new organosoluble aromatic polyamides and polyazomethines containing pendent pentadecyl chains. High Perform Polym 23:494
72. Ghatge ND, Maldar NN (1984) Polyimides from dianhydride and diamine: structure property relations by thermogravimetric analysis (t.g.a.). Polymer 25:1353
73. Mathew JS, Vernekar SP, Mercier R, Kerboua R (2002) U.S. Patent US6500913B2
74. Mathew JS (2001) Ph.D. thesis, University of Pune, India
75. More AS, Sane PS, Patil AS, Wadgaonkar PP (2010) Synthesis and characterization of aromatic polyazomethines bearing pendant pentadecyl chains. Polym Degrad Stab 95:1727
76. Shingte RD, Wadgaonkar PP (2004) U.S. US6790993B1
77. Sadavarte NV, Avadhani CV, Naik PV, Wadgaonkar PP (2010) Regularly alternating poly(amideimide)s containing pendent pentadecyl chains: synthesis and characterization. Eur Polym J 46:1307
78. Tawade BV, Kulkarni AD, Wadgaonkar PP (2015) Synthesis and characterization of polyetherimides containing multiple ether linkages and pendent pentadecyl chains. Polym Int 64:1770
79. Jadhav AS, Vernekar SP, Maldar NN (1993) Synthesis and characterization of new aromatic sulfone ether polyamides containing pendant pentadecyl groups. Polym Int 32:5
80. More AS, Pasale SK, Wadgaonkar PP (2010) Synthesis and characterization of polyamides containing pendant pentadecyl chains. Eur Polym J 46:557
81. Sadavarte NV, Patil SS, Avadhani CV, Wadgaonkar PP (2013) New organosoluble aromatic poly(esterimide)s containing pendent pentadecyl chains Synthesis and characterization. High Perform Polym 25:735
82. More AS, Patil AS, Wadgaonkar PP (2010) Poly(amideimide)s containing pendant pentadecyl chains: synthesis and characterization. Polym Degrad Stab 95:837
83. More AS, Menon SK, Wadgaonkar PP (2012) New poly(1,3,4-oxadiazole)s bearing pentadecyl side chains: synthesis and characterization. J Appl Polym Sci 124:1281
84. Seo S, Kim Y, You J, Sarwade BD, Wadgaonkar PP, Menon SK, More AS, Kim E (2011) Electrochemical fluorescence switching from a patternable poly(1,3,4-oxadiazole) thin film. Macromol Rapid Comm 32:637
85. Cyriac A, Amrutha SR, Jayakannan M (2008) Renewable resource-based poly (m-phenylenevinylene)s and their statistical copolymers: synthesis, characterization, and probing of molecular aggregation and forster energy transfer processes. J Polym Sci A Polym Chem 46:3241
86. Pillai CKS, Sherrington DC, Sneddon A (1992) Thermotropic liquid crystalline copolyester based on 8-(3-hydroxyphenyl) octanoic acid and p-hydroxybenzoic acid. Polymer 33:3968
87. Abraham S, Prasad VS, Pillai CKS, Ravindranathan M (2002) Copolyesters of hydroxyphenylalkanoic acids: synthesis and thermal properties of poly{(4-oxybenzoate)-co-[8-(3-oxyphenyl)octanoate]} and poly{(3-bromo-4-oxybenzoate)-co-[8-(3-oxyphenyl)octanoate]}. Polym Int 51:475
88. Lee DG, Chang VS (1978) Oxidation of hydrocarbons. 8. Use of dimethyl polyethylene glycol as a phase transfer agent for the oxidation of alkenes by potassium permanganate. J Org Chem 43:1532
89. Grubbs RH (2004) Olefin metathesis. Tetrahedron 60:7117
90. Astruc D (2005) The metathesis reactions: from a historical perspective to recent developments. New J Chem 29:42
91. Grubbs RH (2006) Olefin-metathesis catalysts for the preparation of molecules and materials (Nobel lecture). Angew Chem Int Ed 45:3760

92. Martina F (2008) Ph.D. thesis, University of Salento, Italy
93. Vasapollo G, Mele G, Sole RD (2011) Cardanol-based materials as natural precursors for olefin metathesis. Molecules 16:6871
94. Bloise E, Carbone L, Colafemmina G, D'Accoltil L, Mazzetto SE, Vasapollo G, Mele G (2012) First example of a lipophilic porphyrin-cardanol hybrid embedded in a cardanol-based micellar nanodispersion. Molecules 17:12252
95. Attanasi OA, Ciccarella G, Filippone P, Mele G, Spadavecchia J, Vasapollo G (2003) Novel phthalocyanines containing cardanol derivatives. J Porphyrins Phthalocyanins 7:52
96. Mgaya JE, Bartlett SA, Mubofu EB, Mgani QA, Slawin AMZ, Porgorzelec PJ, Cole-Hamilton DJ (2016) Synthesis of bifunctional monomers by the palladium-catalyzed carbonylation of cardanol and its derivatives. Chem Cat Chem 8:751
97. Cassidy PE (1980) Thermally stable polymers: synthesis and properties. Mercel Dekker Inc, New York
98. Hergenrother PM (2003) The use, design, synthesis, and properties of high performance/high temperature polymers: an overview. High Perform Polym 15:3
99. Dautel OJ, Wantz G, Flot D, Lere-Porte J-P, Moreau JJE, Parneix J-P, Serein-Spirau F, Vignau L (2005) Confined photoactive substructures on a chiral scaffold: the design of an electroluminescent polyimide as material for PLED. J Mater Chem 15:4446
100. Jung M-S, Lee T-W, Hyeon-Lee J, Sohn BH, Jung I-S (2006) Synthesis and characterizations of a polyimide containing a triphenylamine derivative as an interlayer in polymer light-emitting diode. Polymer 47:2670
101. Kausar AS, Zulfiqar S, Ahmad Z, Sarwar MI (2010) Novel processable and heat resistant poly(phenylthiourea azomethine imide)s: Synthesis and characterization. Polym Degrad Stab 95:1826
102. Rubal M, Jr Wilkins CW, Cassidy PE, Lansford C, Yamada Y (2008) Fluorinated polyimide nanocomposites for CO2/CH4separation. Polym Adv Technol 19:1033
103. Zhang B, Li W, Yang J, Fu Y, Xie Z, Zhang S, Wang L (2009) Performance enhancement of polymer light-emitting diodes by using ultrathin fluorinated polyimide modifying the surface of poly(3,4-ethylene dioxythiophene):poly(styrenesulfonate). J Phys Chem C 113:7898
104. Ulmer II CW, Smith DA, Sumpter BG, Noid DI (1998) Computational neural networks and the rational design of polymeric materials: the next generation polycarbonates. Comput Theor Polym Sci 8:311
105. Dove AP, Meier MAR (2014) Step-growth polymerization in the 21st century. Macromol Chem Phys 215:2135
106. Mittal KL (ed) (2001) Polyimides and other high temperature polymers: synthesis, characterisation and applications. VSP, Boston
107. Volksen W (1994) Condensation polyimides: synthesis, solution behavior, and imidization characteristics. Adv Polym Sci 117:111
108. Wallace JS, Arnold FE, Tan LS (1987) In situ rigid-rod aromatic polyimides. Am Chem Soc Polym Prepr 28:316
109. St. Clair TL, Wilson D, Stenzenberger HD, Hergenrother PM (eds) (1990) Polyimides. Chapman and Hall, New York
110. Sato M (1997) Polyimides. In: Olabisi O (ed) Handbook of thermoplastics. Marcel Dekker, New York, p 665
111. Yeganeh H, Tamami B, Ghazi I (2004) A novel direct method for preparation of aromatic polyimides via microwave-assisted polycondensation of aromatic dianhydrides and diisocyanates. Eur Polym J 40:2059
112. Imai Y, Kojima K (1972) Preparation of polyimides from pyromellitic dithioanhydride and aromatic diamines. J Polym Sci A Polym Chem 10:2091
113. Bella SD, Consiglio G, Leonardi N, Failla S, Finocchiaro P, Fragala I (2004) Film polymerization—a new route to the synthesis of insoluble polyimides containing functional nickel(II) schiff base units in the main chain. Eur J Inorg Chem 2004:2701

114. Gheneim R, Perez-Berumen C, Gandini A (2002) Diels–Alder reactions with novel polymeric dienes and dienophiles: synthesis of reversibly cross-linked elastomers. Macromolecules 35:7246
115. Chi JH, Shin GJ, Kim YS, Jung JC (2007) Synthesis of new alicyclic polyimides by Diels-Alder polymerization. J Appl Polym Sci 106:3823
116. Munoz DM, de la Campa JG, de Abajo J, Lozano AE (2007) Experimental and theoretical study of an improved activated polycondensation method for aromatic polyimides. Macromolecules 40:8225
117. Yang HH (1989) Aromatic high-strength fibers. Wiley Interscience, New York
118. Fink JK (2008) High performance polymers. William Andrew Inc, Burlington
119. Volbracht L (1989) Aromatic Polyimides In: Allen G, Bevington B, Eastmond GV, Ledwith-A, Russo S, Sigwald P (eds) Comprehensive polymer science, vol 5. Pergamon Press, Oxford, p 373
120. Shockravi A, Abouzari-Lotf E, Javadi A, Atabaki F (2009) Preparation and properties of new ortho-linked polyamide-imides bearing ether, sulfur, and trifluoromethyl linkages. Eur Polym J 45:1599
121. Sarkar A, More AS, Wadgaonkar PP, Shin GJ, Jung JC (2007) Synthesis and liquid-crystal-aligning properties of novel aromatic poly(amide imide)s bearing n-alkyloxy side chains. J Appl Polym Sci 105:1793
122. Behniafar H, Haghighat S (2006) Preparation and properties of aromatic poly(amide-imide) s derived from N-[3,5-bis(3,4-dicarboxybenzamido)phenyl]phthalimide dianhydride. Eur Polym J 42:3236
123. Yang C-P, Chen Y-P, Woo EM (2004) Thermal behavior of 1,4-bis(4-trimellitimido-2-trifluoromethyl phenoxy)benzene (DIDA) solvated with polar organic solvents and properties of DIDA-based poly(amide-imide)s. Polymer 45:5279
124. Liaw D-J, Chen W-H (2003) Synthesis and characterization of new soluble cardo poly(amide–imide)s derived from 2,2-bis[4-(4-trimellitimidophenoxy)phenyl]norbornane. Polymer 44:3865
125. Lee C, Iyer NP, Min K, Pak H, Han H (2004) Synthesis and characterization of novel poly(amide-imide)s containing 1,3-diamino mesitylene moieties. J Polym Sci A Polym Chem 42:137
126. Behniafar H, Banihashemi A (2004) Synthesis and characterization of new soluble and thermally stable aromatic poly(amide-imide)s based on N-[3,5-bis(N-trimellitoyl)phenyl]phthalimide. Eur Polym J 40:1409
127. Abid S, Gharbi RE, Gandini A (2004) Polyamide-imides bearing furan moieties. 1. Solution polycondensation of aromatic dianhydrydes with 2-furoic acid dihydrazides. Polymer 45:6469
128. Sarkar A, Honkhambe PN, Avadhani CV, Wadgaonkar PP (2007) Synthesis and characterization of poly(amideimide)s containing pendent flexible alkoxy chains. Eur Polym J 43:3646
129. Privalko VP, Mudrak CV, Privalko EG, Usenko AA, Karpova IL (2001) Structure-property relationships for film-forming poly(amide imide)s. Macromol Symp 175:403
130. Tussot CM, Sagnier C, Pham Q-T (2001) Tolylene diisocyanate-based poly(imido-amide)s, 2. Distributions of amide and imide groups in ternary poly(imido-amide)s containing terephthalic acid studied by 1H and 13C NMR. Macromol Chem Phys 202:1071
131. Hong Y-T, Jin MY, Suh DM, Lee J-H, Choi K-Y (1997) New preparation method of poly(amide-imide)s using direct polycondensation with thionyl chloride and their characterization. Angew Makromol Chem 248 :105
132. Kakimoto M-A, Akiyama R, Negi YS, Imai Y (1988) Synthesis and characterization of aromatic polyimide and polyamide-imide from 2,5-bis(4-isocyanatophenyl)-3,4-diphenylthiophene and aromatic tetra- and tricarboxylic acids. J Polym Sci A Polym Chem 26:99
133. Imai Y, Maldar NN, Kakimoto M-A (1985) Synthesis and characterization of aromatic polyamide-imides from 2,5-bis(4-aminophenyl)-3,4-diphenylthiophene and 4-chloroformylphthalic anhydride. J Polym Sci A Polym Chem 23:2077

134. Grigoras M, Catanescu CO, Simionescu CI (2001) Poly(azomethine)s. Rev Roum Chim 46:927
135. Grigoras M, Catanescu CO (2004) Imine oligomers and polymers. J Macromol Sci C Polym Rev C44:131
136. Grigoras M, Cianga I, Farcas A, Nastase G, Ivanoiu M (2000) Fully conjugated and soluble polyazomethines containing 1,1′-binaphtyl groups. Rev Roum Chim 45:703
137. Bagheri M, Entezami A (2002) Synthesis of polymers containing donor–acceptor Schiff base in side chain for nonlinear optics. Eur Polym J 38:317
138. Dutta PK, Jain P, Sen P, Trivedi R, Sen PK, Dutta J (2003) Synthesis and characterization of a novel polyazomethine ether for NLO application. Eur Polym J 39:1007
139. Suh SC, Shim SC (2000) Synthesis and properties of a novel polyazomethine, the polymer with high photoconductivity and second-order optical nonlinearity. Synth Metals 114:91
140. Iwan A, Sek D (2008) Processible polyazomethines and polyketanils: from aerospace to light-emitting diodes and other advanced applications. Prog Polym Sci 33:289
141. Morgan PW, Kwolek SL, Pletcher TC (1987) Aromatic azomethine polymers and fibers. Macromolecules 20:729
142. Rudzinski WE, Guthrie SR, Cassidy PE (1988) Poly(Schiff base) polymers based on substituted biphenyl. J Polym Sci A Polym Chem 26:1677
143. Banerjee S, Gutch PK, Saxena C (1995) Polyether azomethines. I. Synthesis and characterization. J Polym Sci A Polym Chem 33:1719
144. Park KH, Tani T, Kakimoto M, Imai Y (1998) Synthesis and characterization of new diphenylfluorene-based aromatic polyazomethines. Macromol Chem Phys 199:1029
145. Chen Y, Yang Y, Su J, Tan L, Wang Y (2007) Preparation and characterization of aliphatic/aromatic copolyesters based on bisphenol-A terephthalate, hexylene terephthalate and lactide mioties. React Funct Polym 67:396
146. Chern Y-T, Huang C-M (1998) Synthesis and characterization of new polyesters derived from 1,6- or 4,9-diamantanedicarboxylic acyl chlorides with aryl ether diols. Polymer 39:2325
147. Bertini F, Zuev VV (2006) Investigation of the thermal degradation of fully aromatic regular polyesters: Poly(oxy-1,4-phenyleneoxy-fumaroyl-bis-4-oxybenzoate). Polym Degrad Stab 91:3214
148. Mallakpour S, Meratian S (2008) Synthesis and characterization of organosoluble optically active poly(ester-imide)s derived from trimellitic anhydride, L-methionine and bisphenols. High Perform Polym 20:3
149. Maiti S, Das S (1980) Synthesis and properties of a new polyesterimide. Angew Makromol Chem 86:181
150. Lienert KW (1999) Poly(ester-imide)s for Industrial Use. Adv Polym Sci 141:45
151. Yang C-P, Chiang H-C, Chen R-S (2003) Synthesis and characterization of organosoluble poly(arylate-imide)s prepared from direct polycondensation of bis(trimellitimide)-diacids and various bisphenols. J Appl Polym Sci 89:3818
152. Mulvaney JE, Figueroa FR, Wu SJ (1986) Polymers from 4,4′-sulfonyldiphenol. J Polym Sci A Polym Chem 24:613
153. Yang J-W, Wang C-S (1999) Novel perfluorononenyloxy group-containing polyimides. Polymer 40:1411
154. Schmidt K, Wille D (1971) U.S. Patent US3562219A
155. Shen DC (1982) U.S. Patent US4362861
156. Dolui SK, Pal D, Maiti S (1985) Synthesis of a novel polyesterimide. J Appl Polym Sci 30:3867
157. Maiti S, Das S (1981) Synthesis and properties of polyesterimides and their isomers. J Appl Polym Sci 26:957
158. Li C-H, Chen C-C, Chen K-M (1994) Studies on the synthesis and properties of copolyesterimide. J Appl Polym Sci 52:1751
159. Li C-H, Chen C-C, Chen K-M, Chang T-C (1995) Studies on the synthesis and properties of soluble homo- and copolyester–imide derived from imide–diacid. J Appl Polym Sci 55:747

160. Kurita K, Koyama Y, Mikawa N, Kaneda K, Murakoshi H (1989) Polymers based on p-aminophenol. 5. New synthetic route to wholly aromatic polyimide-ester. J Polym Sci C Polym Lett 27:115
161. Kurita K, Mikawa N, Koyama Y, Nishimura S (1990) Polymers based on p-aminophenol. 6. Facile synthesis of the simplest wholly aromatic poly(imide-ester) by pyrolytic polymerization of monomers containing preformed ester linkages. Macromolecules 23:2605
162. Li C-H, Jung A, Liang A-L, Chang T-C (1995) Studies on the synthesis and properties of thermotropic liquid crystalline copoly(amide–ester–imide) derived fromN-(hydroxyphenyl) phthalimide-4-carboxylic acid with amino acid. J Appl Polym Sci 56:1661
163. Johnson RN, Farnham AG (1967) Poly(aryl ethers) by nucleophilic aromatic substitution. III. Hydrolytic side reactions. J Polym Sci A Polym Chem 5:2415
164. Mandal BK, Maiti S (1986) Displacement polymerization—V. Synthesis and characterization of polyarylethers containing furoxan and furazan units. Eur Polym J 22:447
165. Cotter RJ (1995) Engineering thermoplastics, a handbook of poly(aryleneether)s. Gordon and Breach, Postfach
166. Chen M, Gibson HW (1996) Large-sized macrocyclic monomeric precursors of poly(ether ether ketone): Synthesis and Polymerization. Macromolecules 29:5502
167. Xie D, Gibson HW (1996) A 40-membered cyclic arylene ether sulfone from bisphenol-A: improved synthesis and properties. Macromol Chem Phys 197:2133
168. Kricheldorf HR, Delius U, Tonnes KU (1988) New polymer syntheses. 14. New crystalline and amorphous aromatic poly(ether ketone)s. New Polym Mater 1:127
169. Colon I, Kwiatkowski GT (1990) High molecular weight aromatic polymers by nickel coupling of aryl polychlorides. J Polym Sci A Polym Chem 28:367
170. Frazer AH, Wallenberger FT (1964) Poly(1,3,4-oxadiazole) fibers: new fibers with superior high temperature resistance. J Polym Sci A Polym Chem 2:1171
171. Frazer AH, Sarasohn IM (1966) Thermal behavior of polyhydrazides and poly-1,3,4-oxadiazoles. J Polym Sci A Polym Chem 4:1649
172. Frazer AH, Wallenberger FT (1964) Poly(1,3,4-oxadiazolidine). J Polym Sci A General Papers 2:1181
173. Sava I, Schulz B, Zhu S, Bruma M (1995) Synthesis and characterization of new silicon-containing poly(arylene-1,3,4-oxadiazole)s. High Perform Polym 7:493
174. Iwakura Y, Uno K, Hara S (1965) Poly-l,3,4-oxad.iazoles.I.Polyphenylene-1,3,4-oxadiazoles. J Polym Sci A Polym Chem 3:45
175. Hensema ER, Boom JP, Mulder MHV, Smolders CA (1994) Two reaction routes for the preparation of aromatic polyoxadiazoles and polytriazoles: Syntheses and properties. J Polym Sci A Polym Chem 32:513
176. Schulz B, Leibnitz E (1992) Aromatic poly(oxadiazole)s–new aspects of their synthesis, structures and properties. Acta Polym 43:343
177. Mythili CV, Retna AM, Gopalakrishnan S (2004) Synthesis, mechanical, thermal and chemical properties of polyurethanes based on cardanol. Bull Mater Sci 27:235
178. Ionescu M, Wan X, Bilic N, Petrovic ZS (2012) Polyols and rigid polyurethane foams from cashew nut shell liquid. J Polym Environ 20:647
179. Suresh KI (2013) Rigid polyurethane foams from cardanol: synthesis, structural characterization, and evaluation of polyol and foam properties.ACS Sustainable Chem Eng 1:232
180. Kathalewar M, Sabnis A (2014) Preparation of novel CNSL-based urethane polyol via non-isocyanate route: curing with melamine-formaldehyde resin and structure–property relationship. J Appl Polym Sci 131:41391
181. Suresh KI, Kishanprasad VS (2005) Synthesis, structure, and properties of novel polyols from cardanol and developed polyurethanes. Ind Eng Chem Res 44:4504
182. Mohapatra S, Mohanty N, Guru BN, Pal NC (2014) The synthesis and FTIR, kinetics and TG/DTG/dta study of biopolymers derived from polyurethanes of glycerol modified linseed oil and cardanol based dyes. J Chem Pharm Res 6:1493

Chapter 10
Cashew Nut Shell Liquid—Natural Solution for Industrial Problems

Timothy Stonis, Fernanda Tavares and Anbu Natesh

Abstract Renewable resources as substitutes for petrochemical derivatives have attracted considerable attention from polymer researchers worldwide. Cashew nut shell liquid (CNSL), an agricultural renewable resource and a by-product of the cashew industry, holds considerable promise in this direction, being a natural source of unsaturated alkyl-phenols. Its peculiar chemical features have been widely investigated and described in over a Century of scientific literature, and successfully applied in several final application fields, from abrasive materials to coatings. This Chapter highlights the successful experience of Cardolite Corporation, the leader in CNSL technology, applying cardanol and its derivatives at industrial level, validating their benefits and pioneering new applications for such a unique natural occurring molecule.

10.1 CNSL Commercialization History

The cashew (*Anacardium Occidentale*) tree is a large tropical evergreen plant that belongs to the Anacardiaceae family and is native to northeastern Brazil. In the sixteenth century, Portuguese traders introduced cashew trees to Indian shores with the purpose of binding the soil and preventing erosion. It did not take long for cashew trees to prosper and spread throughout the southern part of India, Vietnam, Ceylon, Malaysia, the Andaman Islands, East and West Africa, Mauritius, and the Seychelles [1].

Cashew trees consist of trunk, leafs, apples (fruits), and cashew nuts. The cashew fruit is an edible yellow-red apple, rich in Vitamins A and C. The cashew nut is quite peculiar with respect to other tree nuts as it is located outside the fruit.

T. Stonis · F. Tavares · A. Natesh (✉)
Cardolite Corporation, 11 Deer Park Dr., Suite 124, Monmouth Junction, NJ 08852, USA
e-mail: anatesh@cardolite.com

P. Anilkumar (ed.), *Cashew Nut Shell Liquid*,
DOI 10.1007/978-3-319-47455-7_10

The nut is comprised of an outer shell, inner shell, and the kernel; the thickness of the nutshell is about 0.3–0.4 cm, and it is characterized between outer and inner shells, by a honeycomb structure containing a dark brown, irritant liquid known as cashew nut shell liquid (CNSL). CNSL makes up approximately 25 % by weight of the cashew nut shell [2].

CNSL was originally used in Indian medicine for skin treatments (psoriasis, skin peeling), decorative tattooing in the West Indies, and to protect floors and rafters from termites. CNSL moved to industrial applications thanks to the investigations of M. T. Harvey and S. Caplan of Harvel Research Corporation [3]. Harvel entered into a joint venture with Irvington Varnish and Insulator Company with the purpose of developing commercial uses for CNSL. In the 1930s, they succeeded by developing CNSL-based friction particles for the brake lining industry, which was the first major (and still large) use of CNSL. In 1953, 3M acquired Irvington and in the subsequent years developed new families of products derived from CNSL such as modified cardanol grades, epoxy functional derivatives, liquid resins, and amine functional curing agents called phenalkamines. In 1985, the CNSL business and assets were acquired from 3M through a management buyout led by Anthony Stonis, the company's current Chairman. The new corporation was named "Cardolite" and has since pioneered the development and growth of new CNSL-based derivatives in coatings, adhesives, and friction applications.

10.2 CNSL: Chemical Composition

The first analytical work done on CNSL was published in 1847 [4], and it described the characterization of the oil recovered from the cashew nut through solvent extraction as two different components: anacardic acid (approx. 90 % w/w) and cardol. They were originally assigned formulas $C_{44}H_{30}O_5$ and $C_{42}H_{31}O_4$, respectively, but were later adjusted using combustion data to $C_{44}H_{60}O_5$ (anacardic acid) and $C_{42}H_{61}O_4$ (cardol). Anacardic acid continued to be analyzed and was further characterized as a hydroxycarboxylic acid with formula $C_{22}H_{32}O_3$ [5]. Subsequent investigations were performed by hydrogenation followed by oxidation and distillation over zinc chloride [6], describing anacardic acid as a salicylic acid homolog, with a side hydrocarbon chain of general formula $C_{15}H_{28}$. At the same time, the second constituent of cashew nutshell liquid, cardol, was analyzed using similar approaches [7] and assigned formula $C_{32}H_{50}O_3{\cdot}H_2O$, even if not fully confirmed.

In 1941, the phenolic nature of both anacardic acid and cardol was confirmed by hydrogenation and oxidative degradation experiments [8], and identified as 2-carboxy-pentadecadienylphenol and 5-pentadecadienylresorcinol, respectively, although no explanation on the olefinic linkages' position was provided. During the same experiments, the easy decarboxylation of anacardic acid was highlighted, resulting in a monophenol identified as 3-pentadecadienylphenol after having synthesized the corresponding tetrahydro derivative. This derivative was initially called "anacardol," on the basis of what was previously reported [9], but then changed to "cardanol" [10], in order to avoid any confusion with a different monophenol of empirical formula $C_{18}H_{38}O$ extracted from the kernel oil of the marking nut (*Semecarpus anacardium*) [11].

Posterior studies to determine the composition of cardanol showed its chemical structure was independent from the method utilized to extract CNSL from the shells [12]. Since these initial studies, many further investigations have been accomplished [13], finally defining naturally occurring CNSL as a mixture of four components: cardanol, cardol, anacardic acid, and 2-methyl-cardol.

10.3 CNSL Extraction

Cashew nutshell liquid extraction is the first step of the cardanol recovery process (Fig. 10.1).

Cashew nut shell liquid can be extracted from the nut by heat [14] to recover "technical-grade CNSL" or at low-to-room temperatures to recover "natural-grade CNSL". Several techniques have been described and patented, providing CNSL grades with different compositions, yields, and purity [15]. Some examples include pyrolysis [16], solvent extraction [17], and the use of supercritical fluids like carbon dioxide [18], but the most common extraction methods (thermal, mechanical and solvent-based) are explained below:

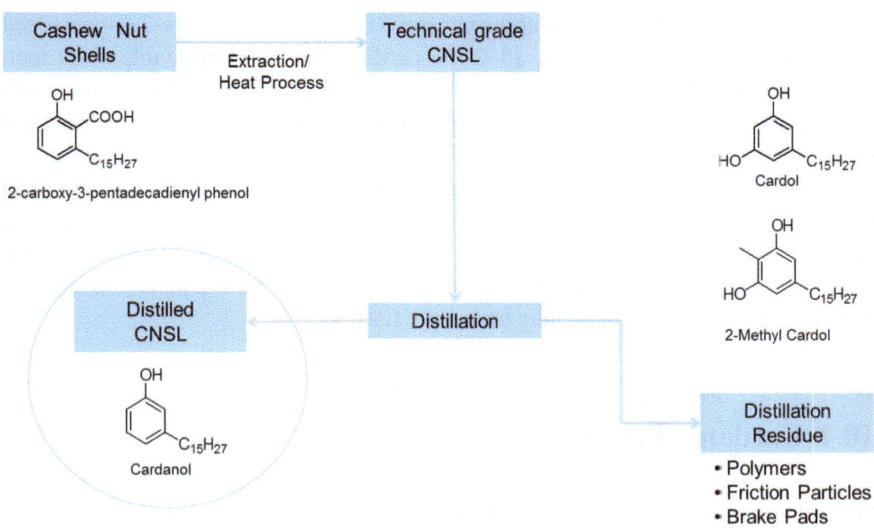

Fig. 10.1 Schematic description of cardanol recovery process

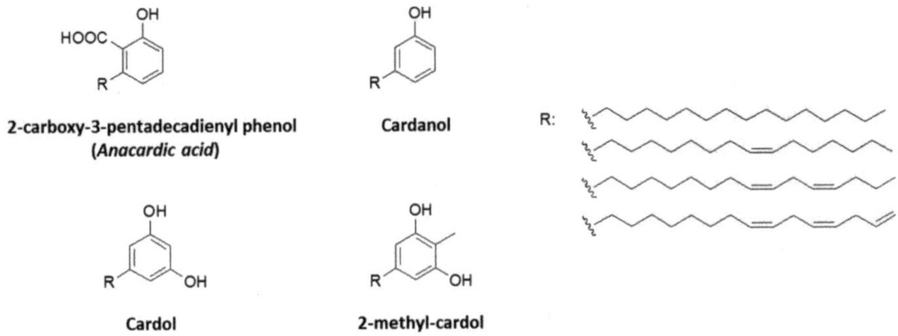

Fig. 10.2 Main components of cashew nut shell liquid

- *Hot oil bath method*: raw nuts are passed through a hot CNSL bath (180–200 °C) releasing CNSL (approx. 50 % oil recovery). This approach can be improved by a first surface wetting step followed by immersion in water at 20–25 °C and subsequent steam treatment prior to exposure to the hot CNSL bath. Another 20 % of CNSL could be recovered passing the spent shells through an expeller, eventually adding a solvent extraction step to maximize final yields.
- *Roasting method*: shells are charred during this process, producing an explosive pressure in the cellular structure, which forces the liquid out of the shell. This technique might be employed in conjunction with an expeller, where the oil is expelled from the shells to an extent of 90 %.
- *Screw press method*: [19] raw shells are screwed by a hydraulic press in order to release an anacardic-rich CNSL.

Natural-grade CNSL contains approximately 60–65 % 2-carboxy-3-3-pentadecadienyl phenol (anacardic acid), 15–20 % cardol, and 10 % cardanol, while technical-grade CNSL, due to the anacardic acid decarboxylation to cardanol under heating conditions, is composed of approximately 60–65 % cardanol, 15–20 % cardol, and 10 % polymeric material (Fig. 10.2) [20].

Each CNSL component is characterized by the presence of a C_{15} side chain that contains up to three non-conjugated double bonds (approx. 5–8 % saturated isomer; 48–49 % monoene isomer; 16–17 % diene isomer, and 29–30 % triene isomer) [21], whose relative percentages within the mixture depends on the geographical area where cashew nuts have been harvested.

10.4 Cardanol Purification

Several methods for cardanol purification have been explored and explained in the literature that target the isolation of the highest boiling fractions (cardol and 2-methyl-cardol) [22], the isolation of anacardic acid in high purity [23], or the separation of different isomers through chromatography techniques (e.g., argentation liquid chromatography) [24]. Alternative selective techniques such as extraction with supercritical carbon dioxide [25] or solvents [26] have been described as well, but are not known to be applied on an industrial scale.

Cardanol is traditionally obtained by distillation of CNSL at relatively high temperatures (210–280 °C) and under vacuum (2–8 mmHg), and it is recovered as a pale yellow liquid that darkens over time [27]. Distillation of cardanol can be done using a batch distillation unit or a thin film distillation. In batch distillation (a discontinuous process), CNSL is charged into the distillation unit and brought to high temperature and vacuum under stirring. Cardanol is recovered as the lowest boiling fraction, while the majority of cardol and 2-methyl-cardol undergo thermal oligomerization turning into a thick black polymer (residol) because of crosslinking of the side chain double bonds. Batch distillation is normally characterized by limited efficiency (yield 65–68 %) [28] with cardanol purity of approx. 80 % due to the presence of residual high boiling fractions; cardanol can be redistilled under similar conditions to improve its purity up to about 90 %.

Thin film distillation (a continuous process) is widely used for the purification of natural oils and thermosensitive substrates. In a typical equipment setup, CNSL is continuously pumped through a heating column in the distillation unit (thus also allowing a complete decarboxylation before distillation, guaranteeing an optimal vacuum control) where a specific stirrer homogenously spreads the oil over the internal heated wall, creating a thin liquid film. Under these conditions (vacuum and heating), cardanol is continuously removed from the top of the distillation unit in high yield and purity, while high boiling fractions (polymeric residol, cardol, and 2-methyl-cardol) are removed from the bottom of the distillation unit.

In both batch and thin film distillations, a certain amount of residol is formed due to double-bond crosslinking [29] and recovered as a dark brown to black viscous liquid. Its chemical composition and molecular weight distribution are strongly related to the composition of the starting CNSL as well as the distillation and recovery parameters. Because of its phenolic character and relative lower cost, residol (by-product) has successfully found use in varnishes, tiles, cements, asphalts [30], and friction products [31], providing flexibility, and improved resistance to impact, corrosion, and abrasion. Alternative chemical routes to producing residol-type derivatives have been described in the literature such as treating cardanol with Lewis acids and targeting a better control of final crosslinked cardanol average physicochemical properties [32].

In spite of advancements in distillation processing of CNSL, the use of cardanol has been historically restricted in some applications due to its darker color and poor color stability over time. It was not until recently that Cardolite developed a proprietary process technology to yield very high purity cardanol grades with ultra-light color (Gardner 1) and good color stability over shelf life. This leap forward in CNSL processing technology is expanding the use of cardanol as an epoxy diluent and modifier and as a building block for other derivatives. As a modifier for epoxy systems, cardanol is a suitable replacement for potentially toxic diluents such as nonyl-phenol as it provides better dilution

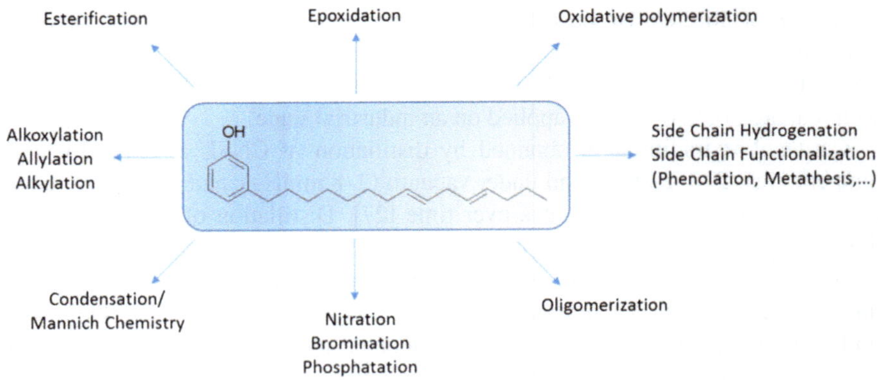

Fig. 10.3 Cardanol average molecule and examples of its potential functionalization

power, higher impact resistance, and similar acceleration effects to epoxy-amine reactions.

10.5 CNSL Technology Platforms

Cardanol is a unique molecule that can be chemically modified in many different sites: functionalization can in fact occur on the phenoxy OH, the aromatic ring and the double bonds on the side chain (Fig. 10.3).

The versatility of this molecule has prompted academic researchers and industries over the last century to investigate cardanol's applicability as building block in many different synthetic routes [33], amounting to more than 3,000 published papers, 2,000 patents, and 30,000 citations.

Cardanol (or crude CNSL) has been successfully employed in polyester resins [34], silanized structures [35], acrylates [36], solvents [37], surfactants [38], nanomaterials [39], and as biodiesel [40]. The main examples of cardanol derivatives widely used in industrial applications are phenolic resins (novolac- and resole-type materials), amines as epoxy curing agents (phenalkamines and phenalkamides), epoxy resins, and diluents, and more recently cardanol-based polyols and benzoxazines [41].

10.5.1 Cardanol-Based Novolacs and Resoles

Phenolic resins are one of the oldest and most well-known thermosetting polymers widely used in adhesives, composites, and insulating materials. These derivatives can be easily obtained by phenol homopolymerization using a suitable

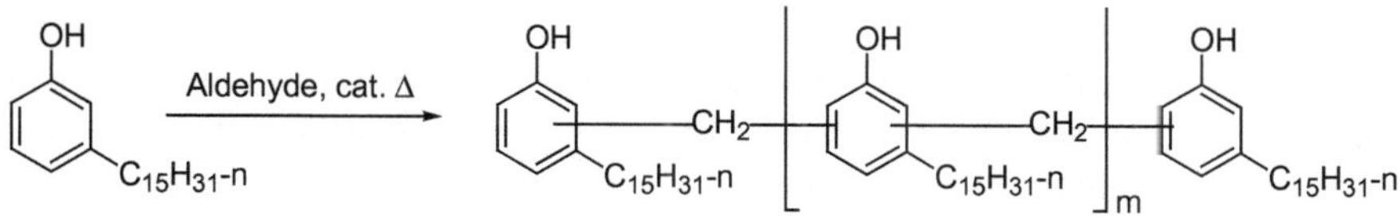

Scheme 10.1 Generic reaction scheme to cardanol-based novolacs and resoles

aldehyde (typically formaldehyde or paraformaldehyde) under both acidic and basic conditions, recovering novolac-type and resole-type materials, respectively (Scheme 10.1).

Phenolic resins' average properties are tuned by adjusting the phenol/aldehyde ratio, unreacted materials content, and reaction time. These resins are generally characterized by high-temperature resistance and fire retardancy, but are also quite brittle. To overcome this limitation, different aldehydes and substituted phenols (like cardanol) are used. Cardanol has been extensively studied as a complete or partial replacement for other phenolic [42] and aromatic materials [43] in novolac and resole resins to impart flexibility, toughness, and lower viscosity while providing high chemical resistance [44].

Due to the presence of the long C15 in the *meta* position, cardanol-based novolacs have a different reaction kinetic than phenol-based derivatives [45], thus requiring adjustment of process conditions as in the case where "*high-ortho*" novolacs are needed [46]. Response surface methodology [47], NMR analysis [48], and thermal degradation [49] have been recently adopted to optimize cardanol-based novolac processes by giving a precise chemical structure characterization of final products and a more comprehensive understanding of all the parameters that might affect their quality and consistency.

Cardanol-based novolacs are commercially available (e.g., Cardolite Novolac Series) and have been successfully used as reinforcement for rubber materials [50], thermosetting resins cured with HEXA for the production of glass laminates [51], curing agents and modifiers for epoxy resins [52], intermediates for the production of cyanate-ester resins [53], and oil pour point depressants [54] among others.

10.5.2 Cardanol-Based Amine Curing Agents (Phenalkamines and Phenalkamides)

Epoxies are widely used in coatings, adhesives, and composites due to their excellent bonding and anticorrosion properties, mechanical strength, and chemical resistance. The most common types of epoxy curing agents for ambient-cure systems include the following unmodified or epoxy-adducted amines: aliphatic polyamines (e.g., unmodified, ketamines, aryl-amines, and Mannich base);

cycloaliphatic amines; fatty acid modified amines (e.g., polyamides, adducts and amido-amines); and aromatic amines. There are pros and cons to formulating with each of these amines. For instance, polyamides provide excellent flexibility and good water resistance, but are very slow to cure especially at lower temperatures. Unmodified polyamines are very reactive, but typically cause poor flexibility and blushing in addition to potential toxicity problems. Mannich-base curing agents derived from petrochemical phenols offer many advantages due to the presence of the phenoxy OH that enhances the rate of nucleophilic ring opening of epoxies by amines (fast cure) and their relatively high molecular weight (400–1000 Da) that improves the physical properties of the cured material. On the other hand, the use of these phenol-derived Mannich bases results in systems with short pot life and low impact resistance. Moreover, there have been increased health and environmental concerns with the use of several phenols.

Mannich-base curing agents obtained from the reaction of cardanol, amines, and an aldehyde (generally formaldehyde or paraformaldehyde [55]), are called "*phenalkamines*" (Scheme 10.2) [56]. Phenalkamines were first developed by 3M in the 1970s and have been advanced over the years by Cardolite Corporation. Phenalkamines share similar advantages to other Mannich-base curing agents, such as extremely fast cure, low temperature cure (even below 0 °C), good chemical resistance, and non-blushing properties. However, due to the long hydrophobic side chain of cardanol, phenalkamines also offer better pot life and flexibility, surface tolerance, and excellent water and corrosion resistance. Adhesion to poorly prepared or wet surfaces, such as water-saturated concrete, is especially good with phenalkamines because of the combination of hydrophobicity and fast cure, which ensure surrounding water does not influence the surface resin bond.

The desirable properties of phenalkamines have made them one of the most commonly used curing agents in the formulation of heavy-duty coatings for the marine and protective markets. Over the years, Cardolite Corporation pioneered new developments with CNSL technology, which have expanded the use of phenalkamines to new markets such as adhesives and sealants, and other industrial coatings applications. First generation phenalkamines were dark in color (Gardner 18+) and high in viscosity (20,000–50,000 cps) whereas latest generation phenalkamines from Cardolite are very light color (Gardner 1–10) and low viscosity and solvent-free (<1000 cps). Other advancements in CNSL curing agent technology championed by Cardolite include faster curing at below zero Celsius temperatures, highly chemical-resistant phenalkamines, and the development of phenalkamides.

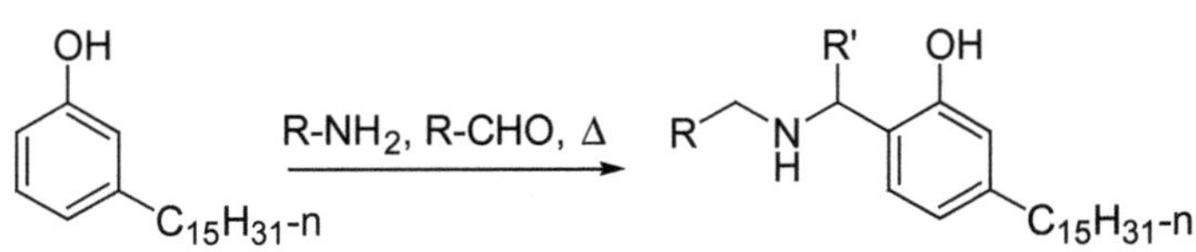

Scheme 10.2 Generic reaction scheme to cardanol-based phenalkamines

Phenalkamides (Cardolite LITE 3000 Series) are a novel class of epoxy curing agents developed to fill a gap between polyamide and phenalkamine chemistries. By chemically combining these technologies, phenalkamides offer the benefits of both while mitigating their limitations [57]. Broader adoption of phenalkamines has been limited due to drawbacks in color stability, long overcoatability, and sometimes flexibility. With phenalkamide technology, epoxy formulations have the desired color stability, overcoat window, and flexibility of polyamide-cured formulations, but with the outstanding corrosion protection and fast low temperature cure and bond strength development of phenalkamines. These balanced properties have enabled the use of CNSL curing agent technology in new coatings and adhesives applications.

10.5.3 Cardanol-Based Epoxy Derivatives

Cardanol has been widely used as an intermediate for the synthesis of epoxidized structures where the epoxy ring is introduced on the phenoxy OH by epichlorohydrin addition under basic pH conditions or through side chain double bonds treatment with peracids (Scheme 10.3).

These derivatives have found broad use in epoxy protective coatings [58] (marine, industrial, flooring, transportation, cans, drums), electrical and electronic applications, adhesives [59], and fire-resistant plasticizers [60].

Cardanol-based epoxies of various functionalities provide improved flexibility, and impact and water resistance, and are compatible with standard petro-based epoxy resins (Bisphenol A- and Bisphenol F-based). These features have been extensively described in the literature, where the curing [61] and

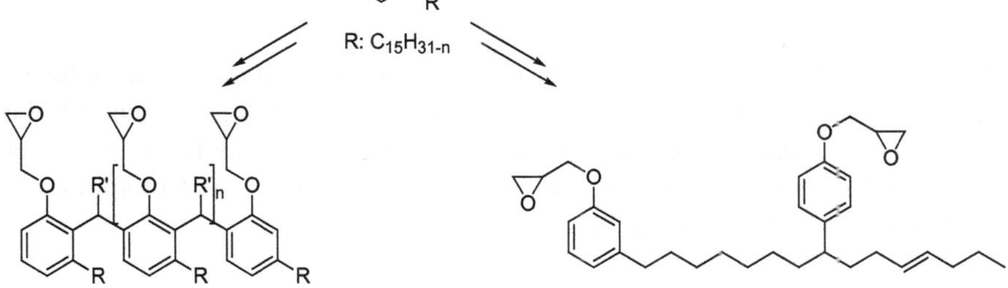

Scheme 10.3 Generic reaction schemes to cardanol-based epoxy derivatives

thermo-mechanical properties of commercial multifunctional cardanol-based epoxy resins (e.g., Cardolite NC-547, NC-514) have been fully characterized. For example [62], thermal and dynamic mechanical thermal analysis of a composite showed a gradual decrease in glass transition temperatures (T_g) with cardanol content increase due to the C15 chain, thus contributing to an increase in overall system flexibility and toughness, as highlighted by the increase in the Izod impact strength. Moreover, similar advantages are observed when epoxy derivatives based on cardanol resoles are used [63] with respect to unmodified epoxy resins that include an increase in tensile strength and impact strength and improved in scratch hardness, adhesion, and flexibility.

In addition to multifunctional resins, mono-functional cardanol-based epoxies (e.g., Cardolite NC-513) are commonly formulated as reactive diluents in epoxy coatings and adhesives due to their excellent dilution power and anticorrosion properties as well as benefits in flexibility and water resistance. Cardolite's proprietary process technology described in the cardanol purification section was applied to manufacture a mono-functional cardanol-based epoxy with maximum color Gardner 1 (Ultra LITE 513), which enables the direct replacement of non-cashew-derived glycidyl ethers that might present health or environmental concerns.

Epoxidized cardanol derivatives (bearing oxiranic ring on the phenoxy OH or on the side chain double bonds or on both positions [64]) have been recently investigated and tested as follows: toughening agents for DGEBA resins [65], in cathodically electrodeposed epoxy coatings [66], in combination with CTPB to improve mechanical and thermal properties [67], as intermediates for the synthesis of biobased UV curable resins [68] with improved hydrophobicity [69], in self-curable epoxy coatings [70], as inhibitors for composite propellants [71] and biobased epoxy coatings [72].

Further chemical derivatization through maleinization [73] of epoxidized cardanol derivatives; the evaluation of cardanol-epoxy materials in novel fields of application (e.g., PVC plasticizers [74], epoxy foams [75], antioxidants [76]); and the use of enzymatic approaches [77] as alternative green routes for its production are a testimony to the versatility and outstanding properties offered by cardanol-based epoxy derivatives.

10.5.4 Cardanol-Based Polyols

Most of the polyols currently used in polyurethane applications, such as foams, coatings, adhesives, and elastomers, derive from petroleum-based feedstock, but the increasing interest in materials from renewable sources has led to the development of biobased polyols with a more favorable environmental footprint than petroleum-based counterparts [78].

The naturally occurring alkyl phenolic structure of cardanol offers tremendous advantages to polyurethanes: the aromatic ring provides good chemical and thermal resistance (advantage over most renewable polyols); the OH group gives good reactivity for fast and low temperature cure and strong adhesion even to damp or poorly prepared surfaces; and the long aliphatic side chain delivers excellent water resistance and anticorrosive properties, good flexibility, low viscosity, and extended pot life.

The possibility of introducing a variety of chemical modifications to create biobased monomers and polymers including polyols makes cardanol an ideal raw material for applications requiring environmental sustainability and easy processability. Several chemical routes to synthesize cardanol-based polyols have been reported in the literature [79] and patented, highlighting the possibility of recovering low-viscous derivatives with different functionalities and hydroxyl values. Cardanol-based polyols can be obtained by cardanol homopolymerization using novolac chemistry, by Mannich condensation with suitable alkanolamines followed by propoxylation [80], and by hydrolysis of cardanol-based epoxy novolacs or cardanol-based epoxy monomers using aqueous acids [81] or nucleophiles (Scheme 10.4) [82].

When used in polyurethane matrices (from coatings [83] to flexible foams [84]), cardanol derivatives have demonstrated to contribute to mechanical properties [85] as well as thermal stability [86] and fire performances [87]. Recent research work has been focused on the evaluation of cardanol-based urethanes that enable isocyanate-free systems, maximum biocontent, reduced environmental impact, and good final product performances [88]. A variety of CNSL-based polyols for coatings, adhesives, and foam applications are commercially available from Cardolite Corporation (NX-9000 Series).

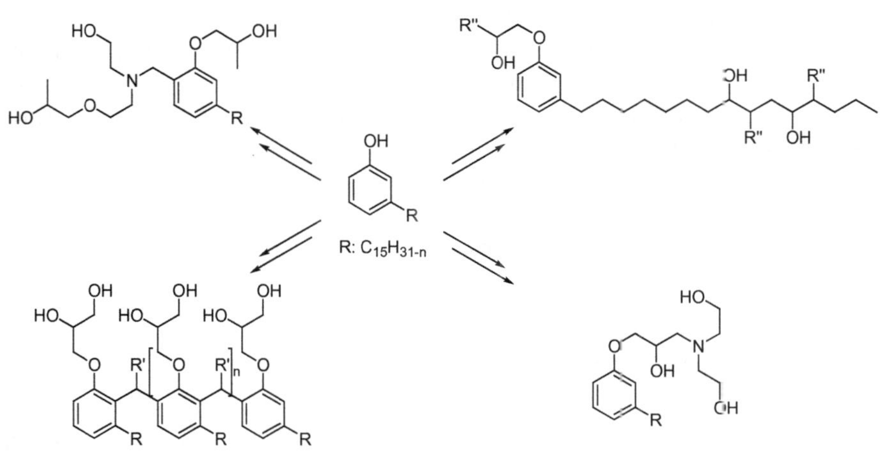

Scheme 10.4 Generic reaction schemes to cardanol-based epoxy derivatives

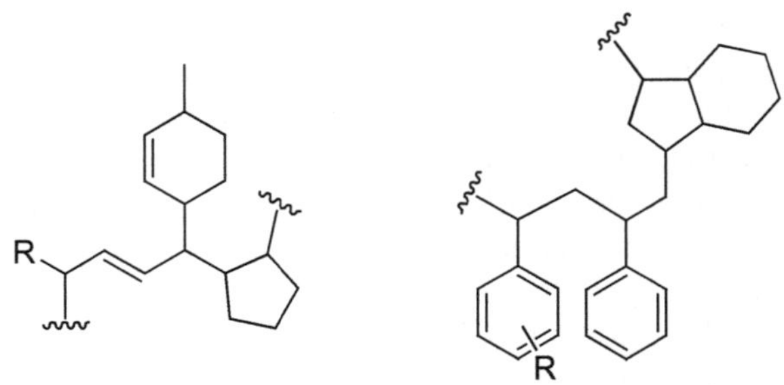

Fig. 10.4 Generic structures of hydrocarbon resins

10.5.5 CNSL-Based Hydrocarbon Resins

Hydrocarbon resins represent a wide family of derivatives generally classified into aliphatic hydrocarbons (typically indicated as "C5" since they are based on monomeric units with 5 carbon atoms), aromatic hydrocarbons (described as "C9" since they are based on aromatic units with 9 carbons), or aliphatic/aromatic hydrocarbons (shortened to "C5/C9" as they are aliphatic-aromatic copolymers) depending on the nature of the starting raw materials [89]. Monomers such as dicyclopentadiene, cylcopentadiene, 1,3-pentadiene (in its *cis* or *trans* configuration), styrene, indene, α-methylstyrene, and tall oil derivatives are cationically polymerized or subjected to Diels–Alder-type cycloadditions; final structures still contain unmodified double bonds that can be potentially further functionalized or hydrogenated to target-specific final properties (Fig. 10.4) [90].

Hydrocarbon resins are commonly used as tackifying polymers (e.g., especially natural rubber), and as compatibilizers or reinforcements in adhesives, printing inks, sealants, and paints. Additionally, re-enforcers lower-cost-type hydrocarbons are often used in packaging HMA based on EVA.

Cardanol has been recently used on an industrial scale by Cardolite Corporation to develop innovative hydrocarbon resins (commercially available as "LITE 2100" and "LITE 2100R") that aim to overcome the limitations of their fully petro-derived phenolic offsets [91]. These novel biobased liquid and low-colored structures are able to improve final systems' flexibility and hydrophobicity, and impart better adhesion properties, corrosion resistance, and faster hardness development than most standard hydrocarbon resins.

10.5.6 CNSL-Based Friction Particles and Resins

CNSL-based friction particles (sometimes referred to as "friction dust") were the first CNSL derivatives to be successfully commercialized in the 1930s. Friction particles and resins are typically made from CNSL polymers such as residol, a by-product of the CNSL distillation, and are still largely used in low steel, ceramic, non-asbestos-organic, semi-metallic and asbestos formulations of brake pads, drum linings, train blocks, clutches, and industrial friction materials.

CNSL friction particles come in different varieties (brown, black, and modified), sizes, organic content, and cross-link density to meet requirements in heat resistance and compressibility among other properties. Their resilient nature reduces the friction material hardness and improves conformance to the mating surface. At elevated brake temperature, friction particle that is exposed at the rubbing surface of the friction material decomposes. The decomposition products help stabilize friction coefficient and reduce friction material wear. The inclusion of friction particles also helps control brake noise.

CNSL-based friction resins are used as a binder for friction products in brake pads, linings, and train blocks and are supplied with or without solvent at various viscosities. Upon heat cure with a cross-linking agent (e.g., hexamine), CNSL resins provide desirable impact resistance, flexibility, and thermal stability making them suitable replacements for petro-based phenolic resins.

A comprehensive product range of friction particles and resins are available from Cardolite Corporation for OEM and aftermarket passenger cars and heavy-duty applications.

10.6 Conclusions

Cashew nutshell liquid and its main component cardanol represent a unique example of naturally occurring non-edible oil, joining the properties of phenolic and alkenyl structures. The different functional groups within its chemical backbone can be either selectively or simultaneously modified, by simply choosing the most suitable chemical approach according to desirable properties. Cardanol's applicability in a broad spectrum of products and formulations, from low added-value to high-performance applications, further confirms its versatility and prompts researchers and industries to look at this molecule with an increasing interest to optimize existing technologies and develop new processes for its purification and chemical functionalization. As a result of the continued advancement in cardanol technology, cardanol remains an "evergreen molecule" suitable as building block for many new products and markets.

References

1. Morton JF (1961) Econ Bot 15(1):57
2. Piyali D, Anuradda G (2003) Biomass Bioenergy 25:113; Venkata Ramanan M, Lakshmanan E, Sethumadhavan R, Renganarayanan S (2008) Braz J Chem Eng 25(3):585; Sengar SH, Mohod AG, Khandetod YP (2012) Int J Energy Eng 2(3):78; Chaudhari AP, Thakor NJ (2015) Asian J Biol Life Sci 4(1):71
3. Harvey MT, Caplan S (1940) Ind Eng Chem 32(10):1306

4. Stadeler A (1847) Ann Chem Pharm 63:137
5. Ruhemann S, Skinner S (1887) J Chem Soc Trans 51:663
6. Smit AJH (1931) Proc Acad Sci Amsterdam 34:165
7. Spiegel L, Correll M (1913) Ber Deut Pharm Ges 23:356; Spiegel L, Correll M (1895) Ber Deut Pharm Ges 5:309
8. Backer HJ, Haack NH (1941) Rev Trav Chim 60:661; Gokhale G, Patel M, Shah R (1940) Current Sci 9:362
9. Pillay PP (1935) J Indian Chem Soc 12:226
10. Harvey MT. US2,098,824
11. Naidu DSJ (1925) Indian Inst Sci 8:129
12. Wasserman D, Dawson CR (1945) Ind Eng Chem 37:396
13. Menon ARR, Pillai CKS, Sudha JD, Malhew AG (1985) Sci Ind Res 44:324; Gedam PH, Sampathkumaran PS (1986) Prog Org Coat 14:115; Tyman JHP (1973) Chem Soc Rev 8:499; Paul V, Yeddanapally LM (1954) Nature 174:604
14. Laurenco Marques. Indian Patent 78612; Siddiqui S, Khan AA. Indian Patent 32237
15. Subbarao ChNV, Krishna Prasad KMM, Prasad VSRK (2011) Pharma Res J 6(1):21
16. Das P, Sreelatha T, Ganesh A (2004) Biomass Bioenergy 27:265; Tsamba AJ, Yang W, Blasiak W (2006) Fuel Process Technol 87:523
17. Gandhi TS, Dholakiya BZ, Patel M (2013) R Polym J Chem Tech 15(4):24; Tyman JHP (1989) J Am Oil Chem Soc 66(4):533
18. Shobha SV, Ravindranath B (1991) J Agric Food Chem 39:2214; Smith RL, Malaluan RM, Setianto WB, Inomata H, Arai K. Bioresour Technol 88:1
19. Rodrigues FHA, França FCF, Souza JRR, Ricardo NMPS, Feitosa JPA (2011) Polímeros 21(2):156
20. Kumar PP, Paramashivappa R, Vithayathil PJ, Rao PVS, Rao AS (2002) J Agric Food Chem 50:4705
21. Lubi MC, Thachil ET (2000) Des Monomers Polym 3:123; Tyman JHP (1975) J Chromatogr 111:285; Tyman JHP, Tychopoulos V, Colenutt BA (1981) J Chromatogr 213:287
22. Tyman JHP, Bruce IE, Payne P (1992) Nat Prod Lett 1:117; Njuku FW, Muturi PM, Thiong'o GT (2014) Chem Mater Res 6(9):65; Mwangi PM, Mbugua BK (2013) Int J Cur Trans Res 2(1):285; Tyman JHP, Patel MS, Manzara AP. US4352944; De Braganca RM, Baird MS WO2011138608
23. Paramashivappa R, Phani Kumar P, Vithayathil PJ, Srinivasa Rao A (2001) J Agric Food Chem 49:2548
24. Sood SK, Tyman JHP, Durrani A, Johnson RA (1986) Lipids 21:241; Lloyd HA, Denny C, Krishna G (1980) J Liquid Chromatogr Related Technol 3(10):1497
25. Shobha SV. Ravindranath B (1991) J Agric Food Chem 39:2214; Marriott R, Gould M WO2013026727
26. Gandhi T, Patel M, Dholakiya BKJ (2012) Nat Prod Plant Resour 2(1):135
27. Durrani AA, Davis GL, Sood SK, Tychopoulos V, Tyman JHP (1982) J Chem Technol Biotechnol 32(7–12):681; Risfaheri, Irawadi TT, Anwar Nur M, Sailah I (2009) Indones J Agric 2(1):11
28. Sheeja KR, Avila Thanga Booshan S (2014) Int J Dev Res 4(11):2246
29. Rodrigues FHA, Souza JRR, França FCF, Ricardo NMPS, Feitosa JPA (2006) E-polymers 81:1
30. Venema JB, Eijkenboom ACJ, Van Weezenbeek KD EP2389415
31. Bijwe J (1997) Polym Compos 18(3):378; Keller EC. US3448071; Prasad VSN, Sudha JNDA, Pillai CKS, Menon AR, Menon R, Damodaran AD. Indian Patent 242267
32. Antony R, Pillai CKS (1990) J Appl Polym Sci 41:1765; Manjula S, Kumar VG, Pillai CKS (1992) J Appl Polym Sci 45:309; Bai W, Xiao X, Chen Q, Y, Zheng S, Lin J (2012) Progr Org Coat 75:184
33. Lubi MC, Thachil ET (2000) Des Monomers Polym 3(2):123; Voirin C, Caillol S, Sadavarte NV, Tawade BV, Boutevin B, Wadgaonkar PP (2014) Polym Chem 5:3142

34. Pillai CKS, Both SGLS, Madhavan NS, Bijymol J, Alathur D. Indian Patent 197453; Mwangi PM, Mbugua BK (2013) Int J Cur Trans Res 2(1):285
35. Balgude D, Sabnis AJ (2014) Renew Mater 2(3):235; Dai Z, Dalal A, Lawson DC III, He J. US20110065947
36. Sultania M, Rai JSP, Srivastava D (2010) J Appl Polym Sci 118:1979–1989; Sultania M, Rai JSP, Srivastava D (2010) Eur Polym J 46:2019; Liu R, Zhang X, Zhu J, Liu X, Wang Z, Yan J (2015) ACS Sustain Chem Eng 3:1313; Li S, Yang X, Huang K, Li M, Xia J (2014) Progr Org Coat 77:388–394; Shukla P, Srivastava D (2014) Int J Plast Technol 18(1):1
37. Njuku FW, Muturi PM, Thiong'o GT (2014) Chem Mater Res 6(9):65
38. Abhijit T, Trissa J, Vineet VS. US8360150; Glos M. US20110028578; Hulkko J, Koskimies S, Kylakoski R, Laine A, Luttikhedde H. WO2008131918; Peungjitton P, Sangvanich P, Pornpakakul S, Petsom A, Roengsumran S (2009) J Surf Deterg 12:85
39. Balachandran VS, Jadhav SR, Vemula PK, John G (2013) Chem Soc Rev 42:427
40. Vedharaj S, Vallinayagam R, Yang WM, Chou SK, Chua KJE, Lee PS (2014) Exp Thermal Fluid Sci 53:259; Sanjeeva SK, Nair CB, Kini GM, Rao PVS, Ramamoorthy S, Barrow C, Pullela PKu (2015) Iran J Energy Environ 6(4):282
41. Rao BS, Palanisamy A (2012) Progr Org Coat 74:427; Lochab B, Varma IK, Bijwe J (2010) J Therm Anal Calorim 102:769; Zhang C, Luo X, Zhu R, Ling H, Gu Y (2015) J Adhesion Sci Technol 29(8):767; Campaner P, D'Amico D, Longo L, Stifani C, Tarzia A, Tiburzio S (2011) Handbook of benzoxazine Resins, Chapter 19, Elsevier, Amsterdam; Calò E, Maffezzoli A, Mele G, Martina F, Mazzetto SE, Tarzia A, Stifani C (2007) Green Chem 9:754; Xu G, Shi T, Liu J, Wang Q (2014) J Appl Polym Sci 131(11):40353; Sharma P, Lochab B, Kumar D, Kumar Roy P (2016) ACS Sustain Chem Eng 4(3):1085; Attanasi OA, Behalo MS, Favi G, Lomonaco D, Mazzetto SE, Mele G, Pio I, Vasapollo G (2012) Curr Org Chem 16:2613; Kotzebue LRV, Ribeiro FWM, Sombra VG, Feitosa JPA, Mele G, Mazzetto SE, Lomonaco D (2016) Polymer 92:189
42. Mahanwar PA, Kale DD (1996) J Appl Polym Sci 61:2107
43. Srivastava R, Srivastava D (2015) J Coat Technol Res 12(2):303; Srivastava R, Srivastava D (2013) Int J ChemTech Res 5(5):2575
44. Parameswaran PS, Abraham BT, Thachil ET (2010) Progr Rubber Plast Recycl Technol 26(1):31
45. Devi A, Srivastava D (2006) J Appl Polym Sci 102:2730; Yadav R, Srivastava D (2007) Mater Chem Phys 106:74; Sultania M, Rai JSP, Srivastava D (2009) Int J Chem Kinet 41(9):559; Shukla P, Yadaw SB, Srivastava D (2010) Int J Chem Kinet 42(6):380
46. Sammbasivan E, Basu P, Roy D. WO2002102867
47. Yadav R, Devi A, Tripathi G, Srivastava D (2007) Eur Polym J 43(8):3531
48. Roy D, Basu PK, Raghunathan P, Eswaran SV (2003) J Appl Polym Sci 89:1959; Sharma M, Raghunathan P, Eswaran SV (2008) J Appl Polym Sci 108:804
49. Natarajan M, Murugavel SC (2013) High Perform Polym 25(6):685
50. Chuayjuljit S, Rattanametangkool P, Potiyaraj PJ (2007) Appl Polym Sci 104:1997
51. Sathiyalekshmi K (1993) Bull Mater Sci 16(2):137
52. Campaner P, D'Amico D, Longo L, Stifani C, Tarzia A (2009) J Appl Polym Sci 114:3585; Unnikrishnan KP, Thachil ET (2006) Int J Polym Mater 55:323; Unnikrishnan KP, Thachil ET (2008) J Elastom Plast 40:271
53. Reghunadhan Nair CP, Bindu RL, Joseph VC (1995) J Polym Sci, Part A: Polym Chem 33:621
54. Huyen QTM, Khanh NV (2014) PetroVietnam J 6:48
55. Liu Y, Wang J, Xu SJ (2014) Polym Sci Part A Polym Chem 52:472
56. Ding C, Matharu AS (2014) ACS Sustain Chem Eng 2:2217
57. Kim Y, Tavares F, Lawson D, Chen M, Dai J (2012) Phenalkamide—new hybrid molecule and its unique coating properties
58. Yadav R, Awasthi P, Srivastava D (2009) J Appl Polym Sci 114:1471
59. Gunasekaran S, Gorczyca TB, Cole HS. US6548189

60. Constantinescu A, Reyes J. US8349924
61. Agrawal JP, Bhide NM, Naidu SR (1993) J Therm Anal 39(3):351
62. Gour RS, Kodgire VV, Badiger MVJ (2016) Appl Polym Sci 133:43318
63. Shukla SK, Srivastava K, Srivastava D (2015) Mater Res 18(6):1217
64. Patel MB, Patel RG, Patel VS (1989) J Therm Anal 35(1):47
65. Natarajan M, Muragavel SC (2016) J Therm Anal Calorim 1
66. Kumar RP (2011) J Coat Technol Res 8(5):563
67. Devi A, Srivastava D (2007) Eur Polym J 43:2422
68. Sultania M, Rai JSP, Srivastava D (2010) Eur Polym J 46:2019
69. Chen Z, Chisholm BJ, Webster DC, Zhang Y, Patel S (2009) Prog Org Coat 65:246
70. Raju Shukla R, Kumar P (2011) Pigment Resin Technol 40/5:311
71. Agrawal JP, Agawane NT (2001) J Propul Power 17:5
72. Darroman E, Durand N, Boutevin B, Caillol S (2015) Prog Org Coat 83:47
73. Kathalewar M, Sabnis A (2014) J Coat Technol Res 11(4):601; Kathalewar M, Sabnis A, Waghoo G (2014) Pigment Resin Technol 43(6):379
74. Chen J, Liu Z, Jiang J, Nie X, Zhouab Y, Murray RE (2015) RSC Adv 5:56171
75. Dworakowska S, Cornille A, Bogdał D, Boutevin B, Caillol S (2015) Eur J Lipid Sci Technol 117(11):1893
76. Liu Z, Chen J, Knothe G, Nie X, Jiang J (2016) ACS Sustain Chem Eng 4(3):901
77. Kim YH, An ES, Park SY, Song BK (2007) J Mol Catal B Enzym 45:39
78. Madbouly S, Zhang C, Kessler MR (2015) Bio-based plant oil polymers and composites. Elsevier, Amsterdam; Liu Z, Kraus G (2014) Green materials from plant oils. Royal Society of Chemistry; Desroches M, Escouvois M, Auvergne R, Caillol S, Boutevin B (2012) Polym Rev 52(1):38; Guner FS, Yagci Y, Erciyes AT (2006) Progr Polym Sci 31:633; Raquez JM, Deleglise M, Lacrampe MF, Krawczak P (2010) Progr Polym Sci 35:487
79. Bhunia HP, Jana RN, Basak A, Lenka S, Nando GB (1998) J Polym Sci Part A Polym Chem 36:391; Bhunia HP, Nando GB, Chakia TK, Basak A, Lenka S, Nayak PL (1999) Eur Polym J 35:1381; Agrawal JP, Satpute RS (1993) J Macromol Sci Part A Pure Appl Chem 30(1):19
80. Ionescu M, Wan X, Bilić N, Petrovic ZS (2012) Polym Environ 20:647; Gandhi TS, Patel MR, Dholakiya BZ (2015) J Polym Eng 35(6):533; Suresh KI (2013) ACS Sustain Chem Eng 1(2):232
81. Suresh KI, Kishanprasad VS (2005) Ind Eng Chem Res 44:4504; Suresh KI, Kishanprasad VS. WO2006003668
82. Prasad VSN, Pillai CKS. WO200707756
83. Mythili CV, Retna AM, Gopalakrishnan S (2004) Bull Mater Sci 27(3):235; Minh Tan TT(1997) J Appl Polym Sci 65(3):507
84. Reese JR, Moore MN., Wardius DS, Hager SL. EP1930355
85. Tan Ton That Minh (1996) Polym Int I 41:13
86. Gopalakrishnan S, Fernando TL (2012) Bull Mater Sci 35(2):243; Mythili CV, Retna AM, Gopalakrishnan S (2005) J Appl Polym Sci 98:284
87. Zhang M, Zhang J, Chen S, Zhou Y (2014) Polym Degrad Stab 110:27
88. Kathalewar M, Sabnis A (2015) J Appl Polym Sci 132:41391; Kathalewar M, Sabnis A, D'Mello D (2014) Eur Polym J 57:99
89. Lowery RD (2001) Kirk-Othmer encyclopedia of chemical technology, hydrocarbon resins. Wiley, Hoboken
90. Naumkin PV, Nesterova TN, Nesterov IA, Toikka AM, Shakun VA (2015) Ind Eng Chem Res 54(35):8629; Allen RP. US4393261; Allen RP, Ruggles AC. US4342850; Kumar KD, Tsou AH, Bhowmick AK (2010) Int J Adhesion Adhes 30:200; Macedo AV, Burgers MHW, Garcia ML, Haluska JL, Jacob LE, Jagisch FC, Lowery RD, Sims CL, Smits JAF. WO1998057999
91. Pinto M, Xu H, Natesh A, Wei J. WO2015054360

FSC
www.fsc.org
MIX
Papier aus verantwortungsvollen Quellen
Paper from responsible sources
FSC® C105338

If you have any concerns about our products, you can contact us on
ProductSafety@springernature.com

In case Publisher is established outside the EU, the EU authorized representative is:
Springer Nature Customer Service Center GmbH
Europaplatz 3, 69115 Heidelberg, Germany

Printed by Libri Plureos GmbH
in Hamburg, Germany